Catalytic Activation of Small Molecules

Catalysis Series

Editor-in-chief:
Justin S. J. Hargreaves, *University of Glasgow, UK*

Series editors:
Harry Bitter, *Wageningen University, The Netherlands*
Jose Rodriguez, *Brookhaven National Laboratory, USA*

For a list of titles in this series see: rsc.li/catalysis-series

How to obtain future titles on publication:
A standing order plan is available for this series. A standing order will bring delivery of each new volume immediately on publication.

For further information please contact:
Book Sales Department, Royal Society of Chemistry, Thomas Graham House, Science Park, Milton Road, Cambridge, CB4 0WF, UK
Telephone: +44 (0)1223 420066, Fax: +44 (0)1223 420247
Email: booksales@rsc.org
Visit our website at books.rsc.org

Catalytic Activation of Small Molecules

Edited by

Mustafa Yasin Aslan
Usak University, Turkey
Email: mustafa.aslan@usak.edu.tr

Angela Daisley
University of Glasgow, UK
Email: Angela.Daisley@glasgow.ac.uk

Justin S. J. Hargreaves
University of Glasgow, UK
Email: Justin.Hargreaves@glasgow.ac.uk

and

José L. Rico
Universidad Michoacana de San Nicolás de Hidalgo, Mexico
Email: jose.rico@umich.mx

Catalysis Series No. 49

Hardback ISBN: 978-1-83767-418-3
EPUB ISBN: 978-1-83767-681-1
PDF ISBN: 978-1-83767-680-4
Print ISSN: 1757-6725
Electronic ISSN: 1757-6733

A catalogue record for this book is available from the British Library

© The Royal Society of Chemistry 2025

All rights reserved

Apart from fair dealing for the purposes of research for non-commercial purposes or for private study, criticism or review, as permitted under the Copyright, Designs and Patents Act 1988 and the Copyright and Related Rights Regulations 2003, this publication may not be reproduced, stored or transmitted, in any form or by any means, without the prior permission in writing of the Royal Society of Chemistry or the copyright owner, or in the case of reproduction in accordance with the terms of licences issued by the Copyright Licensing Agency in the UK, or in accordance with the terms of the licences issued by the appropriate Reproduction Rights Organization outside the UK. Enquiries concerning reproduction outside the terms stated here should be sent to the Royal Society of Chemistry at the address printed on this page.

Whilst this material has been produced with all due care, The Royal Society of Chemistry cannot be held responsible or liable for its accuracy and completeness, nor for any consequences arising from any errors or the use of the information contained in this publication. The publication of advertisements does not constitute any endorsement by The Royal Society of Chemistry or Authors of any products advertised. The views and opinions advanced by contributors do not necessarily reflect those of The Royal Society of Chemistry which shall not be liable for any resulting loss or damage arising as a result of reliance upon this material.

The Royal Society of Chemistry is a charity, registered in England and Wales, Number 207890, and a company incorporated in England by Royal Charter (Registered No. RC000524), registered office: Burlington House, Piccadilly, London W1J 0BA, UK, Telephone: +44 (0)20 7437 8656.

For further information see our website at www.rsc.org

For general enquiries, please contact books@rsc.org

For EU product safety enquiries, please email books@rsc.org or contact Royal Society of Chemistry Worldwide (Germany) GmbH, Römischer Hof, Unter den Linden 10, 10117 Berlin.

Printed in the United Kingdom by CPI Group (UK) Ltd, Croydon, CR0 4YY, UK

Preface

The catalytic activation of small molecules is a very interesting and enduring topic which spans current industrial practices (*e.g.* steam reforming of methane and the Haber–Bosch process for ammonia synthesis) and desirable target processes (*e.g.* the oxidative coupling of methane to produce ethylene). The contents of this book reflect this with different chapters being dedicated to both industrial processes as well as selected target reactions. Its aim is to provide an indication of the state of the art. In this respect, the contents have extended beyond traditional thermal heterogeneous catalysis to encompass the emerging themes in both electrocatalysis and photocatalysis as well as chemical looping.

In bringing this book to fruition, we would like to express our appreciation to all who have contributed chapters. We are most grateful to them for their timely submissions. We also wish to acknowledge the editorial and production staff from the Royal Society of Chemistry and thank them for their very kind assistance throughout the commissioning, preparation and production processes. In this respect we particularly acknowledge the contributions made by Helen Armes and also Amina Headley.

Mustafa Yasin Aslan,
Angela Daisley,
Justin S. J. Hargreaves and
José L. Rico

Contents

Chapter 1 Steam Reforming and Dry Reforming of Methane   1
Mustafa Yasin Aslan

     1.1 Introduction   1
         1.1.1 Importance of H_2 Production and H_2 Production Technologies   1
         1.1.2 Steam Reforming of Methane   2
         1.1.3 Dry Reforming of Methane (DRM)   3
         1.1.4 Partial Oxidation and Autothermal Reforming of Methane and Carbon Formation   4
     1.2 Thermodynamics of Reactions   5
         1.2.1 Steam Reforming of Methane   5
         1.2.2 Dry Reforming of Methane   6
     1.3 Catalysts   7
         1.3.1 Steam Reforming of Methane   8
         1.3.2 Dry Reforming of Methane   13
     1.4 (Micro)kinetics and DFT Studies   21
         1.4.1 Steam Reforming of Methane   21
         1.4.2 Dry Reforming of Methane   23
     1.5 New Generation Reactors and Process Intensification   25
         1.5.1 Sorption-enhanced Reforming of Methane   25
         1.5.2 Oxidative (Steam or Dry) Reforming of Methane   27

Catalysis Series No. 49
Catalytic Activation of Small Molecules
Edited by Mustafa Yasin Aslan, Angela Daisley, Justin S. J. Hargreaves and José L. Rico
© The Royal Society of Chemistry 2025
Published by the Royal Society of Chemistry, www.rsc.org

	1.5.3	Chemical Looping Reforming of Methane	27
	1.5.4	Other New-generation Methane-reforming Routes	28
1.6	General Summary and Outlook		28
References			29

Chapter 2 Methane Cracking: A Comprehensive Review of Catalysts, Solar Reactors, and Integrated System Case Studies   40
Aliya Banu and Yusuf Bicer

2.1	Introduction		40
2.2	Catalysts		41
	2.2.1	Metal Catalyst	42
	2.2.2	Carbon Catalyst	45
	2.2.3	Reaction Kinetics	47
	2.2.4	Co-feeding	49
	2.2.5	Catalyst Regeneration	50
2.3	Solar Reactors		51
	2.3.1	Directly Irradiated Solar Reactors	52
	2.3.2	Indirectly Irradiated Solar Reactors	53
	2.3.3	Catalyzed Solar Reactors	56
	2.3.4	Molten Media Reactors	57
	2.3.5	Solar Energy Collection	58
2.4	Case Studies		58
	2.4.1	System Description	58
	2.4.2	Methodology	62
	2.4.3	Results and Discussion	67
2.5	Conclusion		76
Acknowledgements			77
References			77

Chapter 3 Oxidative Coupling of Methane   84
Gabriel L. Catuzo, Letícia F. Rasteiro, Larissa B. Lopes, Luiz H. Vieira, José M. Assaf and Elisabete M. Assaf

3.1	Introduction	84
3.2	Methane: Characteristics and Environmental Impact	85
3.3	Activation of the C–H Bond and Possibilities for Methane Conversion	87
3.4	Oxidative Coupling of Methane (OCM)	89
3.5	Thermodynamic Aspects	90

	3.6	Reaction Mechanism and Kinetic Aspects	91
	3.7	Impact of Reaction Conditions in OCM	94
	3.8	Alternative Oxidants for OCM	95
	3.9	Active Sites for OCM	100
	3.10	Development of Catalysts for OCM	103
		3.10.1 Alkaline Earth Metal Oxides	103
		3.10.2 Rare Earth Oxides	106
		3.10.3 Perovskites	109
		3.10.4 Mn–Na–W–SiO$_2$ Catalyst	111
	3.11	Conclusion	114
	Acknowledgements		115
	References		115
Chapter 4	Exploring Chemical Looping for Methane Utilisation		122
	A. F. B. Abu Kasim and E. J. Marek		
	4.1	Chemical Looping – Concept, History, and Fundamentals	122
	4.2	Chemical Looping with Methane	125
	4.3	Chemical Looping Combustion (CLC) of CH_4	126
	4.4	Partial Oxidation of Methane with Oxygen Carriers (CL–POM)	129
	4.5	Dry Reforming of Methane with Chemical Looping (CL–DRM)	131
	4.6	Methane Cracking with Chemical Looping (CL–MC)	135
	4.7	Chemical Looping for Oxidative Coupling of Methane (CL–OCM)	137
	4.8	Methane to Other Products *via* Chemical Looping Routes	138
		4.8.1 Chemical Looping Methanol Synthesis	139
		4.8.2 Chemical Looping Benzene Synthesis	140
	4.9	Chemical Looping for CH_4 Conversion – Perspectives	141
	References		142
Chapter 5	Catalytic Dehydroaromatisation of Methane		147
	J. S. J. Hargreaves		
	5.1	Introduction	147
	5.2	Molybdenum-containing Catalysts	148
	5.3	Other Catalytic Systems	154
	5.4	Conclusions	155
	References		156

Chapter 6	**Activation of CO, CO_2, and H_2 Toward Synthetic Fuel Manufacture by Fischer–Tropsch Synthesis**	**158**
	A. N. Akin, O. Ozcan, M. Dogan-Ozcan and D. Uner	
6.1	Overview	158
6.2	Production of Hydrocarbons from Carbon Monoxide Hydrogenation	160
6.3	Thermodynamics of the CO Hydrogenation Reaction	163
	6.3.1 Heats of Reaction	163
	6.3.2 Free Energy of Reaction	164
6.4	Product Distribution	165
6.5	Carbon Monoxide Hydrogenation Catalysts	167
	6.5.1 Active Metals	169
	6.5.2 Supports	175
	6.5.3 Promoters	177
6.6	CO and H_2 Adsorption on Fischer–Tropsch Catalysts	179
	6.6.1 Electronic Configuration of CO	179
	6.6.2 Carbon Monoxide Adsorption	180
	6.6.3 Hydrogen Adsorption	183
	6.6.4 Co-adsorption and Interaction of Carbon Monoxide and Hydrogen	184
6.7	Mechanism of Carbon Monoxide Hydrogenation	186
	6.7.1 Carbide Mechanism	186
	6.7.2 Enolic Mechanism	188
	6.7.3 CO Insertion Mechanism	190
	6.7.4 Alkoxy Mechanism	192
	6.7.5 Other Alternative Mechanisms	192
6.8	Kinetics of Carbon Monoxide Hydrogenation	196
6.9	Fischer–Tropsch Synthesis Through Carbon Dioxide Hydrogenation	199
6.10	Products of Carbon Dioxide Hydrogenation	200
6.11	Thermodynamics of the Carbon Dioxide Hydrogenation Reactions	201
6.12	Product Distribution	202
6.13	Carbon Dioxide Hydrogenation Catalysts	202
	6.13.1 Active Metals	203
	6.13.2 Supports	204
	6.13.3 Promoters	205
6.14	Electronic Configuration of CO_2	206
6.15	Carbon Dioxide Adsorption	207

	6.16	Mechanism of Carbon Dioxide Hydrogenation	207
		6.16.1 Conversion of CO_2 to Methanol: Formate and CO Pathway	208
		6.16.2 Hydrogenation of CO_2 to Methane	209
		6.16.3 Fischer–Tropsch Synthesis Mechanism	210
	6.17	Kinetics of Carbon Dioxide Hydrogenation	211
	6.18	Outlook and Summary	211
	Acknowledgements		212
	References		212
Chapter 7	**Electrocatalytic Carbon Dioxide Activation** *J. L. Rico*		**223**
	7.1	Introduction	223
	7.2	Theoretical Aspects of CO_2	224
	7.3	Electrocatalytic Reduction of CO_2	225
		7.3.1 Electrocatalytic Cells	229
		7.3.2 Electrolytes	232
	7.4	Catalysts	241
		7.4.1 Metal-based Catalysts	241
		7.4.2 Hybrid Catalyst	246
		7.4.3 Metal-free Catalyst	247
	7.5	Computational Studies	256
		7.5.1 Final Remarks	261
	References		262
Chapter 8	**Construction of Active Sites for CO_2 Photoreduction** *Wa Gao, Yong Zhuo and Zhigang Zou*		**271**
	8.1	Introduction	271
	8.2	Basic Principles of Photocatalytic CO_2 Reduction	272
		8.2.1 Photogenerated Charge Carrier Separation	273
		8.2.2 CO_2 Adsorption and Activation	273
		8.2.3 C–C Coupling	275
	8.3	Strategies for Creating Active Sites	276
		8.3.1 Hydro/Solvothermal Route	276
		8.3.2 Thermal Annealing	277
		8.3.3 Etching	278
		8.3.4 Ultrasonication	279
	8.4	Typical Active Sites for Selective Photocatalytic CO_2 Reduction	280
		8.4.1 Metal Active Sites (MASs)	280
		8.4.2 Defect Engineering	282

		8.4.3 Edge Configurations	285
		8.4.4 Facet Engineering	285
	8.5	Conclusions and Perspectives	287
	Acknowledgements		289
	References		289

Chapter 9 Heterogeneous Catalysts in Ammonia Synthesis — 294
Masaaki Kitano and Hideo Hosono

9.1	Introduction	294
9.2	Transition Metal-based Catalysts	296
9.3	Electride-based Catalysts	299
9.4	Hydride-based Catalysts	303
9.5	Nitride-based Catalysts	307
9.6	Non-conventional Ammonia Synthesis Approaches	310
9.7	Outlook	313
Acknowledgements		314
References		314

Chapter 10 Electrocatalytic and Photocatalytic Conversion of Nitrogen — 318
A. Daisley

10.1	Introduction		318
10.2	Electrocatalytic Conversion of N_2		319
	10.2.1	Electrocatalytic Conversion of N_2 by a Metal Cathode	320
	10.2.2	Electrocatalytic Conversion of N_2 by Metal Nitrides and Carbides	323
	10.2.3	Electrocatalytic Conversion of N_2 by Metal Oxides	325
10.3	Photocatalytic Conversion of N_2		326
	10.3.1	Photocatalytic Conversion of N_2 by Metal Oxides	328
	10.3.2	Photocatalytic Conversion of N_2 by g-C_3N_4	331
	10.3.3	Photocatalytic Conversion of N_2 by Oxyhalides	331
	10.3.4	Photocatalytic Conversion of N_2 by Metal Sulphides	332
	10.3.5	Alternative Materials for the Photocatalytic Conversion of N_2	333
10.4	Conclusions		335
References			336

| Chapter 11 | Catalytic Combustion of Methane in Low-concentration Gas Streams | 343 |

M. Bligh, M. Drewery, L. Harvey, E. M. Kennedy and
M. Stockenhuber

11.1	Background		343
	11.1.1	Natural Gas Engines	343
	11.1.2	Natural Gas for Power Generation	344
	11.1.3	Natural Gas in Transport	345
11.2	Methane Slip		345
	11.2.1	Lean *versus* Stoichiometric	345
	11.2.2	Exhaust Methane	346
	11.2.3	Ventilation Air Methane (VAM)	347
11.3	Active Metal Catalysts		348
	11.3.1	Noble Metals	348
	11.3.2	Non-noble Metal Oxides	350
11.4	Catalyst Supports		355
	11.4.1	Alumina	355
	11.4.2	Silicas	356
	11.4.3	Zirconia	356
	11.4.4	Zeolites	356
11.5	Catalyst Preparation		358
	11.5.1	Impregnation	358
	11.5.2	Precipitation	359
	11.5.3	Sol–Gel	360
	11.5.4	Passivation of the Support	360
	11.5.5	Use of Different Precursors	360
11.6	Deactivation		361
	11.6.1	Sintering	361
	11.6.2	Sulphur Dioxide Poisoning	361
	11.6.3	Water Inhibition	363
	11.6.4	Coking	364
11.7	Regeneration		364
	11.7.1	Temperature-based Regeneration	364
	11.7.2	Nitrogen Oxide Regeneration	365
11.8	Stability		365
	11.8.1	Water	365
	11.8.2	Sulphur Dioxide	365
11.9	Reaction Kinetics		366
	11.9.1	Mechanisms	366
	11.9.2	Effect of Water	369
	11.9.3	Effect of Support	371
	11.9.4	Reaction Order	371

	Acknowledgements	373
	References	373
Chapter 12	**Conclusion**	**378**
	Mustafa Yasin Aslan, Angela Daisley, Justin S. J. Hargreaves and José L. Rico	
	References	380
Subject Index		**382**

CHAPTER 1

Steam Reforming and Dry Reforming of Methane

MUSTAFA YASIN ASLAN

Department of Chemical Engineering, Faculty of Engineering and Natural Sciences, Usak University, 64200, Usak, Türkiye
Email: mustafa.aslan@usak.edu.tr

1.1 Introduction

1.1.1 Importance of H_2 Production and H_2 Production Technologies

Hydrogen is one of the key substances of the chemical industry due to its use as a raw material in many areas such as petroleum refining, ammonia production, methanol production and others.[1] The annual hydrogen production was reported as 87 million tonnes in 2020 and estimations show that the hydrogen demand of industrial markets will rise to nearly 400 million tonnes by 2050.[2] In addition, global warming and the regulations to reduce CO_2 emissions are challenges that are directly related to the use of fossil fuels in many areas.[3,4] These facts dictate the need for humankind to transform the use of fossil fuels as the primary source of energy to a more sustainable and environmentally friendly method. From this point of view, the use of hydrogen-based energy technologies is a promising solution for future of life on Earth. When all of these considerations are evaluated collectively, the necessity for hydrogen production *via* sustainable methods in the present and future can be elucidated (from black to green hydrogen production).[5]

There are several methods for hydrogen production. These methods can be classified as (i) mature technologies, (ii) semi-mature technologies and (iii) technologies that are in the R&D phase.[6] Although hydrogen is the most abundant element in the world, only a small portion of it exists as the molecular structure (H_2). In this framework, hydrogen production technologies should be based on the raw materials from which the hydrogen will be produced. The most feasible raw materials have been hydrocarbons and water according to already-developed technologies. Among the hydrocarbons, natural gas is widely used for hydrogen production due to it mostly consisting of methane. Steam reforming of methane (SRM) is a conventional hydrogen production method, which has been using for almost a century.[7] The other methods that are used for hydrogen production starting from methane (hydrocarbons) are listed below:

- CO_2 (dry) reforming of methane (DRM)
- Partial oxidation of methane (POX)
- Autothermal reforming of methane (ATR)

A general outlook on the SRM and DRM processes is given in Sections 1.1.2 and 1.1.3 from an industrial perspective. In addition, brief information is given on the side reactions during the reforming reactions and other hydrogen and/or synthesis gas production methods are also considered in Section 1.1.4.

1.1.2 Steam Reforming of Methane

The steam reforming of methane is a highly endothermic reaction, as shown in eqn (1.1). The reaction conditions under which the SRM takes place are also favourable for the water gas shift reaction given in eqn (1.2), which is a slightly exothermic reaction. Due to the thermodynamics/nature of the reaction, high methane conversion can be achieved at high temperature. A gas mixture of $H_2 : CO = 3 : 1$ or even higher can be obtained during the operation of SRM reaction with the feed ratio of $H_2O : CH_4 \geq 3$ or above, although it is not certain due to side reactions and operating conditions.

$$CH_4 + H_2O \rightleftharpoons CO + 3H_2 \quad \Delta H°_{298K} = +206 \text{ kJ mol}^{-1} \quad (1.1)$$

$$CO + H_2O \rightleftharpoons CO_2 + H_2 \quad \Delta H°_{298K} = -41 \text{ kJ mol}^{-1} \quad (1.2)$$

The initial proposal of the interaction between hydrocarbons and metals was put forth by Davy in the early 1800s.[8] In addition, the production of hydrogen from hydrocarbons by using steam was discovered at the end of the 19th century.[9,10] The industrial practice of SRM was taken into consideration by several companies in the early 20th century.[11,12] One of the first comprehensive studies was performed by Neumann and Jacob, who achieved near equilibrium conversion in the presence of nickel as a catalyst.[13] Since then, SRM has become a mature industrial technology for the production of synthesis gas or hydrogen.

Today, the industrial operation of the SRM reaction has two main challenges. First the SRM reaction is a reversible and highly endothermic reaction. In other words, it is thermodynamically limited. Therefore, higher methane conversions require higher operation temperatures. The second challenge is coke deposition. Although the catalysts, especially supported Ni catalysts, are active even at 400 °C, they are prone to severe coke deposition under reaction conditions at such relatively low temperatures.[14] The coke deposition mechanism will be addressed in Section 1.4. These problems cause deactivation of the catalyst material by coke formation and sintering of the metal particles over the support material.[15] The traditional industrial catalyst for SRM is Ni/Al_2O_3 due to its high activity as well as low cost.[14] Coke formation and sintering of Ni particles are major challenges for the commercial Ni catalysts.

To overcome the above-mentioned problems and to obtain an efficient/intensified SRM process, many studies have been carried out, and these studies can be divided into three groups.[16] The first group of studies have focussed on developing a catalyst that is active and stable under reaction conditions, resistant to coke formation, and durable over an extended operational period. The second part of the studies are related with modifying and/or intensifying the process flow layout to overcome the thermodynamic limitation of the reaction such as orientation to the process of a chemical looping sorption enhanced SRM reactor. The third group of studies have aimed at optimizing the process parameters to achieve more efficient operation.

1.1.3 Dry Reforming of Methane (DRM)

Reaction of methane with CO_2, which is known as dry reforming of methane in the literature, is illustrated in eqn (1.3). Similar to SRM, the DRM reaction has a highly endothermic nature. The reverse water gas shift reaction takes place as a side reaction, which is slightly endothermic, as shown in eqn (1.4). As can be seen from eqn (1.3), a gas mixture with a $H_2:CO$ ratio close to unity can be obtained from the reaction.

$$CH_4 + CO_2 \rightleftharpoons 2CO + 2H_2 \quad \Delta H°_{298K} = +260.5 \text{ kJ mol}^{-1} \quad (1.3)$$

$$CO_2 + H_2 \rightleftharpoons CO + H_2O \quad \Delta H°_{298K} = +41 \text{ kJ mol}^{-1} \quad (1.4)$$

The initial studies for the reaction of methane with CO_2 date to the end of the 19th century.[17] After that, Fischer and Tropsch investigated the DRM reaction in the presence of Ni–Co catalyst.[18]

The importance of the DRM reaction becomes more apparent day by day due to the abundance of CO_2 in the atmosphere and the impacts of the global climate crisis.[19] The annual estimated CO_2 emission reached 38 billion tonnes according to data from the International Energy Agency (IEA).[20] In these circumstances, the elimination of CO_2 and CH_4 through the DRM reaction and production of synthesis gas with a $H_2:CO$ ratio approaching unity would be an opportunity for solving one of the greatest challenges of humankind.[21] The reaction stoichiometry dictates a synthesis gas mixture of $H_2:CO = 1:1$ under reaction conditions, which would be beneficial for the production of synthetic chemicals *via* the Fischer–Tropsch reaction route.[22]

DRM technology is not industrially mature due to similar problems to those faced by SRM, which are deactivation of the catalyst due to sintering of metal particles and severe coke deposition over the catalyst surface.[23] Similar to the SRM, Ni/Al_2O_3 or $Ni/MgAl_2O_4$ are the most appropriate catalysts for the DRM reaction.[24] The reason for this can be explained as comparable catalytic activity under reaction conditions and low cost.[21] Under the reaction conditions, the formation of distinct carbon structures is observed on the catalyst, particularly on Ni particles. The elevated temperature of the reaction, exceeding the Tammann temperature, resulted in the sintering of metal particles. As an economic solution, partially sulphur-passivated Ni catalyst was suggested as a promising catalyst and has recently been taken into consideration.[25]

1.1.4 Partial Oxidation and Autothermal Reforming of Methane and Carbon Formation

Partial oxidation of methane (POX) and autothermal reforming of methane (ATR) processes are mature technologies that have been used by industry for a long time. The characteristic reactions for both are given in eqn (1.5) and (1.6). As can be seen from the formation enthalpies of the reaction, POX is a mild exothermic reaction, while ATR has a mildly endothermic nature. In the first instance, it can be said that both of these processes can be feasible in terms of energy efficiencies with respect to SRM and DRM reactions. In addition, POX and ATR processes, generally, produce a synthesis gas mixture whose $H_2:CO$ ratio is between 1–2, which can be beneficial for production of chemicals *via* the Fischer–Tropsch reaction.[19,22] On the other hand, the main disadvantage of both processes is the requirement of a pure oxygen production unit, which means additional cost. Use of air instead of oxygen can be considered but it is not preferred due to additional volume increase because of the nitrogen content of air and heating issues inside the reformer.

$$CH_4 + 0.5O_2 \rightleftharpoons CO + 2H_2 \quad \Delta H^\circ_{298K} = -35.7 \text{ kJ mol}^{-1} \quad (1.5)$$

$$CH_4 + O_2 + 0.5H_2O \rightleftharpoons CO + 2.5H_2 \quad \Delta H^\circ_{298K} = 85.3 \text{ kJ mol}^{-1} \quad (1.6)$$

$$CH_4 \rightleftharpoons C + 2H_2 \quad \Delta H^\circ_{298K} = 75.0 \text{ kJ mol}^{-1} \quad (1.7)$$

$$2CO \rightleftharpoons C + CO_2 \quad \Delta H^\circ_{298K} = -173.0 \text{ kJ mol}^{-1} \quad (1.8)$$

Two common side reactions for almost all of the hydrogen production technologies are the Boudouard eqn (1.8) and methane decomposition eqn (1.7) reactions, which are thermodynamically favourable at SRM and/or DRM reaction conditions, causing coke formation over the catalyst surface, especially at mild temperatures (400–700 °C).[26]

1.2 Thermodynamics of Reactions

1.2.1 Steam Reforming of Methane

SRM and the accompanying side reactions are reversible reactions with endothermic characteristics. During thermodynamic calculations, SRM and water gas shift reactions are taken into consideration. The equilibrium conversion of methane during the SRM reaction is given (see Figure 1.1) at different pressures and $CH_4:H_2O$ ratio of 1. Due to the endothermic characteristics of the reaction, equilibrium conversion of methane increases with increasing temperature and decreases with increasing pressure at constant temperature. Similarly, the equilibrium compositions of products and reactants at the outlet of the SRM reactor are given (see Figure 1.2). The calculations were performed with a $CH_4:H_2O$ feed ratio of 1 and under atmospheric pressure. As can be seen from Figure 1.2, the $H_2:CO$ ratio increases with increasing temperature. The equilibrium consumption of H_2O is higher than CH_4 due to the water gas shift reaction.

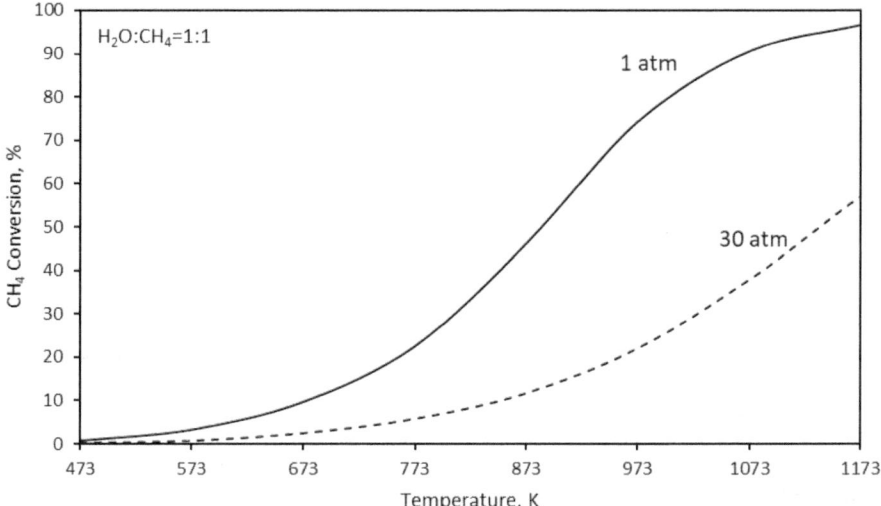

Figure 1.1 Steam reforming of methane reaction equilibrium conversions of methane at different pressures at 200–900 °C and $CH_4:H_2O = 1:1$ feed ratio.

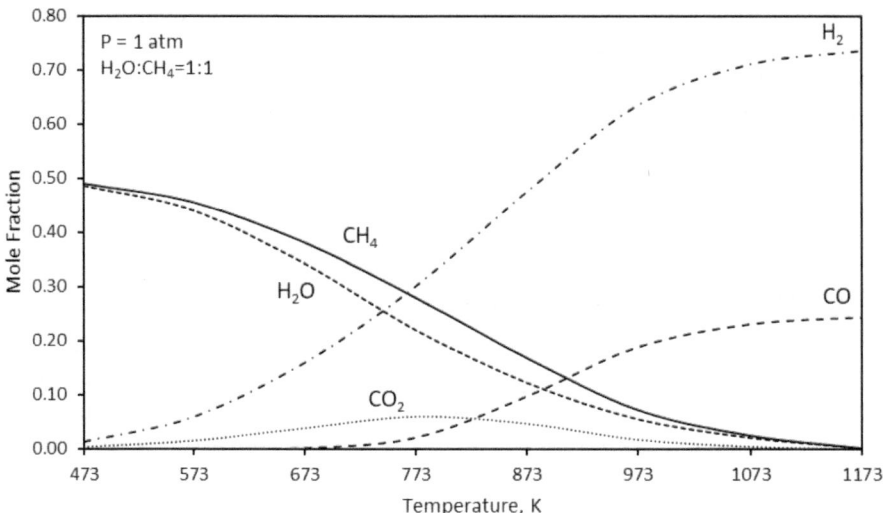

Figure 1.2 Steam reforming of methane reaction equilibrium composition at the reactor outlet at 1 atm and $H_2O:CH_4 = 1:1$.

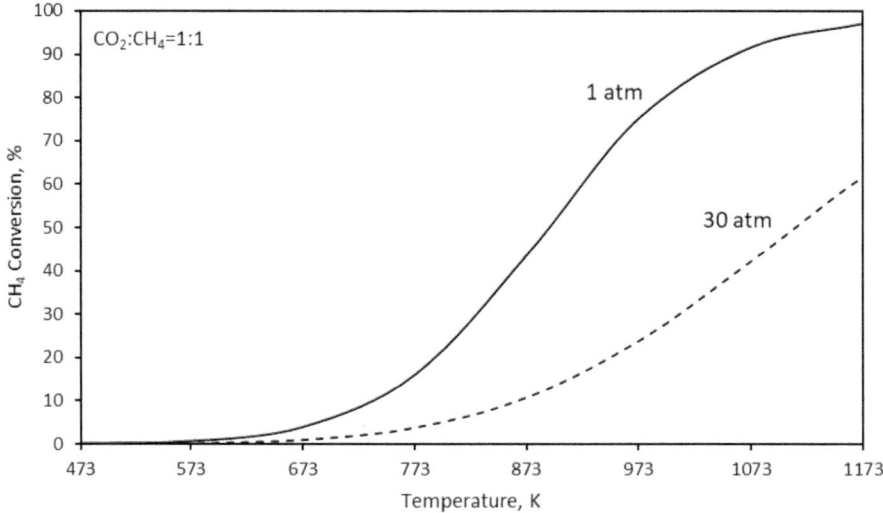

Figure 1.3 Dry reforming of methane reaction equilibrium conversions of methane at different pressures at 200–900 °C and $CO_2:CH_4 = 1:1$ feed ratio.

1.2.2 Dry Reforming of Methane

The thermodynamic analyses of the DRM and side reactions are shown in Figures 1.3 and 1.4. The DRM and the reverse water gas shift reactions are considered in the thermodynamic calculations. The equilibrium conversion

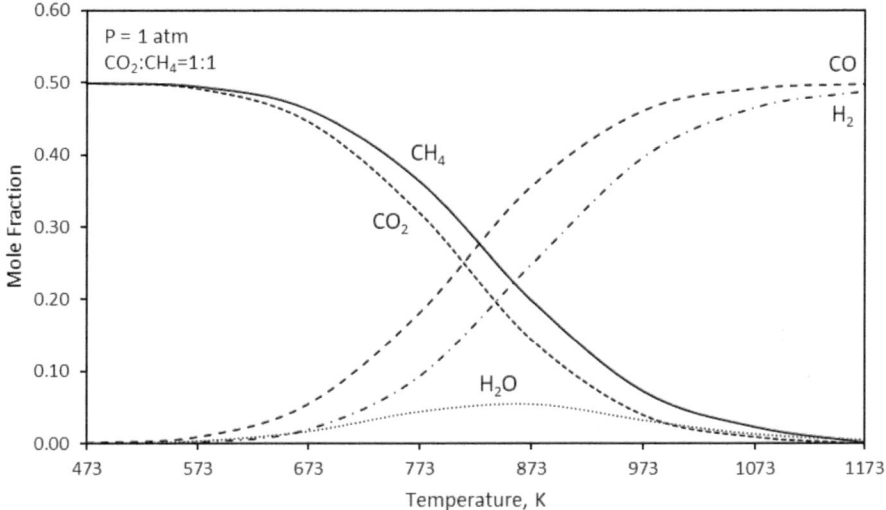

Figure 1.4 Dry reforming of methane reaction equilibrium composition at the reactor outlet at 1 atm and $CO_2 : CH_4 = 1 : 1$.

of methane increases with increasing temperature at constant pressure and decreases with increasing pressure at constant temperature when $CH_4 : CO_2 = 1 : 1$ is fed into the reactor as shown in Figure 1.3. This behaviour can be explained by the reversible endothermic nature of the DRM. Equilibrium compositions at the reactor outlet at 1 atm and a $CH_4 : CO_2$ ratio of 1 are shown in Figure 1.4. The consumption of CH_4 and CO_2 does not follow the same trend due to the occurrence of the reverse water gas shift reaction. According to the reaction stoichiometry of DRM, equal molar amounts of H_2 and CO should be produced. However, this is not possible due to the simultaneous occurrence of the reverse water gas shift reaction, which consumes some of the H_2 produced. On the other hand, the $H_2 : CO$ ratio approaches unity as the temperature increases.

1.3 Catalysts

A number of catalysts have been investigated in the literature for their potential use in SRM and DRM reactions. The objective has been to identify a catalyst that can achieve high activity, long stability and operate the manufacturing process in a cost-effective manner. Despite the fact that noble metal catalysts demonstrate greater activity per active site, the commercial catalyst employed in SRM is Ni/Al_2O_3.[7] The primary reasons for the industrial utilization of the Ni/Al_2O_3 catalyst can be attributed to its high activity and relatively low production cost. As might be expected, industrial catalyst production companies are currently developing Ni/Al_2O_3 SRM catalysts with certain promoters to enhance the catalyst's resistance to deactivation.[7] This section of the chapter presents the results of the literature review on the tested catalysts for both SRM and DRM reactions, based on a Ni/Al$_2$O$_3$ catalyst considered as a benchmark for both reactions.

1.3.1 Steam Reforming of Methane

1.3.1.1 Promoters

The addition of noble metal promoters (Pt, Pd, Ir, Ru and Rh) to Ni/Al$_2$O$_3$-based catalyst has been widely described in the literature.[27–30] In general, it was found that the noble metal promoters improved the dispersion and stability of the Ni metal nanoparticles. In this way, Ni metal nanoparticles are not affected by sintering. For example, Jaiswar *et al.* investigated the effect of the addition of small amounts of Pt to Ni/MgAl$_2$O$_4$ catalyst. The addition of a low concentration of Pt to the catalyst resulted in an enhanced dispersion of Ni nanoparticles through the formation of a Pt–Ni structure and an improvement in the stability under the reaction conditions. Consequently, the 0.1 wt% Pt-loaded Ni/MgAl$_2$O$_4$ catalyst exhibited the optimal performance among the samples studied.[27] In another study, the addition of Rh to Ni/Al$_2$O$_3$ catalyst was investigated. It was found that the addition of rhodium increased the catalytic activity in terms of methane conversion and improved the reduction ability of the catalyst. It was concluded that the addition of rhodium increased the density of active sites over the catalyst.[28] Morales-Cano claimed that while Ni metal dispersion increased with the use of Ir and Rh as promoter, the same observation could not be seen in the presence of Ru. They explained this situation as the crystal structures of Ni and Ru are not similar and thus they cannot interact well with each other.[29] On the contrary, the impact of Au (Ag) addition to the Ni/Al$_2$O$_3$ and Ni/MgAl$_2$O$_4$ catalysts was examined. Au addition resulted in the modification and/or blockage of both low and high coordination sites of Ni atoms/nanoparticles, which led to a decrease in the number/density of active sites of Ni nanoparticles and a reduction in the catalytic activity in terms of CH$_4$ conversion. Furthermore, an increase in the measured activation energy was observed with the addition of increasing amounts of Au. Based on these findings, the authors proposed that two adjacent Ni sites are necessary for the CH$_4$ activation process.[31–33]

The addition of redox metal oxides (CeO$_2$, Nb$_2$O$_5$, La$_2$O$_3$, TiO$_2$, CeO$_2$–ZrO$_2$) to Ni-based catalysts was also investigated. The main observation for the modification of the catalyst surface with a reducible oxide is a decrease in the amount of coke deposition/formation over the catalyst surface.[34–41] The researchers attributed this result to the oxygen transfer ability of reducible oxides, which increase the rate of coke removal from the catalyst surface. Gonçalves and Souza promoted Ni/Al$_2$O$_3$ with Nb$_2$O$_5$. Nb$_2$O$_5$ addition up to an extent improved the activity and stability of the catalyst *via* suppression of coke deposition.[34] Ce addition on Ni/Al$_2$O$_3$ catalyst was also investigated. It was found that Ce promotion increased the Ni metal dispersion and inhibited the coke formation. Furthermore, higher loading of Ce affected the catalyst negatively due to blocking of the Ni metal active sites.[38]

The effect of alkaline and earth alkaline (Ca, Sr, K) metals has also been investigated for the SRM reaction over Ni/Al$_2$O$_3$. The results showed that the addition of alkali and/or earth alkaline promoter increased the coke resistance of the catalyst, thus improving the catalytic activity.[39,42–44] Borowiecki et al. investigated K addition to Ni/Al$_2$O$_3$. The coke resistance ability of the catalyst increases depending on the location of the potassium, which should be in close proximity with the Al$_2$O$_3$ surface.[42] In another study, it was reported that coke resistance and stability characteristics of the Ni/MCM-41 catalyst improved with the addition of Sr.[43] Promotion of Ni/Al$_2$O$_3$ with CaZrO$_3$ nanoparticles was also studied. 15 wt% CaZrO$_3$-loaded Ni/Al$_2$O$_3$ catalyst demonstrated comparable or enhanced performance in terms of H$_2$ yield and CH$_4$ conversion at lower steam/methane (S/C) ratios compared to the bare Ni/Al$_2$O$_3$ catalyst. This was attributed to the presence of oxygen vacancies on CaZrO$_3$ and an increased amount of water adsorption to these sites.[44]

There have been studies on the effect of the addition of non-noble metals to Ni-based catalysts in the literature. For example, Co addition over Ni/Al$_2$O$_3$ catalyst has been studied. Although Co promotion increased the coke resistance ability and stability of the catalyst over a long time on stream conditions, dispersion of metal particles decreased, as well as the catalytic activity. The reason for this is explained by the location of Co atoms over low coordination sites of Ni nanoparticles thus resulting in a decrease in the inhibition effect of coking.[45] Furthermore, the promotion of Ni/zeolite-Y catalyst with Mo$_2$C was investigated. The Ni-Mo$_2$C/zeolite-Y catalyst exhibited enhanced activity and stability in comparison to the Ni/zeolite-Y catalyst. The observed increase in catalyst activity upon addition of Mo$_2$C was attributed to an improvement in methane activation characteristics, rather than an enhancement in water activation.[46]

1.3.1.2 Supports

The existing literature on the improvement of the benchmark Ni/Al$_2$O$_3$ catalyst also focuses on the support materials. The primary function of the support material for a SRM catalyst can be defined as providing a support structure for metal nanoparticles. In this context, Al$_2$O$_3$ is regarded as one of the most effective support materials for the SRM reaction due to its high thermal stability and surface characteristics. In order to enhance the performance of the catalyst, modified alumina supports and new support materials with distinct characteristics have been investigated in the literature.

As evidenced by the extant literature, support modifications or changes have been implemented primarily to enhance the two distinct characteristics of the catalysts: Ni dispersion and coke resistance. In order to enhance the dispersion of Ni in the catalyst, a range of support materials were used. In some of the studies, the support materials such as mesoporous alumina,[47,48] MCM-41,[43] MgO,[49,50] SiO$_2$,[51] Zn–Al,[52] hydrotalcite,[53] Mg–Al,[52] SiO$_2$–Al$_2$O$_3$,[54]

Nb–Al$_2$O$_3$,[55] NiO–MgO,[56,57] MgO–ZrO$_2$,[57] CeO$_2$[58] and Zr–CeO$_2$[58] were used to improve the Ni dispersion.

Different synthesis techniques were also employed for the support materials and metal loading strategies to increase the catalytic activity of the catalyst *via* increasing the metal dispersion by preventing sintering. This involved the use of different support materials, synthesized using varying techniques, and the incorporation of varying Ni loadings. Additionally, the synthesis of distinct Al$_2$O$_3$ structures was undertaken to facilitate the formation of smaller Ni nanoparticles. The synthesis methods differ according to the desired position of Ni metal nanoparticles in the catalyst. The Ni nanoparticles and other additives were incorporated into the crystal structure of the support material in order to prevent the sintering of the Ni particles and the additives that are used to promote the support materials. A variety of synthesis methods were employed to create complex support structures, including precipitation,[55] co-precipitation,[52,53,58] exsolution,[56] microemulsion[51] and solution combustion.[57] Another strategy is loading of the Ni metal nanoparticles and promoters onto the support materials by impregnation, co-impregnation and sol–gel synthesis methods[43,47,48]. The implementation of diverse catalyst synthesis methodologies has permitted researchers to synthesize well-dispersed nickel nanoparticles on support materials, thereby facilitating enhanced SRM reaction performance in terms of methane conversion, hydrogen yield and catalyst stability.

Another important problem of Ni/Al$_2$O$_3$ catalyst is coke formation over the surface under reaction conditions. In order to provide resistance against coke deposition on Ni-based catalysts, generally, support materials capable of oxygen transfer were preferred such as reducible oxides (CeO$_2$, Ce–ZrO$_2$, *etc.*), perovskites (LaFeO$_3$, CaTiO$_3$, *etc.*) and pyrochlores (La$_2$Sn$_2$O$_7$, La$_2$Zr$_2$O$_7$, *etc.*).[59–68] The main characteristics of the support materials listed is their redox capabilities, where the metal reduces to a low oxidation state and oxygen is transferred to another domain over the support surface. Salcedo *et al.* tested Ni/CeO$_2$ catalyst under SRM reaction conditions and reported that a good metal–support interaction and low amount of metal loading are needed for a coke-free operation.[59] Urasaki *et al.* showed that LaAlO$_3$-supported Ni catalyst showed higher stability than Ni/Al$_2$O$_3$ catalyst, which was attributed to the reducibility of LaAlO$_3$ under reaction conditions and it having the capability of oxygen transfer and regeneration characteristics.[67] Another type of material, which is named pyrochlore, was also used as a support. Ma *et al.* reported that the catalytic activity of Ni/La$_2$Zr$_2$O$_7$ was better than that of Ni/Al$_2$O$_3$. Although the catalytic activity of Ni/La$_2$Sn$_2$O$_7$ was not as good as Ni/La$_2$Zr$_2$O$_7$, the long-term stabilities of both of them, with no coke formation, were better than Ni/Al$_2$O$_3$.[65]

Pd-promoted and unpromoted Ni/YSZ catalysts were also synthesized and tested for the SRM reaction. The catalytic activity in terms of CH$_4$ conversion and H$_2$ yield was better for 0.5 Pd–Ni/YSZ catalyst compared to unpromoted Ni/YSZ. The better performance of Pd-promoted Ni/YSZ was attributed to the high water adsorption capacity of the catalyst.[69]

1.3.1.3 Supported Noble Metal Catalysts

It is well known in the literature that supported noble metal catalysts are highly active for SRM reaction as well as being resistant to coke formation. The main drawback of supported noble metal catalysts is their cost. Therefore, the industrial use of supported noble metal catalysts does not look promising in the near future. However, the R&D of supported noble metal catalysts is still evolving, especially on the enlightenment of the surface reaction mechanisms for the SRM reaction. In this manner, the studies that were performed in the presence of supported noble metal catalysts for SRM reaction are summarized.

SRM reaction experiments were performed in the presence of supported Ru catalysts in the literature. The test results were also compared with traditional Ni/Al_2O_3 catalyst. Kikuchi et al. performed activity tests with 0.5 wt% Ru catalyst with various types of support materials such as La_2O_3, Al_2O_3, SiO_2, TiO_2, ZrO_2, Nb_2O_5, V_2O_3, CaO and MgO.[70] They reported that the highest catalytic activity was observed over Ru/Al_2O_3, which was similar to the catalytic activity of 15 wt% Ni/Al_2O_3. Some of the researchers claimed that support materials have no influence on the catalytic activity of Ru catalysts.[71,72] Wei and Iglesia investigated Al_2O_3 and ZrO_2 supported Ru catalysts and concluded that the support materials only influence the dispersion of the metal particles.[71] A comparable investigation was conducted by Amjad et al. on the catalytic activities of Ru/Al_2O_3 and Ru/CeO_2 catalysts.[73] On the other hand, in a study conducted by Amjad et al., the catalytic activity of 1.5 wt% Ru-loaded Ru/MgO and Ru/Nb_2O_5 catalysts were evaluated. The findings revealed that the Ru/Nb_2O_5 catalyst exhibited higher catalytic activity, which was attributed to the presence of Nb_2O_5 in the tetragonal crystal structure and Ru in the metallic state.[73] Carvalho et al. studied the effect of Mg addition to Ru/Al_2O_3. It was noted that Mg addition decreased the acidity, increased the Ru dispersion and stability of the catalyst but did not improve the catalytic activity of the catalyst in terms of CH_4 conversion.[74] Furthermore, some support materials, synthesized using special techniques, were studied, such as Al_2O_3@Al, Ni_6Al, $Co_{6-x}Mg_xAl_2$, etc. For most of these studies, the reaction test results showed that near-equilibrium CH_4 conversion was achieved.[75–77]

The catalytic activities and stabilities of various types of supported Rh catalysts were evaluated for the SRM reaction. The reaction test results of reducible oxide (CeO_2 and/or La_2O_3) supported and/or promoted Rh catalysts exhibited higher activity and stability than Rh/Al_2O_3 catalyst due to their redox abilities, which increased the coke resistance of the catalysts.[78–82] Furthermore, the promotion of Al_2O_3 with CeO_2 led to the formation of $CeAlO_3$, which enhanced the stability and metal–support interaction of the Rh/Sm_2O_3–CeO_2–Al_2O_3 catalyst. The addition of Sm_2O_3 enabled a greater reduction of CeO_2 and, consequently, the dispersion of the Rh nanoparticle remained consistent under reaction conditions.[80] Lanthanum zirconate pyrochlore type supported Rh catalysts were tested for SRM reaction. All of the catalysts demonstrated high catalytic activity and good performance against the formation of coke deposits.[83]

Different synthesis parameters for Rh/Al_2O_3 catalysts were investigated by several research studies. Yu et al. showed the effect of particle size on the catalytic activity of Rh/Al_2O_3 catalysts.[84] Rh particles were distributed within a narrow range of 1–3 nm on Al_2O_3 synthesized by flame spray pyrolysis (FAl), whereas the Rh particle size ranged from 2 to 7 nm on Al particle suspension (SAl). It was proposed that the preformed Al_2O_3 support (SAl) promotes the catalytic activity, possibly enhancing H_2 spillover, which is considered to facilitate the rate-determining steps in SRM. In order to evaluate the potential of noble metal catalysts supported on $MgAl_2O_4$, a series of experiments were conducted. While the noble metals exhibited varying turnover frequencies, with Pd > Ir > Pt ~ Rh > Ru > Ni, the most active catalysts were found to be $MgAl_2O_4$-supported Rh and Ir. This was attributed to the highly dispersed metal nanoparticles on the catalyst surface.[85] The addition of Sr to hexaaluminate-supported Rh catalyst resulted in enhanced strength, which prevented sintering of Rh nanoparticles. As a consequence, the catalyst exhibited superior activity and stability.[86] Hydroxyapatite (HAp)-supported Rh catalysts with different metal loadings were studied and 1 wt% Rh-loaded Rh/HAp showed the highest CH_4 conversion without deactivation over 30 h, which was attributed to its high coke-resistance ability.[87]

Pt- and Ir-based catalysts were also evaluated for SRM reaction by many research studies. Amongst the most used supports for Pt metal are reducible oxides such as CeO_2 and La_2O_3. They are used as either a support material or promoter for Al_2O_3 support. It can be said that platinum-doped reducible oxide powders showed good activity towards the SRM reaction.[88–94] The most commonly proposed explanation for this observation is the oxygen transfer ability of CeO_2 and/or La_2O_3 to the Pt surface and the oxidation of adsorbed C species. Wattanathana et al. have also revealed, according to their experimental results, that an increase in the Pt loading results in a decrease in coke formation.[88] In addition, the test results revealed the presence of $[LaPt_xO]Pt^0$-like species when La_2O_3 was used as a promoter. Apart from the reducible oxides, Al_2O_3–MgO-supported Pt catalyst was synthesized with different Al/Mg ratios. The reaction test results revealed that the highest activity was achieved when Al-rich MgO–Al_2O_3 support was used.[95]

1.3.1.4 Supported Non-noble Metal Catalysts

Supported non-noble transition metal catalysts exhibited significant catalytic activity for the SRM reaction. In most studies, the main problem for non-noble metal catalysts is rapid deactivation due to coke deposition. Additives such as noble metal promoters increase the activity and stability of supported non-noble metal catalysts. Profeti et al. investigated the addition of noble metals as promoters to Co/Al_2O_3 catalyst. Methane conversion increased from 7% to 60% with the addition of Pt.[96] In addition, Lucredio et al.

demonstrated a significant decrease in coke deposition with the addition of La and Ce as promoters to Co/Mg/Al catalysts.[97,98] In another study, superior catalytic activity and high resistance to carbon formation were reported over Pt-, La- and Zr-promoted Co/Al$_2$O$_3$ catalyst.[99] Co$_6$Al$_2$, a double-layered hydroxide, supported Cu catalysts were synthesized and achieved nearly complete conversion at 700°C without any coke formation.[100,101] Al$_2$O$_3$-supported and unsupported Mo-based catalysts were also evaluated. All of the samples were carburized to obtain Mo$_2$C. The highest catalytic activity as well as lowest coke deposition were observed in the presence of an unsupported sample that was carburized at 700°C.[102]

1.3.2 Dry Reforming of Methane

1.3.2.1 Modifications of Ni/Al$_2$O$_3$ Catalyst

In order to increase the efficiency of Ni/Al$_2$O$_3$ catalyst, some research studies have been performed as described in the literature. Generally, these studies focused on changing the surface morphology of Ni/Al$_2$O$_3$ catalyst to increase resistance to coke formation and deposition. In this framework, different types of catalyst synthesis and pre-treatment methods were used to investigate the coke formation mechanism. The test results indicated that the type of pre-treatment influenced the mean particle size of the Ni metal nanoparticles and there was a direct relationship between the degree of coke formation and the particle size of the metal.[103–106] Similarly, Xu et al. reported that while mainly filamentous carbon was observed on the Ni/α-Al$_2$O$_3$ catalyst, a low amount of graphitic carbon was deposited on the Ni/γ-Al$_2$O$_3$ due to the high dispersion and small nanoparticle size of the Ni metals.[107] Yang et al. demonstrated that the type of Ni attachment on the support influences the coke resistance of the Ni/Al$_2$O$_3$ catalyst. They observed that the coke resistance ability of the catalyst increases with an enhanced interaction between the Ni and the support.[108] Zhou et al. observed that the mean particle size of the catalyst decreased with increasing calcination temperature of the catalyst, which resulted in a reduction in the coke deposition.[109]

Conversely, Zhang et al. demonstrated the formation of NiAl$_2$O$_4$ structure over a Ni/Al$_2$O$_3$ surface, as well as the production of small-sized Ni metal nanoparticles through the partial reduction of NiAl$_2$O$_4$. It was observed during the reaction tests that both the Ni particles and the NiAl$_2$O$_4$ phase inhibited coke formation and deposition on the catalyst surface, thereby providing stability to the catalyst for long-time on-stream tests. While NiO nanoparticles were used for CH$_4$ activation, CO$_2$ activation was carried out by oxygen vacancies, which formed over NiAl$_2$O$_4$ during the reaction.[110]

1.3.2.2 Noble and Non-noble Metals as Promoters

1.3.2.2.1 Metal Promoters. Various types of promoters were used in the literature to enhance the catalytic activity and coke resistance of the traditional Ni/Al_2O_3 catalyst. The promoters can be divided into three groups, which are alkaline/earth alkaline metals, metals and noble metals.

The addition of alkaline and earth alkaline metals as promoters to Ni/Al_2O_3 catalysts improved the coke removal from the catalyst surface.[111,112] Dias and Assaf indicated that CaO interacts with CO_2 during the reaction.[112] On the other hand, Juan-Juan et al. reported that potassium migrates from the support surface to Ni metal nanoparticles and improves the removal of coke by oxidation.[111]

Studies in the literature on the use of noble metals as promoters for Ni/Al_2O_3 catalysts have generally shown that the addition of small amounts of noble metals increases the dispersion and reducibility of Ni metal nanoparticles, thereby enhancing the catalytic activity.[113–116] In addition, lower amounts of coke formation were reported.[115,117] Garcia-Dieguez demonstrated that Pt inhibited the formation of $NiAl_2O_4$ and avoided the sintering of Ni metal nanoparticles.[113] Niu et al. expressed that Pt addition improved the $H_2:CO$ ratio by inhibiting the reverse water gas shift reaction. In addition, density functional theory (DFT) studies revealed that oxidation of the C–H bond was more favourable than carbon formation over the Pt–Ni surface, which provides the catalyst with high coke resistance.[117] Damyanova et al. reported that the reducibility and dispersion of Ni metal nanoparticles increased in the presence of Rh promoter.[114] Alvarez Moreno et al. claimed that Ru improved the reducibility and metal dispersion of Ni via hydrogen spillover. The addition of small amounts Ru increased the catalytic activity while decreasing the coke formation.[115]

Transition metals were also used to promote Ni/Al_2O_3-based catalysts. The addition of Co increased both the catalytic activity and resistance to coke formation and sintering, therefore improving the stability of the catalyst for a longer time on stream tests.[118–120] The addition of Cu metal improved significantly the reforming stability due to the formation of a Ni–Cu alloy and a lower rate of coke deposition, although a significant change in the catalytic activity was not observed.[118,121] The addition of V also inhibited both $NiAl_2O_4$ formation and coke formation over Ni/Al_2O_3 catalyst.[122] On the other hand, the addition of Mn and Zr adversely affected the catalytic activity by covering the Ni metal particles, causing inaccessibility of Ni metal and formation of carbon filaments depending on the coke deposition.[118]

1.3.2.2.2 Support Promoters. Coating the surface of Al_2O_3 with alkaline and earth alkaline metals/oxides (K, MgO, CaO, SrO, BaO) has been shown to increase the catalytic efficiency of Ni/Al_2O_3 catalyst. Coating of Al_2O_3 with MgO doubled the activity of the catalyst via preventing the formation of $NiAl_2O_4$. SEM and TEM analyses of the used $Ni/MgAl_2O_4$ catalyst showed the formation of whisker type carbon on the surface. EPR spectroscopy revealed the existence of oxygen vacancies, which are beneficial for coke removal. In situ XRD showed the stability of the Ni metal and $MgAl_2O_4$ phase during the reaction conditions. When an ordered mesoporous Al_2O_3 support was used and promoted by MgO or CaO, two different promotional effects were observed: (i) stabilization of Ni metal nanoparticles and thus preventing sintering and (ii) MgO or CaO prevented the deposition of coke and improved the catalytic activity by increasing the number of basic sites. Coke suppression was also achieved by limiting diffusion within the pores.[123–132] One of the reasons for the activity and stability enhancement can be explained by the similar crystalline of NiO and MgO.[133] The promotion of ZSM-5-supported Ni catalyst with K and Ca showed that alkaline metals react with CO_2 to form carbonate species and with the help of Ni at the adjacent site facilitate the dissociate adsorption of CO_2.[134,135] The addition of Sn to Ni/MgO catalyst helped the catalyst to inhibit the coke deposition reaction through CH_4 decomposition.[136] The addition of K in the presence of CeO_2 increased the strength of coke resistance of the catalyst. Furthermore, K facilitated the reverse Boudouard reaction, which promoted the gasification of carbon-like species and reduced the rate of CO_2 adsorption on the catalyst surface.[137,138] The effects of additions of Co and MgO or CaO were investigated and they was reported to have an impact on the reducibility of the catalyst.[139]

CeO_2- and/or La_2O_3-promoted Ni/Al_2O_3 catalyst showed better performance compared to Ni/Al_2O_3. The increase in the catalytic activity and low carbon deposition were attributed to the oxidative properties of CeO_2/La_2O_3 and the increase in Ni metal dispersion. Furthermore, La increased the catalytic activity and stability of the catalyst by enhancing the basic sites, which was used to inhibit coke deposition by forming $La_2O_2CO_3$.[140–145]

Xu et al. investigated the effect of MgO and CeO_2 addition on Al_2O_3 support. They put forth that while MgO inhibits the methane dehydrogenation to produce coke, CeO_2 is responsible for increasing the carbon resistance of the catalyst.[146]

Li and Wang studied the effect of ZrO_2 as a promoter and observed a catalytic activity increase and suppression of coke deposition over Ni/Al_2O_3–ZrO_2 catalyst. They explained that ZrO_2 addition increased the dispersion of the metal and decreased the mean particle size of Ni. In addition, ZrO_2 prevents the formation of $NiAl_2O_4$ structure. Therefore, better catalytic activity and low carbon deposition were obtained.[147,148]

Nandini et al. investigated the influence of CeO_2 or MnO as a promoter for Ni/Al_2O_3. The results exhibited that CeO_2 and/or MnO addition inhibited the coke formation without affecting CH_4 conversion and H_2 yield. CeO_2 or MnO promoters were located over the Al_2O_3 surface as patches and therefore the metal dispersion was increased and the mean particle size of Ni decreased.[137,138]

1.3.2.3 Supports

MgO has been used as a support. XPS revealed that there is an electron transfer between Ni and Mg, which increases the strength of the bond, therefore preventing sintering and coke deposition. Pulse MS study also showed that CO_2 and CH_4 were responsible for CO and H_2 production *via*

CO_2 and CH_4 decompositions, respectively.[149] Isotopic exchange experiments showed that lattice oxygen of the support first oxidizes Ni metal then reacts with coke.[150] One possible explanation for the activity and stability enhancement is the similar crystalline structures of NiO and MgO.[133] The addition of Sn to Ni/MgO catalyst helped the catalyst to inhibit the coke deposition reaction through CH_4 decomposition.[136] Wei and Iglesia put forth that CH_4 activation was carried out over Ni metal particles.[151] The effect of the precursor, which was used to synthesize MgO, was investigated. The test results revealed that Ni/MgO catalyst that was prepared using $[MgCO_3]_4Mg(OH)_2$ showed the best performance amongst the other precursors.[152]

A Ni/ZrO_2 catalyst was prepared and tested for the DRM reaction. The results demonstrated that ZrO_2 had weak basic sites. The catalyst exhibited an effective reaction performance, which can be attributed to the presence of smaller Ni metal nanoparticles, whose agglomeration was hindered by the high interaction of Ni particles with the support and by the oxygen transfer ability of ZrO_2.[153] On the other hand, some researchers argued that there is a weak interaction between Ni and ZrO_2, thus deactivation was observed due to agglomeration.[154,155] Han et al. reported that Ni/ZrO_2 catalyst had five times lower initial TOF values than Ni/MgO or Ni/A_2O_3 catalysts.[156] MgO addition to ZrO_2 catalyst improved the stability and performance of the catalyst while minimizing the coke depositions.[157]

Ce–ZrO_2 has also been used as a support material for the DRM reaction. Several synthesis methods were used. Better activity and stability were achieved. It was found that $Ce_{0.8}Zr_{0.2}O$ crystalline support had a strong interaction with highly dispersed Ni metal particles, which was highly resistant to sintering. Besides, the redox ability of CeO_2 enhanced the coke oxidation through oxygen transfer from Ce^{4+}–Ce^{3+} transformation. While Ce–ZrO_2 shows promise as a support for DRM reactions, the synthesis type of the catalyst is of significant importance with regard to the location and stability of Ni metal particles.[158–163] The addition of MgO to the Ce–ZrO_2 improved the sintering resistance of the catalyst.[164–168] WO_3 and CeO_2 addition to Ni/ZrO_2 catalyst was studied. The oxidation ability of the catalyst increased with the addition of Ce and W redox couples and the density of basic sites increased with WO_3 and CeO_2 promotion, which facilitated the removal of coke.[169]

CeO_2 has been used as a support material for the DRM reaction due to the redox properties of the material. Laosiripojana and Assabumrungrat reported that lattice oxygen of CeO_2 could react with CH_4 and CO to produce CO and CO_2, respectively.[62] CeO_2-supported Ni–Co bimetallic catalyst was examined and compared with Ni/CeO_2 under reaction conditions. It was revealed that increasing the calcination temperature of the catalyst decreased the carbon deposition on the support surface.[170]

La_2O_3 was also used as a support for the DRM reaction. La_2O_3 reacted with CO_2 to form $La_2O_2CO_3$ and this caused the material to exhibit better performance without deactivation.[171] When Cr_2O_3 and La_2O_3 metal oxides were added to Ni/MgO catalyst together, lattice oxygen could easily transfer through the Ni metal where carbon species tended to be deposited. Lattice oxygen that migrated to Ni metal reacted with carbon; therefore, this mechanism resulted in inhibiting the coke formation *via* methane dehydrogenation.[172]

In most of the studies, SiO_2 and TiO_2 were tried as support materials for the DRM reaction but no promising results have been obtained up to date.[173–175] A high amount of carbon deposition was measured over various amounts of Ni-loaded SiO_2 and TiO_2 catalysts.[175] Ni/TiO_2 under photo-thermal reaction conditions showed better performance.[176] On the other hand, it was demonstrated that Ni/SiO_2 (Ni/Silicalite) catalyst with a metal particle size of *ca.* 2.0–3.0 nm had almost excellent catalytic activity and stability due to the high dispersion of Ni metal nanoparticles, strong metal–support interaction and the existence of basic sites on the support coming from the synthesis technique.[177–179] The material was prepared by a special synthesis method, which resulted in a very small particle size of Ni.

Different types of structure silica-based materials were used as a support material. Ni/MCM-41 catalyst was tested and it was found that improved catalytic activities could be achieved with an optimum amount of Ni loading due to the fact that highly dispersed Ni metal nanoparticles can be stabilized in the pore walls of MCM-41 ordered structure.[180] Y_2O_3–Ni/MgO–MCM-41 catalyst was synthesized, characterized and tested for the DRM reaction. The addition of MgO and Y_2O_3 increased the basicity of support and reducibility and dispersion of Ni metal, respectively. Tip type carbon nanotubes were detected over used catalyst, which deactivated the catalyst.[181] The addition of Ce on Ni/MCM-41 catalyst was also used. A small amount of CeO_2 loading covered the Ni metal particles with a very thin layer such that CeO_2 can transfer oxygen easily.[182] Indium addition to Ni/SiO_2 improved the coking resistance of the catalyst such that the electron density over Ni metal increased due to In–Ni alloy formation which decreased the C–H bond activation ability. Deep coke formation can be eliminated by inhibiting the C–H bond activation over Ni metal.[183] In another study, a fibrous silica was used as the support material, which had basic sites and achieved higher reaction rates and good stability. Fibrous silica encapsulated the Ni particles, thus no sintering and high dispersion were observed. With the help of basic sites located on the support, coke deposition was prevented.[184] Amin *et al.* compared MCM-41, TMS, KIT-6, SBA-15 and MCF as support for Ni metal for the DRM reaction. The most promising one was reported as TMS due to the thickness of the pore wall and pore size opening of the material.[185] Ce addition to SiO_2 catalyst was also investigated. When the catalyst was treated under Ar flow, high lattice transfer ability of CeO_2 become apparent and oxidized the reactive carbon species. On the other hand, when H_2 was used for pre-treatment, the oxygen transfer ability of CeO_2 diminished due to reduction, therefore coke deposition cannot be removed.[186] CeO_2-modified silicate nanotubes showed improved catalytic activity. The good performance of the catalyst was attributed to the existence of defect sites at the interface of Ni and CeSiO$_{2-x}$ and oxygen transfer capability of CeO$_2$. This property facilitated dissociative CO$_2$ adsorption to form CO* and O*.[187]

Bimetallic catalysts were also tested for the DRM reaction. Generally, Ni metal was promoted with a (noble) metal in the presence of a support material apart from Al$_2$O$_3$. The addition of Pt or Pd to Ni/ZrO$_2$, preferably by co-impregnation, prevents coke formation.[188] The presence of Pd on a mesoporous silica-supported Ni catalyst surface affected mainly the Ni metal distribution of the catalyst such that accessible and smaller Ni metal particles are formed and avoid sintering of Ni nanoparticles.[189] Pt addition to Ni/CeO$_2$ catalyst improved the anti-coking ability of the catalyst as well as increased the metal dispersion and stabilized them on the surface.[190,191] Co–Ni/CeO$_2$ exhibited better catalytic activity and stability than Ni/CeO$_2$, which was attributed to the formation of Ni–Co alloy during reduction, thus hydrogen can be weakly adsorbed to the surface and the reverse water–gas shift reaction can be suppressed. In addition, the oxygen transfer ability of CeO$_2$ was promoted with the presence of Co, hence the carbon removal was improved.[192] Ru addition to Ni/Al$_2$O$_3$ catalyst blocked the formation of NiAl$_2$O$_4$ and thus increased the activity and stability of the catalyst.[116] Promotion of Ni/MgO catalyst with Mo increased the activity and stability of the catalyst for longer times with no carbon deposition. It was elucidated that Ni-Mo alloy has the potential to migrate to the edge and corner sites of MgO crystalline during a temperature ramp, where it can be situated with 17 nm particle size. In addition to this, Mo–C interaction was observed.[193,194] Co-Ni/TiO$_2$ was tested and characterized. It was shown that a homogeneous Co–Ni alloy was obtained and this exhibited resistance to coke formation and metal oxidation.[195]

Perovskites are also a candidate as support materials for the DRM reaction and are used widely. LaNiO$_4$ perovskite type material was used since it can be transformed into La$_2$O$_3$-supported Ni catalyst under reaction conditions and showed similar activity with pristine La$_2$O$_3$ support material.[196–198] Lima et al. used La$_{1-x}$Ce$_x$NiO$_3$ as a catalyst since LaNiO$_3$ deactivated slowly with time due to coke deposition. A low amount of Ce addition influenced the catalytic activity and stability positively due to improved oxygen transfer characteristics of the catalyst.[199] Valderrama et al. performed the reaction tests starting from LaNi$_{1-x}$Co$_x$O$_3$ perovskite-type oxide. XRD results showed that LaNiO$_3$ and LaCoO$_3$ phases were present. During the reaction tests, Ni^0 and Co^0 were in the metallic phase and La was in the form of La$_2$O$_2$CO$_3$.[200] In another study, they added Sr to the perovskite matrix and prepared La$_{1-x}$Sr$_x$CoO$_3$ catalyst. The presence of highly dispersed metallic Co, SrO and La$_2$O$_3$ was demonstrated under reaction conditions. They reported that Sr addition improved the stability of nanosized Co metal particles.[201] The effect of the addition of Cr to the B site of La$_{0.8-x}$Sr$_x$Cr$_{0.85}$Ni$_{0.15}$O$_3$ perovskite structure was tested and Cr was shown to increase the stability of the catalyst.[202]

Boron nitride (BN) was also used as a support for the DRM reaction. Dong et al. prepared hexagonal-BN=supported metal (Ni, Fe, Co, Ru) nanoparticles. The oxidative etching pretreatment was applied to the catalyst to have a BO_x layer over the metal nanoparticles. It was observed that the BO_x layer prevents the sintering of Ni nanoparticles.[203] Bu *et al.* prepared a defect-confined BN-supported Ni catalyst and Ni metal particles were located in the defects of the support, which avoided sintering of particles due to SMSI. In addition, B-O-H sites were generated over the BN support, which facilitated CH_4 decomposition and consecutively hydrogenation of adsorbed CO_2.[204]

The promotion of ZSM-5 supported Ni catalyst with K and Ca showed that alkaline metals react with CO_2 to form carbonate species and with the help of Ni at the adjacent site facilitate the dissociative adsorption of CO_2.[134,135] On the other hand, Ni/ZSM-5 catalyst with Si/Al = 30 exhibited a good reaction performance for 30 h without deactivation, which was attributed to the high acidity of the catalyst.[205] Ni/HZSM-5 enhanced the hydrogen spillover from metal to support and improved the CO_2 reduction ability of the catalyst.[206] Ni–Co/Attapulgite-derived MFI (ADM) zeolite was tested. ADM support not only prevents the adsorption of excess amounts of CO_2 but also avoids sintering of the Ni metal particles.[207] Cr addition to dealuminated β-supported Ni catalyst exhibited lower carbon deposition and inhibited CH_4 decomposition.[208]

1.3.2.4 Supported Noble Metal Catalysts

One of the main problems for the supported Ni-based catalyst is deactivation due to coke deposition under DRM reaction conditions. In this framework, the use of supported metal catalysts has been taken into consideration. Their strength of resistance to coke deposition compared to the nickel-based catalyst has been proven but they have not been given a chance industrially due to their high cost.[14,209] On the other hand, the use of supported metal catalysts in R&D studies have been beneficial in terms of providing information on the surface reaction mechanism of a dry methane reforming reaction. In this manner, there have been many studies in the literature to find a supported metal catalyst that is active under reaction conditions, resistant to coke deposition and stable for a long time on stream reaction tests.

Rostrup-Nielsen and Hansen tested MgO-supported noble metal catalysts for DRM reaction and reported the activity of the metals as the following:

$$Ru > Rh \approx Ni > Ir > Pt > Pd$$

They also showed that carbon deposition over noble metal catalyst is about 0–0.5 wt% at 500 °C, while it is 3.5 wt% over Ni catalyst.[24] On the other hand, a similar study was carried out by Kikuchi and Chen[210] over Al_2O_3-supported noble metal catalysts and the order of the activity of noble metals was as follows:

$$Rh \approx Pt > Pd > Ru \approx Ir$$

Wei and Iglesia carried out a study on ZrO_2–CeO_2-supported noble metal catalysts (Ru, Ir, Pt, Rh and Ni) and reported that Pt/ZrO_2–CeO_2 catalyst had the highest activity and coke resistance ability.[71,151,211–214] They claimed that CH_4 activated on the metal surface irreversibly, CO_2 activation and water gas shift reaction should be in quasi equilibrium.

Bitter et al. performed kinetic experiments on Pt/ZrO_2 catalyst. The results revealed that while CH_4 is activated over a metal surface, CO_2 is activated over either a support surface or metal–support interface.[215] Carrara et al. also achieved the same conclusion with Bitter et al. over Ru/La_2O_3 catalyst, although severe coke deposition and metal agglomeration were observed.[216] Nagaoka et al. showed that coke resistance of Pt/ZrO_2 was better than Pt/Al_2O_3.[217] Hou et al. compared the activities and coke resistance abilities of Al_2O_3-supported noble metal (Ru, Rh, Pt, Pd, Ir) and Ni catalysts. The results revealed that noble metal catalysts had better resistance to coke deposition than Ni catalyst and Rh/Al_2O_3 showed the best performance among the others.[218] Richardson et al. also obtained a similar result when they compared Rh/Al_2O_3 and Pt–Re/Al_2O_3 catalyst.[219] A series of reducible (CeO_2, ZrO_2, Nb_2O_5, Ta_2O_5, TiO_2) and irreducible (Al_2O_3, La_2O_3, MgO, SiO_2, Y_2O_3) metal oxide-supported Rh catalysts were investigated and it was found that Rh/MgO and Ru/Al_2O_3 were the most promising catalysts. While SiO_2-, Y_2O_3-, TiO_2- and Ta_2O_5-supported catalysts rapidly deactivated, ZrO_2 and CeO_2 catalysts needed longer times to be activated.[220]

Apart from the conventional support materials, different types of materials were used for DRM reaction. Gour et al. synthesized pyrochlore type Rh-substituted catalysts and tested their activity for the DRM reaction. They reported that Rh-substituted and Ca-Rh-substituted pyrochlore catalysts exhibited the best performance. The carbon resistance of Ca-Rh-substituted catalyst was better than the Rh-substituted one.[221] In another study, the catalytic activities of $CePr_{0.1}O_2$-, $CeZr_{0.1}O_2$- and CeO_2-supported Ir catalyst were investigated. The reaction test results indicated that the $CePr_{0.1}O_2$- and $CeZr_{0.1}O_2$-supported catalysts exhibited enhanced catalytic activity, which was attributed to their ability to stabilize the Ir metal, preventing agglomeration.[222]

1.3.2.5 Supported Non-noble Metal Catalysts

The main disadvantage of supported noble metal catalysts is their cost. In this framework, supported non-noble metals were also evaluated for the DRM reaction, which has been mainly Co. It was revealed that supported Co catalysts rapidly deactivate due to severe coke deposition when supports such as Al_2O_3, ZrO_2 and TiO_2 are used.[223–225] The mechanism of deactivation was explained to be due to the formation of carbon over the metal and oxidation of the metal.[223] On the contrary, it was revealed that the loading amount of Co and calcination temperature of the catalyst directly determined the activity and stability of the catalyst. Ruckenstein and Wang showed that a highly active and stable Co/Al_2O_3 catalyst was tested for DRM reaction with an optimum Co loading (9 wt%) and a calcination temperature of 500 °C.[223] A similar conclusion was achieved for Co/activated carbon.[226]

1.4 (Micro)kinetics and DFT Studies

1.4.1 Steam Reforming of Methane

Steam reforming of methane is a mature technology for hydrogen production, however gaining an insight into the reaction mechanism over the catalyst surface is an important issue as this could improve the efficiency of the process. In this manner, many research studies have been carried out to learn more information on the surface reaction mechanism of SRM.

One of the earliest studies on the microkinetics of the SRM reaction was performed by Temkin.[230] Temkin's model takes into account both the SRM and the water gas shift reactions, as shown in eqn (1.9)–(1.15):

$$CH_{4(g)} + * \rightleftharpoons CH_2* + H_2 \tag{1.9}$$

$$CH_2* \rightleftharpoons C* + H_2 \tag{1.10}$$

$$C* + H_2O_{(g)} \rightleftharpoons CO* + H_2 \tag{1.11}$$

$$CO* \rightleftharpoons CO + * \tag{1.12}$$

$$H_2O_{(g)} + * \rightleftharpoons O* + H_2 \tag{1.13}$$

$$OH* + * \rightleftharpoons O* + H* \tag{1.14}$$

$$CO + O* \rightleftharpoons CO_2 + * \tag{1.15}$$

and the corresponding reaction rate expression for the SRM reaction is given in eqn (1.16):

$$r = \frac{k_1 P_{CH_4} P_{H_2O} \left[1 - \left(\frac{1}{K}\right)\left(\frac{P_{CO} P_{H_2}^3}{P_{CH_4} P_{H_2O}}\right)\right]}{\left(P_{H_2O} + I_1 P_{H_2O} P_{H_2} + I_2 P_{H_2}^2 + I_3 P_{H_2}^3\right)\left[1 + K_5\left(\frac{P_{H_2O}}{P_{H_2}}\right)\right]} \tag{1.16}$$

where k_i and K_i represent the rate and equilibrium constant, respectively, and I_1, I_2, I_3 are the parameters related with the rate constants of elementary steps which can be found in detail in ref. 230.

Later, Ross and Stell performed kinetic experiments over Ni/α-Al$_2$O$_3$ to determine the order of the reactants according to power law kinetics. The results revealed that the orders of the SRM reaction are 1 and −0.5 with respect to partial pressures of methane and steam, respectively, at 500–700 °C and under vacuum conditions.[231]

Xu and Froment also proposed a microkinetic model that was compatible with experimental results.[232] Although their experimental results are similar to those obtained by Temkin, they added some elementary steps to the reaction mechanism such as the formation of the CH_2O intermediate which was proposed early by Menon et al.[233]

Aparicio studied the microkinetic modelling of the SRM over a $Ni/MgAl_2O_4$ catalyst and separately investigated the dissociation of methane and steam and the formation of CO reactions on the catalyst surface.[234] Aparicio proposed the dissociation of methane to adsorbed carbon (C*) species and desorption of molecular hydrogen from the surface. It was concluded that the availability of surface oxygen was an important parameter for the reaction mechanism and could not be considered as a single rate-determining step for the reaction.

In light of previous studies and additional experimental results, Avetisov et al. proposed a microkinetic model which included the dissociation of methane up to C* and formation of CO through COH intermediate.[235] On the other hand, Wei and Iglesia studied a series of noble metals and Ni catalysts to gain an insight into the reaction mechanism for the SRM.[115,211,212] The test results demonstrated that dissociation of methane is the rate-determining step for SRM reaction independent of where the reaction proceeds. In addition, the reaction rate is first order with respect to methane partial pressure under the conditions that experiments were performed. No reaction order was reported for partial pressure of steam. The other steps should be in either equilibrium or quasi-equilibrium.

Berman et al. carried out a kinetic study of Ru/Al_2O_3 and reported that the reaction order with respect to partial pressure of methane changed with temperature, which was determined as below 1 at 450 and 500 °C and as nearly equal to 1 at 700–900 °C.[236] In addition, steam had a negative reaction order. They concluded that methane dissociates over Ru metal and steam adsorbs on the support. In the next step, steam migrates and dissociates over the Ru surface. Finally, the adsorbed carbon species oxidize with adsorb oxygen to form CO. Similarly, Jacobsen reported that the reaction was first order with respect to the methane partial pressure and that the kinetics of the reaction were negatively affected by the adsorption of CO and H_2 on Rh/ZrO_2 catalyst under the experimental reaction conditions.[237]

Jones et al. published a study on the reaction mechanism of the SRM reaction, where the elementary steps are given in eqn (1.17)–(1.25), and DFT calculations of the relevant elementary reaction steps over different metal surfaces.[238] Their study eliminated many unknowns and put forth the activity–noble metal relationship both experimentally and theoretically by testing the supported noble metal catalysts with different nanoparticle sizes and constructing volcano plots as a function of C and O adsorption energies. They concluded that the formation of CO is kinetically the most relevant step over noble metals at low temperatures. At higher temperatures, the most important step for the reaction mechanism is determined as the dissociation of methane due to changes in the characteristics of noble metals. This finding explains the different outcomes of many research studies in the literature.

$$CH_{4(g)} + 2* \rightleftharpoons CH_3* + H* \quad (1.17)$$

$$CH_3* + * \rightleftharpoons CH_2* + H* \quad (1.18)$$

$$CH_2* + * \rightleftharpoons CH* + H* \quad (1.19)$$

$$CH* + * \rightleftharpoons C* + H* \quad (1.20)$$

$$H_2O_{(g)} + 2* \rightleftharpoons OH* + H* \quad (1.21)$$

$$OH* + * \rightleftharpoons O* + H* \quad (1.22)$$

$$C* + O* \rightleftharpoons CO* + * \quad (1.23)$$

$$H* \rightleftharpoons 0.5H_{2(g)} + * \quad (1.24)$$

$$CO* \rightleftharpoons CO_{(g)} + * \quad (1.25)$$

Potential energy diagrams obtained from DFT calculations can be used to guide the selection of the optimal catalyst for the SRM reaction. There are several studies in the literature that calculate the reaction energetics either for all the elementary steps or for a part of the elementary steps that are considered to be rate limiting. Jones et al. carried out a DFT study for whole steps over noble and transition metal catalysts at 500 °C.[238] The energy diagrams suggested that Ru and Ni were the most feasible metals for the reaction. Yu et al. carried out DFT calculations over the Ni_4/CeO_2 cluster to determine the thermodynamically most feasible pathway for the SRM reaction.[239] The results revealed that the optimum reaction pathway was dissociation of methane up to *CH radical and oxidation to obtain *CHO radical and synthesis of CO by cleavage of H atom. Zhu et al. investigated the activation of methane over Ni/Al_2O_3 catalyst in the presence or absence of OH radical over Ni metal theoretically.[240] The calculations showed that the reaction barrier for the dissociation of the methane step decreased from 2.15 eV to 0.46 eV in the presence of two OH radicals compared to the absence of OH radicals.

1.4.2 Dry Reforming of Methane

DRM is not yet a mature technology for either H_2 or synthesis gas production. One of the main drawbacks of the technology is the deactivation of the catalyst due to coke formation, as mentioned earlier. Therefore, many studies have been carried out to gain an insight into the surface reaction mechanism of the DRM reaction in order to manage problematic reaction steps. Kathiraser et al. compiled many of the kinetic studies on DRM including the rate expressions based on power law, Eley–Rideal and LHHW kinetics.[241] Although the results of most studies have reached a consensus on the rate-determining step(s) and the coke formation mechanism for DRM reactions over mainly Ni-based catalysts, some debates are still awaiting an answer to explain the whole picture.

One of the earliest studies on the detailed reaction mechanism of DRM over Ni catalyst was carried out by Bradford and Vannice.[242] The reaction mechanism consisted of the decomposition of CH_4, dissociation of CO_2 *via* reverse water gas shift reaction and reaction of surface hydroxyl with CH_x species that was obtained by CH_4 decomposition as given in eqn (1.26)–(1.33). In this mechanism, the decomposition of CH_4 and the reaction of CH_xO radicals with surface hydroxyl were determined to be the rate-determining steps for the reaction. In addition, a corresponding rate expression was given in eqn (1.34);

$$CH_{4(g)} + * \rightleftharpoons CH_x* + \left(\frac{4-x}{2}\right)H* \tag{1.26}$$

$$CO_{2(g)} + * \rightleftharpoons CO_2* \tag{1.27}$$

$$H_2 + 2* \rightleftharpoons 2H* \tag{1.28}$$

$$CO_2* + H* \rightleftharpoons CO* + OH* \tag{1.29}$$

$$OH* + H* \rightleftharpoons H_2O_{(g)} + 2* \tag{1.30}$$

$$CH_x* + OH* \rightleftharpoons CH_xO* + H* \tag{1.31}$$

$$CH_xO* \rightarrow CO* + \left(\frac{x}{2}\right)H_2 \tag{1.32}$$

$$CO* \rightleftharpoons CO_{(g)} + * \tag{1.33}$$

$$r_{CH_4} = \frac{k_1 P_{CH_4} P_{CO_2}}{\frac{k_{-1}K}{k_7} P_{CO} P_{H_2}^{(4-x)/2} + \left(1 + \frac{k_1}{k_7} P_{CH_4}\right) P_{CO_2}} \tag{1.34}$$

where k_i and K were the rate constants and equilibrium constant, respectively.

Later, Iglesia[151,211] and Rostrup-Nieslen[14] and their co-workers discussed the similarity of the SRM and DRM reactions in terms of their kinetics. Wei and Iglesia proposed that the CH_4 dissociation step was the rate-determining step and took place on the metal surface and the other steps were in equilibrium including the dissociation of CO_2.[151] Some other researchers have also come to similar conclusions as a result of their microkinetic modelling and DFT studies.[243–245]

Recent studies have revealed that some different dominating pathways and/or rate-determining steps would occur during the DRM reaction.[241,246] Tsipouria and Verykios reported that the rate-determining steps were CH_4 decomposition and reaction of *C with oxycarbonates originated from

La$_2$O$_2$CO$_3$ over Ni/La$_2$O$_3$ catalyst.[243] Zhu *et al.* showed that the oxidation of *C species, which was produced by the dissociation of CH$_4$, was the rate-determining step for the reaction.[247] Fan *et al.* reported that the dominant pathways for the DRM reaction were the oxidation of *CH and *C by surface *O or *OH.[270] The most likely step between them was determined to be the oxidation of *C by surface *O. They also added that the rate-determining step may change with respect to the reaction pressure, such that while the rate-determining step at high pressures was either CH$_4$ dissociation or *CH/*C oxidation, the rate-determining step at low pressures was the *CH/*C oxidation step. Omran also carried out a mechanistic study and reached a similar conclusion.[248] On the contrary, Wang *et al.* indicated that oxidation of *CH species was energetically the most favourable step according to DFT calculations over a Ni catalyst.[249]

Although many modifications to the surface reaction mechanism for the DRM reaction have been made according to experimental and theoretical research studies, a consensus has almost been achieved, which states that the reaction mechanism goes through synthesis of *CHO radical by hydrogenation of *CO as shown in refs. 250–252. Some minor changes can be seen in different papers. For example, Dehimi *et al.* claimed that the main oxidizer is *OH instead of *O.[250]

1.5 New Generation Reactors and Process Intensification

Two types of conventional reactors have been used for H$_2$ production *via* the SRM route. These reactors are named the tubular reformer and heat exchange reformer. In the operation of the tubular reformer, the reactants enter the reactor at 450–650 °C and leave the reactor at 700–950 °C.[253] The reactor operates at around 25 bar.[254] Fifty percent of the heat that is generated by fired hydrocarbon can be transferred inside to the tubular reforming reactor. On the other hand, approximately 80% of the heat generated can be transferred inside the reactor due to the special design of the reactor. Conventional reactors are almost optimized to produce H$_2$ or synthesis gas as efficiently as possible; however, they are limited in terms of yield due to the reversible nature of the reactions. Therefore, different types of reactors have been investigated and implemented in laboratory and pilot scales to overcome the thermodynamic barrier of the steam and DRM reaction. In the upcoming section, brief information is given on the most investigated reactor/process intensification developments.

1.5.1 Sorption-enhanced Reforming of Methane

Sorption-enhanced SRM is based on the adsorption of CO$_2$ on a sorbent to shift the reaction to the product side. The reactions that take place during the SRM process are given in eqn (1.1) and (1.2) neglecting the other side reactions. The sum of eqn (1.1) and (1.2) is given in eqn (1.35);

$$CH_4 + H_2O \rightleftharpoons CO_2 + 4H_2 \quad \Delta H^\circ_{298K} = +165 \text{ kJ mol}^{-1} \quad (1.35)$$

Eqn (1.35) implies that removal of CO_2 from the reaction medium directly shifts the equilibrium reaction to the product side. This improvement adds many advantages to the process. First, almost pure-grade CO_2 can be produced, which can either be used as a chemical or sequestrated easily.[255] Secondly, the per pass conversion rate of the SRM reaction can increase beyond the thermodynamic limitation.[256] Finally, nearly pure-grade H_2 can be produced even if it is not pure enough for some customers.

A continuous and fluidized bed type SRM process is depicted in Figure 1.5. There are two different beds, which are the reformer and regenerator. In the reformer part, the SRM reaction takes place at relatively low temperatures (500–600 °C) and CaO captures the produced CO_2. Then, $CaCO_3$ is transferred to the regenerator part to decompose into CaO and CO_2 at high temperature (900 °C).[257] The regenerated sorbent is fed again into the reformer part to complete the cycle. Therefore, H_2 can be produced at a high purity level even if it is not pure enough for some customers.[256]

In this process, one of the most important parameters is the stability of the combination of the catalytic and sorbent materials. Numerous types of catalytic and sorbent materials have been tried in the literature.[15] The sorbent/catalytic materials can be evaluated in two groups, which are natural sorbents and synthetic sorbents. The main advantage of the natural sorbents such as CaO and dolomite is their low price. On the other hand, the synthetic sorbents can be regenerated <100 times, which means that their lifetimes are much more than those of the natural substitutes,[256] but they are costly. In order to increase the lifespan of the natural type sorbents, stabilizers such as Al_2O_3, $Ca_9Al_6O_{18}$ and $CaZrO_3$ are used, and promising results have been obtained.[258]

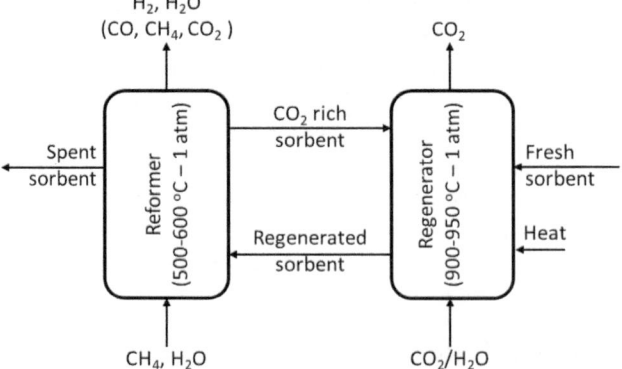

Figure 1.5 Simplified schematics of a continuous–fluidized bed-based sorption-enhanced steam reforming of methane process.

Although the experimental and preliminary results of sorption-enhanced SRM technology are promising, some challenges are still waiting to be solved. One of the main problems is discovering a catalyst and sorbent couple that has a long lifespan without affecting the severe operation conditions of the process. The other issue is the optimization of the energy demand and heat recovery of the process under industrial operating conditions.[259]

1.5.2 Oxidative (Steam or Dry) Reforming of Methane

Oxidative reforming of methane is a beneficial process intensification to produce H_2 or synthesis gas without deactivation of the catalyst due to coke formation at relatively low temperature.[260] As a result, sintering and safety problems are eliminated by operating the process at low temperature. The oxidative reforming of methane can be defined as the simultaneous execution of partial oxidation of methane and (steam/dry) reforming of methane reactions. A general reaction is given for oxidative SRM in eqn (1.6), which is based on the addition of small amounts of O_2 to the reactor to remove deposited coke on the catalyst surface and to facilitate some energy saving by the oxidation of coke. In addition to the prevention of coke formation on the catalyst surface, there are some other benefits to the oxidative reforming of methane. The oxidative (autothermal) SRM process can be operated at lower steam to methane ratios compared to current industrial applications in order to obtain a product composition that fulfils customer requirements.[253] A similar advantage of the oxidative DRM is that the product composition can be changed by tuning the feed composition.[261]

1.5.3 Chemical Looping Reforming of Methane

Chemical looping is one of the popular methods of intensifying the chemical processes, which have challenges such as reforming of methane technologies. Chemical looping has been tried for both SRM and DRM processes.[262] Simplified schematics of chemical looping SRM and chemical looping DRM are given in Figure 1.6. In this process, methane and the oxidizer (H_2O or CO_2) react in different reactors and an oxygen carrier cycles between the two reactors. In the methane (fuel) reactor, methane converts to H_2 in the SRM process and to synthesis gas ($H_2 + CO$) in the DRM process by reacting with a metal oxide. Then, the metal oxide is transferred to the oxidizer or CO_2 reactor to re-oxidize in the presence of H_2O or CO_2. In some cases, an air reactor is included in the process layout to ensure complete oxidation of the metal oxide. As a result, high-purity H_2 and synthesis gas can be produced at the outlet of the chemical looping SRM and DRM processes, respectively. The looping continues by transferring the oxidized metal oxide into the fuel reactor.

The oxygen carrier is generally a metal oxide that can release and capture its lattice oxygen in its structure in different reactor mediums.[263] There are many studies in the literature to develop or discover an oxygen carrier that

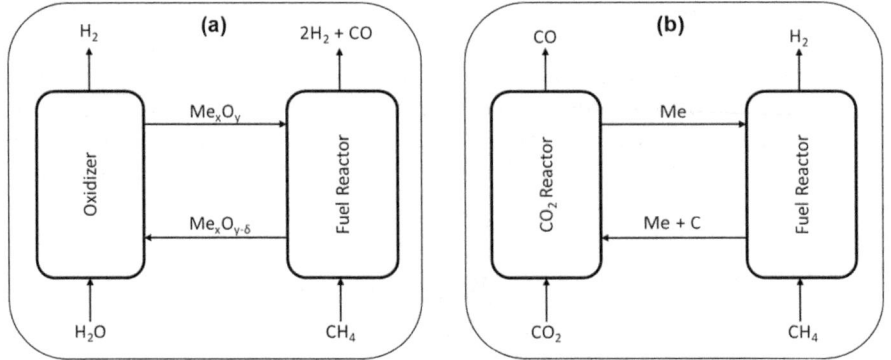

Figure 1.6 Simplified schematics of (a) chemical looping of steam reforming of methane, (b) chemical looping of dry reforming of methane.

can be operated for longer times.[15] The ones mostly used are Fe_2O_3, NiO, supported NiO, La–Fe-containing perovskites and some others, and their detailed analyses can be found elsewhere.[15,262]

In recent years, sorption-enhanced chemical looping SRM has gained the attention of researchers.[15] In this method, the sorbent, catalytic material and oxygen carrier are circulated together to simultaneously facilitate methane conversion and hydrogen production. Higher methane conversion rates and H_2 production can be observed at 650 °C.[264] The main problem with this system is the regeneration of the catalytic materials, which can only be achieved under pure oxygen.[15]

1.5.4 Other New-generation Methane-reforming Routes

There are some other emerging technologies to convert methane to H_2 and/or synthesis gas more effectively such as the use of membrane reactors,[265,266] microreactors,[267] photocatalytic SRM,[268] and electrical field-enhanced and plasma-enhanced SRM.[15,269] All of these technologies have some advantages but are still in the early R&D phase. Therefore, detailed information can be found elsewhere for readers with further curiosity.[15]

1.6 General Summary and Outlook

A general summary of the SRM and DRM processes was conducted within this chapter. The importance of sustainable hydrogen production is mentioned and the related production methods are evaluated from an industrial point of view. It is well known that both SRM and DRM reactions with their side reactions are thermodynamically limited, therefore the equilibrium compositions at the reactor outlet have been calculated. The main focus of attention has been given to compiling the recent advances in the development of new catalytic materials for SRM and DRM processes. Furthermore, surface reaction kinetics and process intensifications for the reactor operations are discussed. From a technological point of view, it can be said that the main challenge for the production of sustainable (green) hydrogen is directly related to the thermodynamic limitations of both SRM and DRM processes, where operation of the reaction at lower temperatures causes many problems such as coke formation. On the contrary, high-temperature operation causes sintering of the catalyst and higher energy consumption. In these circumstances, the development of new-generation reactor operations based on unsteady-state operation may result in more efficient hydrogen production technologies.

References

1. DOE Hydrogen and Fuel Cells Program Record, https://www.hydrogen.energy.gov/docs/hydrogenprogramlibraries/pdfs/19002-hydrogen-market-domestic-global.pdf?Status=Master. [Accessed: 10.10.2024].
2. Global hydrogen demand in the Net Zero Scenario, 2022–2050, https://www.iea.org/data-and-statistics/charts/global-hydrogen-demand-in-the-net-zero-scenario-2022-2050. [Accessed: 10.10.2024].
3. S. Solomon, G.-K. Plattner, R. Knutti and P. Friedlingstein, *Proc. Natl. Acad. Sci. U. S. A.*, 2009, **106**, 1704.
4. J. DeAngelo, I. Azevedo, J. Bistline, L. Clarke, G. Luderer, E. Byers and S. Davis, *Nat. Commun.*, 2021, **12**, 6096.
5. A. Ganguli and V. Bhatt, *Front. Therm. Eng.*, 2023, **3**, 143987.
6. C. Kalamaras and A. Efstathiou, *Conf. Pap. Sci.*, 2013, **2013**, 690627.
7. J. Rostrup-Nielsen, *Catalysis: Science and Technology*, Springer, Berlin, Heidelberg, 1984, vol. 5.
8. H. Davy, *Philos. Trans. R. Soc. London*, 1817, (107), 77.
9. M. Tessie Du Motay and M. Marechal, *Bull. Soc. Chim. Paris*, 1868, **334**, 9.
10. L. Mond and C. Langer, Pat., DRP 51572, 1889.
11. O. Dieffenbach and W. Moldenhauer, Pat. DRP 229406, 1909.
12. A. Mittasch and C. Schneider, Pat. DRP 296866, 1912.
13. B. Neumann and K. Jacob, *Z. Elektrochem. Angew. Phys. Chem.*, 1924, **30**, 557.
14. J. R. Rostrup-Nielsen and J. Sehested, *Adv. Catal.*, 2002, **47**, 65.
15. H. Zhang, Z. Sun and Y. H. Hu, *Renewable Sustainable Energy Rev.*, 2021, **149**, 111330.
16. L. Chen, Z. Qi, S. Zhang, J. Su and G. A. Somorjai, *Catalysts*, 2020, **10**, 858.
17. J. Lang, *Z. Phys. Chem.*, 1888, **2**, 161.
18. F. Fischer and H. Tropsch, *Brennst.-Chem.*, 1928, **3**, 39.
19. H. Zhu, H. Chen, M. Zhang, C. Liang and L. Duan, *Catal. Sci. Technol.*, 2024, **14**, 1712.
20. IEA, CO2 Emissions in 2023, https://www.iea.org/reports/co2-emissions-in-2023/executive-summary. [Accessed: 10.10.2024].

21. I. V. Yentekakis, P. Panagiotopoulou and G. Artemakis, *Appl. Catal., B*, 2021, **296**, 120210.
22. C. Zhang, Y. Li, Z. Chu, Y. Fang, K. Han and Z. He, *Sep. Purif. Technol.*, 2024, **329**, 125109.
23. W.-J. Jang, J.-O. Shim, H.-M. Kim, S.-Y. Yoo and H.-S. Roh, *Catal. Today*, 2019, **324**, 15.
24. J. R. Rostrup-Nielsen and J. H. B. Hansen, *J. Catal.*, 1993, **144**, 38.
25. J. R. Rostrup-Nielsen, J.-H. Bak Hansen and L. M. Aparicio, *Bull. Jpn. Pet. Inst.*, 1997, **40**, 366.
26. C. Ávila-Neto, S. C. Dantas, F. Silva, T. Franco, L. Romanielo, C. Hori and A. Assis, *J. Nat. Gas Sci. Eng.*, 2009, **1**, 205.
27. V. K. Jaiswar, S. Katheria, G. Deı and D. Kunzru, *Int. J. Hydrogen Energy*, 2017, **42**, 18968.
28. E. C. Luna, A. M. Becerra and M. I. Dimitrijewits, *React. Kinet. Catal. Lett.*, 1999, **67**, 247.
29. F. Morales-Cano, L. F. Lundegaard, R. R. Tiruvalam, H. Falsig and M. S. Skjøth-Rasmussen, *Appl. Catal., A*, 2015, **498**, 117.
30. S.-C. Baek, K.-W. Jun, Y.-J. Lee, J. D. Kim, D. Y. Park and K.-Y. Lee, *Res. Chem. Intermed.*, 2012, **38**, 1225.
31. K. de Oliveira Rocha, C. M. P. Marques and J. M. C. Bueno, *Chem. Eng. Sci.*, 2019, **207**, 844.
32. Y.-H. Chin, D. L. King, H.-S. Roh, Y. Wang and S. M. Heald, *J. Catal.*, 2006, **244**, 153.
33. M. Dan, M. Mihet, A. R. Biris, P. Marginean, V. Almasan, G. Borodi, F. Watanabe, A. S. Biris and M. D. Lazar, *React. Kinet., Mech. Catal.*, 2012, **105**, 173.
34. J. F. Gonçalves and M. M. V. M. Sozua, *Catal. Lett.*, 2018, **148**, 1478.
35. M. Dan, M. D. Lazar, V. Rednic and V. Almasan, *Rev. Roum. Chim.*, 2011, **56**, 643.
36. F. Salahi, F. Zarei-Jelyani, M. Farsi and M. Rahimpour, *J. Energy Inst.*, 2023, **108**, 101208.
37. F. Zarei-Jelyani, F. Salahi, M. Meshksar, M. Farsi and M. Rahimpour, *J. Energy Inst.*, 2023, **110**, 101363.
38. X. Yang, J. Da, H. Yu and H. Wang, *Fuel*, 2016, **179**, 353.
39. S. Y. Lee, H. Lim and H. C. Woo, *Int. J. Hydrogen Energy*, 2014, **39**, 17645.
40. Z.-W. Liu, K.-W. Jun, H.-S. Roh and S. Park, *J. Power Sources*, 2002, **111**, 283.
41. A. J. de Abreu, A. F. Lucrédio and E. M. Assaf, *Fuel Process. Technol.*, 2012, **102**, 140.
42. T. Borowiecki, A. Denis, M. Rawski, A. Golebiowski, K. Stolecki, J. Dmytrzyk and A. Kotarba, *Appl. Surf. Sci.*, 2014, **300**, 191.
43. M. Hasani Estalkhi, M. Yousefpour, H. Koohestan and Z. Taherian, *Int. J. Hydrogen Energy*, 2024, **68**, 1344.
44. K. Lertwittayanon, W. Youravong and W. J. Lau, *Int. J. Hydrogen Energy*, 2017, **42**, 28254.

45. X. You, X. Wang, Y. Ma, J. Liu, W. Liu, X. Xu, H. Peng, C. Li, W. Zhou, P. Yuan and X. Chen, *ChemCatChem*, 2014, **6**, 3377.
46. X. Zhang, K. Yim, J. Kim, D. Wu and S. Ha, *Appl. Catal., B*, 2022, **310**, 121250.
47. S. Zolghadri, B. Honarvar and M. Rahimpour, *Fuel*, 2023, **335**, 127005.
48. H. Khosravani, M. Meshksar, M. Koohi-Saadi, K. Taghanam and M. Rahimpour, *J. Energy Inst.*, 2023, **108**, 101203.
49. S. Miura, Y. Umemura, Y. Shiratori and T. Kitaoka, *Chem. Eng. J.*, 2013, **229**, 515.
50. S. Charojrochkul, M. Chanthanumataporn, S. Wongsakulphasatch, S. Assabumrungrat and S. Ratchahat, *Int. J. Hydrogen Energy*, 2024, **86**, 58.
51. B. Han, F. Wang, L. Zhang, Y. Wang, W. Fan, L. Xu, H. Yu and Z. Li, *Res. Chem. Intermed.*, 2020, **46**, 1735.
52. M. Nieva, M. M. Villaverde, A. Monzon, T. F. Garetto and A. J. Marchi, *Chem. Eng. J.*, 2014, **235**, 158.
53. R. Denghan-Niri, J. C. Walmsley, H. Anders, P. A. Midgley, E. Rytter, A. H. Dam, A. B. Hungria, J. C. Hernandez-Garrido and D. Chen, *Catal. Sci. Technol.*, 2012, **12**, 2476.
54. S. Ali, M. J. Al-Marri, A. G. Abdelmoneim, A. Kumar and M. M. Khader, *Int. J. Hydrogen Energy*, 2016, **41**, 22876.
55. W. Zeng, L. Li, M. Song, X. Wu, G. Li and C. Hu, *Int. J. Hydrogen Energy*, 2023, **48**, 6358.
56. Y. S. Park, M. Kang, P. Byeon, S.-Y. Chung, T. Nakayama, T. Ko and H. Hwang, *J. Power Sources*, 2018, **397**, 318.
57. P. Sintuya, S. Charojrochkul, M. Chanthanumataporn, S. Wongsakulphasatch, S. Assabumrungrat and S. Ratchahat, *Int. J. Hydrogen Energy*, 2024, **86**, 58.
58. I. Iglesias, M. Forti, G. Baronetti and F. Mariño, *Int. J. Hydrogen Energy*, 2019, **44**, 8121.
59. A. Salcedo, P. Lustemberg, N. Rui, R. Palomino, Z. Liu, S. Nemsak, S. Senanayake, J. Rodriguez, M. Rodriguez and B. Irigoyen, *ACS Catal.*, 2021, **11**, 8327.
60. X. Fang, X. Zhang, Y. Guo, M. Chen, W. Liu, X. Xu, H. Peng, Z. Gao, X. Wang and C. Li, *Int. J. Hydrogen Energy*, 2016, **41**, 11141.
61. K. Kusakabe, K.-I. Sotowa, T. Eda and Y. Iwamoto, *Fuel Process. Technol.*, 2004, **86**, 319.
62. N. Laosiripojana and S. Assabumrungrat, *Appl. Catal., B*, 2005, **60**, 107.
63. S. Lee, M. Monai, K. Shen, J. Chang, J. Vohs and R. Gorte, *J. Phys. Chem. C*, 2022, **126**, 11619.
64. X. Zhang, L. Peng, X. Fang, Q. Cheng, W. Liu, H. Peng, Z. Gao, W. Zhou and X. Wang, *Int. J. Hydrogen Energy*, 2018, **43**, 8298.
65. Y. Ma, X. Wang, X. You, J. Liu, J. Tian, X. Xu, H. Peng, W. Liu, C. Li, W. Zhou, P. Yuan and X. Chen, *ChemCatChem*, 2014, **6**, 3366.

66. R. Thalinger, M. Gocyla, M. Heggen, R. Dunin-Borkowski, M. Grünbacher, M. Stöger-Pollach, D. Schmidmair, B. Klötzer and S. Penner, *J. Catal.*, 2016, **337**, 26.
67. K. Urasaki, Y. Sekine, S. Kawabe, E. Kikuchi and M. Matsukata, *Appl. Catal., A*, 2005, **286**, 23.
68. Z. Ou, Z. Zhang, C. Qin, H. Xia, T. Deng, J. Niu, J. Ran and C. Wu, *Sustainable Energy Fuels*, 2021, **5**, 1845.
69. S. Lee and S. Hong, *Int. J. Hydrogen Energy*, 2014, **39**, 21037.
70. E. Kikuchi, S. Uemiya, A. Koyama, A. Machino and T. Matsuda, *J. Jpn. Pet. Inst.*, 1990, **33**, 152.
71. J. Wei and E. Iglesia, *J. Phys. Chem. B*, 2004, **108**, 7253.
72. J. Jacobsen, T. Jørgensen, I. Chorkendorff and J. Sehested, *Appl. Catal., A*, 2010, **377**, 158.
73. U. Amjad, G. Gonçalves Lenzi, N. Camargo Fernandes-Machado and S. Specchia, *Catal. Today*, 2015, **257**, 122.
74. L. Carvalho, A. Martins, P. Reyes, M. Oportus, A. Albonoz, V. Vicentini and M. Rangel, *Catal. Today*, 2009, **142**, 52.
75. H. Lee, Y. Potapova and D. Lee, *J. Power Sources*, 2012, **216**, 256.
76. M. Nawfal, C. Gennequin, M. Labaki, B. Nsouli, A. Aboukaïs and E. Abi-Aad, *Int. J. Hydrogen Energy*, 2015, **40**, 1269.
77. D. Homsi, S. Aouad, C. Gennequin, A. Aboukaïs and E. Abi-Aad, *Int. J. Hydrogen Energy*, 2014, **39**, 10101.
78. U. Amjad, A. Vita, C. Galletti, L. Pino and S. Specchia, *Ind. Eng. Chem. Res.*, 2013, **52**, 15428.
79. R. Duarte, M. Nachtegaal, J. Bueno and J. van Bokhoven, *J. Catal.*, 2012, **296**, 86.
80. R. Duarte, O. Safanova, F. Krumeich, M. Makosch and J. van Bokhoven, *ACS Catal.*, 2013, **3**, 1956.
81. R. Duarte, F. Krumeich and J. van Bokhoven, *ACS Catal.*, 2014, **4**, 1279.
82. R. Duarte, M. Olea, E. Iro, T. Sasaki, K. Itako and J. van Bokhoven, *ChemCatChem*, 2014, **6**, 2898.
83. Y. Zhou, D. Haynes, J. Baltrus, A. Roy, D. Shekhawat and J. Spivey, *Appl. Catal., A*, 2020, **606**, 117802.
84. J. Yu, Z. Zhang, F. Dallmann, J. Zhang, D. Miao, H. Xu, A. Goldbach and R. Dittmeyer, *Appl. Catal., B*, 2016, **198**, 171.
85. D. Mei, V.-A. Glezakou, V. Lebarbier, L. Kovarik, H. Wan, K. O. Albrecht, M. Gerber, R. Rousseau and R. A. Dagle, *J. Catal.*, 2014, **316**, 11.
86. N. McGuire, N. Sullivan, R. Kee, H. Zhu, J. Nabity, J. Engel, D. Wickham and M. Kaufman, *Chem. Eng. Sci.*, 2009, **64**, 5231.
87. Z. Boukha, M. Gil-Calvo, B. de Rivas, J. González-Velasco, J. Gutiérrez-Ortiz and R. López-Fonseca, *Appl. Catal., A*, 2018, **556**, 191.
88. W. Wattanathana, N. Nootsuwan, C. Veranitisagul, N. Koonsaeng, N. Laosiripojana and A. Laobuthee, *J. Mol. Struct.*, 2015, **1089**, 9.
89. U. Amjad, C. Quintero, G. Ercolino, C. Italiano, A. Vita and S. Specchia, *Catal. Today*, 2015, **257**, 122.

90. P. Prieto, A. Ferreira, P. Haddad, D. Zanchet and J. Bueno, *J. Catal.*, 2010, **276**, 351.
91. J. Araujo, D. Zanchet, R. Rinaldi, U. Schuchardt, C. Hori, J. Fierro and J. Bueno, *Appl. Catal., B*, 2008, **84**, 552.
92. K. Rocha, J. Santos, D. Meira, P. Pizani, C. Marques, D. Zanchet and J. Bueno, *Appl. Catal., A*, 2012, **431–432**, 79.
93. V. Mortola, S. Damyanova, D. Zanchet and J. Bueno, *Appl. Catal., B*, 2011, **107**, 221.
94. S. Cheah, L. Massin, M. Aouine, M. Steil, J. Fouletier and P. Gélin, *Appl. Catal., B*, 2018, **234**, 279.
95. A. Martins, L. Carvalho, P. Reyes, J. Grau and M. Rangel, *Mol. Catal.*, 2017, **429**, 1.
96. L. Profeti, E. Ticianelli and E. Assaf, *Fuel*, 2008, **87**, 2076.
97. A. Lucrédio and E. Assaf, *J. Power Sources*, 2006, **159**, 667.
98. A. Lucrédio, G. Filho and E. Assaf, *Appl. Surf. Sci.*, 2009, **255**, 5851.
99. S. Itkulova, Y. Boleubayev and K. Valishevskiy, *J. Sol–Gel Sci. Technol.*, 2019, **92**, 331.
100. D. Homsi, S. Aouad, C. Gennequin, J. Nakat, A. Aboukaïs and E. Abi-Aad, *C. R. Chim*, 2014, **17**, 454.
101. D. Homsi, S. Aouad, C. Gennequin, A. Aboukaïs and E. Abi-Aad, *Adv. Mater. Res.*, 2011, **324**, 453.
102. T. Christofoletti, J. Assaf and E. Assaf, *Chem. Eng. J.*, 2005, **106**, 97.
103. J. Juan-Juan, M. Román-Martínez and M. Illán-Gómez, *Appl. Catal., A*, 2009, **355**, 27.
104. S. Ali, M. Khader, M. Almarri and A. Abdelmoneim, *Catal. Today*, 2020, **343**, 26.
105. Y. Xu, X. Du, L. Shi, T. Chen, H. Wan, P. Wang, S. Wei, B. Yao, J. Zhu and M. Song, *Int. J. Hydrogen Energy*, 2021, **46**, 14301.
106. W. Kim, Y. Lee, H. Park, Y. Choi, M. Lee and J. Lee, *Catal. Sci. Technol.*, 2016, **6**, 2060.
107. J.-K. Xu, Z.-J. Li, J.-H. Wang, W. Zhou and J.-X. Ma, *Acta Phys.-Chim. Sin.*, 2009, **25**, 253.
108. B. Yang, J. Deng, H. Li, T. Yan, J. Zhang and D. Zhang, *iScience*, 2021, **24**, 102747.
109. L. Zhou, L. Li, N. Wei, J. Li and J.-M. Basset, *ChemCatChem*, 2015, **7**, 2508.
110. S. Zhang, M. Ying, J. Yu, W. Zhan, L. Wang, Y. Guo and Y. Guo, *Appl. Catal., B*, 2021, **291**, 120074.
111. J. Juan-Juan, M. Román-Martínez and M. Illán-Gómez, *Appl. Catal., A*, 2006, **301**, 9.
112. J. Dias and J. Assaf, *Catal. Today*, 2003, **85**, 59.
113. M. García-Diéguez, I. Pieta, M. Herrera, M. Larrubia and L. Alemany, *J. Catal.*, 2010, **270**, 136.
114. S. Damyanova, I. Shtereva, B. Pawelec, L. Mihaylov and J. Fierro, *Appl. Catal., B*, 2020, **278**, 119335.

115. A. Álvarez Moreno, T. Ramirez-Reina, S. Ivanova, A.-C. Roger, M. Centeno and J. Odriozola, *Front. Chem.*, 2021, **9**, 694976.
116. S. Andraos, R. Abbas-Ghaleb, D. Chlala, A. Vita, C. Italiano, M. Laganà, L. Pino, M. Nakhl and S. Specchia, *Int. J. Hydrogen Energy*, 2019, **44**, 25706.
117. J. Niu, Y. Wang, S. E. Liland, S. K. Regli, J. Yang, K. Rout, J. Luo, M. Rønning, J. Ran and D. Chen, *ACS Catal.*, 2021, **11**, 2398.
118. X. Yu, F. Zhang and W. Chu, *RSC Adv.*, 2016, **6**, 70537.
119. J. Zhang, H. Wang and A. Dalai, *J. Catal.*, 2007, **249**, 300.
120. J. Zhang, H. Wang and A. Dalai, *Appl. Catal., A*, 2008, **339**, 121.
121. A. Chatla, M. Ghouri, O. El Hassa, N. Mohamed, A. Prakash and N. Elbashir, *Appl. Catal., A*, 2020, **602**, 117699.
122. A. Valentini, N. Carreño, L. Probst, P. Lisboa-Filho, W. Schreiner, E. Leite and E. Longo, *Appl. Catal., A*, 2003, **255**, 211.
123. V. Choudhary, B. Uphade and A. Mamman, *Catal. Lett.*, 1995, **32**, 387.
124. N. Habibi, Y. Wang, H. Arandiyan and M. Rezaei, *ChemCatChem*, 2016, **8**, 3600.
125. N. Habibi, Y. Wang, H. Arandiyan and M. Rezaei, *Int. J. Hydrogen Energy*, 2017, **42**, 24159.
126. E. Akbari, S. Alavi and M. Rezaei, *Fuel*, 2017, **194**, 171.
127. Z. Bao, Y. Zhan, J. Street, W. Xu, F. To and F. Yu, *Chem. Commun.*, 2017, **53**, 6001.
128. B. Jin, S. Li and X. Liang, *Fuel*, 2021, **284**, 119082.
129. L. Xu, H. Song and L. Chou, *Appl. Catal., B*, 2011, **108–109**, 177.
130. L. Xu, H. Song and L. Chou, *ACS Catal.*, 2012, **2**, 1331.
131. J. Guo, H. Lou, H. Zhao, D. Chai and X. Zheng, *Appl. Catal., A*, 2004, **273**, 75.
132. J.-E. Min, Y.-J. Lee, H.-G. Park, C. Zhang and K.-W. Jun, *J. Ind. Eng. Chem.*, 2015, **26**, 375.
133. E. Ruckenstein and Y. Hu, *Appl. Catal., A*, 1995, **133**, 149.
134. S.-E. Park, J.-S. Chang, H.-S. Roh, M. Anpo and H. Yamashita, *Stud. Surf. Sci. Catal.*, 1998, **114**, 395.
135. J.-S. Chang, S.-E. Park and H. Chon, *Appl. Catal., A*, 1996, **145**, 111.
136. K. Tomishige, Y. Himeno, Y. Matsuo, Y. Yoshinaga and K. Fujimoto, *Ind. Eng. Chem. Res.*, 2000, **39**, 1891.
137. A. Nandini, K. Pant and S. Dhingra, *Appl. Catal., A*, 2005, **290**, 166.
138. L. Azancot, L. Bobadilla, M. Centeno and J. Odriozola, *Appl. Catal., B*, 2021, **285**, 119822.
139. S. Sengupta and G. Deo, *J. CO2 Util.*, 2015, **10**, 67.
140. S. Wang and G. Lu, *Appl. Catal., B*, 1998, **19**, 267.
141. E. Akbari, S. Alavi and M. Rezaei, *J. CO2 Util.*, 2018, **24**, 128.
142. E. Akiki, D. Akiki, C. Italiano, A. Vita, R. Abbas-Ghaleb, D. Chlala, G. Drago Ferrante, M. Laganà, L. Pino and S. Specchia, *Int. J. Hydrogen Energy*, 2020, **45**, 21392.
143. A. Marinho, F. Toniolo, F. Noronha, F. Epron, D. Duprez and N. Bion, *Appl. Catal., B*, 2021, **281**, 119459.

144. X. Yu, N. Wang, W. Chu and M. Liu, *Chem. Eng. J.*, 2012, **209**, 623.
145. K. Li, C. Pei, X. Li, S. Chen, X. Zhang, R. Liu and J. Gong, *Appl. Catal., B*, 2020, **264**, 118448.
146. G. Xu, K. Shi, Y. Gao, H. Xu and Y. Wei, *J. Mol. Catal. A: Chem.*, 1999, **147**, 47.
147. H. Li and J. Wang, *Chem. Eng. Sci.*, 2004, **59**, 4861.
148. N. Rahemi, M. Haghighi, A. A. Babaluo, M. Jafari and P. Estifaee, *J. Ind. Eng. Chem.*, 2013, **19**, 1566.
149. Y. Hu and E. Ruckenstein, *Catal. Lett.*, 1997, **43**, 71.
150. E. Ruckenstein and Y. Hu, *Catal. Lett.*, 1998, **51**, 183.
151. J. Wei and E. Iglesia, *J. Catal.*, 2004, **224**, 370.
152. E. Ruckenstein and Y. Hu, *Appl. Catal., A*, 1997, **154**, 182.
153. Y. Wang, Q. Zhao, Y. Wang, C. Hu and P. D. a Costa, *Ind. Eng. Chem. Res.*, 2020, **59**, 11441.
154. R. Zhang, G. Xia, M. Li, Y. Wu, H. Nie and D. Li, *J. Fuel Chem. Technol.*, 2015, **43**, 1359.
155. X. Li, J.-S. Chang and S.-E. Park, *Chem. Lett.*, 1999, **28**, 1099.
156. J. Han, J. Park, M. Choi and H. Lee, *Appl. Catal., B*, 2017, **203**, 625.
157. B.-J. Kim, H.-R. Park, Y.-L. Lee, S.-Y. Ahn, K.-J. Kim, G.-R. Hong and H.-S. Roh, *J. CO2 Util.*, 2023, **68**, 102379.
158. H. Potdar, H. Roh, K.-W. Jun, M. Ji and Z.-W. Liu, *Catal. Lett.*, 2002, **84**, 95.
159. H.-S. Roh, K.-W. Jun, W.-S. Dong, J.-S. Chang, S.-E. Park and Y.-I. Joe, *J. Mol. Catal. A: Chem.*, 2002, **181**, 137.
160. H.-S. Roh, H. Potdar and K.-W. Jun, *Catal. Today*, 2004, **93–95**, 39.
161. H.-S. Roh, H. Potdar, K.-W. Jun, J.-W. Kim and Y.-S. Oh, *Appl. Catal., A*, 2004, **276**, 231.
162. Y. Lyu, J. Jocz, R. Xu, C. Stavitski and C. Sievers, *ACS Catal.*, 2020, **10**, 11235.
163. F. Zhang, Z. Liu, X. Chen, N. Rui, L. Betancourt, L. Lin, W. Xu, C. Sun, A. Abeykoon, J. Rodriguez, J. Teržan, K. Lorber, P. Djinović and S. Senanayake, *ACS Catal.*, 2020, **10**, 3274.
164. W.-J. Jang, D.-W. Jeong, J.-O. Shim, H.-S. Roh, I.-H. Son and S. Lee, *Int. J. Hydrogen Energy*, 2013, **38**, 4508.
165. D.-W. Jeong, W.-J. Jang, J.-O. Shim, H.-S. Roh, I. Son and S. Lee, *Int. J. Hydrogen Energy*, 2013, **38**, 13649.
166. W.-J. Jang, H.-M. Kim, J.-O. Shim, S.-Y. Yoo, K.-W. Jeon, H.-S. Na, Y.-L. Lee, D.-W. Jeong, J. Bae, I. Nah and H.-S. Roh, *Green Chem.*, 2018, **20**, 1621.
167. W.-J. Jang, D.-W. Jeong, J.-O. Shim, H.-M. Kim, W.-B. Han, J. Bae and H.-S. Roh, *Renewable Energy*, 2015, **79**, 91.
168. A. Al-Fatesh, R. Kumar, A. Fakeeha, S. Kasim, J. Khatri, A. Ibrahim, R. Arasheed, M. Alabdulsalam, M. Lanre, A. Osman, A. Abasaeed and A. Bagabas, *Sci. Rep.*, 2020, **10**, 13861.
169. R. Patel, A. Al-Fatesh, A. Fakeeha, Y. Arafat, S. Kasim, A. Ibrahim, S. Al-Zahrani, A. Abasaeed, V. Srivastava and R. Kumar, *Int. J. Hydrogen Energy*, 2021, **46**, 25015.

170. H. Ay and D. Üner, *Appl. Catal., B*, 2015, **179**, 128.
171. Z. Zhang and X. Verykios, *Appl. Catal., A*, 1996, **138**, 109.
172. P. Chen, H.-B. Zhang, G.-D. Lin and K.-R. Tsai, *Appl. Catal., A*, 1998, **166**, 343.
173. J. Han, J. Park, M. Choi and H. Lee, *Appl. Catal., B*, 2017, **203**, 625.
174. R. Zhang, G. Xia, M. Li, Y. Wu, H. Nie and D. Li, *J. Fuel Chem. Technol.*, 2015, **43**, 1359.
175. E. Ruckenstein and H. Hu, *J. Catal.*, 1996, **162**, 230.
176. T. Xie, Z.-Y. Zhang, H.-Y. Zheng, K.-D. Xu, Z. Hu and Y. Lei, *Chem. Eng. J.*, 2022, **429**, 132507.
177. S. Das, A. Jangam, S. Xi, A. Borgna, K. Hidajat and S. Kawl, *ACS Appl. Energy Mater.*, 2020, **3**, 7719.
178. X. Gao, K. Hidajat and S. Kawi, *J. CO2 Util.*, 2016, **15**, 146.
179. J. Wang, Y. Fu, W. Kong, F. Jin, J. Bai, J. Zhang and Y. Sun, *Appl. Catal., B*, 2021, **282**, 119546.
180. D. Liu, R. Lau, A. Borgna and Y. Yang, *Appl. Catal., A*, 2009, **358**, 110.
181. Z. Taherian, A. Khataee and Y. Orooji, *Renewable Sustainable Energy Rev.*, 2020, **134**, 110130.
182. A. Al-Fatesh, R. Kumar, S. Kasim, A. Ibrahim, A. Fakeeha, A. Abasaeed, H. Atia, U. Armbruster, C. Kreyenschulte, H. Lund, S. Bartling, Y. Ahmed Mohammed, Y. Albaqma, M. Lanre, M. Chaudary, F. Almubaddel and B. Chowdhury, *Ind. Eng. Chem. Res.*, 2022, **61**, 164.
183. W. Liu, L. Li, S. Lin, Y. Luo, Z. Bao, Y. Mao, K. Li, D. Wu and H. Peng, *J. Energy Chem.*, 2022, **65**, 34.
184. A. Abdulrasheed, A. Jalil, M. Hamid, T. Siang, N. Fatah, S. Izan and N. Hassan, *Int. J. Hydrogen Energy*, 2020, **45**, 18549.
185. M. Amin, *Catalysts*, 2020, **10**, 51.
186. J. Zhu, X. Peng, L. Yao, X. Deng, H. Dong, D. Tong and C. Hu, *Int. J. Hydrogen Energy*, 2013, **38**, 117.
187. D. Guo, Y. Lu, Y. Ruan, Y. Zhao, Y. Zhao, S. Wang and X. Ma, *Appl. Catal., B*, 2020, **277**, 119278.
188. F. Menegazzo, M. Signoretto, F. Pinna, P. Canton and N. Pernicone, *Appl. Catal., A*, 2012, **439–440**, 80.
189. C. Pan, Z. Guo, H. Dai, R. Ren and W. Chu, *Int. J. Hydrogen Energy*, 2020, **45**, 16133.
190. D. Araiza, D. Arcos, A. Gómez-Cortés and G. Díaz, *Catal. Today*, 2021, **360**, 46.
191. M. Vasiliades, C. Damaskinos, K. Kyprianou, M. Kollia and A. Efstathiou, *Catal. Today*, 2020, **355**, 788.
192. Y. Turap, I. Wang, T. Fu, Y. Wu, Y. Wang and W. Wang, *Int. J. Hydrogen Energy*, 2020, **45**, 6538.
193. Y. Song, E. Ozdemir, S. Ramesh, A. Adishev, S. Subramanian, A. Harale, M. Albuali, B. Fadhel, A. Jamal, D. Moon, S. Choi and C. Yavuz, *Science*, 2020, **367**, 777.

194. X. Zhang, J. Deng, M. Pupucevski, S. Impeng, B. Yang, G. Chen, S. Kuboon, Q. Zhong, K. Faungnawakij, L. Zhen, G. Wu and D. Zhang, *ACS Catal.*, 2021, **11**, 12087.
195. K. Takanabe, K. Nagaoka, K. Nariai and K.-I. Aika, *J. Catal.*, 2005, **232**, 268.
196. C. Batiot-Dupeyrat, G. Valderrama, A. Meneses, F. Martinez, J. Barrault and J. Tatibouët, *Appl. Catal., A*, 2003, **248**, 143.
197. C. Batiot-Dupeyrat, G. Gallego, F. Mondragon, J. Barrault and J.-M. Tatibouët, *Catal. Today*, 2005, **107–108**, 474.
198. N. Bonmassar, M. Bekheet, L. Schlicker, A. Gili, A. Gurlo, A. Doran, Y. Gao, M. Heggen, J. Bernardi, B. Klötzer and S. Penner, *ACS Catal.*, 2020, **10**, 1102.
199. S. Lima, J. Assaf, M. Peña and J. Fierro, *Appl. Catal., A*, 2006, **311**, 94.
200. G. Valderrama, A. Kiennemann and M. Goldwasser, *Catal. Today*, 2008, **133–135**, 142.
201. G. Valderrama, C. Urbina de Navarro and M. Goldwasser, *J. Power Sources*, 2013, **234**, 31.
202. T. Wei, L. Jia, J.-L. Luo, B. Chi, J. Pu and J. Li, *Appl. Surf. Sci.*, 2020, **506**, 144699.
203. J. Dong, C. Fu, H. Li, J. Xiao, B. Yang, B. Zhang, Y. Bai, T. Song, R. Zhang, L. Gao, J. Cai, H. Zhang, Z. Liu and X. Bao, *J. Am. Chem. Soc.*, 2020, **142**, 17167.
204. K. Bu, J. Deng, X. Zhang, S. Kuboon, T. Yan, H. Li, L. Shi and D. Zhang, *Appl. Catal., B*, 2020, **267**, 118692.
205. G. Moradi, F. Khezeli and H. Hemmati, *J. Nat. Gas Sci. Eng.*, 2016, **33**, 657.
206. Q. Zhu, H. Zhou, L. Wang, C. Wang, H. Wang, W. Fang, M. He, Q. Wu and F.-S. Xiao, *Nat. Catal.*, 2022, **5**, 1030.
207. D. Liang, Y. Wang, M. Chen, X. Xie, C. Li, J. Wang and L. Yuan, *Appl. Catal., B*, 2023, **322**, 122088.
208. K. Tamura, D. Murata, T. Sumi, S. Kokuryo, H. Kitamura, S. Tsubota, K. Miyake, Y. Uchida, M. Miyamoto and N. Nishiyama, *Energy Fuels*, 2023, **37**, 18945.
209. J. Rostrup-Nielsen, J. B. Hansen and L. M. Aparicio, *J. Jpn. Pet. Inst.*, 1997, **40**, 366.
210. E. Kikuchi and Y. Chen, *Stud. Surf. Sci. Catal.*, 1997, **107**, 547.
211. J. Wei and E. Iglesia, *J. Phys. Chem. B*, 2004, **108**, 4094.
212. J. Wei and E. Iglesia, *J. Catal.*, 2004, **225**, 116.
213. J. Wei and E. Iglesia, *Angew. Chem., Int. Ed.*, 2004, **43**, 3685.
214. J. Wei and E. Iglesia, *Phys. Chem. Chem. Phys.*, 2004, **6**, 3754.
215. J. Bitter, K. Seshan and J. Lercher, *Top. Catal.*, 2000, **10**, 295.
216. C. Carrara, J. Múnera, E. Lombardo and L. Cornaglia, *Top. Catal.*, 2008, **51**, 98.
217. K. Nagaoka, K. Seshan, J. Lercher and K.-I. Aika, *Stud. Surf. Sci. Catal.*, 2001, **136**, 129.

218. Z. Hou, P. Chen, H. Fang, X. Zheng and T. Yashima, *Int. J. Hydrogen Energy*, 2006, **31**, 555.
219. J. Richardson, M. Garrait and J.-K. Hung, *Appl. Catal., A*, 2003, **255**, 69.
220. H. Wang and E. Ruckenstein, *Appl. Catal., A*, 2000, **204**, 143.
221. S. Gaur, D. J. Haynes and J. Spivey, *Appl. Catal., A*, 2011, **403**, 142.
222. F. Wang, L. Xu, J. Yang, J. Zhang, L. Zhang, H. Li, Y. Zhao, H. Li, K. Wu, G. Xu and W. Chen, *Catal. Today*, 2017, **281**, 295.
223. E. Ruckenstein and H. Y. Wang, *J. Catal.*, 2002, **205**, 289.
224. Ş. Özkara-Aydınoğlu and A. E. Aksoylu, *Catal. Commun.*, 2010, **11**, 1165.
225. K. Takanabe, K. Nagaoka, K. Nariai and K.-I. Aika, *J. Catal.*, 2005, **230**, 75.
226. G. Zhang, A. Su, Y. Du, J. Qu and Y. Xu, *J. Colloid Interface Sci.*, 2014, **433**, 149.
227. N. Wang, W. Chu, L. Huang and T. Zhang, *J. Nat. Gas Chem.*, 2010, **19**, 117.
228. S. Zeng, L. Zhang, X. Zhang, Y. Wang, H. Pan and H. Su, *Int. J. Hydrogen Energy*, 2012, **37**, 9994.
229. G. Valderrama, C. Urbina de Navarro and M. Goldwasser, *J. Power Sources*, 2013, **234**, 31.
230. M. I. Temkin, *Adv. Catal.*, 1979, **28**, 173.
231. J. R. H. Ross and M. C. F. Steel, *J. Chem. Soc., Faraday Trans. 1*, 1973, **69**, 10.
232. J. Xu and G. F. Froment, *AIChE J.*, 1989, **35**, 88.
233. P. G. Menon, J. C. de Deken and G. F. Froment, *J. Catal.*, 1985, **95**, 313.
234. L. M. Aparicio, *J. Catal.*, 1997, **165**, 262.
235. A. K. Avetisov, J. R. Rostrup-Nielsen, V. L. Kuchaev, J.-H. Bak Hansen, A. G. Zyskin and E. N. Shapatina, *J. Mol. Catal. A: Chem.*, 2010, **315**, 155.
236. A. Berman, R. K. Karn and M. Epstein, *Appl. Catal., A*, 2005, **282**, 73.
237. J. G. Jakobsen, M. Jakobsen, I. Chorkendorff and J. Sehested, *Catal. Lett.*, 2010, **140**, 90.
238. G. Jones, J. G. Jacoben, S. S. Shim, J. Kleis, M. P. Andersson, J. Rossmeisl, F. Abild-Pedersen, T. Bligaard, S. Helveg, B. Hinnemann, J. R. Rostrup-Nielsen, I. Chorkendorff, J. Sehested and J. K. Nørskov, *J. Catal.*, 2008, **259**, 147.
239. W. Yu, Y. Wang, D. Tian and C. Zeng, *Comput. Mater. Sci.*, 2024, **244**, 113217.
240. L. Zhu, T. Lei, X. Liu, X. Yang, B. Zhang, H. Jiao, W. Guo, B. Teng and X. Wen, *ACS Catal.*, 2024, **14**, 12342.
241. Y. Kathiraser, U. Oemar, E. Saw, Z. Li and S. Kawi, *Chem. Eng. J.*, 2015, **278**, 62.
242. M. Bradford and M. Vannice, *Appl. Catal., A*, 1996, **142**, 97.
243. V. Tsipouriari and X. Verykios, *Catal. Today*, 2001, **64**, 83.
244. J. Niu, X. Du, J. Ran and R. Wang, *Appl. Surf. Sci.*, 2016, **376**, 79.
245. B. Sawatmongkhon, K. Theinnoi, T. Wongchang, C. Haoharn and A. Tsolakis, *Int. J. Hydrogen Energy*, 2017, **42**, 24697.
246. D. Pakhare and J. Spivey, *Chem. Soc. Rev.*, 2014, **43**, 7813.

247. Y.-A. Zhu, D. Chen, X.-G. Zhou and W.-K. Yuan, *Catal. Today*, 2009, **148**, 260.
248. A. Omran, S. Yoon, M. Khan, M. Ghouri, A. Chatla and N. Elbashir, *Catalysts*, 2020, **10**, 1043.
249. Z. Wang, X.-M. Cao, J. Zhu and P. Hu, *J. Catal.*, 2014, **311**, 469.
250. L. Dehimi, Y. Benguerba, M. Virginie and H. Hijazi, *Int. J. Hydrogen Energy*, 2017, **42**, 18930.
251. Y.-X. Yu, G. Wang, Y.-A. Zhu and X.-G. Zhou, *Chem. Eng. J.*, 2024, **479**, 146959.
252. L. Foppa, M.-C. Silaghi, K. Larmier and A. Comas-Vives, *J. Catal.*, 2016, **343**, 196.
253. J. Rostrup-Nielsen, I. Dybkjaer and L. Christiansen, *Chemical Reactor Technology for Environmentally Safe Reactors and Products*, Springer Netherlands, Dordrecht, 1992, p. 249.
254. J. Rostrup-Nielsen and T. Rostrup-Nielsen, *Cattech*, 2002, **6**, 150.
255. K. Johnsen, H. Ryu, J. Grace and C. Lim, *Chem. Eng. Sci.*, 2006, **61**, 1195.
256. L. Barelli, G. Bidini, F. Gallorini and S. Servili, *Energy*, 2008, **33**, 554.
257. A. Di Giuliano and K. Gallucci, *Chem. Eng. Process.*, 2018, **130**, 240.
258. X. Chen, L. Yang, Z. Zhou and Z. Cheng, *Chem. Eng. Sci.*, 2017, **163**, 114.
259. S. Masoudi Soltani, A. Lahiri, H. Bahzad, P. Clough, M. Gorbounov and Y. Yan, *Carbon Capture Sci. Technol.*, 2021, **1**, 10003.
260. P. Chaudhary and G. Deo, *Colloids Surf., A*, 2022, **646**, 128973.
261. B. Nematollahi, M. Rezaei and M. Khajenoori, *Int. J. Hydrogen Energy*, 2011, **36**, 2969.
262. R. Ramezani, L. Felice and F. Gallucci, *J. Phys.: Energy*, 2023, **5**, 024010.
263. Y. Sun, J. Li and H. Li, *Chem. Eng. J.*, 2022, **431**, 134173.
264. A. Antzaras, E. Heracleous and A. Lemonidou, *Fuel Process. Technol.*, 2020, **208**, 106513.
265. T. Chompupun, S. Limtrakul, T. Vatanatham, C. Kanhari and P. Ramachandran, *Chem. Eng. Process.*, 2018, **134**, 124.
266. Y. Chen, Y. Wang, H. Xu and G. Xiong, *Appl. Catal., B*, 2008, **81**, 283.
267. R. Rajasree, V. Kumar and B. Kulkarni, *Energy Fuels*, 2006, **20**, 463.
268. D. Simakov, M. Wright, S. Ahmed, E. Mokheimer and Y. Román-Leshkov, *Catal. Sci. Technol.*, 2015, **5**, 1991.
269. R. Abiev, D. Sladkovskiy, K. Semikin, D. Murzin and E. Rebrov, *Catalysts*, 2020, **10**, 1358.
270. C. Fan, Y.-A. Zhu, M.-L. Yang, Z.-J. Sui and D. Chen, *Ind. Eng. Chem. Res.*, 2015, **54**, 5901.

CHAPTER 2

Methane Cracking: A Comprehensive Review of Catalysts, Solar Reactors, and Integrated System Case Studies

ALIYA BANU AND YUSUF BICER*

Division of Sustainable Development, College of Science and Engineering, Hamad Bin Khalifa University, Qatar Foundation, Doha, Qatar
*Email: ybicer@hbku.edu.qa

2.1 Introduction

Methane cracking represents a more sustainable method for hydrogen production from natural gas, as it generates no emissions of carbon monoxide (CO) or carbon dioxide (CO_2). This route has the smallest carbon footprint compared to other hydrogen production routes from natural gas. Additionally, it yields carbon black (CB) as a by-product, which has applications in various processes for the production of plastics, electronics, and inks.[1] When effectively marketed, this by-product can enhance the economic viability of the process. Methane cracking can be viewed as a transitional technology toward a sustainable future[2] as it is one of the most cost-effective short-term solutions for producing hydrogen with low CO_2 emissions.[3] The literature widely supports the benefits of transitioning to a hydrogen economy, emphasizing the importance of clean hydrogen production and the necessity of moving away from traditional sources like liquefied natural gas (LNG) in the pursuit of decarbonization. Economically, the implementation

Catalysis Series No. 49
Catalytic Activation of Small Molecules
Edited by Mustafa Yasin Aslan, Angela Daisley, Justin S. J. Hargreaves and José L. Rico
© The Royal Society of Chemistry 2025
Published by the Royal Society of Chemistry, www.rsc.org

of the cracking of methane is deemed more appropriate for small- to medium-scale on-site hydrogen production.[4–6] Furthermore, the reliance of methane cracking on the existing natural gas utilization infrastructure positions it as a favorable option for near-term clean hydrogen production.

$$CH_4 \rightarrow C + 2H_2 \quad \Delta H° = 74.8 \text{ kJ mol}^{-1} \quad (2.1)$$

One of the key advantages of methane cracking is its relatively low energy requirement compared to steam methane reforming (SMR) and other hydrogen production processes. Specifically, SMR requires an energy input of 63.3 kJ mol^{-1} of hydrogen, while this reaction only needs 37.8 kJ mol^{-1}.[7] Moreover, the methane cracking process is comparatively simpler, not involving additional stages like the water gas shift reaction (WGS) and the stages for removal of CO_2 needed in SMR. Additionally, the economic viability of methane thermal decomposition may be better than that of SMR, dependent of the price of the co-product.[4]

Owing to the strength of the C–H bonds present in methane, the thermal reaction occurs at temperatures exceeding 1473.15 K.[8] The implementation of catalysts can reduce the temperature requirements for this reaction. Various carbon and metal catalysts have been employed to decrease the energy required for the process.[9] A significant portion of the research in this field has concentrated on the development of an effective catalyst. Additionally, CB is generated as a co-product, which obstructs the active sites of the catalysts, resulting in its quick deactivation.[10]

Solar energy also provides a sustainable means to produce hydrogen as it uses a renewable energy source that is abundant to meet the high-energy requirements of methane cracking, producing a carbon-free fuel. Compared to natural gas, using renewable energy for producing hydrogen provides better utilization and conversion of energy. Mitigation of climate change and reduction of greenhouse gases are both benefits of using renewable energy sources.[11] Solar energy possesses significant potential as a renewable energy source for hydrogen production.[12] To combat the intermittent availability of solar energy, it can be transformed and stored as chemical energy.[13] Firstly, this chapter reviews the different catalysts developed for methane cracking. The second section explores the solar reactors developed for this reaction in order to integrate solar energy into the process. The last section outlines two case studies of proposed integrated systems with solar methane cracking for the sustainable utilization of the hydrogen and carbon product.

2.2 Catalysts

The production of hydrogen from methane cracking necessitates temperatures exceeding 1473.15 K.[8] Utilizing a catalyst can reduce this temperature. Extensive research has sought effective catalysts for this reaction, yet rapid catalyst deactivation remains a significant challenge due to carbon by-product formation that obstructs active sites.[10] Additionally, natural gas

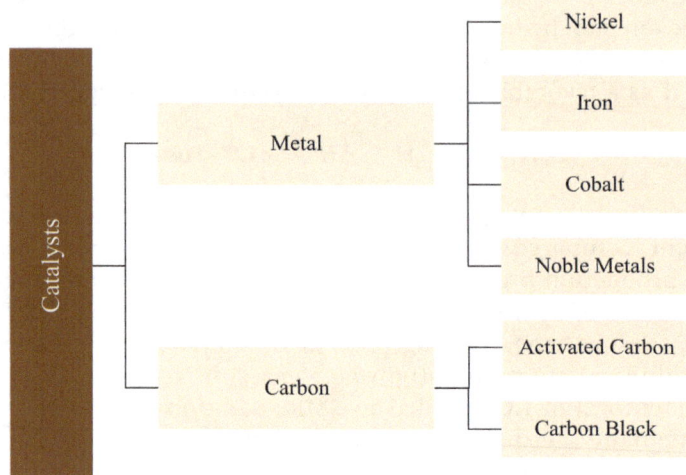

Figure 2.1 Categorization of the various catalyst groups tested for the methane cracking process. Reproduced from ref. 16, https://doi.org/10.1016/j.ecmx.2021.100117, under the terms of the CC BY 4.0 license, https://creativecommons.org/licenses/by/4.0/.

impurities, such as sulfur, can diminish metal catalyst activity over time.[14] Nevertheless, employing an appropriate catalyst can greatly enhance reaction kinetics; at 1093.15 K hydrogen production through catalysis was 13 times greater compared to direct methane cracking.[15]

The primary catalysts used in the reaction are carbon and metal catalysts, as Figure 2.1 illustrates. These catalysts exhibit distinct performances; the activity of metal catalysts arises from surface-reduced metal, while carbon catalysts depend on structural defects within the graphene layer.[17]

Subsequent sections will address the development and testing of carbon and metal catalysts reported for methane cracking. A comparison and analysis of the effects of catalyst supports and the addition of promoters will be conducted. Additionally, various regeneration studies addressing deactivation issues will be examined.

2.2.1 Metal Catalyst

Nickel (Ni), cobalt (Co), and iron (Fe) are the most commonly studied metals for catalysis for methane cracking because of their abundance and affordability.[18] Various supports such as Al_2O_3, MgO, and SiO_2, are utilized to improve the catalytic performance of these metal catalysts.

Ni catalysts exhibit high activity toward the decomposition of methane.[19] In a study by Zhang *et al.*[20] on the variance of reaction temperature and its effect on the catalytic activity of non-supported Ni catalysts, it was found that an increase in the temperature leads to higher activity (Figure 2.2). Ni with magnesium aluminate support was evaluated for methane cracking in a

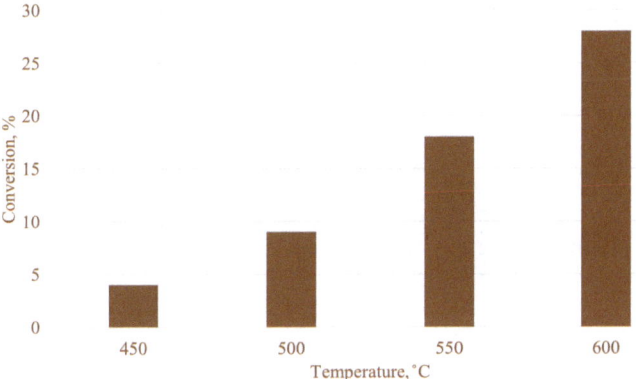

Figure 2.2 The rate of conversion of the nickel catalyst, synthesized *via* co-precipitation, at varying reaction temperatures. Reproduced from ref. 16, https://doi.org/10.1016/j.ecmx.2021.100117, under the terms of the CC BY 4.0 license, https://creativecommons.org/licenses/by/4.0/.

fixed bed reactor. Optimal conditions were identified as a lower $N_2:CH_4$ feed ratio and 773.15 K. Reduction of the catalyst at 973.15 K for 1 hour demonstrated superior performance, indicating that the preparation method affects catalyst activity.[21] In another study,[22] it was reported that lanthanum (La) addition to alumina-supported Ni catalysts improved methane decomposition performance, as characterized by homogeneous Ni particle dispersion and high surface area of the metal. Furthermore, the addition of copper (Cu) to Ni supported on SiO_2 was found to enhance the activity of the catalyst.[23] Cu addition to Ni/Al_2O_3 reduced the initial activity while significantly enhancing stability.[24] Ce addition to Ni supported on SiO_2 prolonged the deactivation time and increased the initial rate of conversion.[25] Table 2.1 presents various nickel catalyst modifications tested and reported in the literature.

Fe-based catalysts are increasingly favored due to their cost-effectiveness compared to Ni.[36] Testing of Fe/Al_2O_3 at 973.15 K resulted in a H_2 yield of 77.2%.[37] Pinilla *et al.* researched Fe and Ni catalysts with various textural promoters, achieving high reaction temperatures with the Fe-based catalyst and an 82% conversion rate. However, Ni-based catalysts demonstrated superior hydrogen production due to higher activity. Al_2O_3-supported catalysts produced more hydrogen than those supported by MgO,[36] with analogous results observed for Fe with different promoters.[38] Additionally, the amount of promoter loading significantly impacts catalytic activity (Figure 2.3). Mo doping with Fe enhanced performance in MgO but had a lesser effect in Al_2O_3. A subsequent study[39] examined the influence of operating conditions with Fe-based catalysts, revealing that as the temperature increases and the space velocity decreases, higher conversion can be achieved. Table 2.2 summarizes various reported Fe catalyst combinations.

Table 2.1 Selected Ni-based catalysts reported in the literature.

Catalyst	T (K)	Mass (g)	Method[a]	Max. X_{CH_4} (%)	Y_{H_2} (%)	Feed	Ref.
Ni/Al$_2$O$_3$	923.15	—	A	55	—	—	26
Ni/Al$_2$O$_3$	873.15	0.03	B	36	—	Pure CH$_4$	22
La–Ni/Al$_2$O$_3$				50	—		21
Ni/MgAl$_2$O$_4$	823.15	0.1	C	37	—	N$_2$:CH$_4$ = 7:1	
				12	—	N$_2$:CH$_4$ = 1:3	
				18	—	N$_2$:CH$_4$ = 1:1	
0.1%Pt–15%Ni/MgAl$_2$O$_4$	973.15	0.1	C	45	—	N$_2$:CH$_4$ = 7:1	27
	823.15			15	—	N$_2$:CH$_4$ = 7:1	
	823.15			8	—	N$_2$:CH$_4$ = 1:3	
50%Ni/SiO$_2$	1023.15	1	B	77	—	N$_2$:CH$_4$ = 1:1	23
10%Cu–50%Ni/SiO$_2$				83	—		
10% Ni/TiO$_2$	973.15	1	D	—	43	Pure CH$_4$	28
50%Ni/TiO$_2$	973.15			—	56		
	923.15			—	53		
	873.15			—	45		
12.5%Ni–12.4%Co/La$_2$O$_3$	973.15	0.3	B	82	—	N$_2$:CH$_4$ = 1:9	29
	873.15			54	—		
	823.15			21	—		
25%Ni–25%Co/Al$_2$O$_3$	973.15	0.3	B	74	—	—	30
50%Ni–25%Ce/Al$_2$O$_3$	973.15	0.5	C	—	53	—	31
40%Ni-ZSM-5(25)	973.15	0.5	C	—	77	Pure CH$_4$	32
40%Ni-ZSM-5(400)				—	77		
40%Ni-AS[a]				—	77		
50%Ni–10%Pd/Al$_2$O$_3$	948.15	0.05	C	75	—	N$_2$:CH$_4$ = 7:3	33
	1023.15			90	—		
50%Ni–10%Fe/Al$_2$O$_3$	948.15	0.05	C	68	—	N$_2$:CH$_4$ = 7:3	34
	1023.15			85	—		
50%Ni–10%Fe–10%Cu/Al$_2$O$_3$	948.15	0.05	C	56	—	N$_2$:CH$_4$ = 7:3	35
	1023.15			81	—		

[a] A = fusion, B = co-precipitation, C = wet impregnation, D = sol–gel.

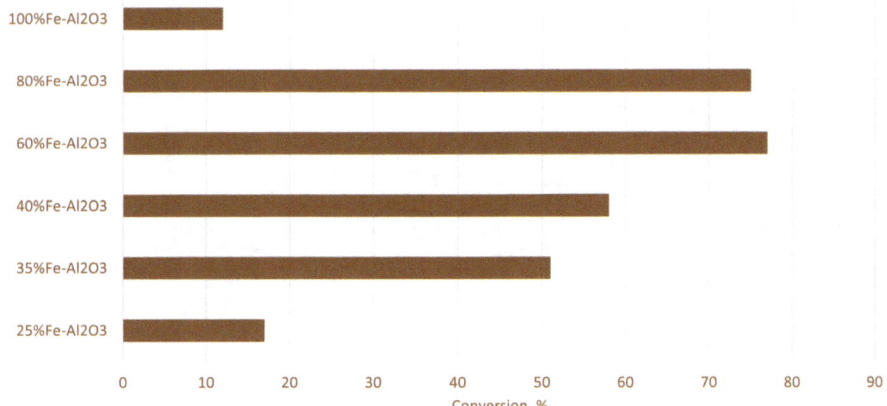

Figure 2.3 The conversion rate (maximum) of Fe/Al$_2$O$_3$ with varying % support loadings at a $T = 973.15$ K, using a N$_2$:CH$_4$ = 1:10 feed. Reproduced from ref. 16, https://doi.org/10.1016/j.ecmx.2021.100117, under the terms of the CC BY 4.0 license, https://creativecommons.org/licenses/by/4.0/.

2.2.2 Carbon Catalyst

Carbon catalysts such as CB and activated carbon (AC), when supported differently, exhibit significant catalytic activity in methane cracking. It was shown by Muradov[46] that AC demonstrates superior initial activity, whereas CB offers enhanced resistance to deactivation, making its activity more stable.

To leverage the benefits of both carbon and metal, carbon was evaluated with metals like Fe and Ni. Prasad *et al.*[47] found that carbon supported with Ni exhibited improved catalytic activity. Various loadings of Ni were assessed, revealing that a 23.3% loading provided optimal performance, achieving 74% conversion at 1123.15 K (Figure 2.4). It was observed that the cracking rate initially decreased for the first 2 h before increasing, attributed to surface properties. Further research is necessary to evaluate the catalyst's sustainability. Carbon doped with Ni is synthesized by incorporating nickel nitrate.[48] The synthesized catalyst demonstrated superior stability and activity compared to Ni/SiO$_2$, Ni/Al$_2$O$_3$, and other carbon catalysts. Additionally, in ref. 49, Pt, Pd, and Cr were added to AC and the performance was examined. The inclusion of Cr metal diminished activity, whereas Pd enhanced both stability and activity.

Wei *et al.*[50] researched methane cracking in a fixed bed reactor using coal chars, where four forms of char were derived through parent coal pyrolysis. The study revealed that catalytic activity diminished with higher coal ranks, resulting in reduced reaction rates and yields due to deactivation over time. Additionally, tests indicated that increased mineral matter enhanced catalytic activity,[51] as it facilitates mesopore formation during KOH activation.[52]

Zhang *et al.*[53] prepared AC from direct coal liquefaction residue (CLR) using KOH activation, with some samples incorporating SBA-15 or SiO$_2$.

Table 2.2 Selected iron catalysts reported in the literature.

Catalyst	T (K)	Mass (g)	Method[a]	Max. X_{CH_4} (%)	Y_{H_2} (%)	Feed	Ref.
Fe/Al$_2$O$_3$	1073.15	0.15	A	77	—	Pure CH$_4$	38
Fe–Mo/Al$_2$O$_3$				81	—		
Fe/MgO				38	—		
Fe–Mo/MgO				68	—		
10Fe–10Al$_2$O$_3$/AC	1132.15	0.1	C	28	—	N$_2$:CH$_4$ = 1:4	40
15Fe–15Al$_2$O$_3$/AC				21	—		
20Fe–20Al$_2$O$_3$/AC				27	—		
30Fe–30Al$_2$O$_3$/A.C.				24	—		
Fe–5.1%Mo/MgO	1023.15	—	A	75	—	Pure CH$_4$	41
Fe–5.1%Mo/Al$_2$O$_3$				69	—		
Ni-Fe/SBA-15	973.15	3	C	35	52	Pure CH$_4$	42
Co–Fe/SBA-15				34	51		
Fe–SiO$_2$	1073.15	3	D	—	58	Pure CH$_4$	43
20%Fe/MgO	973.15	—	B	47	—	N$_2$:CH$_4$ = 1:9	44
20%Fe/Al$_2$O$_3$				35	—		
20%Fe/TiO$_2$				14	—		
30%Fe/MgO				46	—		
30%Fe/Al$_2$O$_3$				43	—		
30%Fe/TiO$_2$				15	—		
40%Fe/MgO				38	—		
40%Fe/Al$_2$O$_3$				42	—		
40%Fe/TiO$_2$				16	—		
Fe/ZrO$_2$	1073.15	0.3	C	60	—	N$_2$:CH$_4$ = 7:13	45
Fe/La$_2$O$_3$–ZrO$_2$				80	—		
Fe/WO$_2$–ZrO$_2$				91	—		

[a] A = fusion, B = co-precipitation, C = wet impregnation, D = sol–gel.

Methane Cracking

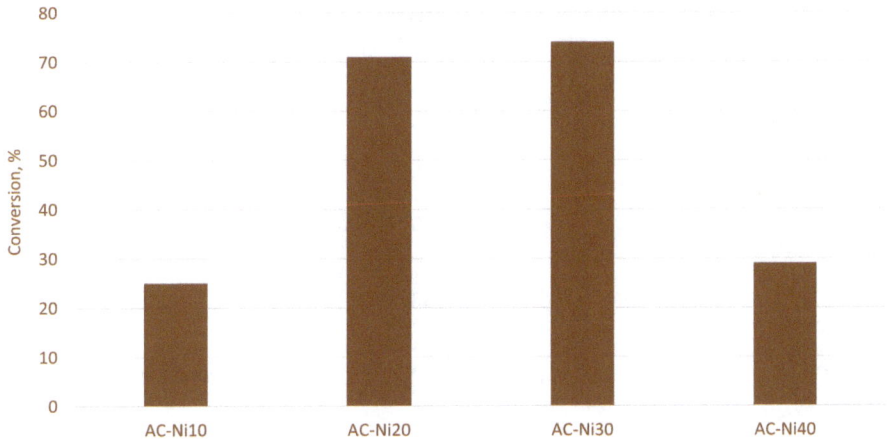

Figure 2.4 The maximum conversion rate of AC with varying loading of nickel (wt%) ($T = 1123.15$ K, pure methane feed). Reproduced from ref. 16, https://doi.org/10.1016/j.ecmx.2021.100117, under the terms of the CC BY 4.0 license, https://creativecommons.org/licenses/by/4.0/.

The catalytic activity of these samples was evaluated against commercial CB (BP2000) and AC revealing that the developed carbons exhibited improved performance. The inclusion of SBA-15 and SiO_2 enhanced catalytic performance compared to catalysts activated directly by KOH. Additionally, incorporating Al_2O_3 into CLR-derived carbon achieved a methane conversion rate of 61% over 10 hours, attributed to the formation of fibrous carbon.[54]

In another study,[55] carbon texture was altered by varying synthesis procedures, resulting in different catalytic activity. An optimum ratio of solvent, KOH activation, and carbonization was identified. Mesoporous carbon outperformed catalysts with microporous structure. Due to their higher specific surface area, narrow pore size distribution, and increased pore volume, ordered nano- and mesoporous carbons possess superior qualities for catalytic activity.[56] Additionally, mesoporous AC exhibited greater resistance to deactivation.[57] A summary of various reported carbon catalysts is provided in Table 2.3. Nishii *et al.*[58] found that activated carbon exhibited superior initial methane conversion compared to other tested carbon structures (Figure 2.5). Recent research by Tokunaga *et al.* indicated that fullerene facilitates catalytic methane decomposition at 673.15 K, a temperature lower than that required for graphite and CB. This process yields hydrogen along with concentric spherical carbon and amorphous carbon. It was found that spherical carbon enhances catalytic activity and increases hydrogen production over time.[59]

2.2.3 Reaction Kinetics

This section analyzes the reported reaction kinetics of the methane cracking process, including previous studies that have investigated both catalytic and non-catalytic systems. Steinberg[61] investigated the reaction kinetics in a

Table 2.3 Different carbon catalysts reported in the literature.

Catalyst	T (K)	Mass (g)	Method[a]	Max. X_{CH_4} (%)	Y_{H_2} (%)	Feed	Ref.
Shengli lignite char	1123.15	10	A	86	88	$N_2:CH_4 = 19:1$	50
Xiaolongtan lignite char				80	82		
Binxian bituminous char				70	68		
Jincheng anthracite char				35	27		
nCLR-SiO$_2$	1123.15	0.2	B	30	—	Pure CH_4	53
CLR-SBA-15				27	—		
AC			C	28	—		
BP2000				21	—		
CLR-Al$_2$O$_3$	1123.15	0.2	B	61	—	Pure CH_4	54
CLR				20	—		
AC	1213.15	20	C	52	—	Pure CH_4	15
AC	1173.15			37	—		
AC	1133.15			25	—		
AC	1093.15			16	—		
2C-K$_2$O$_3$-2Ni	1123.15	0.5	D	80	—	$N_2:CH_4 = 1:3$	60
2C-K$_2$O$_3$-5Ni				88	—		
2C-K$_2$O$_3$-8Ni				85	—		
2C-K$_2$O$_3$-10Ni				83	—		

[a] A = pyrolysis, B = KOH activation, C = commercial, D = precipitation.

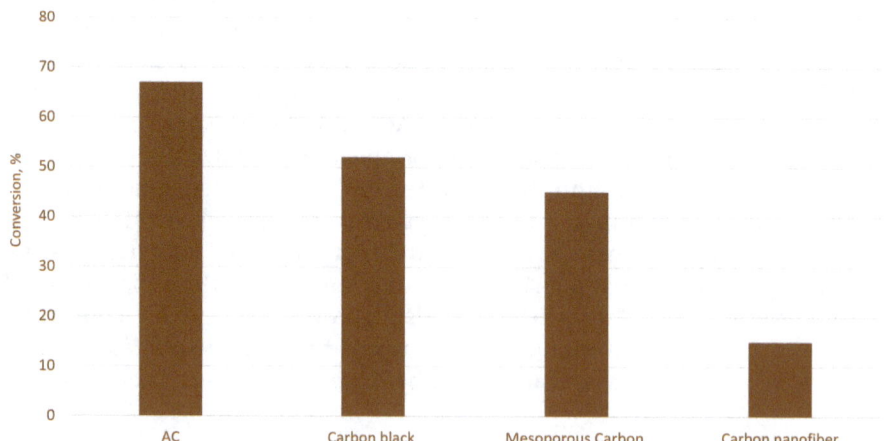

Figure 2.5 The maximum methane conversion rate of various commercial carbon catalysts at 900 °C in a pure methane feed. Reproduced from ref. 16, https://doi.org/10.1016/j.ecmx.2021.100117, under the terms of the CC BY 4.0 license, https://creativecommons.org/licenses/by/4.0/.

Table 2.4 Activation energies observed in previous research.

Ref.	Catalyst	E_a (kJ mol^{-1})	Temp. range (°C)
Steinberg et al.[61]	Non-catalytic	131	700–900
Dahl et al.[62]	Non-catalytic	208	1260–1870
Caballero et al.[64]	Pd/Al$_2$O$_3$	35.5	400–750
	Pd/1%Nd–Al$_2$O$_3$	37.9	400–750
	Pd/Nd10%–Al$_2$O$_3$	33	400–750
Saraswat et al.[63]	60% Ni–5% Cu–5% Zn/Al$_2$O$_3$	73.2	600–800
Zein et al.[65]	Ni/TiO$_2$	60	550–900
Alstrup and Tavares[66]	Ni/SiO$_2$	90	450–570

tubular type reactor, reporting an activation energy (E_a) of 131 kJ mol^{-1} between 973.15 and 1173.15 K and pressures of 28–56 atm, concluding that the reaction is self-catalyzed at 1173.15 K. Dahl et al.[62] performed modeling of an aerosol flow reactor, determining the E_a to be 208 kJ mol^{-1}.

In a study by Saraswat et al.,[63] it was demonstrated that the methane cracking reaction rate is significantly influenced by reactor temperature, with flow rate and partial pressure also playing a role. Ni doped with Zn and Cu on Al$_2$O$_3$ support yielded an E_a of 73.2 kJ mol^{-1}. Another study[64] synthesized Pd-based catalysts with Al$_2$O$_3$ supports and Nd promoter, and a fixed bed quartz reactor was used to analyze the kinetics. The reaction order was identified as 1, with activation energies between 33 and 37.9 kJ mol^{-1}. 10 wt% Nd catalyst exhibited the lowest E_a, indicating that promoters enhance reaction kinetics.

Table 2.4 summarizes the E_a and the corresponding temperature range. A notable reduction in activation energy occurs with catalyst use; however, no consistent trend is observed. Various catalyst systems exhibit a wide range of reported activation energies. Further research is necessary to enhance understanding of the kinetics involved with different catalysts.

2.2.4 Co-feeding

A key challenge in methane cracking with catalysts is the fast deactivation. Co-feeding can enhance the rate of conversion. Incorporating an appropriate hydrocarbon in the feed extends catalyst activity by generating more catalytically active carbon.[67] Carbon derived from hydrocarbons like ethylene and benzene exhibits greater activity than that produced from methane.[68]

Muradov[67] conducted tests with methane with 70% propane at 1123.15 K using an activated Al$_2$O$_3$ catalyst, yielding no enhancement in reaction. Conversely, testing 70% acetylene using quartz wool at the same temperature resulted in the production of more catalytically active carbon and increased hydrogen output. Malaika and Kozlowski[68] demonstrated that co-feeding 10% propylene in the reactor with AC improved catalyst stability. Additionally, another study by the same group[69] showed that *in situ* ethylene production *via* the oxidative coupling of methane achieved a conversion rate 250% higher than the catalytic route. While *in situ* ethylene production decreases process efficiency, it enhances economic viability due to the high cost of ethylene.

Rechnia *et al.*[70] demonstrated that adding ethanol enhances methane cracking by generating more catalytically active carbon, resulting in reduced deactivation rates and increased stability. A follow-up study[71] utilized carbon from hazelnut shells as a catalyst, which, however, experienced rapid deactivation. The addition of ethanol, which decomposes into ethylene, mitigates carbon catalyst deactivation. Ethanol and methanol were alternately introduced into a reactor with a fixed bed, with variations in temperature and ethanol dosing times. The cycle with the highest temperature of 1223.15 K resulted in decreased activity after the ethanol run, likely because of carbon catalyst graphitization. Conversely, performing decomposition at 1223.15 and ethanol decomposition at 1123.15 K slightly enhanced the cracking rate. Decomposition of ethanol is complex and yields various products beyond ethylene.

2.2.5 Catalyst Regeneration

Regeneration allows continuous operation.[72] It involves restoring the activity of the catalyst through three oxidation processes using: CO_2, steam, and air.[73] Usually performed in cycles, upon deactivation of the catalyst, the feed is shifted from the methane stream to the regeneration stream, facilitating removal of deposited carbon from the surface of the catalyst.

During oxidation with air, deposited carbon reacts with O_2 and is then removed from the surface of the catalyst. In a study by Amin *et al.*[74] supported Ni catalysts were evaluated in a reactor with a fluidized bed. Regeneration of the catalyst was undertaken through carbon gasification in air at 823.15 K. The Ni/Al_2O_3 catalysts deactivated after the first cycle of regeneration because of sintering. In contrast, the nickel supported on silica exhibited thermal stability during multiple cycles of regeneration, although there was a decrease in the size of the particles, affecting the quality of fluidization. The mechanical strength of catalyst pellets was negatively affected by regeneration, but this issue can be reduced through partial regeneration. In ref. 75, partial regeneration showed superior performance, maintaining stable activity throughout the studied cycles of regeneration.

Amin *et al.*[73] examined non-porous and porous Al_2O_3-supported Ni catalysts during regeneration cycles with air. Regeneration at 773.15 K was ineffective due to slow carbon loss from the catalyst, whereas regeneration at 823.15 K was significantly quicker. After six cycles, the non-porous catalyst became deactivated, whereas the other catalyst stayed active even after 24 cycles. Additionally, the porous catalyst's capacity for carbon deposition increased from 45 $gC\,gNi^{-1}$ in the initial cycle to 80 $gC\,gNi^{-1}$ by the third cycle. Another study[76] evaluated supported Co and Ni catalysts with SiO_2 and Al_2O_3 in a multilayer reactor. The tested Co/Al_2O_3 was identified as optimal during the regeneration cycle, as its material characteristics remained unchanged throughout the cycles. Successful regeneration requires a well-defined particle size distribution of the catalyst and a strong interaction of the metal and support.

Qian et al.[77] evaluated a 40% Fe/Al_2O_3 in a reactor with fluidized bed with a CO_2 regeneration stream at 1023.15 K and 200 mL min^{-1} for 11 hours. The regenerated catalysts exhibited rapid activation compared to fresh catalysts due to reduced Fe_2O_3 particle size and improved distribution over the alumina support. Catalytic activity also showed consistent performance across tested cycles.

Aiello et al.[10] regenerated 15% Ni/SiO_2 catalyst using steam at 923.15 K in a reactor with a fixed bed. Hydrogen production was achieved during the regeneration cycle, although not all carbon was eliminated. Despite this, catalytic activity and methane conversion remained stable over 10 cycles. XRD analysis indicated insignificant structural changes of the catalyst surface post-regeneration.

Regeneration studies indicate that regeneration cycles may reduce catalyst activity over time. Additionally, certain regeneration methods can produce unwanted CO_2 emissions. The reaction and removal of carbon from the catalyst surface result in the loss of a potentially valuable product. Therefore, this approach may not be preferable for the clean production of hydrogen fuel.

2.3 Solar Reactors

As depicted in Figure 2.6, the two main types of solar reactors for methane cracking are directly and indirectly irradiated reactors. In directly irradiated reactors, solar radiation enters by way of quartz windows, heating the reactor walls and the gas flowing within. On the other hand, heat transfer in indirectly irradiated reactors occurs by conduction through the reactor walls in the reactor's sealed cavity. In addition, some reactors use carbon as a catalyst bed or catalytic seeding. The particles of carbon black absorb radiation and provide sites for nucleation in the reactor, allowing for a more consistent distribution of temperature, resulting in enhanced conversion rates.[78]

Each type has been shown to have its advantages and limitations. In an experimental study,[79] it was demonstrated that indirectly irradiated reactors exhibit homogeneous temperatures throughout. Modeling the temperature profile inside directly irradiated reactors has revealed that the highest temperatures are found along the walls, which is where the reaction primarily takes place,[80,81] leading to overall lower methane cracking rates inside the reactor. Another drawback of these types of reactors is the potential for the window to break due to carbon build-up. Additionally, the walls of the reactor require materials which have high temperature specifications for heat transfer.[79] The following sections will explore the advancements in the research on integration of solar energy into methane cracking. Various solar reactors that have been developed and tested for the process will be explored by highlighting the findings from selected articles.

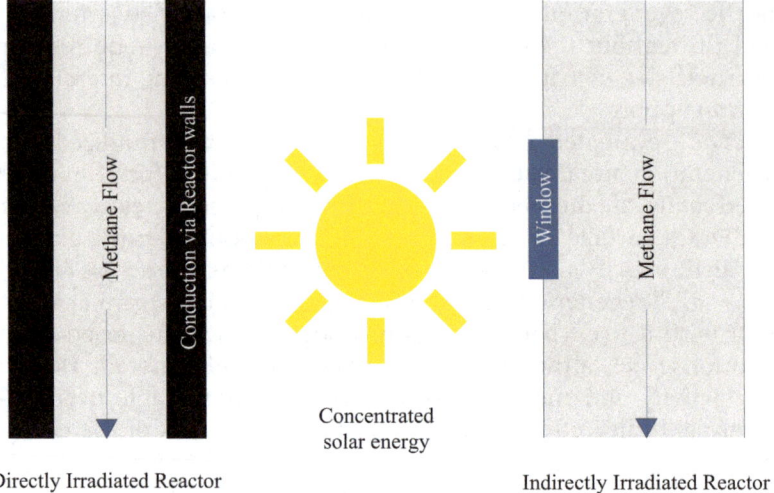

Figure 2.6 Diagram illustrating the two types of solar reactors developed for methane cracking. Reproduced from ref. 16, https://doi.org/10.1016/j.ecmx.2021.100117, under the terms of the CC BY 4.0 license, https://creativecommons.org/licenses/by/4.0/.

2.3.1 Directly Irradiated Solar Reactors

Kogan *et al.* at the Weizmann Institute of Science were the first to suggest and design a solar reactor for methane cracking[82–85] initially constructing and testing a small unseeded reactor. The accumulation of carbon particles on the window of the reactor was a significant limitation that posed a risk of breakage. To combat this issue and remove the settled particles, a tornado type flow of secondary gas was implemented.[86] A 28% conversion at 1050 °C was accomplished by testing two distinct reactor configurations.[82] Albeit effective in clearing the window, the tornado flow did nothing to clear the build-up of carbon at the exit of the reactor.

The reactor design was further developed by seeding with radiation-absorbing particles to address the challenge of inconsistent temperature and low conversion. As a result, the methane flow was heated more evenly, especially in regions far from the reactor walls that had previously received insufficient heating. This development may also mitigate the issue of carbon build-up, as volumetric heating prevents carbon formation on the reactor walls. Various designs for the seeding stream were evaluated, and a configuration featuring a flow inlet that is at least 22% of the length of the reaction chamber demonstrated the highest efficacy in maintaining a clear window.[83] Ozalp and Kanjirakat conducted computational fluid dynamics (CFD) studies on this setup[87] and the experimental outcomes corroborated that the introduction of carbon seeding helps elevate temperatures within the reactor chamber, enhancing the reaction conversion. Additionally, this

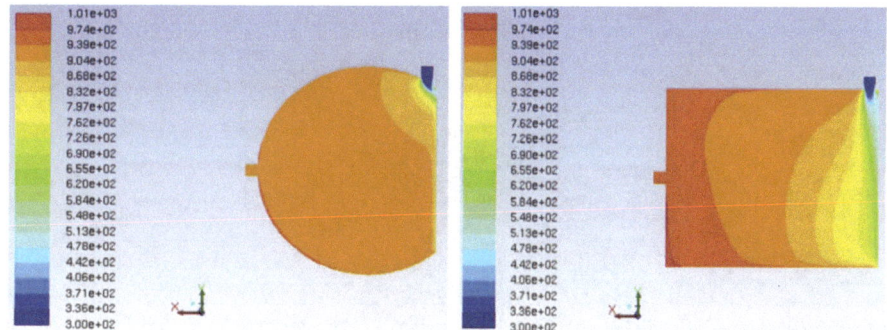

Figure 2.7 Results of the temperature profile in spherical (a) and cylindrical (b) shaped reactors. Reproduced from ref. 89 with permission from Elsevier, Copyright 2012.

reactor configuration was advanced by Yeheskel and Epstein[88] through further CFD analysis, achieving elevated temperatures of up to 1773.15 K, which facilitated high methane conversion rates.

Reactor geometry plays a crucial role in optimizing heat transfer. Through CFD analysis, Costandy *et al.*[89] demonstrated that temperature profiles heavily depend on reactor geometry, effecting the reaction rates. Cylindrical and spherical reactor shapes were evaluated: as illustrated in Figure 2.7 spherical reactors exhibited more homogeneous temperature profiles.

Hirsch and Steinfeld constructed and studied a 5 kW prototype utilizing carbon particles suspended in a vortex methane flow.[78] The reactor geometry (Figure 2.8) achieved a 67% conversion at 1325 °C. A subsequent study[90] verified the experimental results by modeling the reactor setup.

Significant efforts have been made in directly radiated solar reactors for methane cracking at PROMES-CRNS.[80,81,91,92] Figure 2.9 depicts a lab-scale reactor that was set-up for testing at this facility. Various parameters were assessed, revealing that increasing the size of the solar concentrator to 2 m from 1.5 m enhanced conversion rates from 20% to 95%. Additionally, reactor geometry was found to be critical for conversion efficiency, with nozzle geometries exhibiting larger surface areas demonstrating superior radiation absorption resulting in an increased methane cracking rate. While the CB co-product was effectively removed, the reactor walls formed an unwanted carbon layer that could potentially decrease conversion rates over time and clog the reactor.[91]

2.3.2 Indirectly Irradiated Solar Reactors

Dahl *et al.*[93] designed an aerosol flow reactor with a double tube utilizing 10 kW solar concentrator and pure methane feed. Their study assessed the impact temperature and residence time had on conversion rate, revealing that temperature significantly influences reaction extent more than residence time. Despite achieving high temperatures, methane conversion was

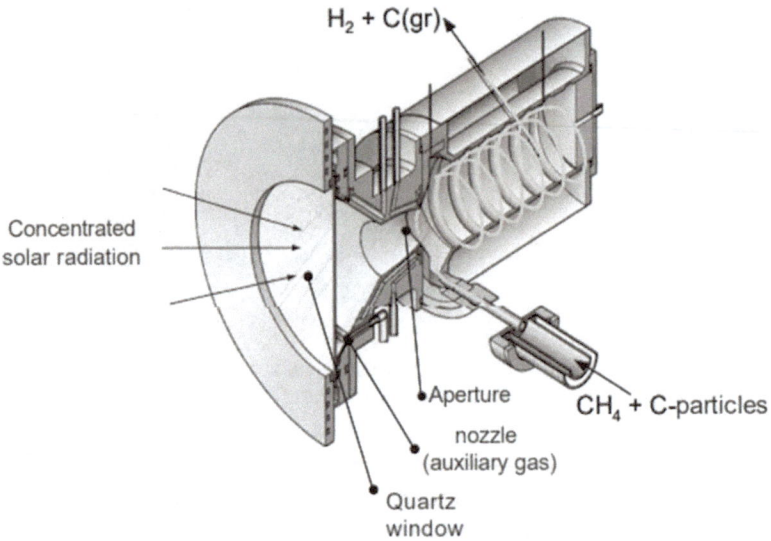

Figure 2.8 A vortex flow type directly radiated reactor designed by Hirsch and Steinfeld. Reproduced from ref. 78 with permission from Elsevier, Copyright 2004.

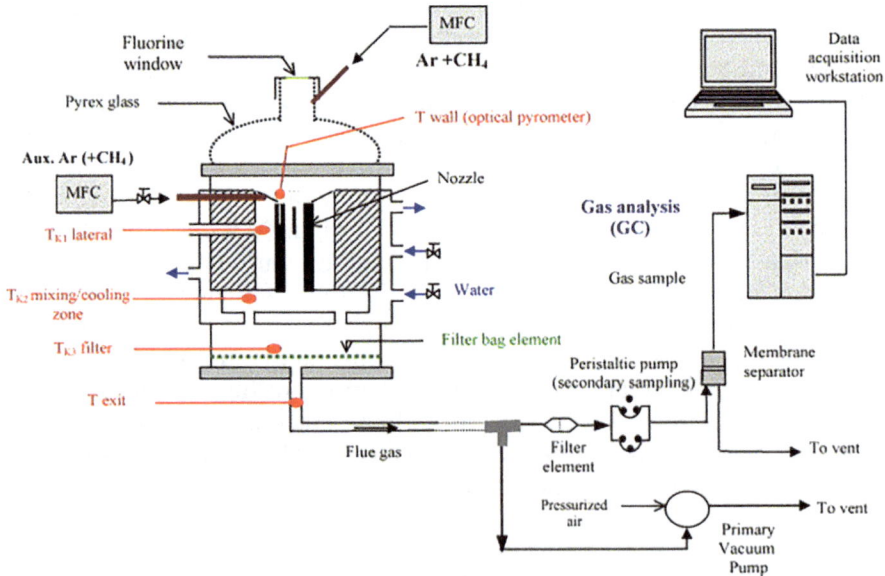

Figure 2.9 Lab-scale directly radiated reactor designed at PROMES-CRNS. Reproduced from ref. 92 with permission from Elsevier, Copyright 2008.

limited to 70% due to inadequate heat transfer. An enhanced reactor design, featuring an additional tube (Figure 2.10),[94] was tested for the effects of temperature of the reactor wall and flow rate of methane, resulting in a

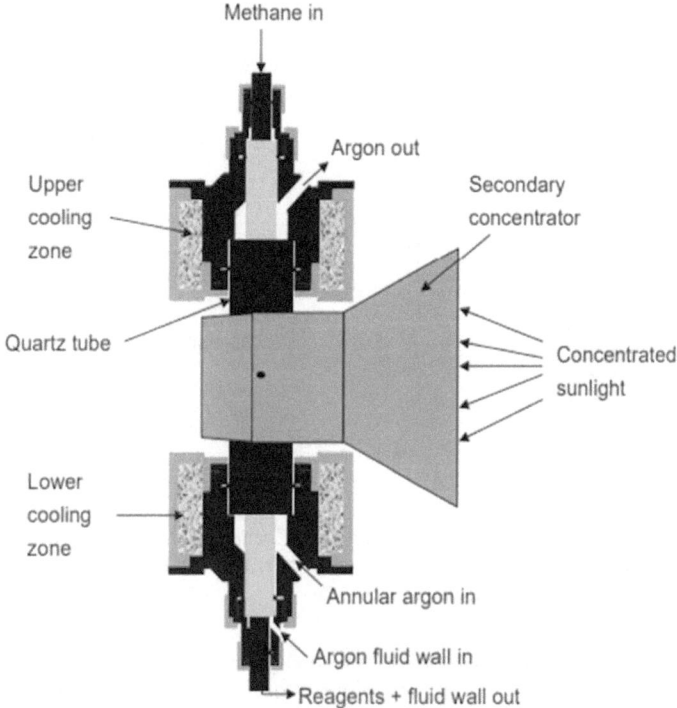

Figure 2.10 Solar reactor developed by Dahl et al. with a fluid wall. Reproduced from ref. 94 with permission from Elsevier, Copyright 2003.

conversion rate of 90% at 2133.15 K with approximately 0.1 seconds residence time. Carbon black seeding was also evaluated, yielding no notable performance improvement, possibly due to the small size of the reactor. Future scaling of this setup may require carbon black seeding as Dahl et al.'s kinetic model confirmed its beneficial impact.[95]

Abanades et al.[96] introduced a double-wall reactor tube concept, demonstrating that methane conversion exceeding 90% can be attained at temperatures of 1849.15 K or higher. Experimental tests on a four-double-walled tube reactor resulted in complete methane conversion and a hydrogen yield of up to 87%. In another study by Rodat et al.[97–99] a 10 kW solar reactor, featuring four indirectly irradiated tubes within a cubic cavity receiver (Figure 2.11), achieved 98% conversion and a 90% hydrogen yield at 1769.15 K The effect on the reaction extent by the residence time, temperature, and methane concentration was researched. Testing using a natural gas feed for this setup was also performed, yielding comparable results.[99] A numerical simulation conducted in ref. 100 indicated that the majority of reactions occur at the reactor's lower section, suggesting the potential for conversion enhancement. Further studies by Rodat et al.[101–104] focused on additional improvements and parameter effects. Another study[103] suggested a 55 MW solar reactor plant accompanied by technoeconomic evaluations to determine its scalability.

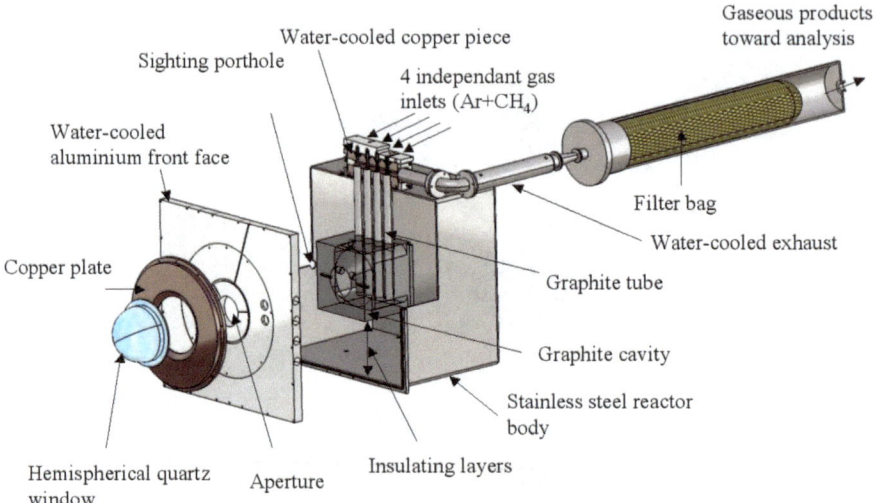

Figure 2.11 Solar reactor developed by Rodat et al. (indirectly irradiated, 19 kW). Reproduced from ref. 99 with permission from American Chemical Society, Copyright 2009.

CFD simulations were performed Abanades et al.[105] on a tube type solar reactor, where the temperature profiles were analyzed to identify optimal parameters for high efficiency. A bench-scale 1 kW reactor was tested at 1673.15 K and was able to achieve a 90% CH_4 conversion and 85% H_2 yield.

2.3.3 Catalyzed Solar Reactors

A study examined the decomposition of methane utilizing solar energy with CB catalysts in a solar reactor using a packed-bed.[106] The reactor, designed for indirect irradiation, ensures homogeneous heating, enhancing temperature control. The catalyst was regenerated using a particle injection mechanism. Various parameters affecting reactor performance were analyzed, revealing that flow rate and temperature significantly impact yield and conversion rates.

The kinetics of carbon catalysts in solar reactors was evaluated by Abanades et al.[107] using a solar-powered thermogravimetric device. Various carbon catalysts were examined for their deactivation kinetics, including activated carbon and CB in pellet and powder forms. The results indicated that activated carbon exhibited greater activity than CB, albeit with faster deactivation. CB powders outperformed pellets, maintaining catalytic activity for an extended period despite high carbon coking on the surface of the catalyst, which did not lead to deactivation, suggesting the carbon deposits act as precursors for the catalyst. Another study[108] introduced CB particles with argon, yet negligible enhancement in the CH_4 conversion was observed, likely caused by insufficient residence time.

A one-dimensional model was developed and simulated by Patrianakos et al.[109] to investigate carbon particle seeding effects.[110] Their findings indicate that particle feeding enhances hydrogen and carbon production rates and reduces carbon growth on reactor walls, leading to improved performance and decreased maintenance. Nezzari and Gomri[111] demonstrated that incorporating a mixture of carbon catalyst particles with varying radius enhances the conversion rate.

Pinilla et al.[26] conducted a detailed study comparing a catalytic solar reactor and a non-catalytic one. The solar reactor without a catalyst yielded negligible conversion rates, with 0.15% at 650 °C and 0.8% at 800 °C. The following materials were tested at their respective optimal temperatures: Ni – 650 °C, Fe – 800 °C, and commercial carbon black – 950 °C, resulting in CH_4 conversion ranging from 34% to 67%. The study concluded that catalysts enhance reaction rates, improve process efficiency, and increase H_2 selectivity by reducing the formation of other hydrocarbons. The use of metal catalysts also leads to higher quality carbon, positively impacting process economics. A key advantage of catalysts is the significant reduction in operating temperatures, which enhances process efficiency and promotes homogeneous heating. This temperature reduction allows for the use of less expensive materials in reactor construction, facilitating easier operation. However, further development is necessary to design and optimize suitable solar reactors for catalyst integration. Current studies primarily focus on carbon catalysts; there is a need to explore and test more metal-based catalyst systems.

2.3.4 Molten Media Reactors

The use of molten media heated *via* concentrated solar energy is another viable option for integrating solar energy into the methane cracking process. Testing has been done on molten metals or salts for this process and shown to be an effective heat transfer medium and also as catalytic materials. Molten media can be used as a catalyst to prevent carbon clogging and catalyst deactivation associated with traditional fixed catalyst beds. Methane flows along the column, facilitating the methane cracking reaction while the carbon particles float above the molten salts medium.[112]

Metals that are catalytically active and possess high melting points are employed as alloys mixed with inert metals[113] A Ni–Bi alloy (27 : 73) was tested in a bubble column by Upham et al.[114] and a cracking rate of 95% was achieved along with pure hydrogen production. Catalan and Rezaei[115] developed a combined kinetic model with hydrodynamics, proposing various designs for liquid bubble reactors. Their findings indicate that higher temperatures facilitate smaller reactor volumes and enhance CH_4 conversion. Recent research by Rowe et al.[116] examined solar reactor designs, revealing a receiver with a beam-down configuration, which achieved an efficiency of only 9%. However, low hydrogen production in these configurations may stem from heat transfer limitations. Further investigation into receiver configurations appropriate for molten media reactors is necessary.

Concentrated solar energy can be harnessed to produce green hydrogen, with molten materials facilitating uniform heat distribution in solar reactors. Zheng et al.[117] introduced a design featuring liquid bubble technology for methane cracking, powered entirely by solar energy. In this system, solar-heated tin is employed in a liquid bubble column to produce hydrogen. The Rankine cycle is integrated to capture heat from the liquid metal to generate electricity or mechanical work which is utilized toward powering internal processes. After the first cycle, the tin metal is then recycled and stored for further use. In another study, a $MgCl_2$–KCl mixture was employed by Boretti et al.[118] as a heat transfer medium from concentrated solar energy. The development of appropriate reactors is required for this system and further research is needed on efficiently removing carbon particles from the column's top and identifying an optimal molten metal mix.

2.3.5 Solar Energy Collection

The economics of hydrogen produced from solar energy is influenced by the cost of the setup for solar energy collection,[119] necessitating the need for affordable and efficient technologies for solar energy collection. This section briefly examines the setup used for the collection of solar energy detailed in previous sections.

At PSI, a high-flux solar furnace was fabricated[120] utilizing sun-tracking heliostats and a paraboloidal concentrator with 8.5 m diameter, achieving 40 kW with a 5000 suns peak concentration, providing input of 5 kW to a reactor operating between 1000–1600 K.[78] Similarly, adjustments in concentrator diameter from 1.5 m to 2 m were made to enhance methane conversion.[91] At CRNS-PROMES, a 1 MW solar furnace featuring 63 sun-tracking heliostats, sized 45 m^2 each, and a parabolic concentrator sized 1830 m^2 delivered up to 9000 suns.[97,99] Smaller numbers of heliostats were used in studies that required reduced solar input. Most studies utilized a heliostat field using a paraboloid concentrator, which is recognized as the most efficient optical system for solar furnaces.[121]

2.4 Case Studies

This section presents two integrated systems proposed for methane cracking using solar energy. A detailed thermodynamic modeling will be conducted, and the efficiency of the systems will be determined.

2.4.1 System Description

2.4.1.1 Integrated System 1

The proposed integrated process uses a solar-powered methane cracker to synthesize dimethyl ether (DME) and methanol as the final products. The system initially stores hydrogen and carbon acting as an energy storage.

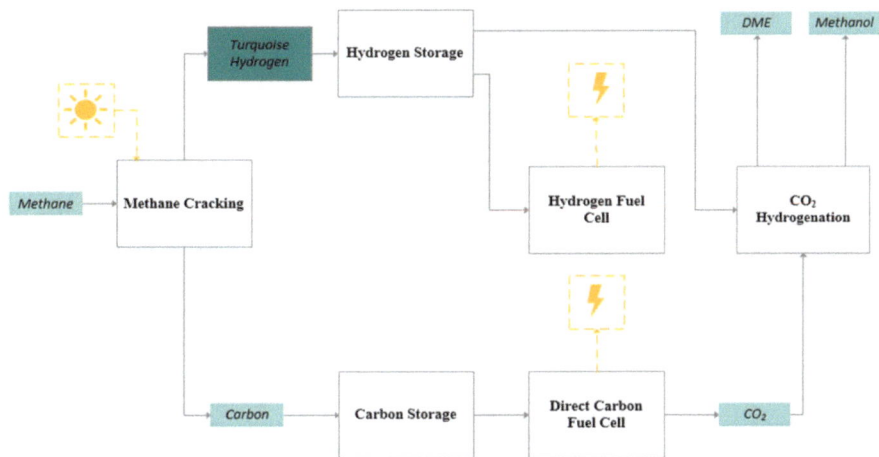

Figure 2.12 Proposed integrated system 1 flow diagram. Reproduced from ref. 122 with permission from Elsevier, Copyright 2022.

Subsequently, part of the hydrogen is utilized in a proton-exchange membrane fuel cell (PEMFC), while carbon is directed to a direct carbon fuel cell (DCFC) for electricity generation. The CO_2 produced by the DCFC reacts with H_2 to synthesize methanol fuel, which is dehydrated to produce DME. This approach enables the generation of sustainable hydrogen from solar input, converting it into a liquid H_2 energy carrier from CO_2 captured in the system. Figure 2.12 presents the system diagram of this integrated process. The system utilizes solar energy, methane, and air, to produce electricity, methanol, and DME. The subsystems included are solar methane cracker, direct carbon fuel cell, hydrogen fuel cell, and CO_2 hydrogenation.

2.4.1.2 Integrated System 2

The proposed system consists of a methane cracking unit producing syngas which is then converted into methanol. By-product CB is used in a DCFC and the produced H_2 is utilized in a hydrogen fuel cell. The CO_2 stream is co-electrolyzed with steam in the presence of solar energy to produce synthetic gas, which is then used toward the synthesis of methanol. Figure 2.13 illustrates the system diagram for this process. This system produces methanol and electricity using solar energy, methane, and air. The subsystems included are solar methane cracker, direct carbon fuel cell, hydrogen fuel cell, co-electrolysis, and methanol synthesis.

2.4.1.3 Subsystems

The included subsystems within both integrated systems are briefly discussed in the subsequent paragraphs.

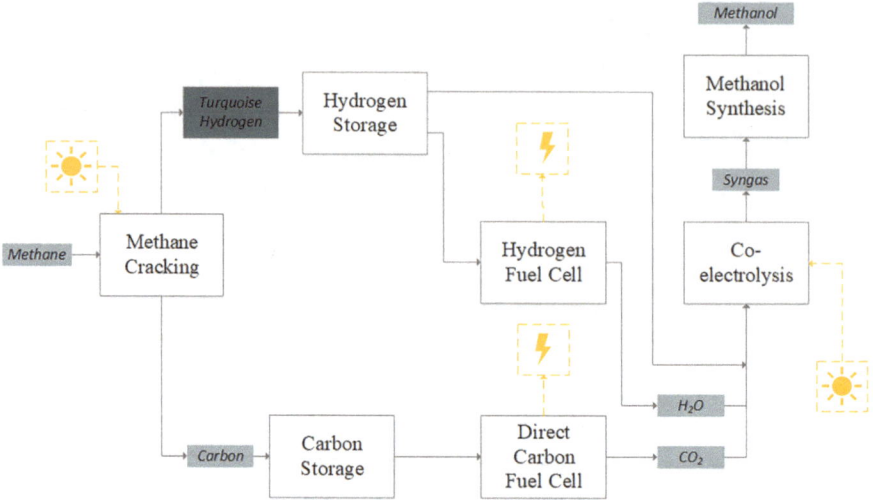

Figure 2.13 Proposed integrated system 2 flow diagram. Reproduced from ref. 123 with permission from Elsevier, Copyright 2024.

2.4.1.3.1 DCFC – Direct Carbon Fuel Cell.
The methane cracking system produces carbon by-product that is directed to the DCFC stack for electricity generation. CO_2 is extracted from the exiting stream of the cathode, captured, and forwarded to the next process. The cell utilizes a molten carbonate anode, enabling an exhaust stream rich in CO_2, making it favorable for capturing the emissions.[124] The reactions occurring at the anode and cathode are detailed in eqn (2.2) and (2.3). The cell functions at approximately 973.15 K with $E = 1.02$ V.[124]

$$O_2 + 2CO_2 + 4e^- \rightarrow 2CO_3^{2-} \tag{2.2}$$

$$C + 2CO_3^{2-} \rightarrow 3CO_2 + 4e^- \tag{2.3}$$

2.4.1.3.2 PEMFC – Hydrogen Fuel Cell.
A PEMFC stack generates electricity from produced H_2. This process is carried out at low temperatures, where H_2 from the methane cracking unit is supplied to the anode. The anode reaction is represented in eqn (2.4). At the cathode, electrons, and protons react with oxygen to produce water, as detailed in eqn (2.5). The overall fuel cell reaction is outlined in eqn (2.6). The cell function at 333.15–353.15 K under atmospheric pressure.

$$H_2 \rightarrow 2H^+ + 2e^- \quad E = 0.000 \text{ V} \tag{2.4}$$

$$1/2 O_2 + 2H^+ + 2e^- \rightarrow H_2O \quad E = 1.229 \text{ V} \tag{2.5}$$

$$H_2 + 1/2 O_2 \rightarrow H_2O \quad \Delta H_{298K} = -285.8 \text{ kJ mol}^{-1} \tag{2.6}$$

Methane Cracking

2.4.1.3.3 MEOH – Methanol Production. Syngas serves as a versatile feedstock for producing various chemicals and fuels. In the proposed integrated system, produced syngas is converted into methanol, which is a cleaner and more transportable fuel than hydrogen. Methanol production occurs at a temperature of 493–553 K and 50–100 bar pressure,[125] as indicated by the following chemical reactions.

$$CO + 2H_2 \rightarrow CH_3OH \tag{2.7}$$

$$CO_2 + 3H_2 \rightarrow CH_3OH + H_2O \tag{2.8}$$

2.4.1.3.4 CO_2 Hydrogenation (CO_2–H). Methanol is produced traditionally through the conversion of H_2 derived from SMR, in the presence of catalysts,[126] and this process has high carbon emissions. Alternatively, methanol can be produced *via* CO_2 hydrogenation, utilizing CO_2 as a feedstock, which is a more favorable option.[127] This route, using CO_2 captured from other systems and sustainably produced H_2, yields a negative carbon footprint. DME, a clean and renewable fuel, is produced by dehydrating methanol. Unlike methanol, DME is non-toxic and non-carcinogenic.[128] DME can be produced through a two-step process with methanol as an intermediate or directly in a single reactor by using hybrid catalysts,[129] which may offer thermodynamic advantages.[130] Captured CO_2 from DCFC reacts with hydrogen gas from methane cracking for one-step synthesis of DME. This reaction first forms methanol, which is then dehydrated to yield DME. Additionally, the water–gas shift reaction occurs, producing some CO in the output stream. The relevant reactions are outlined below.

$$CO_2 + 3H_2 \rightarrow CH_3OH + H_2O \quad \Delta H_{298K} = -49 \text{ kJ mol}^{-1} \tag{2.9}$$

$$CO_2 + H_2 \rightarrow CO + H_2O \quad \Delta H_{298K} = +42 \text{ kJ mol}^{-1} \tag{2.10}$$

$$2CH_3OH \rightarrow CH_3OCH_3 + H_2O \quad \Delta H_{298K} = -91 \text{ kJ mol}^{-1} \tag{2.11}$$

The produced water, unreacted gases, methanol, and DME are separated in the purification section to yield pure methanol and DME streams.

2.4.1.3.5 Co-electrolysis (SOEC). CO_2 can be utilized in co-electrolyzers with water for sustainable syngas production. Although CO_2 and water electrolysis is generally carried out independently, co-electrolysis offers greater energy efficiency and mitigates issues like carbon deposition.[131] This method enables the storage of excess renewable energy, addressing its intermittent availability.[132] The syngas produced through co-electrolysis has diverse applications, including the synthesis of DME and methanol. Utilizing CO_2 for fuel production reduces reliance on fossil fuel sources.[133] At the anode, air rich in oxygen is generated which can be used toward oxy-fuel combustion.[134]

This subsystem contains a solid oxide electrolyzer cell functioning at 1073.15 K. The CO_2 and water streams from the DCFC and PEMFC are heated to the electrolysis temperature and introduced to the cathodic side. Upon supplying heat and electricity, oxygen and hydrogen are produced from water as per eqn (2.12). Carbon monoxide is produced from the reduction of CO_2 according to eqn (2.13).

$$H_2O + 2e^- \rightarrow H_2 + O^{2-} \quad (2.12)$$

$$CO_2 + 2e^- \rightarrow CO + O^{2-} \quad (2.13)$$

A reverse water–gas shift (RWGS) reaction takes place at the cathode (eqn (2.14)).[135]

$$CO_2 + H_2 \rightarrow CO + H_2O \quad (2.14)$$

At the anode, oxygen is also formed *via* eqn (2.15)

$$O^{2-} \rightarrow \frac{1}{2}O_2 + 2e^- \quad (2.15)$$

H_2 is also supplied to the cell, which is required for maintaining optimum conditions for reduction and avoiding cell material oxidation.[136]

2.4.2 Methodology

To evaluate the two proposed systems, a detailed thermodynamic assessment is performed. Firstly, the thermodynamic properties of the streams are obtained, and mass, energy, entropy, and exergy balances are applied to each component. Energy and exergy efficiencies are computed for each subsystem and the overall systems. Modeling of the systems is performed on Aspen Plus®,[137] employing the MIXCIPSD stream class due to solid carbon presence. The Peng–Robinson equation of state serves as the primary property method for calculations of the system thermodynamics. The NRTL-RK method is used for blocks involving DME, methanol, and streams with high pressure.

Pure methane is utilized at ambient conditions, compressed in a multistage process, and pre-heating is done using an output stream from a methane cracker. The methane cracking reactor is modeled using an RGibbs block with an input temperature and pressure of 1473.15 K and 10 bar. H_2 product at the outlet is removed using a membrane separator modeled using a separator block, which is set to achieve 90% separation of the hydrogen. The unreacted methane from the separator is recycled back and mixed with the feed to the methane cracker.

Aspen Plus®[137] lacks built-in blocks for fuel cell simulation, necessitating the use of a separator and RGibbs block for modeling the reactions. This method has been documented in the literature by employing built-in blocks in Aspen plus®.[138,139] In the hydrogen fuel cell, a separator block is used to represent the anode side. This separator block splits and directs a heated stream to the cathode. Cell reactions take place inside an RGibbs reactor representing the cathode. The carbon fuel cell is simulated in a similar manner with the anode represented by an RGibbs reactor at 873.15 K.

In System 1 (Figure 2.14), an RGibbs reactor operating at 373.15 K and 60 bar is used to represent the CO_2 hydrogenation reactor. CO and H_2 feeds are cooled and compressed to the necessary operating conditions. A 3:1 H_2/CO_2 ratio in the feed stream is employed, typical for direct DME synthesis.[140] As mentioned earlier, the property method used here is NRTL. A flash drum is used to separate the unreacted gases and purification of the liquid streams is carried out using two distillation columns to give pure DME and methanol as the final outputs.

In System 2 (Figure 2.15), the CO_2 stream from the carbon fuel cell is conditioned to be fed to the electrolyzer. It is combined with water and H_2, achieving a 45:45:10 ratio ($CO_2:H_2O:H_2$).[141] The model for the electrolyzer is adapted from ref. 142, where the system is divided into three different stages using three blocks, as illustrated in Figure 2.16. This model incorporates the electrochemical and RWGS reactions. 'R1' is the first reactor using an 'REquil' block where the feed enters and the RWGS reaction occurs at 573.15 K. 'R2' is an RStoic block operating at 1073.15 K that is utilized with a conversion rate of 0.05 for CO_2 and 0.98 for H_2O for the below stoichiometric reactions.

$$H_2O \rightarrow H_2 + \frac{1}{2}O_2 \quad (2.16)$$

$$CO_2 \rightarrow CO + \frac{1}{2}O_2 \quad (2.17)$$

A separator block is also used, where oxygen is eliminated from the stream and subsequently enters R3. In this reactor, the second step of the RWGS reaction takes place at 1073.15 K. The output stream includes water which goes to SEP2, a flash drum where it is separated to produce syngas. The NTRL-RK property method is used for modeling the electrolyzer section.

Syngas is compressed to 50 atm in C2 before entering the MEOH conversion reactor, where the fractional conversions for reactions in eqn (2.17) and (2.18) are 0.35 and 0.17, respectively.[143] SEP3, a flash drum, is used to separate the unreacted gases to produce methanol.

$$CO + 2H_2 \rightarrow CH_3OH \quad (2.18)$$

$$CO_2 + 3H_2 \rightarrow CH_3OH + H_2O \quad (2.19)$$

2.4.2.1 Balance Equations

Eqn (2.20) and (2.21) represent the general equations of mass and energy balance.

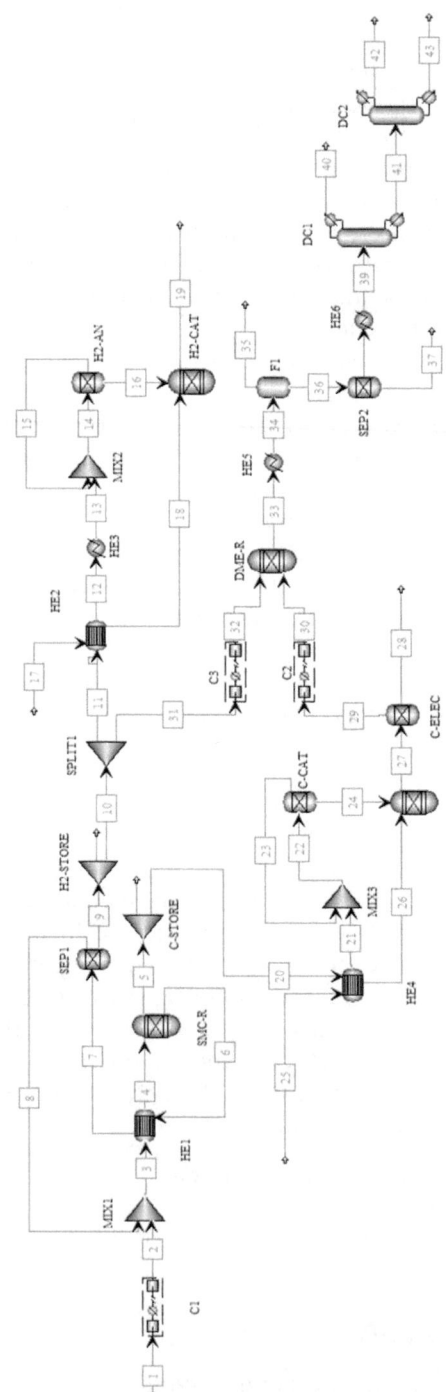

Figure 2.14 Process flowsheet of integrated system 1 from Aspen Plus modeling. Reproduced from ref. 122 with permission from Elsevier, Copyright 2022.

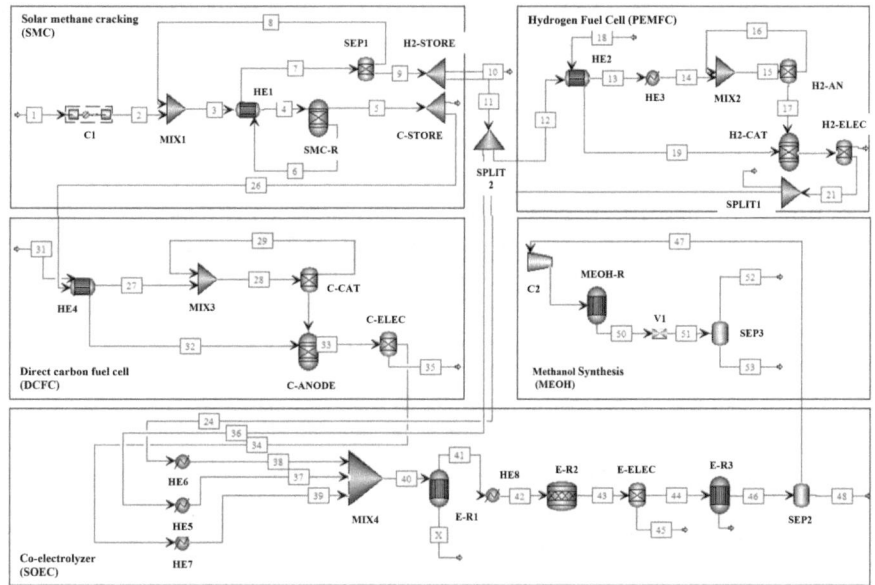

Figure 2.15 Process flowsheet of integrated system 2 from Aspen Plus modeling. Reproduced from ref. 123 with permission from Elsevier, Copyright 2024.

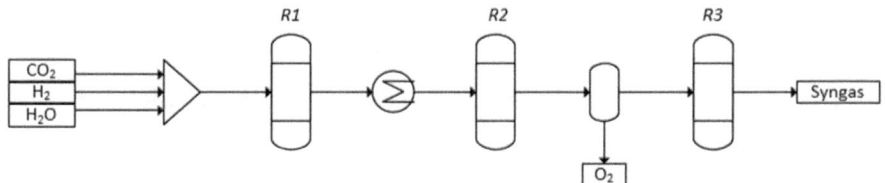

Figure 2.16 Aspen Plus® model of the co-electrolyzer. Reproduced from ref. 123 with permission from Elsevier, Copyright 2024.

$$\sum_{\text{in}} \dot{m}_i = \sum_{\text{out}} \dot{m}_i \tag{2.20}$$

$$\dot{Q}_{\text{in}} + \dot{W}_{\text{in}} + \sum_{\text{in}} \dot{m}_i h_i = \dot{Q}_{\text{out}} + \dot{W}_{\text{out}} + \sum_{\text{out}} \dot{m}_i h_i \tag{2.21}$$

m_i denotes the mass flow rate (kg s^{-1}), h_i the specific enthalpy (kJ kg^{-1}), and $\dot{Q}$ and $\dot{W}$ signify the heat and work rates (kW). Neglecting potential and kinetic energy terms, eqn (2.22) is used for the entropy balance.

$$\frac{\dot{Q}_{\text{in}}}{T_s} + \dot{S}_{\text{gen}} + \sum_{\text{in}} \dot{m}_i s_i = \frac{\dot{Q}_{\text{out}}}{T_s} + \sum_{\text{out}} \dot{m}_i s_i \tag{2.22}$$

s_i denotes the specific entropy (kJ kg^{-1} K^{-1}) and $\dot{S}_{gen}$ denotes the rate of entropy generation of the component (kW K^{-1}). Eqn (2.23) shows the general exergy balance equation.

$$\dot{Ex}_{Q_{in}} + \dot{Ex}_{W_{in}} + \sum_{in} \dot{m}_i ex_i = \dot{Ex}_{Q_{out}} + \dot{Ex}_{W_{out}} + \sum_{out} \dot{m}_i ex_i + \dot{Ex}_D \quad (2.23)$$

ex_i denotes the specific exergy (kJ kg^{-1}). $\dot{Ex}_D$ denotes the rate of exergy destruction of the component (kW). Using eqn (2.24) and (2.25), the exergy rates linked to work and heat can be calculated.

$$\dot{Ex}_Q = \left(1 - \frac{T_0}{T_i}\right)\dot{Q} \quad (2.24)$$

$$\dot{Ex}_W = \dot{W} \quad (2.25)$$

Eqn (2.26) is used for the calculation of specific exergy (neglecting kinetic and potential terms). Eqn (2.27) and (2.28) are used for the calculation of chemical and physical exergies.

$$ex_i = ex_{ph} + ex_{ch} \quad (2.26)$$

$$ex_{ph,i} = (h_i - h_0) - T_0(s_i - s_0) \quad (2.27)$$

$$ex_{ch,i} = \sum x_i ex_{ch}^0 + RT_0 \sum x_i \ln(x_i) \quad (2.28)$$

2.4.2.2 Efficiency

Eqn (2.29) shows the first law efficiency, energy efficiency, which is the ratio of the sum of all useful energy output and sum of all energy inputs.

$$\eta_{en} = \frac{\sum \dot{E}_{useful}}{\sum \dot{E}_{input}} \quad (2.29)$$

Eqn (2.30) shows the second law efficiency, exergy efficiency, which is the ratio of the sum of all useful exergy output to the sum of all exergy inputs, as shown in eqn (2.30).

$$\eta_{ex} = \frac{\sum \dot{Ex}_{useful}}{\sum \dot{Ex}_{input}} \quad (2.30)$$

The overall energy and exergy efficiency equations are as follows for system 1:

$$\eta_{en,overall,I} = \frac{\dot{W}_{elec,PEMFC} + \dot{W}_{elec,DCFC} + \dot{m}_{DME} ex_{DME} + \dot{m}_{MeOH} ex_{MeOH}}{\dot{Q}_{in,Solar}\left(1 - \frac{T_0}{T_S}\right) + \dot{W}_{C1} + \dot{W}_{C2} + \dot{W}_{C3} + \dot{m}_{CH_4} ex_{CH_4} + \dot{Ex}_{Q_{In,TOTAL}}}$$

$$(2.31)$$

$$\eta_{\text{ex,overall},I} = \frac{\dot{W}_{\text{elec,PEMFC}} + \dot{W}_{\text{elec,DCFC}} + \dot{m}_{\text{DME}}\text{ex}_{\text{DME}} + \dot{m}_{\text{MeOH}}\text{ex}_{\text{MeOH}}}{\dot{Q}_{\text{in,Solar}}\left(1 - \frac{T_0}{T_S}\right) + \dot{W}_{C1} + \dot{W}_{C2} + \dot{W}_{C3} + \dot{m}_{\text{CH}_4}\text{ex}_{\text{CH}_4} + \dot{\text{Ex}}_{Q_{\text{In,TOTAL}}}}$$

(2.32)

The overall energy and exergy efficiency equations are as follows for system 2:

$$\eta_{\text{en, overall,II}} = \frac{\dot{m}_{\text{MEOH}}\text{LHV}_{\text{MEOH}} + \dot{W}_{\text{elec,PEMFC}} + \dot{W}_{\text{elec,DCFC}}}{\dot{m}_{\text{CH}_4}\text{LHV}_{\text{CH}_4} + \dot{W}_{\text{SOEC}} + \dot{Q}_{\text{in,total}} + \dot{W}_{\text{in,total}}}$$

(2.33)

$$\eta_{\text{ex,overall,II}} = \frac{\dot{m}_{\text{MEOH}}\text{ex} + \dot{W}_{\text{elec,PEMFC}} + \dot{W}_{\text{elec,DCFC}}}{\dot{m}_{\text{CH}_4}\text{ex} + \dot{\text{Ex}}_{Q_{\text{in,total}}} + \dot{W}_{\text{in,total}}}$$

(2.34)

2.4.3 Results and Discussion

2.4.3.1 Integrated System 1

In this system, methane feed with a flow rate of 0.5 kg s^{-1} is considered. The feed is firstly compressed, cooled, and then directed to the methane cracker integrated with solar energy. The reactor operates at 1473.15 K and 10 bar, achieving a 95% conversion of methane, consistent with literature values for solar methane reactors reported in the literature. This subsystem generates 5.43 tonnes per day of hydrogen and 16.1 tonnes per day of carbon.

The CO_2 hydrogenation reactor was modeled at 60 bar and 373.15 K, achieving an 82% conversion of CO_2 and 61% DME selectivity. The separation and purification of unreacted gases and products yielded 13.79 tonnes per day of DME and 1.56 tonnes per day of methanol. DC1 and DC2 had reboiler and total condenser duties of approximately 300 and 205 kW, respectively.

Exergy destruction rates were calculated for each unit, with totals of the subsystem illustrated in Figure 2.17. The CO_2-H section experiences the highest exergy destruction, accounting for approximately 38% of the system's total. This is primarily due to the compressors needed for high-pressure gas synthesis of DME and methanol. The PEMFC section shows the lowest exergy destruction due to comparatively lower flow rates. The solar reactor contributes about 250 kW of exergy destruction, nearly half of the SMC subsystem's total.

The system generates 3.21 MW of electricity, with overall energy efficiency of 41.9% and overall exergy efficiency of 52.3%, respectively. Subsystem results are presented in Figure 2.18, and the study's outputs are summarized in Table 2.5.

The effect of varying input parameters on the output and performance of the system was evaluated. The reactor temperature is a key parameter in the

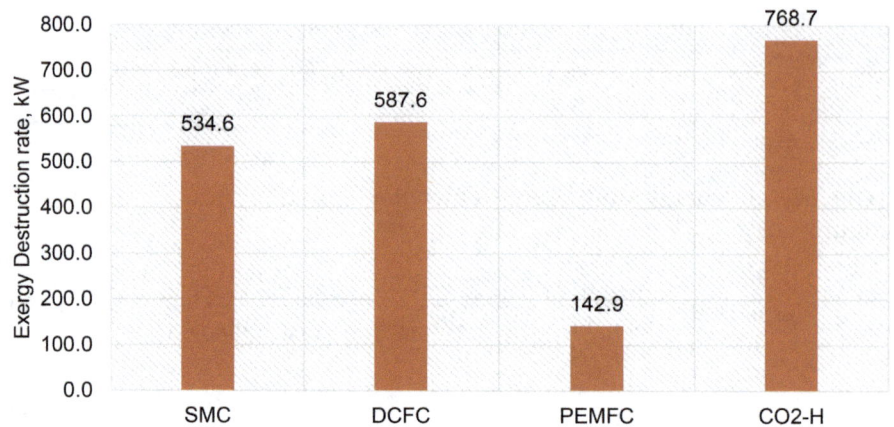

Figure 2.17 Exergy destruction rate results for the integrated system 1. Adapted from ref. 122 with permission from Elsevier, Copyright 2022.

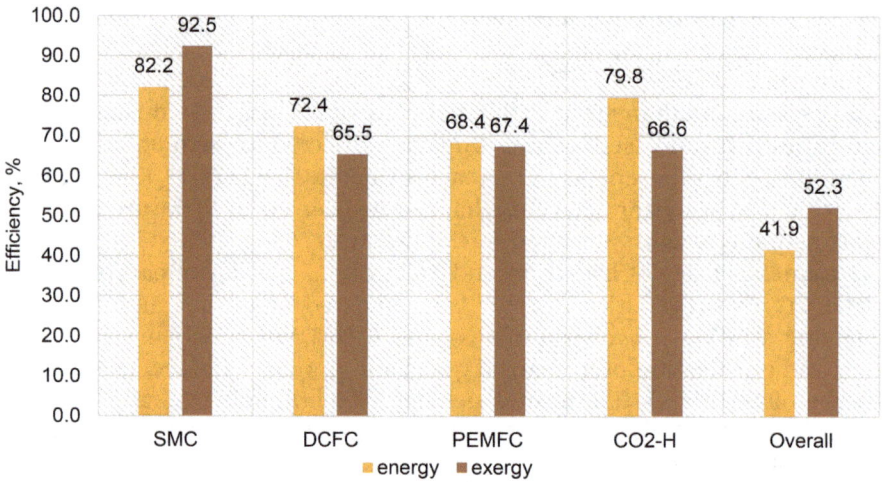

Figure 2.18 Efficiency results for integrated system 1. Adapted from ref. 122 with permission from Elsevier, Copyright 2022.

SMC subsystem, influencing solar energy requirements. The analysis indicates that methane conversion rates are enhanced as reactor temperatures increase, as illustrated in Figure 2.19. Figure 2.20 shows that increased reactor temperatures also lead to a slight decline in energy and exergy efficiencies of the subsystem. Methane conversion rates plateau at temperatures above 1000 °C, resulting in stable system outputs. While higher solar energy inputs are needed for elevated temperatures, this results in decreased efficiency rates. The decrease in exergy efficiency is less pronounced than that of energy efficiency, as solar exergy input constitutes a fraction of the solar energy input.

Table 2.5 Output summary for system 1.

Parameter	Value	Unit
Electricity generated	3.21	MW
Carbon production rate	16.17	Tonnes per day
Hydrogen production rate	5.43	Tonnes per day
DME production rate	13.79	Tonnes per day
Methanol production rate	1.56	Tonnes per day
Methane conversion	95	%
CO_2 conversion	82	%
MEOH selectivity	10	%
DME selectivity	61	%
Solar power requirement	5319.73	kW
Total hot utilities	335.14	kW
Total cold utilities	2108.34	kW
Total work input rate	697.26	kW

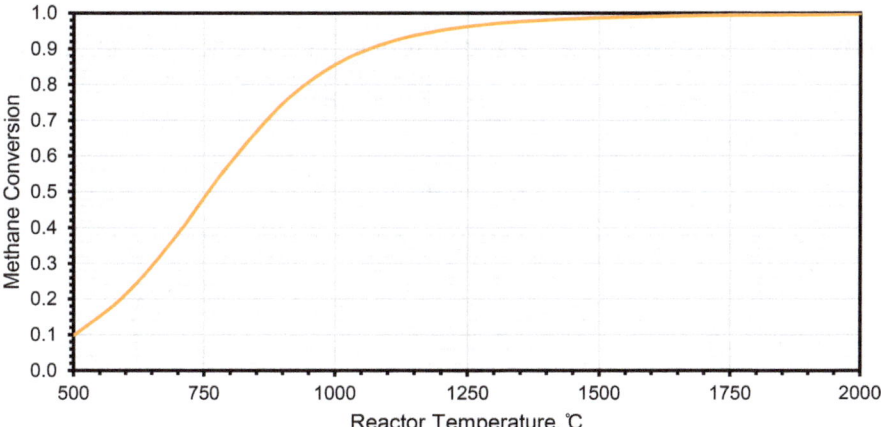

Figure 2.19 Effect of reaction temperature of the methane cracker on its conversion rate. Adapted from ref. 122 with permission from Elsevier, Copyright 2022.

PEMFC operating temperature was raised from 60 to 120 °C, and the energy efficiency increased slightly (Figure 2.21). This temperature increase also reduced exergy destruction rates in the subsystem. The system requires cooling of streams at high temperatures, from methane cracker to the operating temperature of the cell. Despite heat integration to warm the air input, an additional cooler was necessary. Thus, energy input is slightly reduced by higher operating temperatures. However, low-temperature PEMFCs function below 100 °C, as elevated temperatures may impair membrane performance.[144]

To assess the impact of reference temperature on the calculated exergy destruction rates, temperatures were increased from 2 °C to 42 °C. Figure 2.22 illustrates the relationship between reference temperature and exergy

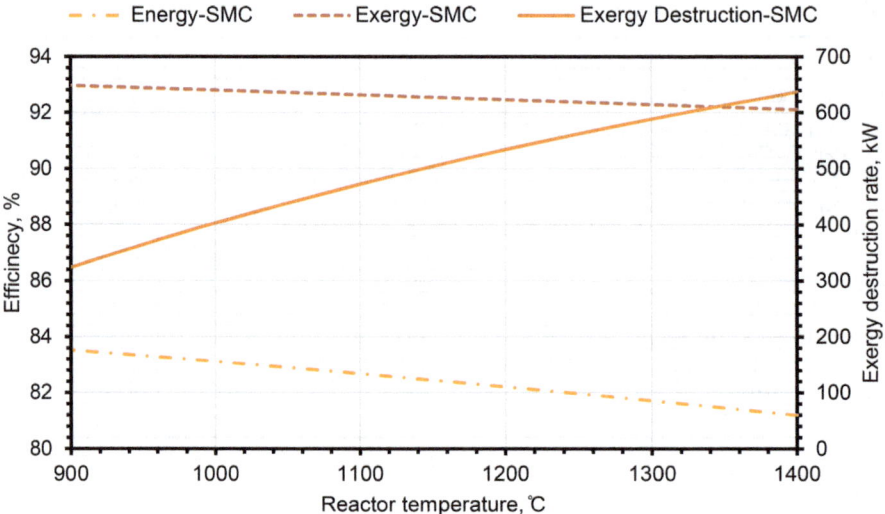

Figure 2.20 Sensitivity of the exergy destruction rate and efficiencies of the SMC subsystem to the reactor temperature. Adapted from ref. 122 with permission from Elsevier, Copyright 2022.

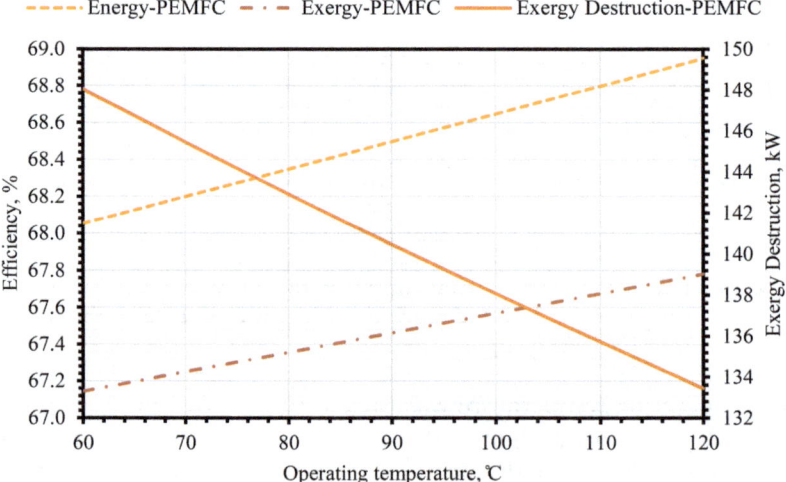

Figure 2.21 Sensitivity of the exergy destruction rates and efficiencies of the PEMFC subsystem to the operating temperature of the fuel cell. Adapted from ref. 122 with permission from Elsevier, Copyright 2022.

destruction rates. In the PEMFC, DCFC and CO_2-H, subsystems, exergy destruction rates increase with temperature. Conversely, in the SMC subsystem, as temperature rises from 2 °C to 42 °C, exergy destruction rates decrease from approximately 597 kW to 489 kW.

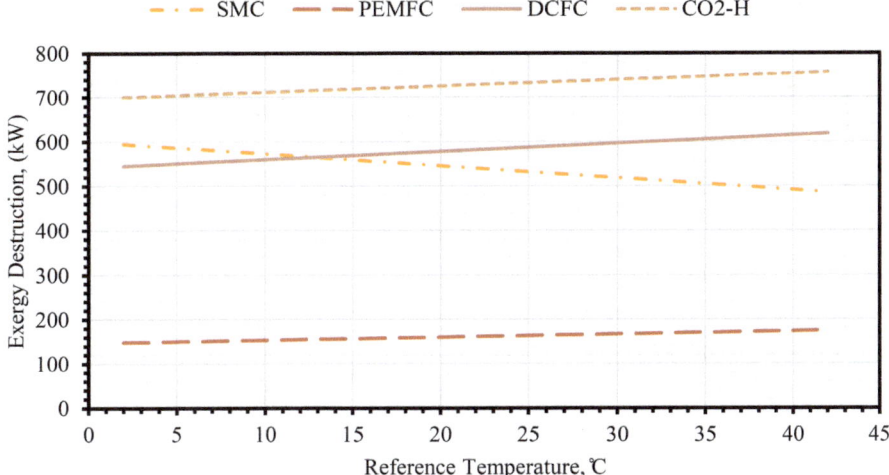

Figure 2.22 Sensitivity of the exergy destruction rates to the reference temperature. Adapted from ref. 122 with permission from Elsevier, Copyright 2022.

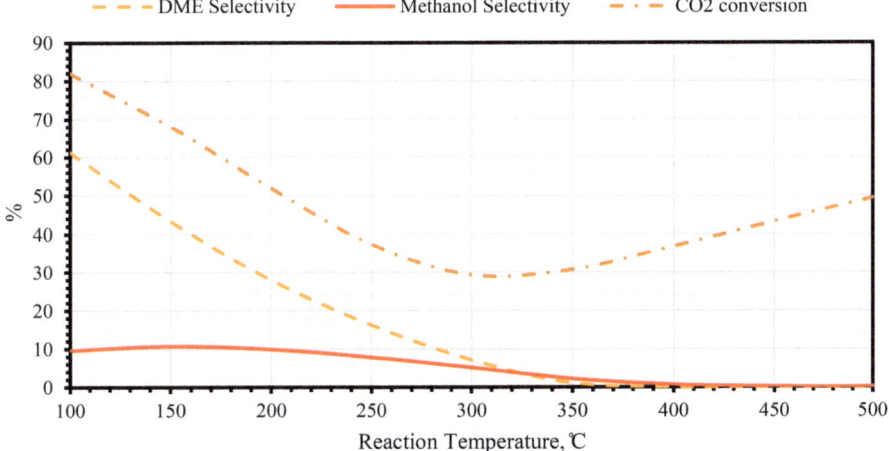

Figure 2.23 Sensitivity of the conversion rate of CO_2, and product selectivity to the CO_2 hydrogenation reaction temperature. Adapted from ref. 122 with permission from Elsevier, Copyright 2022.

Figure 2.23 shows that increasing temperature in the CO_2 hydrogenation reactor initially decreases CO_2 conversion and selectivity for DME and methanol. However, further temperature increases enhance CO_2 conversion due to the water–gas shift reaction. Figure 2.24 illustrates the relationship between varying methane feed and the rates of DME output and electricity generation.

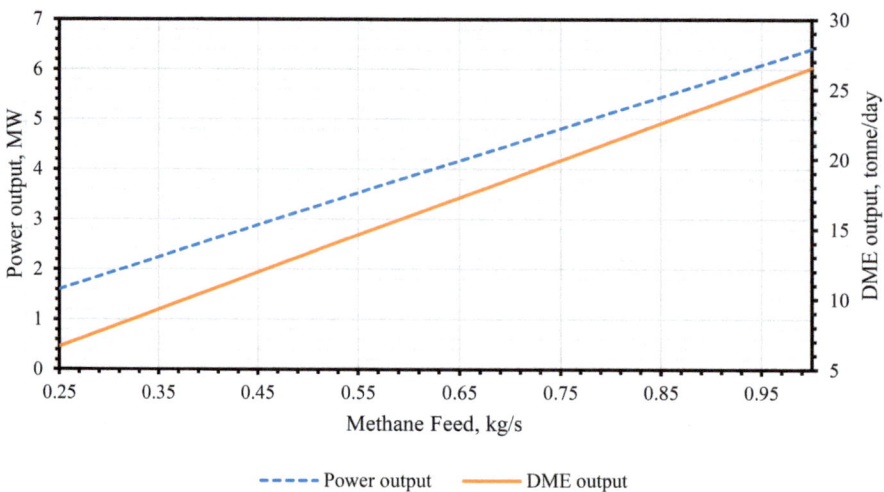

Figure 2.24 Methane and power output dependence on the flow rate of the methane feed. Adapted from ref. 122 with permission from Elsevier, Copyright 2022.

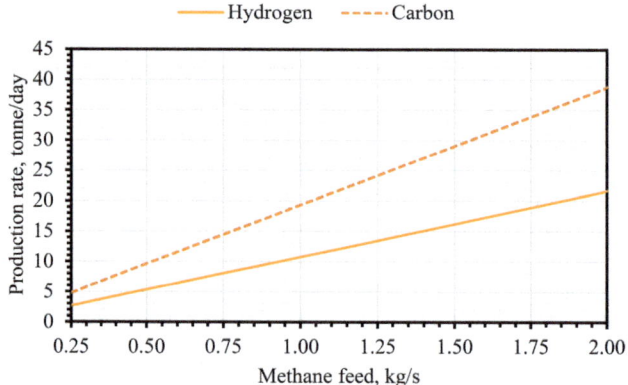

Figure 2.25 Hydrogen and carbon output based on the feed flow rate of methane. Adapted from ref. 123 with permission from Elsevier, Copyright 2024.

2.4.3.2 Integrated System 2

A methane conversion rate of 95% was achieved at 1473.15 K and 10 bar, aligning with literature values for solar reactors. A solar thermal power requirement of 5.3 MW, with 75% efficiency, was determined for methane cracking at the specified conditions. The methane cracking subsystem processes a 0.5 kg s^{-1} feed of pure methane to produce hydrogen and carbon for fuel cell electricity generation. Due to solar energy intermittency, produced carbon and hydrogen during the day are stored. The methane cracking subsystem yields 9.7 tonnes per day of carbon and 5.4 tonnes per day of hydrogen. Figure 2.25 illustrates the flow rate of the products when the methane feed flow rate is varied. It should be noted that bigger heliostat fields will be required for the collection of solar energy for higher flow rates.

The SMC's energy efficiency is 82.2% and exergy efficiency is 92.5%. Literature reports an exergy efficiency of 46.9% for methane cracking.[145] The by-product carbon has a high exergy content but is often deemed a non-useful product, resulting in significant exergy destruction and reduced efficiency. The system's exergy efficiency drops to 49.4% when only hydrogen is considered a useful product. Utilizing the carbon by-product can enhance efficiency, as demonstrated by integrating a DCFC. Additionally, the inclusion of solar energy (using CSP) contributes to higher exergy destruction. The exergy efficiency without CSP is 98.3%. Figure 2.26 provides a summary of efficiency results.

A total of 7.7 MW of electricity is generated from the two fuel cells. CO_2 is co-electrolyzed with water in the SOEC, utilizing electricity from a solar photovoltaic (PV) system with 20% efficiency. This system generates 3.9 tonnes of syngas per day, which is then used toward methanol synthesis. The calculated efficiency of the systems is illustrated in Figure 2.27. The system achieves an energy efficiency of 40.6% and an exergy efficiency of 37.5%. Methanol synthesis experiences low efficiency due to poor conversion rate and sensitivity to the syngas ratio. Figure 2.28 shows the different subsystems' exergy destruction rates. The highest exergy destruction occurs in PV panels, attributed to the low efficiency of conversion. Table 2.6 details the total entropy generation rate calculated for each subsystem.

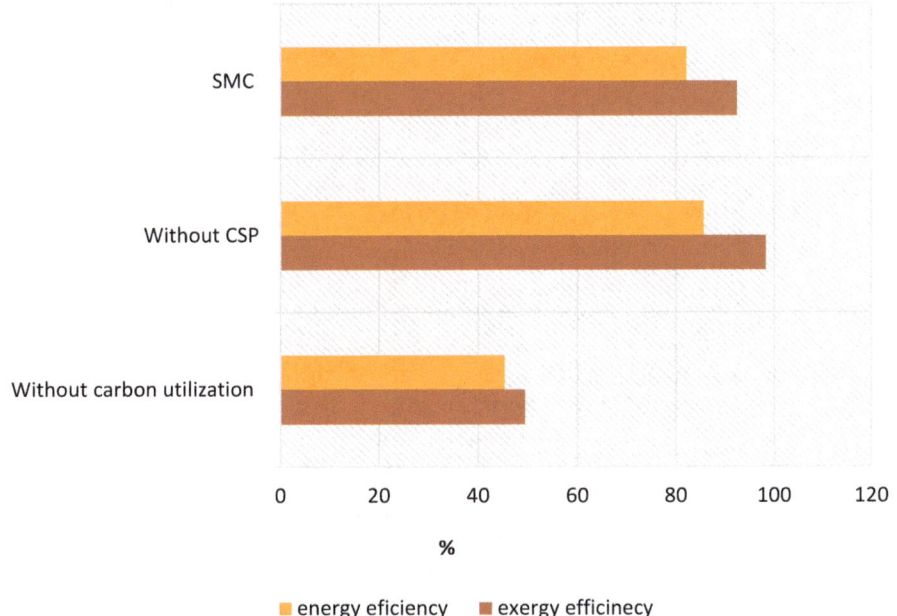

Figure 2.26 SMC subsystem efficiencies in different scenarios.

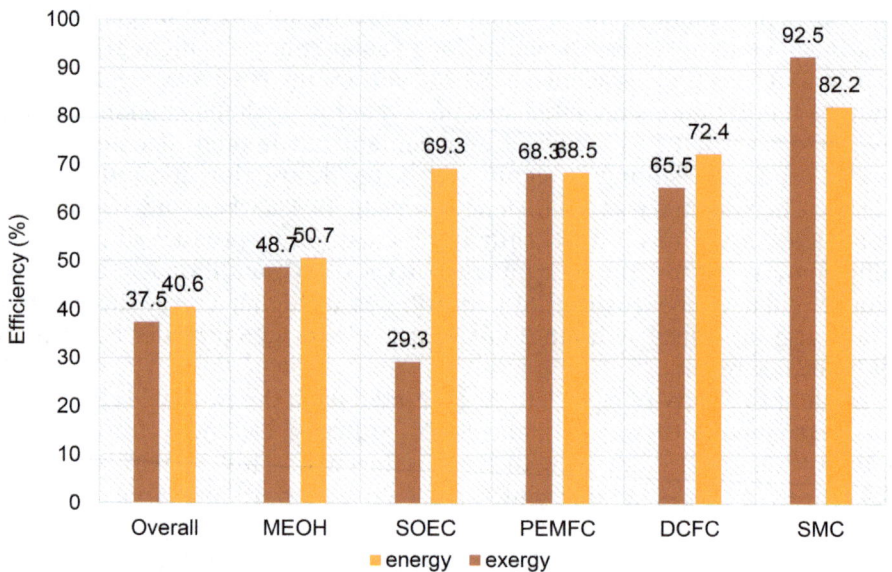

Figure 2.27 Efficiency results for integrated system 2. Adapted from ref. 123 with permission from Elsevier, Copyright 2024.

Figure 2.28 Exergy destruction rate results for the integrated system 2. Adapted from ref. 123 with permission from Elsevier, Copyright 2024.

The reference temperature was varied from 2 °C to 42 °C. Figure 2.29 shows the evaluations and results. The study analyzed the effect of this variation on exergy destruction across different subsystems. The results indicate that exergy destruction rates in the MEOH, SOEC, DCFC, and PEMFC subsystems increase with temperature. Conversely, the SMC subsystem experiences a decrease from approximately 593 kW to 491 kW with rising temperature. Figure 2.30 illustrates the electricity inputs and outputs based

Table 2.6 Entropy generation rate results for the integrated system 2.

System	Entropy generation rate (kW K^{-1})
MEOH	0.902
SOEC	1.008
DCFC	3.719
PEMFC	5.804
CSP	6.251
SMC	8.133
PV	22.064

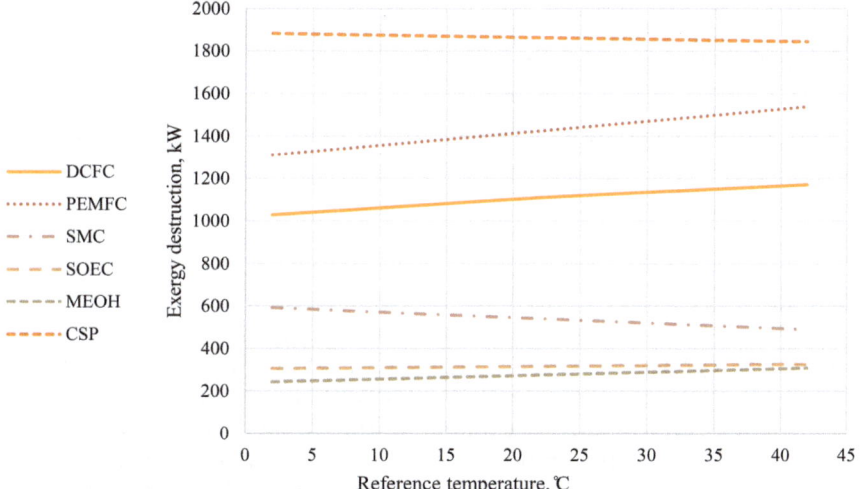

Figure 2.29 Sensitivity of the exergy destruction rates to the reference temperature for system 2. Adapted from ref. 123 with permission from Elsevier, Copyright 2024.

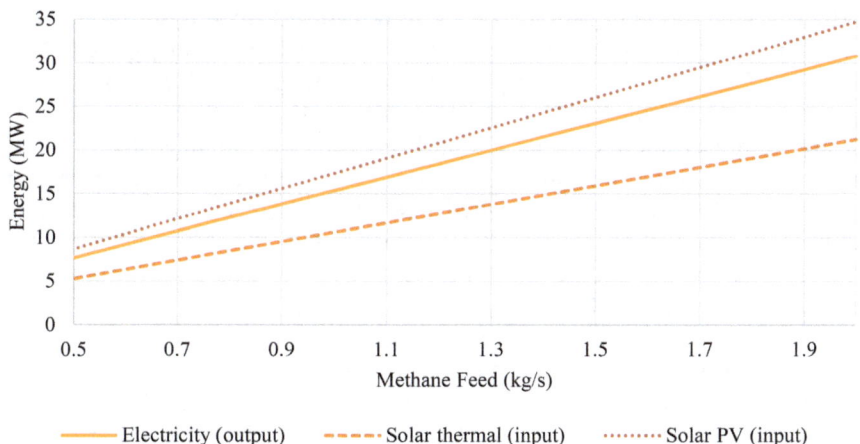

Figure 2.30 The different energy input with varying methane feed flow rate. Adapted from ref. 123 with permission from Elsevier, Copyright 2024.

Table 2.7 System 2 output summary.

Parameter	Value	Unit
Methane conversion	95	%
PEMFC electrical power output	5040	kW
DCFC electrical power output	2667	kW
SOEC heat rate demand	680	kW
SOEC power demand	1737	kW
Solar power requirement (SOEC)	8687.80	kW
Solar power requirement (SMC)	5319.73	kW
Total compressor work input rate	468.02	kW
Syngas production rate	14.8	Tonnes per day
Carbon production rate	9.7	Tonnes per day
Hydrogen production rate	5.4	Tonnes per day
Methanol production rate	3.9	Tonnes per day

on methane feed variations. Table 2.7 summarizes the output flow rates from the system with varying methane feed flow rates.

2.5 Conclusion

This study presents a comprehensive review of the methane cracking process. As this is a high-temperature process, two approaches can enhance its feasibility: integration with renewable solar energy and the use of catalysts. This review examines the integration of solar energy using various types of solar reactors. Additionally, it explores different catalyst systems that have been synthesized and tested for methane cracking. Two integrated systems based on solar methane cracking are proposed, and a detailed thermodynamic assessment is conducted.

For the integrated system 1, the by-products – carbon and hydrogen from the solar methane cracking process – are used toward electricity generation in fuel cells. The DCFC produces a pure CO_2 stream that is hydrogenated to produce DME and methanol. The system showed an energy efficiency of 41.9%, an exergy efficiency of 52.3%, and an overall electricity generation of 3.21 MW. The methane cracking unit using solar energy obtained an energy efficiency of 82.8% and an exergy efficiency of 92.5%. The carbon fuel cell has an energy efficiency of 72.4% and an exergy efficiency of 65.5%. The hydrogen fuel cell showed an energy efficiency of 68.4% and an exergy efficiency of 67.4%. Lastly, the CO_2-hydrogenation section had an energy efficiency of 79.8% and an exergy efficiency of 66.6%. The CO_2 hydrogenation process used for producing methanol and DME is more sustainable compared to current methods.

In integrated system 2, photovoltaic energy is used to power the H_2O and CO_2 co-electrolyzer to produce syngas. The syngas is used toward the synthesis of methanol fuel, a clean fuel that is easily transportable. The system showed an energy efficiency of 40.6%, an exergy efficiency of 37.5%, and an overall electricity generation of 7.71 MW. The methane cracking unit using solar energy obtained an energy efficiency of 82.8% and an exergy efficiency of 92.5%. The carbon fuel cell had an energy efficiency of 72.4% and an exergy efficiency of 65.5%. The hydrogen fuel cell showed an energy efficiency of 68.5% and an exergy efficiency of 68.3%. The co-electrolyzer had an energy efficiency of 69.3% and an exergy efficiency of 29.3%. Lastly, the methanol synthesis subsystem had an energy efficiency of 50.7% and an exergy efficiency of 48.7%.

The hydrogen generated through the cracking of methane has lower emissions compared to conventional processes. The proposed systems also manage the carbon by-product toward electricity generation and subsequently for fuel synthesis.

Acknowledgements

This research was made possible by a Graduate Sponsorship Research Award (GSRA7-2-0427-20025) from the Qatar National Research Fund (a member of Qatar Foundation). The findings herein reflect the work, and are solely the responsibility, of the authors.

References

1. N. Muradov, F. Smith and A. T-Raissi, *Catal. Today*, 2005, **102-103**, 225–233.
2. L. Weger, A. Abánades and T. Butler, *Int. J. Hydrogen Energy*, 2017, **42**, 720–731.
3. B. Parkinson, P. Balcombe, J. F. Speirs, A. D. Hawkes and K. Hellgardt, *Energy Environ. Sci.*, 2019, **12**, 19–40.
4. T. Keipi, H. Tolvanen and J. Konttinen, *Energy Convers. Manage.*, 2018, **159**, 264–273.
5. J. X. Qian, T. W. Chen, L. R. Enakonda, D. Bin Liu, J. M. Basset and L. Zhou, *Int. J. Hydrogen Energy*, 2020, **45**, 15721–15743.
6. S. Timmerberg, M. Kaltschmitt and M. Finkbeiner, *Energy Convers. Manage.*, 2020, **7**, 100043.
7. N. Muradov, *Int. J. Hydrogen Energy*, 2001, **26**, 1165–1175.
8. H. F. Abbas and W. M. A. Wan Daud, *Int. J. Hydrogen Energy*, 2010, **35**, 1160–1190.
9. A. M. Amin, E. Croiset and W. Epling, *Int. J. Hydrogen Energy*, 2011, **36**, 2904–2935.
10. R. Aiello, J. E. Fiscus, H. C. Zur Loye and M. D. Amiridis, *Appl. Catal., A*, 2000, **192**, 227–234.
11. I. Dincer and C. Acar, *Int. J. Energy Res.*, 2015, **39**, 585–606John Wiley and Sons Ltd.
12. C. Acar and I. Dincer, *J. Cleaner Prod.*, 2019, **218**, 835–849.
13. S. Dang, H. Yang, P. Gao, H. Wang, X. Li, W. Wei and Y. Sun, *Catal. Today*, 2019, **330**, 61–75.
14. N. Ozalp, M. Epstein, R. Davis, C. Ophoff and I. Vinck, *Curr. Opin. Chem. Eng.*, 2018, **21**, 111–115.

15. A. A. Al-Hassani, H. F. Abbas and W. M. A. Wan Daud, *Int. J. Hydrogen Energy*, 2014, **39**, 14783–14791.
16. A. Banu and Y. Bicer, *Energy Convers. Manage.:X*, 2021, **12**, 100117.
17. R. Guil-Lopez, J. A. Botas, J. L. G. Fierro and D. P. Serrano, *Appl. Catal., A*, 2011, **396**, 40–51.
18. U. P. M. Ashik, W. M. A. Wan Daud and H. F. Abbas, *Renewable Sustainable Energy Rev.*, 2015, **44**, 221–256.
19. N. S. N. Hasnan, S. N. Timmiati, K. L. Lim, Z. Yaakob, N. H. N. Kamaruddin and L. P. Teh, *Mater. Renewable Sustainable Energy*, 2020, **9**, 8.
20. W. Zhang, Q. Ge and H. Xu, *J. Nat. Gas Chem.*, 2011, **20**, 339–344.
21. G. D. B. Nuernberg, E. L. Foletto, C. E. M. Campos, H. V. Fajardo, N. L. V. Carreño and L. F. D. Probst, *J. Power Sources*, 2012, **208**, 409–414.
22. C. Anjaneyulu, V. V. Kumar, S. K. Bhargava and A. Venugopal, *J. Energy Chem.*, 2013, **22**, 853–860.
23. S. K. Saraswat and K. K. Pant, *J. Nat. Gas Sci. Eng.*, 2013, **13**, 52–59.
24. J. Chen, Y. Qiao and Y. Li, *J. Solid State Chem.*, 2012, **191**, 107–113.
25. K. Tapia-Parada, G. Valverde-Aguilar, A. Mantilla, M. A. Valenzuela and E. Hernańdez, *Fuel*, 2013, **110**, 70–75.
26. J. L. Pinilla, D. Torres, M. J. Lázaro, I. Suelves, R. Moliner, I. Cañadas, J. Rodríguez, A. Vidal and D. Martínez, *Int. J. Hydrogen Energy*, 2012, **37**, 9645–9655.
27. G. D. B. Nuernberg, H. V. Fajardo, E. L. Foletto, S. M. Hickel-Probst, N. L. V. Carreño, L. F. D. Probst and J. Barrault, *Catal. Today*, 2011, **176**, 465–469.
28. M. Pudukudy, Z. Yaakob, A. Kadier, M. S. Takriff and N. S. M. Hassan, *Int. J. Hydrogen Energy*, 2017, **42**, 16495–16513.
29. W. U. Khan, A. H. Fakeeha, A. S. Al-Fatesh, A. A. Ibrahim and A. E. Abasaeed, *Int. J. Hydrogen Energy*, 2016, **41**, 976–983.
30. A. H. Fakeeha, W. U. Khan, A. S. Al-Fatesh, A. E. Abasaeed and M. A. Naeem, *Int. J. Hydrogen Energy*, 2015, **40**, 1774–1781.
31. W. Ahmed, A. E. Awadallah and A. A. Aboul-Enein, *Int. J. Hydrogen Energy*, 2016, **41**, 18484–18493.
32. A. E. Awadallah, D. S. El-Desouki, N. A. K. Aboul-Gheit, A. H. Ibrahim and A. K. Aboul-Gheit, *Int. J. Hydrogen Energy*, 2016, **41**, 16890–16902.
33. N. Bayat, M. Rezaei and F. Meshkani, *Int. J. Hydrogen Energy*, 2016, **41**, 5494–5503.
34. N. Bayat, M. Rezaei and F. Meshkani, *Int. J. Hydrogen Energy*, 2016, **41**, 1574–1584.
35. N. Bayat, F. Meshkani and M. Rezaei, *Int. J. Hydrogen Energy*, 2016, **41**, 13039–13049.
36. J. L. Pinilla, R. Utrilla, M. J. Lázaro, R. Moliner, I. Suelves and A. B. García, *Fuel Process. Technol.*, 2011, **92**, 1480–1488.
37. A. A. Ibrahim, A. H. Fakeeha, A. S. Al-Fatesh, A. E. Abasaeed and W. U. Khan, *Int. J. Hydrogen Energy*, 2015, **40**, 7593–7600.

38. J. L. Pinilla, R. Utrilla, R. K. Karn, I. Suelves, M. J. Lázaro, R. Moliner, A. B. García and J. N. Rouzaud, *Int. J. Hydrogen Energy*, 2011, **36**, 7832–7843.
39. D. Torres, S. De Llobet, J. L. Pinilla, M. J. Lázaro, I. Suelves and R. Moliner, *J. Nat. Gas Chem.*, 2012, **21**, 367–373.
40. L. Jin, H. Si, J. Zhang, P. Lin, Z. Hu, B. Qiu and H. Hu, *Int. J. Hydrogen Energy*, 2013, **38**, 10373–10380.
41. D. Torres, J. L. Pinilla, M. J. Lázaro, R. Moliner and I. Suelves, *Int. J. Hydrogen Energy*, 2014, **39**, 3698–3709.
42. M. Pudukudy, Z. Yaakob and Z. S. Akmal, *Appl. Surf. Sci.*, 2015, **330**, 418–430.
43. M. Pudukudy and Z. Yaakob, *Chem. Eng. J.*, 2015, **262**, 1009–1021.
44. A. A. Ibrahim, A. S. Al-Fatesh, W. U. Khan, M. A. Soliman, R. L. Al Otaibi and A. H. Fakeeha, *J. Chin. Chem. Soc.*, 2015, **62**, 592–599.
45. A. S. Al Fatesh, S. O. Kasim, A. A. Ibrahim, A. S. Al-Awadi, A. E. Abasaeed, A. H. Fakeeha and A. E. Awadallah, *Renewable Energy*, 2020, **155**, 969–978.
46. N. Muradov, *Catal. Commun.*, 2001, **2**, 89–94.
47. J. Sarada Prasad, V. Dhand, V. Himabindu and Y. Anjaneyulu, *Int. J. Hydrogen Energy*, 2011, **36**, 11702–11711.
48. J. Zhang, L. Jin, Y. Li and H. Hu, *Int. J. Hydrogen Energy*, 2013, **38**, 3937–3947.
49. M. Szymańska, A. Malaika, P. Rechnia, A. Miklaszewska and M. Kozłowski, *Catal. Today*, 2015, **249**, 94–102.
50. L. Wei, Y. S. Tan, Y. Z. Han, J. T. Zhao, J. Wu and D. Zhang, *Fuel*, 2011, **90**, 3473–3479.
51. L. Wei, M. Zhu, Y. Ma, Z. Zhang and D. Zhang, *Fuel*, 2016, **183**, 345–350.
52. Y. Wang, H. Yang, L. Jin, Y. Li, H. Hu, H. Ding and X. Bai, *Fuel*, 2019, **258**, 116138.
53. J. Zhang, L. Jin, X. He, S. Liu and H. Hu, *Int. J. Hydrogen Energy*, 2011, **36**, 8978–8984.
54. J. Zhang, L. Jin, Y. Li, H. Si, B. Qiu and H. Hu, *Int. J. Hydrogen Energy*, 2013, **38**, 8732–8740.
55. J. Zhang, L. Jin, S. Liu, Y. Xun and H. Hu, *Carbon*, 2012, **50**, 952–959.
56. V. Shilapuram, N. Ozalp, M. Oschatz, L. Borchardt and S. Kaskel, *Carbon*, 2014, **67**, 377–389.
57. A. A. Al-Hassani, H. F. Abbas and W. M. A. Wan Daud, *Int. J. Hydrogen Energy*, 2014, **39**, 7004–7014.
58. H. Nishii, D. Miyamoto, Y. Umeda, H. Hamaguchi, M. Suzuki, T. Tanimoto, T. Harigai, H. Takikawa and Y. Suda, *Appl. Surf. Sci.*, 2019, **473**, 291–297.
59. T. Tokunaga, K. Kuno, T. Kawakami, T. Yamamoto and A. Yoshigoe, *Int. J. Hydrogen Energy*, 2020, **45**, 14347–14353.
60. J. Zhang, X. Li, W. Xie, Q. Hao, H. Chen and X. Ma, *J. Anal. Appl. Pyrolysis*, 2018, **136**, 53–61.

61. M. Steinberg, *Int. J. Hydrogen Energy*, 1998, **23**, 419–425.
62. J. K. Dahl, V. H. Barocas, D. E. Clough and A. W. Weimer, *Int. J. Hydrogen Energy*, 2002, **27**, 377–386.
63. S. K. Saraswat, B. Sinha, K. K. Pant and R. B. Gupta, *Ind. Eng. Chem. Res.*, 2016, **55**, 11672–11680.
64. M. Caballero, G. Del Angel, I. Rangel-Vázquez and L. Huerta, *Catal. Today*, 2020, **349**, 106–116.
65. S. H. S. Zein, A. R. Mohamed and P. S. T. Sai, *Ind. Eng. Chem. Res.*, 2004, **43**, 4864–4870.
66. I. Alstrup and M. T. Tavares, *J. Catal.*, 1993, **139**, 513–524.
67. N. Z. Muradov, *Energy Fuels*, 1998, **12**, 41–48.
68. A. Malaika and M. Kozłowski, *Int. J. Hydrogen Energy*, 2010, **35**, 10302–10310.
69. A. Malaika, B. Krzyzyńska and M. Kozłowski, *Int. J. Hydrogen Energy*, 2010, **35**, 7470–7475.
70. P. Rechnia, A. Malaika, B. Krzyzyńska and M. Kozłowski, *Int. J. Hydrogen Energy*, 2012, **37**, 14178–14186.
71. P. Rechnia, A. Malaika, L. Najder-Kozdrowska and M. Kozłowski, *Int. J. Hydrogen Energy*, 2012, **37**, 7512–7520.
72. R. Koç, E. Alper, E. Croiset and A. Elkamel, *Turk. J. Chem.*, 2009, **33**, 825–841.
73. A. M. Amin, E. Croiset, C. Constantinou and W. Epling, *Int. J. Hydrogen Energy*, 2012, **37**, 9038–9048.
74. A. M. Amin, E. Croiset, Z. Malaibari and W. Epling, *Int. J. Hydrogen Energy*, 2012, **37**, 10690–10701.
75. R. Koç, E. Alper, E. Croiset and A. Elkamel, *Turk. J. Chem.*, 2008, **32**, 157–168.
76. F. Frusteri, G. Italiano, C. Espro, C. Cannilla and G. Bonura, *Int. J. Hydrogen Energy*, 2012, **37**, 16367–16374.
77. J. X. Qian, L. R. Enakonda, W. J. Wang, D. Gary, P. Del-Gallo, J. M. Basset, D. B. in Liu and L. Zhou, *Int. J. Hydrogen Energy*, 2019, **44**, 31700–31711.
78. D. Hirsch and A. Steinfeld, *Int. J. Hydrogen Energy*, 2004, **29**, 47–55.
79. S. Abanades and G. Flamant, *Chem. Eng. Commun.*, 2008, **195**, 1159–1175.
80. S. Abanades and G. Flamant, *Sol. Energy*, 2006, **80**, 1321–1332.
81. S. Abanades and G. Flamant, *Int. J. Hydrogen Energy*, 2007, **32**, 1508–1515.
82. M. Kogan and A. Kogan, *Int. J. Hydrogen Energy*, 2003, **28**, 1187–1198.
83. A. Kogan, M. Kogan and S. Barak, *Int. J. Hydrogen Energy*, 2004, **29**, 1227–1236.
84. A. Kogan, M. Kogan and S. Barak, *Int. J. Hydrogen Energy*, 2005, **30**, 35–43.
85. A. Kogan, M. Israeli and E. Alcobi, *Int. J. Hydrogen Energy*, 2007, **32**, 4800–4810.
86. A. Kogan and M. Kogan, *J. Solar Energy Eng. Trans. ASME*, 2002, **124**, 206–214.

Methane Cracking

87. N. Ozalp and A. Kanjirakat, *J. Heat Mass Transfer*, 2010, **132**, 122901.
88. J. Yeheskel and M. Epstein, *Carbon*, 2011, **49**, 4695–4703.
89. J. Costandy, N. El Ghazal, M. T. Mohamed, A. Menon, V. Shilapuram and N. Ozalp, *Int. J. Hydrogen Energy*, 2012, **37**, 16581–16590.
90. D. Hirsch and A. Steinfeld, *Chem. Eng. Sci.*, 2004, **59**, 5771–5778.
91. S. Abanades and G. Flamant, *Int. J. Hydrogen Energy*, 2005, **30**, 843–853.
92. S. Abanades and G. Flamant, *Chem. Eng. Process.*, 2008, **47**, 490–498.
93. J. K. Dahl, K. J. Buechler, R. Finley, T. Stanislaus, A. W. Weimer, A. Lewandowski, C. Bingham, A. Smeets and A. Schneider, *Energy*, 2004, **29**, 715–725.
94. J. K. Dahl, K. J. Buechler, A. W. Weimer, A. Lewandowski and C. Bingham, *Int. J. Hydrogen Energy*, 2004, **29**, 725–736.
95. J. K. Dahl, A. W. Weimer and W. B. Krantz, *Int. J. Hydrogen Energy*, 2004, **29**, 57–65.
96. S. Abanades, S. Tescari, S. Rodat and G. Flamant, *J. Nat. Gas Chem.*, 2009, **18**, 1–8.
97. S. Rodat, S. Abanades, J. L. Sans and G. Flamant, *Sol. Energy*, 2009, **83**, 1599–1610.
98. S. Rodat, S. Abanades, J. Coulié and G. Flamant, *Chem. Eng. J.*, 2009, **146**, 120–127.
99. S. Rodat, S. S. Abanades and G. Flamant, *Energy Fuels*, 2009, **23**, 2666–2674.
100. F. J. Valdés-Parada, H. Romero-Paredes and G. Espinosa-Paredes, *Int. J. Hydrogen Energy*, 2011, **36**, 3354–3363.
101. S. Rodat, S. Abanades, J. L. Sans and G. Flamant, *Int. J. Hydrogen Energy*, 2010, **35**, 7748–7758.
102. S. Rodat, S. Abanades and G. Flamant, *Int. J. Chem. React. Eng.*, 2010, **8**, A25.
103. S. Rodat, S. Abanades and G. Flamant, *Sol. Energy*, 2011, **85**, 645–652.
104. S. Rodat, S. Abanades and G. Flamant, *J. Solar Energy Eng. Trans. ASME*, 2011, **133**, 031001.
105. S. Abanades, H. Kimura and H. Otsuka, *Fuel Process. Technol.*, 2014, **122**, 153–162.
106. S. Abanades, H. Kimura and H. Otsuka, *Int. J. Hydrogen Energy*, 2014, **39**, 18770–18783.
107. S. Abanades, H. Kimura and H. Otsuka, *Int. J. Hydrogen Energy*, 2015, **40**, 10744–10755.
108. S. Abanades, H. Kimura and H. Otsuka, *Fuel*, 2015, **153**, 56–66.
109. G. Patrianakos, M. Kostoglou and A. Konstandopoulos, *Int. J. Hydrogen Energy*, 2011, **36**, 189–202.
110. G. Patrianakos, M. Kostoglou and A. G. Konstandopoulos, *Int. J. Hydrogen Energy*, 2012, **37**, 16570–16580.
111. B. Nezzari and R. Gomri, *Int. J. Hydrogen Energy*, 2020, **45**, 135–148.
112. M. Msheik, S. Rodat and S. Abanades, *Energies*, 2021, **14**, 3107.
113. C. Palmer, M. Tarazkar, H. H. Kristoffersen, J. Gelinas, M. J. Gordon, E. W. McFarland and H. Metiu, *ACS Catal.*, 2019, **9**, 8337–8345.

114. D. C. Upham, V. Agarwal, A. Khechfe, Z. R. Snodgrass, M. J. Gordon, H. Metiu and E. W. McFarland, *Science*, 2017, **358**, 917–921.
115. L. J. J. Catalan and E. Rezaei, *Int. J. Hydrogen Energy*, 2020, **45**, 2486–2503.
116. S. C. Rowe, T. A. Ariko, K. M. Weiler, J. T. E. Spana and A. W. Weimer, *Energies*, 2020, **13**, 6229.
117. Z. J. Zheng and Y. Xu, *Energy Convers. Manage.*, 2018, **157**, 562–574.
118. A. Boretti, *Int. J. Energy Res.*, 2021, **45**, 21497–21508.
119. S. E. Hosseini and M. A. Wahid, *Int. J. Energy Res.*, 2020, **44**, 4110–4131.
120. P. Haueter, T. Seitz and A. Steinfeld, *J. Sol. Energy Eng. Trans. ASME*, 1999, **121**, 77–80.
121. J. C. McVeigh, *Sun Power: An Introduction to the Applications of Solar Energy*, Elsevier, 2nd edn, 1983, pp. 132–160.
122. A. Banu and Y. Bicer, *Int. J. Hydrogen Energy*, 2022, **47**, 19502–19516.
123. A. Banu and Y. Bicer, *Int. J. Hydrogen Energy*, 2024, **52**, 580–593.
124. S. Campanari, M. Gazzani and M. C. Romano, *J. Eng. Gas Turbines Power*, 2013, **135**, 011701.
125. F. J. Gutiérrez Ortiz, A. Serrera, S. Galera and P. Ollero, *Fuel*, 2013, **105**, 739–751.
126. A. Rafiee, *J. Environ. Chem. Eng.*, 2020, **8**, 104314.
127. R. Zevenhoven, S. Eloneva and S. Teir, *Catal. Today*, 2006, **115**, 73–79.
128. E. Catizzone, C. Freda, G. Braccio, F. Frusteri and G. Bonura, *J. Energy Chem.*, 2021, **58**, 55–77.
129. X. An, Y. Z. Zuo, Q. Zhang, D. Z. Wang and J. F. Wang, *Ind. Eng. Chem. Res.*, 2008, **47**, 6547–6554.
130. K. Ahmad and S. Upadhyayula, *Environ. Prog. Sustainable Energy*, 2019, **38**, 98–111.
131. C. Stoots, J. O'Brien and J. Hartvigsen, *Int. J. Hydrogen Energy*, 2009, **34**, 4208–4215.
132. S. W. Kim, H. Kim, K. J. Yoon, J. H. Lee, B. K. Kim, W. Choi, J. H. Lee and J. Hong, *J. Power Sources*, 2015, **280**, 630–639.
133. H. R. M. Jhong, S. Ma and P. J. Kenis, *Curr. Opin. Chem. Eng.*, 2013, **2**, 191–199.
134. J. P. Stempien, O. L. Ding, Q. Sun and S. H. Chan, *Int. J. Hydrogen Energy*, 2012, **37**, 14518–14527.
135. V. Menon, Q. Fu, V. M. Janardhanan and O. Deutschmann, *J. Power Sources*, 2015, **274**, 768–781.
136. G. Cinti, A. Baldinelli, A. Di Michele and U. Desideri, *Appl. Energy*, 2016, **162**, 308–320.
137. Aspen Plus|Leading Process Simulation Software, AspenTech.
138. S. Kasemanand, K. Im-orb, P. Tippawan, W. Wiyaratn and A. Arpornwichanop, *Energy Convers. Manage.*, 2017, **149**, 485–494.
139. W. Doherty, A. Reynolds and D. Kennedy, *Energy*, 2010, **35**, 4545–4555.
140. K. Stangeland, H. Li and Z. Yu, *Ind. Eng. Chem. Res.*, 2018, **57**, 4081–4094.

141. C. Graves, S. D. Ebbesen and M. Mogensen, *Solid State Ionics*, 2011, **192**, 398–403.
142. Y. Redissi and C. Bouallou, *Energy Procedia*, 2013, **37**, 6667–6678.
143. M. Puig-Gamero, J. Argudo-Santamaria, J. L. Valverde, P. Sánchez and L. Sanchez-Silva, *Energy Convers. Manage.*, 2018, **177**, 416–427.
144. R. E. Rosli, A. B. Sulong, W. R. W. Daud, M. A. Zulkifley, T. Husaini, M. I. Rosli, E. H. Majlan and M. A. Haque, *Int. J. Hydrogen Energy*, 2017, **42**, 9293–9314.
145. A. Martínez-Rodríguez and A. Abánades, *Entropy*, 2020, **22**, 1286.

CHAPTER 3
Oxidative Coupling of Methane

GABRIEL L. CATUZO,[a] LETÍCIA F. RASTEIRO,[b] LARISSA B. LOPES,[a] LUIZ H. VIEIRA,*[a] JOSÉ M. ASSAF[c] AND ELISABETE M. ASSAF*[a]

[a] São Carlos Institute of Chemistry, University of São Paulo, Av. Trabalhador Sãocarlense, 400, 13560-970, São Carlos, Brazil; [b] School of Chemical & Biomolecular Engineering, Georgia Institute of Technology, Atlanta, Georgia, 30332, USA; [c] Department of Chemical Engineering, Federal University of São Carlos, Rod. W. Luiz, km 235, São Carlos, SP, Brazil
*Emails: lhvieira@iqsc.usp.br; eassaf@iqsc.usp.br

3.1 Introduction

Methane, the primary component of natural gas, is one of the most abundant hydrocarbons on Earth. As global energy demands rise and concerns about environmental sustainability intensify, there is a growing interest in finding efficient ways to utilize methane beyond its conventional use as a fuel. One promising area of research is the oxidative coupling of methane (OCM), a chemical process that aims to convert methane directly into more valuable hydrocarbons, such as ethylene and ethane. These compounds are fundamental building blocks in the petrochemical industry, used to produce a wide range of materials and chemicals.

The concept of OCM emerged in the early 1980s as scientists sought new methods to valorize methane. Traditional processes for converting methane to valuable products, such as steam reforming and Fischer–Tropsch synthesis, are often energy-intensive and involve multiple steps. In contrast, OCM offers a potentially more direct and efficient pathway.

OCM has garnered significant interest because it addresses two key challenges: the efficient utilization of methane and reducing greenhouse gas emissions. Methane is a potent greenhouse gas, and its effective conversion to useful products can help mitigate its environmental impact. Furthermore, producing ethylene directly from methane could reduce reliance on traditional feedstocks derived from oil, offering a more sustainable approach to chemical manufacturing.

Research in OCM has primarily focused on understanding the reaction mechanism and developing effective catalysts. The reaction involves the formation of methyl radicals, which couple to form ethane and ethylene. However, achieving high selectivity and yield while minimizing the production of unwanted byproducts, such as carbon monoxide and carbon dioxide, remains a significant challenge. Various catalysts, including metal oxides and mixed metal oxides, have been explored to optimize the process. Recent advancements in catalyst design, such as nanostructured materials, have shown promise in improving the efficiency and selectivity of OCM.

Despite progress, several hurdles remain before OCM can be commercially viable. These include achieving consistent performance at an industrial scale, ensuring catalyst stability over prolonged periods, and developing cost-effective processes for catalyst synthesis and regeneration. Addressing these challenges requires a multidisciplinary approach, combining experimental research with computational modeling to gain deeper insights into the reaction mechanisms and optimize catalyst formulations.

From an economic perspective, the successful implementation of OCM could enhance the value of natural gas resources by providing a new route for methane utilization. This could lead to the development of new industries and contribute to energy security. Environmentally, OCM offers a cleaner alternative for methane conversion, with the potential to reduce carbon emissions and contribute to more sustainable industrial practices.

The oxidative coupling of methane represents a frontier in the field of catalysis and chemical engineering, with the potential to transform how methane is used. This chapter will provide a comprehensive overview of OCM, including the characteristics and environmental impact of methane, the properties and thermodynamics of the reaction, and the current state of research.

3.2 Methane: Characteristics and Environmental Impact

Methane (CH_4) is the simplest hydrocarbon in the alkane series, formed by one atom of carbon and four hydrogen atoms, with four C–H bond lengths of 1.087 Å and an angle of 109.5° between them. It is a colorless, odorless gas with a molar weight of 16.043 g mol^{-1}. Methane is a stable molecule characterized by a Gibbs free energy of formation ($\Delta G°$) of -50.5 kJ mol^{-1}.[1] Methane is a highly abundant gas on Earth, found in natural gas (comprising over 90% methane), biogas, coal mines, methane hydrates in the ocean, and the atmosphere. Human activities such as rice cultivation, cattle raising, waste decomposition, biomass burning, and fossil fuel production and usage also contribute significantly to methane emissions (Figure 3.1A).[2] The main natural sink for methane is the atmosphere itself, removing 600 Tg CH_4 year^{-1}, where methane reacts with hydroxyl species (OH^-) in the troposphere, leading to the formation of carbon monoxide (CO), water vapor (H_2O), and ozone (O_3).[3,4] Another sink is soil, where methane is oxidized by bacteria, removing around 30 Tg CH_4 year^{-1}.[3] Similar to CO_2, human activities are causing CH_4 concentrations to rise faster than natural sinks can offset them. Methane concentrations in the atmosphere have more than doubled since pre-industrial times, and nowadays, ~60% of methane emissions are the result of human activities.[5] The average concentration of methane in 2023 was approximately 1920 ppb,[6] as shown in Figure 3.1B.

Methane is a potent greenhouse gas and ranks as the second-largest contributor to climate warming, following carbon dioxide (CO_2).[7] Besides being less abundant and having a shorter lifetime than CO_2, methane is more efficient at trapping radiation, making it 32 times greater than CO_2 over a 100-year period.[7] The effects of CH_4 on climate and atmospheric chemistry constitute a significant concern due to its rapid growth rate.[5] In response, researchers have been increasingly exploring its application as a feedstock in recent years. However, its utilization remains underrepresented. Approximately 90% of methane production is used for combustion to generate energy.[8,9] One reason for this under-utilization is that methane reserves are often located far from populated areas, making transportation challenging. The energy density per volume of methane is about three times lower than that of petroleum, making transportation *via* pipelines or ships expensive.[9] To address this issue, converting methane into higher value-added products can justify the transportation costs and provide a better use for methane than simply burning it to form CO_x.[10]

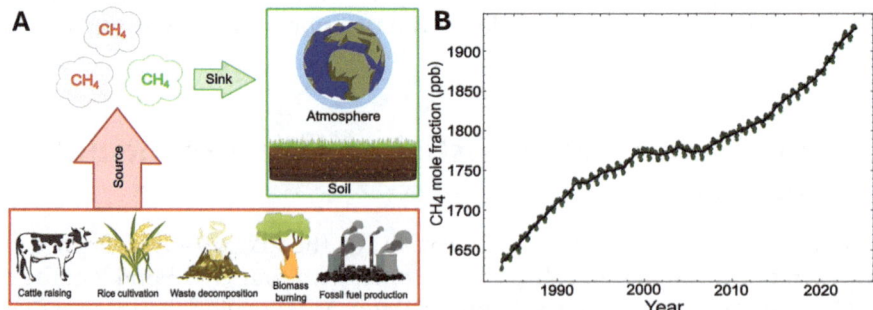

Figure 3.1 (A) Methane sources and sink. (B) Global average methane concentration in the atmosphere for the last 40 years. Data obtained from the NOAA Global Monitoring Laboratory.

3.3 Activation of the C–H Bond and Possibilities for Methane Conversion

Conversions from methane have been considered the best alternatives for producing chemicals and fuels among the potential clean carbon sources, often regarded as the 'holy grail' of catalysis in the 21st century.[9,10] However, some physicochemical properties hinder its activation and conversion. One challenge in reactions using methane as a reagent lies in its high stability. Methane has four very stable C–H bonds ($\Delta_d H = 439$ kJ mol^{-1}), making it the most stable alkane,[10] as seen in Table 3.1.[11,12]

However, compared to other bond energies, the task does not seem too complicated. The energy required to dissociate the C–H bond in methane is only 1% higher than the H–H bond in hydrogen, a molecule relatively easy to activate *via* catalysis. Additionally, the C–H bond is significantly weaker than the O–O bond in molecular oxygen. Despite the energy required to break the C–H bond not seeming very high, other factors also contribute to increasing the stability of this molecule. Methane has four equivalent C–H bonds with a bond length of 1.087 Å and an H–C–H bond angle of 109.5°.[13] The C–H bond in methane is weakly polarized (2.84×10^{-40} C^2 m^2 J^{-1}), resulting in a zero-dipole moment due to the symmetry of its tetrahedral geometry. Difficulties in methane activation can be interpreted in terms of molecular orbitals. Since the lowest unoccupied molecular orbital (LUMO) level is high and the highest occupied molecular orbital (HOMO) level is low, it is difficult to donate an electron to the LUMO and remove an electron from the HOMO.[13,14] Thus, it is an extremely weak acid (p$K_a = 40$) with a high ionization potential (12.6 eV) and low electron affinity (-1.9 eV). Lastly, methane lacks functional groups that make it more labile and is resistant to nucleophilic and electrophilic attacks. Therefore, reactions involving methane require drastic conditions for C–H bond activation, such as high temperatures and aggressive reagents (*e.g.*, superacids and/or radicals).[9] As a consequence of this strategy, the products, which are naturally more reactive

Table 3.1 The bond dissociation energy of several simple molecules.

Molecule	Bond	$\Delta_d H_{298}$ (kJ mol^{-1})
Methane	H$_3$C–H	439
Ethane	H$_3$CH$_2$C–H	423
Ethene	H$_2$CHC–H	464
Acetylene	HCC–H	558
Propane	(CH$_3$)$_2$CH–H	413
Propylene	CH$_2$CHCH$_2$–H	372
Hydrogen	H–H	436
Oxygen	O–O	497
Nitrogen	N–N	945
Carbon monoxide	C–O	1076
Carbon dioxide	OC–O	532

than methane, are even more easily activated than the reagent. Thus, these conditions result in a medium where it is difficult to control the product selectivity.[14] Due to the difficulties in methane activation, there are few known routes for its conversion. Generally, methane conversion can occur either directly or indirectly, as shown in Figure 3.2.[12]

Indirect methane conversion requires the formation of synthesis gas (CO and H_2) as the first step, either through steam reforming, dry reforming, partial oxidation, or a combination of these processes. From these new reagents, it is possible to obtain olefins, gasoline, diesel, and other oxygenated products through Fischer–Tropsch synthesis.[15] Additionally, synthesis gas can be converted to methanol, widely used as an intermediate for producing olefins, gasoline, and aromatic compounds.[13,15] Another important process based on synthesis gas is the Haber–Bosch reaction for ammonia production and hydroformylation (oxo synthesis or oxo process) for producing long-chain alcohols and aldehydes.[14] Except for partial oxidation (mildly exothermic), all processes for obtaining synthesis gas are highly endothermic, requiring high temperature and pressure to achieve reasonable methane conversion (15–40 atm, 630–930 °C).[15,16] These reactions also face many problems related to coke formation and catalyst sintering, leading to their deactivation over time.[13,15]

Direct methane conversion routes can be employed under various conditions depending on the desired product. These routes are economically more advantageous and simpler than indirect conversion reactions because they do not require producing synthesis gas as an intermediate.[15,17] Non-oxidative routes, such as methane dehydroaromatization (or pyrolysis), are operationally attractive because they require fewer reagents and present lower risks of explosions and CO_x emissions.[15] However, the low methane conversion does not justify the costs associated with product separation stages, as well as the high coke formation and catalyst stability problems for this reaction.[11,15] Among the potential methane conversion routes, oxidative pathways are more advantageous due to more favorable thermodynamics and the current prevalence of oxidative processes in industrial applications.[10] Moreover, these routes offer greater versatility regarding

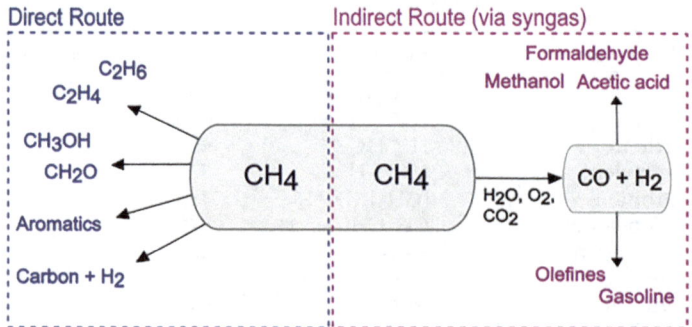

Figure 3.2 Methane conversion routes. Data from ref. 12.

operational conditions, require lower reaction temperatures, and yield better methane conversion results.[15] Some examples of oxidative reactions include partial oxidation of methane to methanol or formaldehyde and oxidative coupling of methane.

3.4 Oxidative Coupling of Methane (OCM)

The oxidative coupling of methane is a promising way to convert methane into higher added-value products, like ethane and ethylene.[18] Given its desirability as a primary chemical reagent in various industrial processes, ethylene typically has a much higher commercial value than other C_1-oxygenated compounds like methanol. Oxidative coupling of methane offers the tantalizing prospect of producing ethylene through a single methane conversion process. This is expected to lead to significant reductions in operational costs and, consequently, economic benefits for olefin production.[10]

In the OCM process, methane molecules interact in the presence of oxidants like O_2 in a temperature range of 500–1000 °C to form C_2 hydrocarbons[9,14] The following equations show the reactions to ethane (eqn (3.1)) and ethylene (eqn (3.2)) production, as well as for the global reaction (eqn (3.3)):

$$2CH_4 + \frac{1}{2}O_2 \rightarrow C_2H_6 + H_2O \quad \Delta H^\circ_{1073K} = -1465 \text{ kJ mol}^{-1} \quad (3.1)$$

$$C_2H_6 + \frac{1}{2}O_2 \rightarrow C_2H_4 + H_2O \quad \Delta H^\circ_{1073K} = 302 \text{ kJ mol}^{-1} \quad (3.2)$$

$$2CH_4 + O_2 \rightarrow C_2H_4 + 2H_2O \quad \Delta H^\circ_{1073K} = -1163 \text{ kJ mol}^{-1} \quad (3.3)$$

The oxidant environment results in a thermodynamically favorable combustion of the products, leading to the undesirable production of CO_x (CO_2, CO) as the main byproduct of the reaction (eqn (3.4)–(3.7)).

$$CH_4 + 2O_2 \rightarrow CO_2 + 2H_2O \quad \Delta H^\circ_{298K} = -802 \text{ kJ mol}^{-1} \quad (3.4)$$

$$CH_4 + O_2 \rightarrow CO + H_2 + H_2O \quad \Delta H^\circ_{298K} = -278 \text{ kJ mol}^{-1} \quad (3.5)$$

$$CH_4 + \frac{1}{2}O_2 \rightarrow CO + 2H_2 \quad \Delta H^\circ_{298K} = -36 \text{ kJ mol}^{-1} \quad (3.6)$$

$$C_2H_4 + 2O_2 \rightarrow 2CO + 2H_2O \quad \Delta H^\circ_{298K} = -757 \text{ kJ mol}^{-1} \quad (3.7)$$

The OCM was first reported by Keller and Bhasin in 1982[19] as represented by eqn (3.3). Their work proposed OCM as an alternative to direct C_2 production through dehydrogenative coupling, which requires temperatures above 800 °C. This high temperature demands significant energy input, operates within narrow conditions, and results in low methane conversion. In contrast, OCM allows the reaction to occur at lower temperatures, with broader operating conditions and achieves higher methane conversions.

Keller and Bhasin tested numerous catalysts and found that metal oxides from groups IIIA, IVA, and VA were the most active. They also concluded that the redox properties of the catalyst are crucial for its activity in the reaction. Since then, the interest in the OCM route has increased, attracting attention as a technological way to suppress the need for C_2 products by an ecofriendly route. However, the OCM route remains limited after several decades of study due to thermodynamic constraints. As already mentioned, CO_x products are thermodynamically more stable than the desired C_2 hydrocarbons, which will be discussed in greater detail in Section 3.5. This thermodynamic challenge explains why OCM has not yet been adopted by industry.

Ethylene is an important product of the OCM reaction, which has gained great commercial interest and is used to produce various chemical intermediates and polymers.[20] For example, it is used for the synthesis of polyethylene $((CH_2-CH_2)_n)$, and its oligomerization can contribute to the production of gasoline (C_5-C_{10}) and diesel $(C_{10}-C_{20})$.[21] Currently, ethylene is mainly produced *via* steam cracking of ethane and naphtha. In these processes, the thermal conversion of the reagents (>750 °C) occurs through a radical mechanism in the gas phase, producing ethylene, short-chain olefins, and H_2.[20] This process is highly endothermic and, combined with the product purification methodologies, emits large amounts of CO_2 into the atmosphere, about 2–3 times more than the amount of ethylene produced.[20,22]

3.5 Thermodynamic Aspects

The energy aspect is the most significant advantage of using OCM compared to other methane conversion routes. The enthalpy for the formation of the water molecule ($DH_f = -285.8$ kJ mol^{-1}) counterbalances the bond dissociation energy of the methane molecule. Combined with the contributions from the coupling reactions of methyl radicals in the gas phase, this results in a negative change in enthalpy. This allows the reaction to be conducted at lower temperatures (around 800 °C) than other methane reactions and reduces the need for external heat input.[12]

However, thermodynamics also presents one of the most significant challenges associated with the OCM reaction: achieving higher C_2 yields. Low C_2 yields have been obtained in OCM for almost all tested catalysts, still far from the thermodynamic limit of 28–30%.[20] Figure 3.3 reveals the reason for this challenge.[15] The formation of undesirable products (CO and CO_2) is thermodynamically more favorable than the formation of ethane and ethylene, thus affecting the reaction's selectivity. Additionally, increasing the temperature further favors the formation of CO compounds, while the formation of ethane becomes even less favorable.[18]

In addition to these factors, catalytic processes involving many sequential chemical conversions are also complicated. Consequently, the undesired product can be formed from the desired product, and being thermodynamically more stable, it results in issues between conversion and

Oxidative Coupling of Methane

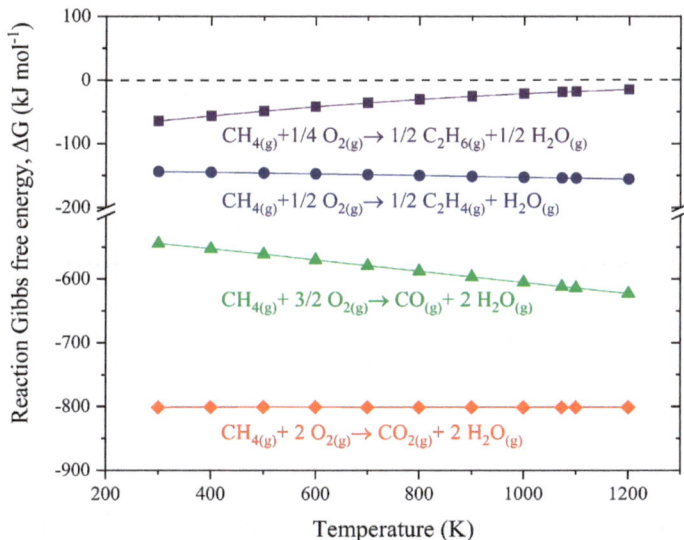

Figure 3.3 Evolution of Gibbs free energy (ΔG) as a function of temperature for selected reactions occurring in the OCM process at 1 atm. Reproduced from ref. 15 with permission from Elsevier, Copyright 2021.

selectivity. Therefore, any catalysts operating at high methane conversions exhibit low C_2 selectivity values. Only by operating at low methane conversion values is it possible to achieve high C_2 selectivity.[18]

3.6 Reaction Mechanism and Kinetic Aspects

Despite discrepancies regarding the details of the OCM reaction mechanism, most studies indicate that the catalysts are metal oxides with active sites that homolytically remove hydrogen from methane, releasing methyl radicals in the gas phase. During this stage, the radical couples with another methyl radical, leading to the formation of ethane. Since the active sites can remove hydrogen from any hydrocarbon, ethane undergoes the same abstraction steps as methane on the catalyst surface and, through oxidative dehydrogenation, forms ethylene,[12] as shown in Figure 3.4.[23] CO_x compounds can originate from various reaction pathways. Methyl radicals or C_2 compounds can be reactivated on the catalyst surface or interact with gas-phase oxygen species.[24]

Studying the OCM mechanism and its kinetic limitations is challenging due to the complex reaction pathways involved.[15] Firstly, the reaction mechanism involves a heterogeneous step (on the catalyst surface) and a homogeneous step (in the gas phase) that strongly influence each other. Additionally, the radical reactions occurring in the gas phase are fast, preventing the identification of the intermediate species formed.[15] Finally, the number of radicals formed can be exorbitant, reaching up to 269 radical

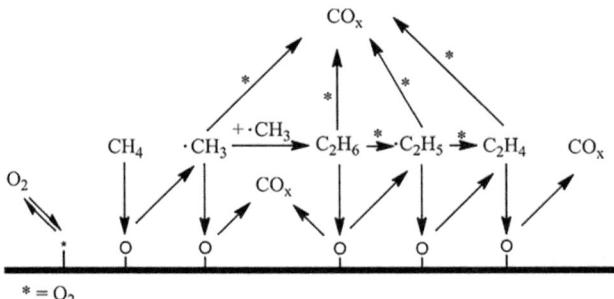

Figure 3.4 Simplified model for the OCM reaction. Reproduced from ref. 23 with permission from Elsevier, Copyright 2014.

species formed *via* 1582 parallel reactions, according to the microkinetic model proposed by Dooley and collaborators.[25]

Other factors also contribute to the lack of rigorous kinetic experiments in OCM. One example is the difficulty in controlling the local temperature due to the formation of hot spots created by highly exothermic over-oxidation reactions. Other factors, such as carbon deposition and the low thermal stability of some catalysts, result in a rapid change in the nature and number of active sites, modifying the reaction pathways.[18]

There is much debate in the literature regarding the C–H bond activation mechanism, especially concerning activation routes *via* homolytic or heterolytic cleavage.[21] Generally, electrophilic oxygen species typically activate methane *via* a homolytic pathway, while nucleophilic oxygens form methoxy groups that lead to less selective reaction pathways.[13] The formation of the methyl radical *via* the heterogeneous route occurs through an Eley–Rideal reaction mechanism. It is initiated by the reaction with the oxygen on the catalyst surface ($O_{(s)}$), as depicted in eqn (3.8).

$$CH_4 + O_{(s)} \rightleftharpoons CH_3^\bullet + OH_{(s)} \tag{3.8}$$

In the gas phase, homolytic dissociation of the C–H bond occurs preferentially. Due to the high ionization potential and low electron affinity of CH_4, the enthalpy for heterolytic dissociation is high. For example, the non-catalytic deprotonation of methane (in the gas phase) has a standard reaction enthalpy of approximately 1745 kJ mol^{-1} to form the methyl anion and proton. The dissociation of CH_4 into a methyl cation and a hydride is slightly less endo-energetic, with an enthalpy of around 1460 kJ mol^{-1}.[13] Gas-phase chemistry contributes to the formation of the methyl radical through both oxidative and non-oxidative mechanisms. Methane can be attacked in the gas phase by both O_2 and oxygen radicals.[21] Examples of gas-phase methane activation routes are highlighted in eqn (3.9)–(3.14).

$$CH_4 + O_2 \rightleftharpoons CH_3^\bullet + HO_2^\bullet \tag{3.9}$$

$$CH_4 + O_2 \rightleftharpoons CH_3O^\bullet + OH^\bullet \tag{3.10}$$

$$CH_4 + O^{\bullet} \rightleftharpoons CH_3^{\bullet} + OH^{\bullet} \tag{3.11}$$

$$CH_4 + HO_2^{\bullet} \rightleftharpoons CH_3^{\bullet} + H_2O_2 \tag{3.12}$$

$$CH_4 \rightleftharpoons CH_3^{\bullet} + H^{\bullet} \tag{3.13}$$

$$CH_4 + H^{\bullet} \rightleftharpoons CH_3^{\bullet} + H_2 \tag{3.14}$$

Radical reactions follow three main stages: initiation, propagation, and termination. After the initiation of radical reactions (formation of the methyl radical), propagation occurs, forming new radicals. These radicals have low selectivity control and can follow hundreds of reaction pathways. Finally, methyl radicals are recombined in the gas phase in the termination stage to form ethane, as shown in eqn (3.15). Through the homogeneous route, ethylene is formed by a dehydrogenation process involving reactions with $H^{\bullet}$, $OH^{\bullet}$, and $CH_3^{\bullet}$, with the ethyl radical ($C_2H_5^{\bullet}$) as an intermediate. The ethyl radical can also be formed from the reaction of ethane with the catalyst surface.

$$CH_3^{\bullet} + CH_3^{\bullet} \rightleftharpoons C_2H_6 \tag{3.15}$$

In the kinetic model proposed by Karakaya and collaborators,[21] CO_x byproducts and C_{2+} hydrocarbon formation occur almost simultaneously. Both homogeneous and heterogeneous routes include the direct and indirect formation of CO_x compounds. Direct oxidation in the gas phase is dominated by the reaction of the methyl radical with O_2, forming $CH_3O^{\bullet}$ and $CH_2O^{\bullet}$ species. At the same time, the indirect route involves the reaction of ethylene with O_2, $H^{\bullet}$, $OH^{\bullet}$, and $CH_3^{\bullet}$, forming the $C_2H_3^{\bullet}$ species. On the catalyst surface, direct oxidation occurs through the reaction of the methyl radical with $O_{(s)}$, forming the $CH_3O_{(s)}$ species, and indirectly through the reaction of ethylene with $O_{(s)}$, forming $C_2H_3^{\bullet}$ or $C_2H_4O_{(s)}$ species.

The contribution of each route depends on the oxygen concentration and reaction temperature. Direct oxidation routes in the gas phase dominate at relatively low temperatures (660 °C). At this temperature, direct and indirect contributions *via* heterogeneous routes are minimal. As the temperature increases to 760 °C, the contribution of indirect routes at the gas phase and direct routes on the catalyst surface increases. At 790 °C, the direct oxidative mechanism in the gas phase predominates. Therefore, gas-phase mechanisms seem to significantly influence the formation of byproducts more than heterogeneous routes.[21] A kinetic model was proposed by Stansch, Mleczko, and Baerns[26] using the La_2O_3/CaO catalyst, as shown in Figure 3.5.

Although relatively simple, this model has proven sufficiently descriptive of the experimental data observed for other OCM catalysts based on oxides. According to the proposed model, the reaction can be described by 10 steps, with CH_4 being converted *via* three parallel reactions. Step 1 is a nonselective route of complete CH_4 oxidation to CO_2, similar to step 3, which

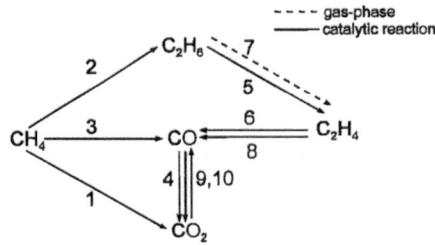

Figure 3.5 Reaction scheme of the kinetic model proposed by Stansch, Mleczko, and Baerns in 1997. Reproduced from ref. 26 with permission from American Chemical Society, Copyright 1997.

forms CO. In step 2, CH_4 is selectively converted to ethane. In subsequent steps, ethane can be converted to ethylene through two parallel routes: oxidative dehydrogenation *via* a heterogeneous route (step 5) or thermal dehydrogenation in the gas phase (step 7). This model neglects the direct conversion of ethane to CO and CO_2. Ethylene oxidation can occur in two parallel steps: reacting with O_2 (step 6) or H_2O *via* steam reforming of ethylene to form CO (step 8). According to the proposed model, CO can react with O_2 to form CO_2 (step 4) or through the water–gas shift reaction (steps 9 and 10, as it is reversible).[26] Finally, it is noted that the model also disregards possible routes for coke formation.

From this model, it can be observed that the reactions leading to CO_x formation exhibit approximately first-order dependence on the partial pressure of O_2, while the reactions for C_2 product formation show $\sim\frac{1}{2}$ order dependence.[18] Therefore, there is a thermodynamic and kinetic preference for CO_x formation.

Thus, there are still some obstacles in the quest for a more efficient OCM process, including the challenge of achieving optimal yield and selectivity, which are constrained by the reaction's thermodynamics and kinetics.[27] Other issues are associated with the difficulty of finding selective catalysts resistant to deactivation over time. This issue is primarily due to the aggregation of catalyst particles under reaction conditions, decreasing their specific surface area.[28]

3.7 Impact of Reaction Conditions in OCM

The reaction conditions under which OCM is employed also significantly affect the performance of the catalysts used in this reaction. Suzuki and collaborators[29] developed a statistical approach (machine learning) to evaluate compositions and reaction conditions from the literature. Based on the adjustment of the statistical model predicting catalyst performance, the authors observed a strong dependence on experimental conditions for obtaining good catalytic predictions.[29] Thus, reaction parameters such as the CH_4/O_2 ratio, reaction temperature, and use of diluents, among others, are essential for achieving better C_2 selectivity.

High O_2/CH_4 ratios lead to excessive oxidation of the products, resulting in low reaction selectivity.[28] Conversely, high CH_4/O_2 ratios result in higher C_2 selectivity values but are accompanied by lower yields of these hydrocarbons. This occurs due to reduced regeneration of oxygen vacancies formed on the surface during the reaction, depleting the active sites for methane activation (resulting in lower methane conversion).[28]

As increased conversion leads to decreased selectivity, controlling selectivity is critical since unreacted methane can be recycled. Additionally, OCM can be coupled with a secondary process to convert unreacted reagents and by-products. For example, OCM can be paired with a reforming reaction of unconverted methane using the byproduct CO_2 as a reagent to produce syngas.[18]

Temperature variations also affect the distribution of the products formed. Increasing the temperature can favor the formation of ethylene over CO_2. This occurs mainly due to the enhanced desorption of ethyl radicals formed on the catalyst surface, reducing the formation of precursors responsible for CO_2 formation.[30] Among the temperature-related issues is the presence of hot spots, primarily due to the overall exothermic reaction, which increases the temperature at specific reaction points. This process makes thermal control difficult and leads to more significant deactivation, mainly through sintering or sublimation of the active phase. Using diluents, such as inert gases (N_2 and He) or ground quartz mixed with the catalyst in the reactor, presents a way to circumvent this problem.[28] On the other hand, highly diluted catalytic systems have higher operational costs related to purification processes.

From this initial analysis, it can be observed that modifications in reaction conditions present both advantages and disadvantages for each parameter variation. Using statistical optimization tools is an excellent option for maximizing the potential of each item. A multivariate method allows for varying the parameters that affect the reaction and understanding the extent of their influence using the fewest possible experiments.[31] All parameters can be varied simultaneously in a set of predetermined experiments. The results describing the system behavior can be investigated by constructing a mathematical model. These experiments offer many advantages over univariate approaches, such as investigating the interaction effects between variables and systematizing the work, giving equal weight to all studied variables.[31] Additionally, the mathematical treatment of the obtained results allows the construction of a response surface that describes the property of interest as a function of the varied factors. This enables the prediction of results within a significant range of experimental conditions, not only the specific points where the experiments were conducted.[31]

3.8 Alternative Oxidants for OCM

An interesting strategy to increase selectivity in OCM is to modify the oxidizing species used. Replacing O_2 with less aggressive oxidants, such as CO_2, N_2O, or S_2, has optimized C_2 selectivity by reducing over-oxidation processes.

Improved catalytic performance can result from changes in reaction thermodynamics, variations in the oxygen species at the active site, or a combination of both factors.[24]

Few studies have explored this parameter, as shown in Figure 3.6, which summarizes the publications over the years, subdividing them by the oxidant used.[24] OCM began to be studied around 1980 and peaked between 1990 and 1995, with interest declining after that. Around 2005, attention related to this reaction grew again due to the shale gas revolution.[24] Since the early years, alternative oxidants have been investigated, but in a much lower volume of publications than traditional OCM using O_2 as the oxidant.

The regeneration reaction of the active site with CO_2, as well as the overall OCM reaction using CO_2, are presented in eqn (3.16) and (3.17), respectively.[32] CO_2 is a cheap, non-toxic, abundant oxidant that is environmentally interesting. As one of the main components of biogas, along with methane, using both gases during the OCM reaction would be advantageous to avoid purification costs. Additionally, CO is expected to be the only byproduct of OCM using CO_2 as the oxidant. An additional benefit of this oxidant is that CO_2 will not induce gas-phase radical reactions, and therefore, a heterogeneous reaction mechanism is expected to dominate.[10]

However, a recurring issue with using CO_2 as an oxidant for OCM is that it can preferentially follow a mechanism that leads to carbon deposits on the catalyst surface, as seen in eqn (3.18)[17] which could result in catalyst deactivation.

$$CO_2 + (\ ) \rightarrow (O) + CO \tag{3.16}$$

$$CH_4 + CO_2 \rightarrow \frac{1}{2} C_2H_4 + CO + H_2O \tag{3.17}$$

$$nCH_4 + 2nCO_2 \rightarrow nC + 2nCO + 2nH_2O \tag{3.18}$$

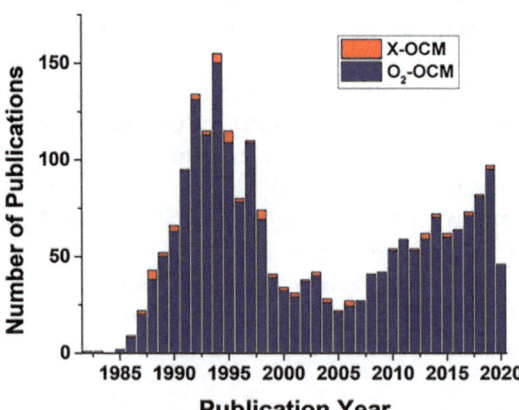

Figure 3.6 Number of publications for OCM over the years. The blue bars represent publications using O_2 as the oxidant, and the red bars represent publications using alternative oxidants. Reproduced from ref. 24 with permission from Wiley-VCH, Copyright 2021.

Another possibility is the application of N_2O as an oxidant for the reaction. In addition to the environmental advantages of using N_2O, its mechanism for regenerating active sites creates more selective oxygen surfaces. The superior performance with N_2O can be explained by the formation of monoatomic oxygen from the dissociation of N_2O on the catalyst surface. In contrast, diatomic oxygen species are formed by the adsorption of O_2.[33] The formation of monoatomic species is facilitated for N_2O compared to O_2 since the O–N bond energy is much lower (167.4 kJ mol^{-1}) than the O–O bond energy in O_2 (498.4 kJ mol^{-1}).[33] Monoatomic species are more effective for methane activation, while diatomic species have a more significant influence on the formation of CO_x compounds.[33]

Most catalysts tested so far indicate that using N_2O results in a decrease in methane conversion and an increase in C_2 selectivity compared to O_2.[24,34,35] Table 3.2 summarizes the variations in Gibbs free energies for the OCM reaction and over-oxidation reactions using O_2 and N_2O as oxidants.[24] Differences in the thermodynamics of the two processes are initial indicators of changes in the product distribution. According to Table 3.3, the reaction carried out with N_2O appears to be more spontaneous than that with O_2. However, variations in reaction kinetics and oxidant activation also affect the reaction mechanisms of these two routes.[24]

Many similarities can be highlighted between OCM reactions using O_2 and N_2O as oxidants. Surface basicity and the quantity and lability of oxygen species are essential factors affecting reactivity and selectivity for both systems. Additionally, the role of both O_2 and N_2O in replenishing/forming active oxygen species on the surface has been emphasized.[24] However, the high cost of this oxidant challenges its practical application. Furthermore, since N_2O activation leads to the formation of N_2, which dilutes the products, this process would require additional separation steps. Nevertheless, OCM with N_2O is a useful process for studying the role of adsorbed oxygen species and highlights that high C_2 yields are possible if the presence of non-selective oxygen species is controlled.[24]

Table 3.2 Gibbs free energy (ΔG) for the oxidative coupling of methane and the complete oxidation of methane and ethylene using O_2 and N_2O as oxidants.

Reactions with O_2	ΔG_{rxn} at 800 °C (kJ mol^{-1})
$2CH_4 + O_2 \rightarrow C_2H_4 + 2H_2O$	-307
$2CH_4 + \frac{1}{2}O_2 \rightarrow C_2H_6 + H_2O$	-114
$CH_4 + 2O_2 \rightarrow CO_2 + 2H_2O$	-792
$C_2H_4 + 3O_2 \rightarrow 2CO_2 + 2H_2O$	-1294
Reactions with N_2O	ΔG_{rxn} at 800 °C (kJ mol^{-1})
$2CH_4 + 2N_2O \rightarrow C_2H_4 + 2H_2O + 2N_2$	$-622,7$
$2CH_4 + N_2O \rightarrow C_2H_6 + H_2O + N_2$	-279
$CH_4 + 4N_2O \rightarrow CO_2 + 2H_2O + 4N_2$	-1432
$C_2H_4 + 6N_2O \rightarrow 2CO_2 + 2H_2O + 6N_2$	-2242

Table 3.3 The OCM performance of some representative catalysts.

Catalyst	Temperature (°C)	CH_4/O_2 ratio	GHSV (mL g^{-1} h^{-1})	CH_4 conversion	C_2 selectivity	Ref.
$MgCO_3$	740 (850)	5	240 000	2.1 (5.3)	35 (70)	82
$CaCO_3$	740 (850)	5	240 000	4.4 (7.8)	41 (75)	82
$SrCO_3$	740 (850)	5	240 000	1.6 (9.5)	47 (79)	82
$BaCO_3$	740 (850)	5	240 000	2.2 (7.0)	54 (84)	82
CaO	800	8	51 360	13	49	86
La–CaO	800	8	51 360	20	67	86
Ce–CaO	800	8	51 360	18	65	86
Sm–CaO	800	8	51 360	19	63	86
Nd–CaO	800	8	51 360	20	71	86
Yb–CaO	800	8	51 360	20	66	86
Li/MgO	650	21	5976	2.6	76	39
$Mg_{0.5}Ti_{0.5}O_y$	800	2	30 000	41	43	87
$BaBr_2:SnO_2=1:1$	800	4	18 000	29	63	50
La_2O_3–CeO_2 nanofiber	570	5	240 000	~29	~62	106
$Sr_{0.1}LaO_x$-H	550	2.6	60 000	39	49	75
$Nd_{0.9}Ce_{0.1}O_{1.55}$	700	4	18 000	~27	~53	103
$La_2Ce_2O_7$	750	4	18 000	~29	~57	73
$La_2Zr_2O_7$	750	4	18 000	~24	~50	73
$La_2Ti_2O_7$	750	4	18 000	~17	~31	73

Oxidative Coupling of Methane

Catalyst	Temp	GHSV	Conv	Sel	Ref
$La_2Ce_2O_7$	700	18 000	4	~38	~58	112
$Y_2Ce_2O_7$	700	18 000	4	~30	~51	112
$Sm_2Ce_2O_7$	700	18 000	4	~30	~51	112
$Sm_2Ce_2O_7$	550	18 000	4	~22	~47	112
$Pr_2Ce_2O_7$	700	18 000	4	~14	~29	112
$La_2Ce_2O_7$–GNC	600	18 000	4	~34	~52	53
$La_{1.7}Ca_{0.3}Ce_2O_{7-\delta}$	650	15 000	4	31	76	64
$LaAlO_3$	775	10 000	3	31	45	126
$LaGaO_3$	775	10 000	3	32	40	126
$SrZrO_3$	775	10 000	3	31	45	126
$LaTi_{0.5}Mg_{0.5}O_{3+\delta}$	800	30 000	8	17	45	118
Na-$BaSrTiO_3$	800	6000 h^{-1}	2	47	51	123
$BaO/CaTiO_3$	650	10 000 h^{-1}	3	~26	~40	121
$SrO/BaTiO_3$	725	10 000 h^{-1}	3	~29	~61	121
$Mn/Na_2WO_4/SiO_2$	850	2700	4.5	33	80	36
5% La_2O_3-Li-Mn/WO_3/TiO_2	750	1140	2.5	~30	~64	68
3 wt% Na_2WO_4/TiO_2	800	10 000	3	49	56	37
1.9% Mn-5% Na_2WO_4-5% Ce/TiO_2	775	4800	2	49	53	38
Core-shell SiO_2@MnO_x@Na_2WO_4@SiO_2	770	20 000 h^{-1}	3	18	54	36
$MnTiO_3$–Na_2WO_4/SBA-15	700	18 500 h^{-1}	3	39	69	128

3.9 Active Sites for OCM

Since Keller and Bhasin's pioneering work, research focused on OCM in the past 40 years has predominantly focused on developing catalysts that maximize C_2 hydrocarbon yields while ensuring long-term stability. Numerous studies have shown that mixed metal oxide catalysts, such as Li/MgO and the multicomponent Mn/Na$_2$WO$_4$/SiO$_2$, achieve the highest C_2 yields, making them the benchmark materials.[36–40] However, the Li/MgO catalyst has low thermal stability, leading to deactivation due to Li$^+$ vaporization in the oxygen atmosphere at elevated temperatures.[41,42] Mn/Na$_2$WO$_4$/SiO$_2$ catalyst is regarded as the most effective for OCM, offering high C_2 yields (14–30%) and stability for up to 500 h. However, the reaction requires high temperatures (above 700 °C),[43] and the catalyst's active sites are not fully understood due to its complex elemental composition.[44] Thus, exploring active sites using simpler structures, such as pure metal oxides, could be highly beneficial for fundamental research.

OCM catalysts must possess cost-effectiveness, and chemical, thermal, and hydrothermal stability, along with a high concentration of basic sites and oxygen vacancies to enhance electrophilic oxygen species. It has also been suggested that low specific surface areas are advantageous in this reaction,[45–47] although this remains a point of debate.[48] The active sites responsible for activating the C–H bond in CH$_4$ molecules during the OCM reaction are generally attributed to reactive oxygen species available on the catalyst surface.[49] The literature agrees that the production of structural defects in the form of oxygen vacancies promotes the adsorption and activation of molecular oxygen, leading to a large number of surface-active oxygen sites, such as superoxide (O_2^-) and peroxide (O_2^{2-}) species. These species activate methane to produce methyl radicals (CH$_3^•$), which are highly favorable for selectively converting methane into C_2 compounds.[50–53]

Numerous theoretical and experimental studies have emerged in an attempt to clarify the primary roles of these oxygen species in the OCM reaction. The work of Osada and colleagues[54] investigated Y$_2$O$_3$–CaO catalysts, finding that superoxide species (O_2^-) could play a significant role in the selective conversion of methane. Yang and colleagues[55] agreed with this interpretation and associated superoxide species as selective active sites for methane activation. Overall, the extent of the contribution of each oxygen species depends on their availability to participate in the C–H bond cleavage. Therefore, the concentration of oxygen vacancies, especially at high temperatures, can be correlated with the abundance of these oxygen species and their availability as active sites. Consequently, efficient catalysts should have a high concentration of oxygen vacancies at the reaction temperature.[49] Oxygen vacancies are also formed during the reaction when the active site oxygen leaves the structure as water and can be replenished by gas-phase oxygen or by bulk oxide oxygen migrating to the surface. Although the first option is kinetically favored, migration occurs when there is insufficient oxygen supply during the reaction.[56]

Variations in C_2 selectivity depend on the ability (or rather, the inability) of the methyl radical to be adsorbed and oxidized by the catalyst surface. There is a balance between the activation energy of the C–H bond, which determines conversion, and the adsorption energy of methyl radicals, which determines the reaction selectivity.[57] Kumar and colleagues[57] studied various pure and doped oxides using density functional theory (DFT). The authors observed that higher activities in OCM were achieved with increased reducibility of the oxide surface and established a correlation between the activation energy of the C–H bond and the energy required for forming oxygen vacancies associated with the catalyst reducibility. However, above a particular concentration of oxygen vacancies, the methane activation barrier becomes independent of its distribution, and beyond a certain degree of reducibility, methane oxidation is favored, forming more significant amounts of CO_x compounds. Thus, reducibility is a factor that needs to be optimized to achieve better C_2 yields.[57]

Various characterization techniques have been employed to detect the presence of oxygen vacancies and/or mobile oxygen species, directly or indirectly, including electron paramagnetic resonance (EPR),[53,58–60] temperature-programmed reduction with hydrogen (TPR-H)$_2$,[61–63] temperature-programmed desorption of oxygen (TPD-O)$_2$,[53,58,60,64] methane temperature-programmed surface reaction (CH_4-TPSR),[58,65,66] Raman spectroscopy (both *in situ* and *ex situ*),[50,59,60,64,67] *in situ* diffuse reflectance infrared Fourier-transform spectroscopy (DRIFTS),[53,64] and X-ray photoelectron spectroscopy (XPS),[58,59,64,67,68] although the effectiveness of *ex situ* XPS analysis for detecting oxygen vacancies is strongly contested.[69,70]

The presence of lattice oxygen (O^{2-}) is commonly associated with the deep oxidation of hydrocarbon reactants to CO and CO_2,[64,71–73] and the ($O_2^- + O_2^{2-}/O^{2-}$) ratio is used as a parameter of comparative analyses of catalyst performance.[50,53,64,74,75] However, this topic is still under discussion. For instance, Gordienko and colleagues[76] studied the contribution of lattice oxygen in catalysts. They revealed that the synthesized catalysts had two types of lattice oxygen species bound to the surface (weakly and strongly bound). Additionally, they established correlations between reaction selectivity and reduction temperature. Since the reduction temperature refers to the activation temperature of each oxygen species, this discovery indicated that each species was responsible for a different pathway in the OCM reaction. Although the contribution of strongly bound species could not be fully accessed, their study revealed that weakly bound oxygen species contributed significantly to better activity and stability in the performance of catalysts during reaction.[76]

Several studies[59,68,73,77] have demonstrated that electrophilic oxygen species become reactive at significantly lower temperatures than lattice oxygen. Consequently, these species could be conducive to the formation of C_2 compounds at lower temperatures, whereas lattice oxygen could be more suitable at higher temperatures.[78] Additionally, the migration of these oxygen species appears to play a crucial role.[44,65] It has also been claimed[79,80] that the role of each species is determined by the catalyst's reducibility: in reducible catalysts, surface lattice oxygen species (O^{2-}) selectively promote the formation of C_2, while O_2^-, O_2^{2-}, and O^- species drive deep oxidation. Conversely, the opposite effect is observed in irreducible metal oxide catalysts.

Feng et al.[65] investigated the impact of surface-chemisorbed oxygen species, formed in oxygen vacancies, along with both bulk and surface lattice oxygen species, following the incorporation of Ce atoms into the $La_2O_2CO_3$ structure. These materials were applied in OCM at 500 °C. Through TPSR-CH_4 and TPSR-CO experiments, the authors associated the formation of C_2 compounds with the presence of chemisorbed oxygen species, while surface lattice oxygen was found responsible for further oxidation of these compounds to CO_x. Interestingly, upon activation of these catalysts with O_2, they observed that a portion of the adsorbed oxygen can migrate into the bulk and convert into bulk La–Ce–O lattice oxygen species *via* a vacancy-mediated mechanism, as seen in Figure 3.7. However, this bulk oxygen can migrate back to the surface, forming surface lattice oxygen. These findings led the authors to conclude that, when more chemisorbed oxygen species form at a certain temperature, the lower will be the OCM reaction temperature.

On the other hand, the study by Kim and colleagues[56] investigated $LaXO_3$ (X = Al, Fe, or Ni) catalysts. Through tests in the absence of oxidants, they demonstrated that methane could react directly with lattice oxygen species in these catalysts. Depending on the cation in X, the properties of these surface lattice oxygens varied, leading to the formation of different products. Variations in the binding energies of lattice oxygens were observed for the three catalysts, resulting in different methane conversion rates and C_2 selectivity. $LaAlO_3$ exhibited the highest binding energy, indicating that the active site had a more electrophilic character, followed by $LaFeO_3$ and $LaNiO_3$, which showed a more nucleophilic character. Correlating with the catalytic results, lattice oxygen species with higher electrophilic character were responsible for selectively producing C_2 hydrocarbons from methane, moderate species led to CO, and nucleophilic species produced predominantly CO_2. Finally, based on their findings, the authors attributed this

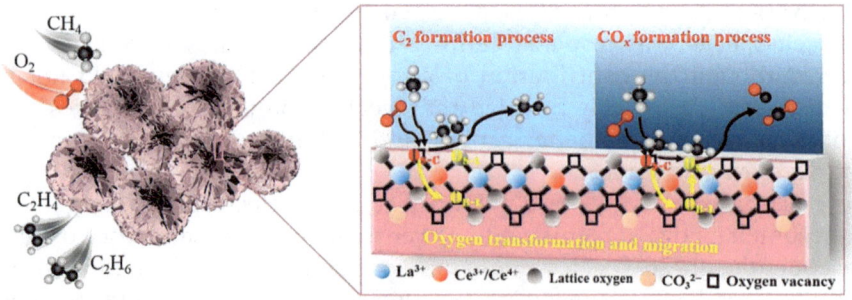

Figure 3.7 Participation of oxygen species in the OCM reaction over $La_xCe_{1-x}O_{1.5+\delta}$ catalyst. Reproduced from ref. 65 with permission from Elsevier, Copyright 2022.

difference to the greater ease with which electrophilic oxygen species on the surface favored the formation of methyl radicals, the rate-limiting step of the reaction.[56]

The acidic/basic nature of the active sites is also an important characteristic of catalysts in the OCM. In addition to the influence of oxygen vacancies, basicity is determined by the binding strength between the cationic and anionic species in the oxide lattice. Theoretical results indicate that basic catalysts exhibit higher C_2 selectivity than acidic ones.[49] Appropriate substituents can alter the arrangement of cations in the crystalline lattice and, consequently, the electronic structure of the oxide. Adding lower valence cations to the oxide structure results in lower oxygen coordination values, weakening its bond and making it more basic.[49] Recently, theoretical studies (using DFT) and experimental studies with a more fundamental approach have provided significant insights into the correlation between catalyst properties and their activity in OCM, as well as discoveries in the reaction mechanism.[49] In general, the active sites of catalysts typically facilitate chemical reactions through electronic interactions with the reactant molecules, which strongly depends on their electronic arrangement.[49] Some studies have linked moderate basic sites with oxygen vacancies.[51,53,74,79] However, since CO_2 is formed as a reaction product, strong basic sites form carbonate species that poison active sites and can only be desorbed at very high temperatures, making them useful only at elevated temperatures.[71,81,82]

The main catalysts used in this reaction can be categorized into four groups: (1) alkaline earth metal oxides; (2) rare earth metal oxides; (3) perovskites; and (4) $Mn/Na_2WO_4/SiO_2$ catalysts. The OCM performance of some representative catalysts is presented in Table 3.3. These materials will be discussed in detail in the next section.

3.10 Development of Catalysts for OCM

3.10.1 Alkaline Earth Metal Oxides

Alkaline earth metals are used in this reaction mostly due to their high alkalinity. Maitra et al.[82] studied the effect of basicity on the catalytic activity of alkaline earth oxides derived from their respective carbonates in OCM. The authors measured basicity by determining the decomposition temperature of carbonates ($MCO_3 \rightarrow MO + CO_2$) using TGA. They observed that basicity increased with the element's period, as shown by the decomposition temperatures of Mg, Ca, Sr, and Ba oxides at 422, 670, 863, and 995 °C, respectively. The authors found a linear correlation between basicity and hydrocarbon selectivity, but only within a reaction temperature range that allows the catalyst to exist, at least partially, as an oxide. Consequently, the catalytic performance of SrO and BaO catalysts was limited by the high stability of the CO_2 bond with the oxides, which is therefore favored at high temperatures, where CO_2 is desorbed. In line with this work, Carreño et al.[45] compared SrO and MgO catalysts in OCM at 750 °C. Although SrO's surface was more active than that of MgO, it was less active under steady-state conditions due to carbonate formation.

Doping one alkaline earth metal into another also proves beneficial. For instance, Ca-promoted MgO catalysts were studied by Cho et al.[83] While pure MgO catalysts exhibited low OCM performance, introducing just 1% Ca increased the C_2 yield to 14.1%. This improvement is ascribed to the creation of additional oxygen vacancies, which in turn, increased the number of moderate basic sites within the framework. Matras et al.[84] synthesized La–Sr/CaO catalysts using the sol–gel method for the OCM. Their study, employing X-ray diffraction computed tomography, pointed out the formation of Sr-doped CaO as a crucial factor in the catalyst's effectiveness. They observed an even distribution of La_2O_3 and CaO–SrO mixed oxide and $SrCO_3$ components, which remained stable at the beginning of the reaction, when using a GHSV (gas hourly space velocity) of 36 000 $mL\,g^{-1}\,h^{-1}$. However, when the GHSV was doubled, decreasing the partial pressure of CO_2, $SrCO_3$ decomposed into SrO, resulting in the formation of a second CaO–SrO mixed oxide (Figure 3.8). Following experiments with a GHSV of 64 000 $mL\,g^{-1}\,h^{-1}$, a final measurement at a reduced GHSV of 36 000 $mL\,g^{-1}\,h^{-1}$ was conducted. Under these conditions, both methane conversion and selectivity toward ethane and ethylene were higher than those observed under the initial reaction conditions, prior to the phase transformation. Such second CaO–SrO mixed oxides have a different extent of Sr incorporation into their unit cell in comparison with the first CaO–SrO mixed oxide, possibly leading to an enrichment of the lattice oxygen diffusion and basicity.

The high basicity of rare earth oxides makes them promising potential dopants for alkaline catalysts in OCM. Choudhary and co-workers[85] studied MgO catalysts promoted with rare earths (La, Ce, Nd, Eu, and Yb). The use of these dopants increased the C_2 selectivity, particularly for the La–MgO, Ce–MgO, and Eu–MgO catalysts (69.8%, 65.7%, and 68.2%, respectively, compared to 37.6% for unpromoted MgO). Rane and colleagues[86] investigated the impact of doping CaO catalysts with various alkali metals, including La, Ce, Sm, Nd, and Yb, on their surface properties and catalytic performance. Their findings indicated that apart from the Sm-promoted CaO catalyst, the incorporation of these alkali metals decreased the number of basic sites, especially the strong ones. It significantly lowered the temperature needed to achieve similar conversion levels compared to the unmodified CaO catalyst and enhanced the selectivity toward C_{2+} hydrocarbons. Among these materials, the Nd-doped CaO catalyst (at a Nd/Ca ratio of 0.05) exhibited the highest catalytic efficiency, achieving a CH_4 conversion of 19.5% and a C_{2+} selectivity of 70.8% in the OCM reaction.

Studies on mixed oxides of Mg and Ti were conducted by Jeon et al.[87] The authors demonstrated that catalysts supported on $Mg_{0.5}Ti_{0.5}O_y$ exhibited significantly higher C_2 yields (17.9%, at 800 °C) compared to pure Ti (6.0%) and Mg (6.3%) oxides, attributing this to a weaker binding force with the oxygen surface.

It has been reported that alkali metal and alkaline earth metal halides improve C_2 selectivity, as halogen anions inhibit CH_4 deep oxidation.[50,88]

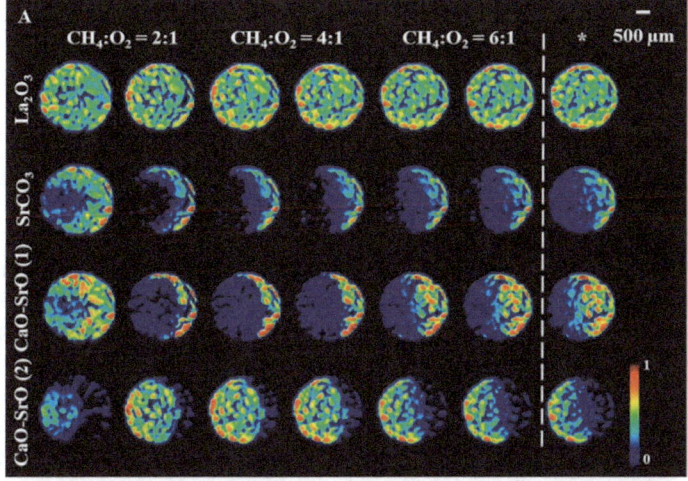

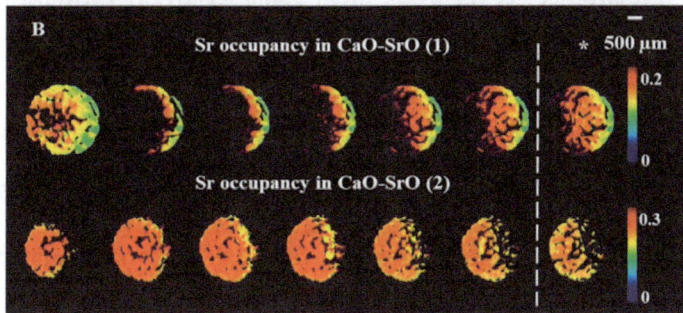

Figure 3.8 (A) Spatial distribution of components during the OCM reaction at a GHSV of 72 000 mL g^{-1} h^{-1}. The * symbol indicates the final measurement conditions with a GHSV of 36 000 mL g^{-1} h^{-1} and a CH$_4$/O$_2$ ratio of 2 : 1. (B) Spatial distribution of Sr occupancy in two CaO–SrO mixed oxides during the OCM reaction (GHSV of 72 000 mL g^{-1} h^{-1}). For each CH$_4$/O$_2$ ratio tested, two consecutive XRD-CT images were acquired, except for the final measurement noted with *. Reproduced from ref. 84, https://doi.org/10.1021/acs.jpcc.8b09018, under the terms of the CC BY 4.0 license, https://creativecommons.org/licenses/by/4.0/deed.en.

In the study by Xi et al.,[50] the influences of various barium halides (BaF$_2$, BaCl$_2$, and BaBr$_2$) as promoters for SnO$_2$ were explored. The inclusion of each barium halide in SnO$_2$ improved the reaction performance by enhancing the number of moderate basic sites and elevating the number of surface vacancies/defects, thereby boosting the concentration of O$_2^-$ sites (and O$_2^{2-}$ in the case of BaBr$_2$ = 1 : 1). The BaBr$_2$ = 1 : 1 catalyst achieved the highest C$_2$ product yield of 18.1% at 800 °C, with a C$_2$H$_4$/C$_2$H$_6$ ratio of 4.

Natural calcium compounds can be used in OCM. For instance, Lima et al.[89] prepared chicken eggshell-derived CaO catalysts. The authors found that the atmosphere in the muffle furnace (N$_2$, air, and static air) significantly influenced the Brunauer–Emmett–Teller (BET) surface area and basicity of the samples. The catalyst calcined under air flow achieved a 30% CH_4 conversion and ethylene selectivity of 53%, at 800 °C. Kwon et al.[48] prepared CaO catalysts from $CaCO_3$-containing waste eggshell and waste cockle shell materials. These catalysts outperformed commercial CaO due to the presence of Na and Mg, which altered their oxygen properties, resulting in the formation of oxygen vacancies and $O_{lat}(e)$ species.

In the literature, among alkaline-earth metals-based catalysts, alkali dopants are the most frequently studied, with Li-doped MgO being of particular interest due to high CH_4 conversion rates at moderately low temperatures.[90] Choudhary et al.[91] studied alkali-metal-promoted MgO catalysts for the OCM at 700 °C and found a relationship between the surface density of basic sites and the C_2 hydrocarbon formation rate per surface area. The Li–MgO, Na–MgO, K–MgO, Rb–MgO, and Cs–MgO catalysts displayed densities of strong basic sites (CO_2 chemisorbed at 500 °C) of 49.2, 17.6, 10.6, 17.3, and 2.5 µmol CO_2 mcat^{-2}, and C_2 hydrocarbon formation rates of 2.800, 0.078, 0.017, 0.027, and 0.016 mmol m^{-2} h^{-1}, respectively, indicating that Li is the most effective promoter.

In one of the most representative studies, Ito and collaborators[40] studied 1%Li/MgO catalysts and obtained 54.7% selectivity to C_2 and 4.4% methane conversion at 650 °C. In comparison, pure MgO showed no C_2 compound formation and achieved only 1% methane conversion. Adding 0.5–1.0 wt% ceria to this catalyst[80] resulted in the formation of surface defects, creating a new pathway for active site formation through electron transfer between Ce^{4+}/Ce^{3+} and Li/MgO. Using 0.1 wt% ceria enhanced CH_4 conversion from 17% to 23% at 750 °C compared to unpromoted Li/MgO, with little change in C_2 selectivity.

The role of Li in this catalyst remains a topic of debate. While early studies suggested that [Li^+O^-] centers are active sites,[92] more recent research indicates that Li acts as a structural modifier.[42] Qian et al.[93] demonstrated that Li restructures the MgO surface, enhancing the surface density of fourfold-coordinated Mg^{2+} sites (Mg_4c^{2+}) on MgO{110} facets, which are more active for CH_4 activation toward methyl radical intermediates while minimizing further dissociation and subsequent combustion reactions.

3.10.2 Rare Earth Oxides

Rare earth oxides represent an important category of materials for OCM because they match with the necessary reactivity prerequisites. These include excellent thermal stability, a significant presence of basic sites, abundant oxygen vacancies, and high oxygen ion mobility. La-,[72,94–98] Ce-,[46,99] and Sm-based catalysts[100] are the most reported. Among these, lanthanum oxide (La_2O_3) stands out as the most extensively studied, because it exhibits balanced reactivity with methyl radicals, thereby allowing CH_3^* to migrate either into the gas phase or across the surface for coupling.[101]

However, a significant drawback of La_2O_3 is its tendency to adsorb CO_2 during reactions, forming strongly bound carbonates, which can poison the catalysts – especially at lower reaction temperatures.[102] In contrast, metal oxides like CeO_2, known for their multiple cationic oxidation states, tend to react extensively with CH_3^*, making them inefficient for radical formation and subsequent coupling reactions.[101] Thus, while La_2O_3 catalysts have high hydrocarbon selectivity, they suffer from low activity, whereas CeO_2 displays the opposite tendency.[95]

However, several studies have shown that doping La into Ce (and *vice versa*) generates a synergistic promoting effect, significantly enhancing catalytic performance. For instance, Zhang et al.[77] demonstrated that introducing La into CeO_2 to form La–Ce mixed oxides creates non-stoichiometric CeO_{2-x}, which is linked to the formation of oxygen vacancies and superoxide ions. Doping CeO_2 structure with Nd_2O_3[103] also brings similar advantages. La–Ce nanofibers[104–106] have also garnered attention due to their ability to minimize temperature gradients and reduce pressure drops in fixed-bed reactors, owing to their high void fractions or bed porosity.[107,108]

Additionally, rare-earth mixed oxide catalysts with $A_2B_2O_7$ have shown excellent performance in OCM, largely due to their high surface basicity and mobile oxygen species.[73,109,110] Notably, the structure of the $A_2B_2O_7$ phase can be modified by tailoring the ionic radii ratio (rA/rB).[111] Specifically, a ratio below 1.46 results in a disordered cubic defective-fluorite phase (space group $Fm\bar{3}m$), while a ratio between 1.46 and 1.78 leads to the formation of an ordered face-centered cubic pyrochlore phase (space group $Fd\bar{3}m$). A ratio above 1.78, on the other hand, results in a monoclinic layered perovskite crystalline phase (space group $P21$).[73] In line with this, Xu et al.[73] synthesized perovskite ($La_2Ti_2O_7$, rA/rB ratio of 1.90), cubic pyrochlore ($La_2Zr_2O_7$, rA/rB ratio of 1.61), and fluorite ($La_2Ce_2O_7$, rA/rB ratio of 1.33) catalysts. Unlike the $La_2Ti_2O_7$ phase, the latter two catalysts possess 8a oxygen vacancies (Figure 3.9). However, due to the random distribution of cations at the A- and B-sites and the resulting disordered oxygen vacancies in the anion sublattices, $La_2Ce_2O_7$ exhibited a greater quantity of mobile oxygen species compared to $La_2Zr_2O_7$, pyrochlore. Consequently, the production of superoxide species and C_2 yield followed the order: $La_2Ce_2O_7 > La_2Zr_2O_7 > La_2Ti_2O_7$. By incorporating different atoms into the A-site of $Ln_2Ce_2O_7$-type catalysts (Ln = La, Pr, Sm, and Y), $La_2Ce_2O_7$ consistently demonstrates the best performance,[112] achieving 58% of C_2 hydrocarbons selectivity with a 31% CH_4 conversion, at 700 °C, making it the most applicable $A_2B_2O_7$ catalyst in OCM. However, the $Sm_2Ce_2O_7$ catalyst showed a significant C_2 yield of 10.2% at 550 °C, making this material highly promising for future applications targeting low-temperature OCM.[112]

Zhang et al.[53] optimized the $La_2Ce_2O_7$ structure by exploring various synthesis methods, including co-precipitation (CP), hydrothermal (HT), and glycine nitrate combustion (GNC). All resulting materials contained superoxide species. However, the sample produced *via* the glycine nitrate combustion method also displayed peroxide species, leading to superior catalytic performance.

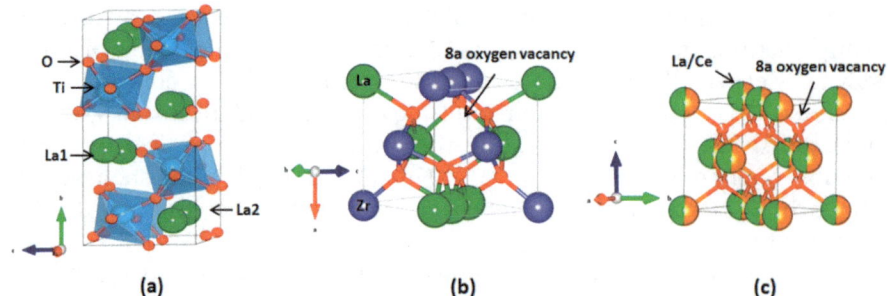

Figure 3.9 Representation of the (a) monoclinic layered perovskite, (b) cubic ordered pyrochlore, and (c) disordered defective cubic fluorite phases. Reproduced from ref. 73 with permission from American Chemical Society, Copyright 2019.

Incorporating low-valent ions, such as alkali or alkaline earth metals, onto the catalyst's structure can enhance its surface basicity and oxygen mobility, thereby enhancing the yield of C_2 compounds. Singh Pal et al.[64] investigated $La_{2-x}M_xCe_2O_{7-\delta}$ (M = Ca, Sr, Ba, and $x = 0.30$) catalysts doped with alkaline earth metals, synthesized *via* a one-step hydrothermal approach for low-temperature OCM. The catalytic performance ranked as follows: $La_{1.7}Ca_{0.3}Ce_2O_{7-\delta}$ (with C_2 yield of 25% at 650 °C) > $La_{1.7}Sr_{0.3}Ce_2O_{7-\delta}$ > $La_{1.7}Ba_{0.3}Ce_2O_{7-\delta}$ > $La_2Ce_2O_7$, which was related with the ratio of reactive oxygen (O_2^- and O_2^{2-}) to lattice oxygen (O^{2-}) and the number of basic sites. Using Raman analysis in a $CH_4 + O_2$ environment, the authors noted that the intensities of peaks associated with superoxide species initially increased as the temperature rose from RT to 600 °C, then decreased with further heating, explaining the reason behind the enhanced OCM activity at 650 °C.

More recently, Wu et al.[113] synthesized macroporous $LaCe_{2-x}Ca_xO_{7-\delta}$ ($A_2B_2O_7$-type) catalysts using a citric acid sol–gel approach. The macroporous structure enhanced the accessibility of reactants (O_2 and CH_4) to the active sites, while the partial substitution of Ce with low-valence Ca ions in the B site led to the creation of surface oxygen vacancies, generating O_2^- species, thereby improving performance compared to the Ca-free $La_2Ce_2O_7$ catalyst. DFT calculations further revealed that the Ca-promoted catalyst exhibited lower oxygen vacancy formation energies than $La_2Ce_2O_7$. Sollier and colleagues[104] fabricated nanofibers of La–Ce and Sr–La–Ce oxides *via* electrospinning, observing that strontium atoms improved CH_4 conversion and C_2 selectivity by creating structural defects that enhanced superoxide species. By incorporating Sr atoms *via* wet impregnation into flower-like La_2O_3 microspheres with hierarchically porous structures, Zhao et al.[75] achieved a C_2 yield of 19.2% at 550 °C. This porous structure allowed the formation of strong interactions between Sr and La_2O_3, which were not observed in the densely structured reference material, resulting in a cooperative effect that enhanced oxygen activation and facilitated methane conversion even at 440 °C.

3.10.3 Perovskites

Perovskite-type oxides are potential catalysts for the OCM reaction because they exhibit good thermal and chemical stability, as well as a suitable structure that allows the creation of electronic defects and oxygen vacancies.[114] Such a class of materials is composed of mixed oxides with a structural formula of ABO_3 (or A_2BO_4). In this structure, the A-site is occupied by a cation with a coordination number of 12, while the B-site is occupied by a cation with a CN of 6. Moreover, multicomponent perovskites can be synthesized by partially substituting the cations at the A and/or B sites, resulting in substituted compounds with the formula $A_{1-x}A'_x B_{1-y}B'_yO_3$, creating electronic defects and oxygen vacancies, which benefits the catalytic performance in OCM.[115] The representation of a partially substituted perovskite structure at the B sites is shown in Figure 3.10. Although most perovskite-type compounds are oxides, some carbides, nitrides, halides, and hydrides also crystallize in this structure.[116–118]

Fung et al.[119] conducted theoretical studies on the effect of A- and B-site cations in perovskite-type oxides for methane activation. Whereas altering the A-site cation has little impact on catalytic performance, it reduces the oxidation state of the B-site cation. The B-site cation, on the other hand, has a strong influence on the reducibility of the catalyst, as the B–O bond is shorter than the A–O bonds, providing a more efficient channel for electron transfer.

Among perovskite-type OCM catalysts, Ti-containing perovskite catalysts have been extensively explored in this reaction. Doping $CaTiO_3$ with Li^+ or Mg^{2+} increases the C_2 yield, related to its p-type conductivity.[59] Ding and coworkers [120] investigated Ti-containing perovskites and alkaline earth metals prepared by two different methods and found that a surface with a high concentration of Ti^{4+} favors full CH_4 oxidation, while a high concentration of alkaline earth metals enhances selectivity toward C_2 compounds. Furthermore, the authors observed that layered perovskites (Sr_2TiO_4 and

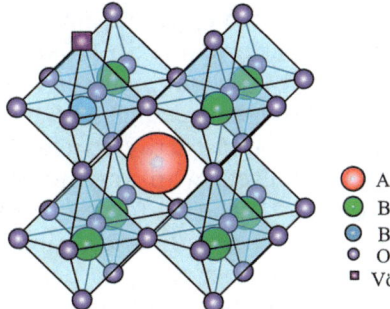

Figure 3.10 The structure of a substituted perovskite oxide with substitution in the B. The purple square represents the oxygen vacancies created by the substitution process. Reproduced from ref. 118 with permission from the Royal Society of Chemistry.

Sr_2SnO_4) exhibit better catalytic performance than their corresponding perovskites ($SrTiO_3$ and $SrSnO_3$) and concluded that O_2^{2-} species are the active sites responsible for the formation of C_2 compounds.

Lim et al.[121] synthesized perovskites such as $CaTiO_3$, $SrTiO_3$, and $BaTiO_3$, in addition to various combinations of CaO, SrO, and BaO oxides deposited on these perovskites, forming complex mixed oxides like Ba–Ca–Ti–O_x, Ba–Sr–Ti–O_x, and Ba_2TiO_4. The interactions between the alkaline-earth metal oxides and the perovskite support enhanced surface basicity, increasing the C_{2+} yields. Although pure perovskite required temperatures above 700 °C to demonstrate significant OCM activity, depositing metal oxides onto these perovskites significantly enhanced their performance below 700 °C. At 650 °C, BaO-deposited perovskites (BaO/$CaTiO_3$ and BaO/$SrTiO_3$) showed the highest C_{2+} yields (10.1% and 10.0%, respectively), whereas the SrO/$BaTiO_3$ catalyst achieved a C_{2+} yield of 17.6% at 725 °C.

Ivanov et al.[114] studied the substitution of $SrTiO_3$ with Mg, Al, Ca, Ba, and Pb, noting an improvement in catalytic activity for OCM, particularly with Mg and Al substitutions. Substitutions in the oxide lattice resulted in a significant increase in selectivity for C_2 compounds, with little change in CH_4 conversion rates. Previous research had demonstrated that $SrTiO_3$ leads to full oxidation of CH_4, resulting in minimal production of C_2 compounds. However, modifications within the oxide lattice significantly increase the selectivity for C_2 compounds, with only minor changes in CH_4 conversion rates.[122]

Doping of $BaSrTiO_3$ catalyst with Mg, Li, and Na was also studied by Fakhroueian et al.,[123] indicating that such doping increased the surface basicity. Kim et al.[56] studied $LaAlO_3$, $LaFeO_3$, and $LaNiO_3$ catalysts and highlighted that changing the B-site cation caused a significant alteration in the electronegativity of the catalytic site. The $LaAlO_3$ catalyst displayed the best catalytic performance, with approximately 25% CH_4 conversion and 40% C_2 selectivity at 800 °C, due to a higher amount of lattice oxygen compared to the other catalysts.

The active site properties of $BaSnO_3$ catalysts promoted with $BaBr_2$ were also studied by Xu et al.[59] Compared to pure $BaSnO_3$, all promoted catalysts exhibited significantly better performance in OCM, especially with an extremely high ethylene selectivity. Several analyses indicated that $BaBr_2$ increases the concentration of oxygen vacancies in $BaSnO_3$, facilitating the conversion of lattice oxygen into more active species for OCM.

A series of Sr/$Sm_2Zr_2O_7$ catalysts with varying Sr concentrations was studied in OCM by Hao et al.[124] The performance of these catalysts was significantly influenced by the synergistic interaction between basicity and increased oxygen vacancies. The best catalytic performance was obtained with 7.6% Sr/$Sm_2Zr_2O_7$, showing 39.1% methane conversion and 47% C_2 selectivity in a reaction carried out at 750 °C.

Lopes et al.[118] investigated the influence of partially substituting Ti with Mg in the $La_2Ti_2O_7$ perovskite structure. The introduction of Mg led to the formation of oxygen vacancies and enhanced the basicity of the catalysts.

Furthermore, they identified an inverse linear correlation between the M–O bond force constants (k) and the selectivity for C_2 compounds, which increased with higher Mg content. The reduction in bond force constant upon Mg incorporation was interpreted as indicative of a greater ionic character in the metal–oxygen bond, with the electronic density shifting toward oxygen, the more electronegative atom, thereby increasing basicity.

Sim and collaborators[125] studied $LaAlO_3$ oxides synthesized under various calcination conditions, such as O_2 fraction, time, and temperature. As calcination time and temperature increased, so did the catalytic activity of $LaAlO_3$, which exhibited good homogeneity and high crystallinity. The $LaAlO_3$ calcined at 1350 °C for 24 h showed the highest C_2 selectivity and yield. According to the authors, the formation of active sites selective for C_2 production is a greater determinant of catalyst performance than the increase in the number and exposure of these sites with the increase in specific surface area. The same group[126] also studied 10 different perovskite compositions in OCM, revealing that the lattice oxygen species play a crucial role in the selective conversion of methane. Specifically, species with moderate binding energies are more selective in OCM for producing C_2 hydrocarbons. Among the tested catalysts, $CaZrO_3$ exhibited the highest C_2 yield, reaching 14.2%.

3.10.4 Mn–Na–W–SiO$_2$ Catalyst

The Mn–Na–W–SiO$_2$ catalyst stands as the state-of-the-art catalyst in OCM due to its remarkable C_2 yield, reaching up to 30%.[36] Typically composed of 0.5–3 wt% Mn, 0.4–2.3 wt% Na, and 2.2–8.9 wt% W – and supported on amorphous SiO_2 – this catalyst undergoes a phase transition to cristobalite upon calcination at around 800 °C. Consequently, the catalyst comprises a mixture of phases, including α-cristobalite, Na_2WO_4, $MnWO_4$, and Mn_2O_3 as major components, with minor phases such as tridymite or quartz, braunite ($MnMn_6SiO_{12}$), Na_4WO_5, and $Na_2W_2O_7$.[127–130]

Each of these components plays a significant role in the OCM reaction. Manganese, particularly in its Mn^{2+} and Mn^{3+} oxidation states, serves as a crucial oxygen source for the reaction and enhances the dispersion of the Na_2WO_4 phase.[131] Dissanayake et al.[132] demonstrated that the specific activity for CH_4 conversion increases linearly with surface Mn concentration, indicating that Mn is responsible for O_2 activation. Fleischer et al.[129] supported this by conducting chemical looping experiments, revealing a Mn^{3+}/Mn^{2+} redox cycle where Mn contributes to oxygen storage, while the tungstate phase enhances the selective activation of methane. During the redox cycle, methane oxidation takes place on the catalyst surface, using lattice oxygen from Mn^{3+}-containing manganese oxide to form C_2 compounds. Subsequently, molecular oxygen from the reaction environment is used to reoxidize the Mn^{2+}-containing manganese oxide catalyst, thereby restoring the cycle.[133]

Concerning the SiO_2 role as a support, Palermo et al.[134,135] reported that the transition from amorphous silica to highly crystalline α-cristobalite is critical for generating ethylene. This phase transition is linked with a noteworthy decrease in BET surface area and the formation of a support inert in OCM. Sodium is essential in lowering the phase transition temperature far below the typical value of 1500 °C, increasing surface basicity, and stabilizing W-oxo species in the appropriate state.[134,135] Ji et al.[127] modeled the transition complex of WO_4 tetrahedron with methane, identifying WO_4 as a key active site due to its proper geometric and energy match with CH_4. They also showed that WO_4 is stabilized in Na- and K-W-Mn/SiO_2 catalysts, resulting in α-cristobalite SiO_2, but not in Li- or Ba-W-Mn/SiO_2, which instead form quartz SiO_2 or remain as amorphous silica.[127]

Although the Mn_2O_3-Na_2WO_4/SiO_2 catalyst is well-established in the literature, ongoing research explores the application of (1) innovative synthesis methods, (2) alternative supports, and (3) the use of promoters to further enhance performance. Zheng and coworkers[136] synthesized Na-W-Mn-Zr-S-P/SiO_2 catalysts using three different approaches: incipient wetness impregnation, sol–gel, and mixture slurry. Among these, the mixture slurry method resulted in a notable enrichment of the MnO_x phase, leading to the highest catalytic performance observed across the synthesis techniques. Ghose et al.[137] achieved one of the highest yields reported in the literature – 27% for C_2 compounds – using Na_2WO_4-Mn/SiO_2 catalysts synthesized via solution combustion synthesis, incorporating 5% La as a promoter and TEOS as the silicon precursor.

Park et al.[131] investigated a core–shell SiO_2@MnO_x@Na_2WO_4@SiO_2 catalyst, where SiO_2 served both as a core and an outer shell, immobilizing Na_2WO_4 and MnO_x (Figure 3.11). This design enhanced the dispersion of the active phases. By using different Mn precursors, they observed that the catalyst synthesized with $Mn(CH_3CO_2)_2 \cdot 4H_2O$ maintained the core–shell structure but resulted in unsatisfactory catalytic performance. Conversely, the catalyst synthesized using $KMnO_4$ led to the formation of α-cristobalite

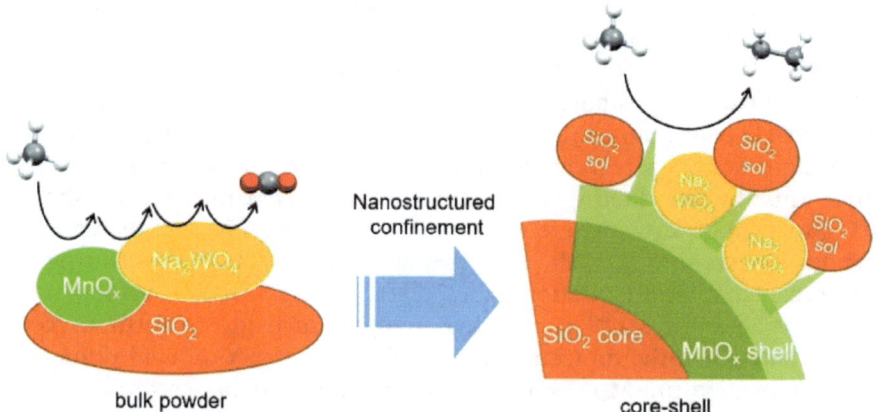

Figure 3.11 Methane activation using core–shell catalysts. Reproduced from ref. 131 with permission from the Royal Society of Chemistry.

SiO_2, which played a stabilizing role in the structure, significantly improving the catalytic results. They also noted that in this configuration, Na permeated the SiO_2 outer shell, facilitating the nucleation of cristobalite and enhancing the interaction between Na_2WO_4 and cristobalite phases.

Researchers have also explored alternative supports, such as SBA-15,[130] COK-12,[129] TiO_2,[138,139] and SiC.[140] Liu[141] employed SiC as a support, noting a significant enrichment of the $MnWO_4$ phase compared to the Mn_2O_3 phase, in contrast to what was observed with silica-supported catalysts. This enhancement was attributed to the reduced surface oxygen availability on the SiC-supported catalysts. Furthermore, the superior thermal conductivity of SiC minimized the formation of hotspots within the catalytic bed. Yildiz et al.[142] conducted a screening of various supports, including La_2O_3, CaO, Al_2O_3, ZrO_2, SiC, MgO, Fe_2O_3, Fe_3O_4, SrO, TiO_2-rutile, and TiO_2-anatase, as candidates to replace SiO_2. Among these, CaO, La_2O_3, SiC, Fe_2O_3, and TiO_2-rutile exhibited catalytic performances comparable to the reference Mn_2O_3–Na_2WO_4/SiO_2 catalyst. However, La_2O_3 and CaO also showed high activity as pure support materials, making it challenging to distinguish between the individual roles of the support and the active phase.

Among Si-based supports, SBA-15 has shown particularly promising results. Yildiz et al.[130] demonstrated that using SBA-15 as the support precursor significantly enhanced catalytic efficiency, achieving a 14% conversion rate with C_2 selectivity exceeding 60%, resulting in a C_2 yield approximately two-fold higher than that of the reference Mn_2O_3–Na_2WO_4/SiO_2 catalyst, at 750 °C. This improvement was attributed to better dispersion of the Mn_xO_y–Na_2WO_4 precursors within the small pores and high surface area of SBA-15, even though the mesostructure of SBA-15 collapsed after thermal treatment. Chukeaw et al.[128] reported similar findings, noting that, compared to fumed-SiO_2 and MCM-41, the Na_2WO_4–$MnTiO_3$ catalyst supported on SBA-15 provided better dispersion of the active phases, enhancing the C_2 yield.

Wang et al.[138,139] demonstrated that the TiO_2-doped Mn_2O_3–$NaWO_4$/SiO_2 catalyst exhibits good performance for OCM at moderately low temperatures. They reported the presence of a chemical cycling between $MnTiO_3$ and Mn_2O_3 that enhanced O_2 activation, thereby improving both the activity and selectivity. Conversely, Kim et al.[38] reported that using TiO_2 as a support increased the enrichment of oxygen species and the proportion of Mn^{2+} and Mn^{3+} species. Moreover, methane conversion increased linearly with the ratio $(Mn^{2+} + Mn^{3+})/Mn_{total}$, highlighting the synergism between Mn^{2+} and Mn^{3+} species in providing oxygen species.

The investigation into the use of promoters has also yielded significant findings. For instance, Zheng et al.[143] investigated Na–W–Mn–Zr/SiO_2 catalysts enhanced with S and P. They discovered that both P and S promoted the migration of key active components, particularly Mn, W, and Zr, to the catalyst surface, and also significantly enhanced the oxygen storage capacity.

Cheng et al.[68] synthesized TiO_2-supported Li–Mn/WO_3/TiO_2 catalysts promoted with La_2O_3 using the wetness impregnation method. By varying the concentration of La_2O_3 from 0 to 7 wt%, they found that methane conversion increased with La content, achieving maximum selectivity for C_2 compounds with the 5% La_2O_3–Li–Mn/WO_3/TiO_2 catalyst. This increase was linked to enhanced oxygen mobility and basic site concentration, as well as a higher fraction of Mn^{2+} and Mn^{3+} species compared to Mn^{4+}, which delayed the anatase-to-rutile transition. While doping increased the BET surface area, the presence of anatase promoted the formation of C_2 compounds over CO_x species. As a result, the catalyst with 7% La_2O_3–Li–Mn/WO_3/TiO_2 exhibited the lowest C_2 selectivity among all the samples tested.

Shahri et al.[144] demonstrated that incorporating Ce as a promoter in Mn/Na_2WO_4/SiO_2 catalysts improved BET area and Na_2WO_4 dispersion, thereby enhancing activity and stability. However, exceeding 5 wt% Ce led to the formation of cerium oxide, which reduced the interaction between active species. Similar results were obtained by Jodaian et al.[37] The authors used different amounts of Ce (from 0 to 8 wt%) in Na_2WO_4/TiO_2 catalysts. They noticed the BET area increased with Ce content, reaching a maximum for the catalyst containing 3 wt% of Ce.

3.11 Conclusion

The oxidative coupling of methane (OCM) presents a promising route for directly converting methane, one of the most harmful greenhouse gases, into value-added hydrocarbons such as ethylene and ethane, which are vital building blocks in the chemical sector. Additionally, this process contributes to climate change mitigation. However, this reaction faces significant thermodynamic and kinetic challenges that have delayed its industrial application. Consequently, over the past four decades, the scientific community has devoted substantial effort to designing catalysts that deliver high yields of these valuable compounds with long-term stability. This chapter highlights recent advancements in OCM, focusing on mechanistic insights, and catalyst developments to enhance C_2 yield, catalysts' durability, and overall viability for industrial application.

To maximize the production of C_2 compounds, catalysts must possess active sites capable of homolytically cleaving methane's C–H bond, thereby generating methyl radicals, which can subsequently be coupled to form C_2 hydrocarbons. However, non-catalytic homogeneous reactions in the gas phase involving molecular oxygen also occur, leading to the formation of CO and CO_2 – which are thermodynamically more stable products – thereby reducing the selectivity toward the desired compounds.

Current literature indicates that effective OCM catalysts must combine high stability with a high concentration of basic sites and electrophilic oxygen species to boost the formation of C_2 compounds. However, there is ongoing debate regarding the roles of adsorbed oxygen species at structural defects versus surface lattice oxygen species. While some studies suggest that the former promotes the formation of C_2 hydrocarbons and the latter favors deep methane oxidation, others report the opposite tendency. Therefore, further investigation is required to clarify the significance of each oxygen species, which could guide the synthesis of more active and selective catalysts. Quantum mechanical calculations involving Density Functional Theory have proven crucial in elucidating the reaction mechanisms and the role of active sites in OCM.

Materials such as alkaline earth metal oxides, rare earth metal oxides, perovskites, and Mn/Na$_2$WO$_4$/SiO$_2$ catalysts have been extensively studied, with the latter achieving the highest C$_2$ yields, reaching up to 30%. However, these yields are typically attained only at elevated reaction temperatures, around 800 °C, which drives operational costs. In contrast, simpler structures like La$_2$O$_3$-based catalysts operate at lower temperatures but produce significantly lower yields. Thus, achieving high productivity at temperatures below 700 °C remains a challenge for all catalyst classes. Overcoming these challenges and enhancing catalyst performance will necessitate customized strategies for each group of catalysts.

To enhance the concentration of basic sites and oxygen species, doping catalysts with additional elements to enhance the number of oxygen vacancies has become a widely adopted strategy. Moreover, ongoing research efforts are increasingly focusing on low-temperature OCM. In this regard, understanding the relationship between each active species with the reaction temperature is essential. Moreover, membrane catalysts have also been explored to mitigate the formation of hot spots during the reaction. Another important area of research involves replacing O$_2$ with mild oxidants, such as CO$_2$, N$_2$O, or S$_2$, to improve C$_2$ selectivity by minimizing overoxidation processes. Additionally, statistical optimization tools can be employed to determine the ideal reaction conditions for each catalyst system, further improving the efficiency and selectivity of OCM processes.

Acknowledgements

The authors acknowledge the financial support provided by the Fundação de Amparo à Pesquisa do Estado de São Paulo (#2023/10582-9, #2021/09394-8, #2022/10615-1, #2022/06419-2, #2023/05285-5) and Conselho Nacional de Desenvolvimento Científico e Tecnológico (#127633/2022-7).

References

1. W. M. Haynes, *Handbook of Chemistry and Physics*, CRC Press, Boca Raton, 26th edn, 1942.
2. M. A. K. Khalil and M. J. Shearer, *Atmospheric Methane*, 2000, **1**, 98.
3. D. S. Reay, P. Smith, T. R. Christensen, R. H. James and H. Clark, *Annu. Rev. Environ. Resour.*, 2018, **43**, 165.
4. A. van Amstel, *J. Integr. Environ. Sci.*, 2012, **9**, 5.
5. D. J. Wuebbles and K. Hayhoe, *Earth-Sci. Rev.*, 2002, **57**, 177.
6. X. Lan, K. W. Thoning and E. J. Dlugokencky, Trends in globally-averaged CH4, N2O, and SF6 determined from NOAA Global Monitoring Laboratory measurements. Version 2024-08.

7. E. G. Nisbet, R. E. Fisher, D. Lowry, J. L. France, G. Allen, S. Bakkaloglu, T. J. Broderick, M. Cain, M. Coleman, J. Fernandez, G. Forster, P. T. Griffiths, C. P. Iverach, B. F. J. Kelly, M. R. Manning, P. B. R. Nisbet-Jones, J. A. Pyle, A. Townsend-Small, A. al-Shalaan, N. Warwick and G. Zazzeri, *Rev. Geophys.*, 2020, **58**, 1.
8. E. V. Kondratenko, T. Peppel, D. Seeburg, V. A. Kondratenko, N. Kalevaru, A. Martin and S. Wohlrab, *Catal. Sci. Technol.*, 2017, **7**, 366.
9. R. Horn and R. Schlögl, *Catal. Lett.*, 2015, **145**, 23.
10. C. Hammond, S. Conrad and I. Hermans, *ChemSusChem*, 2012, **5**, 1668.
11. C. Mesters, *Annu. Rev. Chem. Biomol. Eng.*, 2016, **7**, 223.
12. K. Takanabe, *J. Jpn. Pet. Inst.*, 2012, **55**, 1.
13. P. Schwach, X. Pan and X. Bao, *Chem. Rev.*, 2017, **117**, 8497.
14. A. I. Olivos-Suarez, À. Szécsényi, E. J. M. Hensen, J. Ruiz-Martinez, E. A. Pidko and J. Gascon, *ACS Catal.*, 2016, **6**, 2965.
15. C. A. Ortiz-Bravo, C. A. Chagas and F. S. Toniolo, *J. Nat. Gas Sci. Eng.*, 2021, **96**, 104254.
16. J. H. Lunsford, *Catal. Today*, 2000, **63**, 165.
17. F. C. Muniz, PhD thesis, Universidade Federal do Rio de Janeiro, 2007.
18. B. L. Farrell, V. O. Igenegbai and S. Linic, *ACS Catal.*, 2016, **6**, 4340.
19. G. E. Keller and M. M. Bhasin, *J. Catal.*, 1982, **73**, 9.
20. Y. Gao, L. Neal, D. Ding, W. Wu, C. Baroi, A. M. Gaffney and F. Li, *ACS Catal.*, 2019, **9**, 8592.
21. C. Karakaya, H. Zhu, B. Zohour, S. Senkan and R. J. Kee, *ChemCatChem*, 2017, **9**, 4538.
22. A. Galadima and O. Muraza, *J. Ind. Eng. Chem.*, 2016, **37**, 1.
23. B. Beck, V. Fleischer, S. Arndt, M. G. Hevia, A. Urakawa, P. Hugo and R. Schomäcker, *Catal. Today*, 2014, **228**, 212.
24. A. M. Arinaga, M. C. Ziegelski and T. J. Marks, *Angew. Chem., Int. Ed.*, 2021, **60**, 10502.
25. S. Dooley, M. P. Burke, M. Chaos, Y. Stein, F. L. Dryer, V. P. Zhukov, O. Finch, J. M. Simmie and H. J. Curran, *Int. J. Chem. Kinet.*, 2010, **42**, 527.
26. Z. Stansch, L. Mleczko and M. Baerns, *Ind. Eng. Chem. Res.*, 1997, **36**, 2568.
27. A. I. Olivos-Suarez, À. Szécsényi, E. J. M. Hensen, J. Ruiz-Martinez, E. A. Pidko and J. Gascon, *ACS Catal.*, 2016, **6**, 2965.
28. V. R. Choudhary and B. S. Uphade, *Catal. Surv. Asia*, 2004, **8**, 15.
29. K. Suzuki, T. Toyao, Z. Maeno, S. Takakusagi, K. Shimizu and I. Takigawa, *ChemCatChem*, 2019, **11**, 4537.
30. P. Ciambelli, L. Lisi, R. Pirone, G. Ruoppolo and G. Russo, *Catal. Today*, 2000, **61**, 317.
31. M. C. Breitkreitz, A. M. de Souza and R. J. Poppi, *Quim. Nova*, 2014, **37**, 564.
32. W. Schakel, C. Fernández-Dacosta, M. Van Der Spek and A. Ramírez, *J. CO_2 Util.*, 2017, **22**, 278.
33. K. Langfeld, B. Frank, V. E. Strempel, C. Berger-Karin, G. Weinberg, E. V. Kondratenko and R. Schomäcker, *Appl. Catal., A*, 2012 **417–418**, 145.

34. A. C. Ferreira, T. A. Gasche, J. P. Leal and J. B. Branco, *Mol. Catal.*, 2017, **443**, 155.
35. H. Yamamoto, H. Y. Chu, M. T. Xu, C. L. Shi and J. H. Lunsford, *J. Catal.*, 1993, **142**, 325.
36. J. Liu, J. Yue, M. Lv, F. Wang, Y. Cui, Z. Zhang and G. Xu, *Carbon Resour. Convers.*, 2022, **5**, 1.
37. V. Jodaian and M. Mirzaei, *Inorg. Chem. Commun.*, 2019, **100**, 97.
38. G. J. Kim, J. T. Ausenbaugh and H. T. Hwang, *Ind. Eng. Chem. Res.*, 2021, **60**, 3914.
39. S. Arndt, G. Laugel, S. Levchenko, R. Horn, M. Baerns, M. Scheffler, R. Schlögl and R. Schomäcker, *Catal. Rev.:Sci. Eng.*, 2011, **53**, 424.
40. T. Ito, J. Wang, C. H. Lin and J. H. Lunsford, *J. Am. Chem. Soc.*, 1985, **107**, 5062.
41. P. Myrach, N. Nilius, S. V. Levchenko, A. Gonchar, T. Risse, K. P. Dinse, L. A. Boatner, W. Frandsen, R. Horn, H. J. Freund, R. Schlögl and M. Scheffler, *ChemCatChem*, 2010, **2**, 854.
42. L. Luo, Y. Jin, H. Pan, X. Zheng, L. Wu, R. You and W. Huang, *J. Catal.*, 2017, **346**, 57.
43. J. Xiong, P. Zhang, C. Xie, Q. Lian, T. Wu, D. Han, Y. Yang, Z. Zhao and Y. Wei, *AIChE J.*, 2024, **70**, 1.
44. J. Song, Y. Ren, X. Gao, X. Fan, B. Liu and Z. Zhao, *ACS Catal.*, 2024, **14**, 5116.
45. K.-I. Aika and K. Aono, *J. Chem. Soc., Faraday Trans.*, 1991, **87**, 1273.
46. G. I. Siakavelas, N. D. Charisiou, A. AlKhoori, V. Sebastian, S. J. Hinder, M. A. Baker, I. V. Yentekakis, K. Polychronopoulou and M. A. Goula, *J. Environ. Chem. Eng.*, 2022, **10**, 107259.
47. V. I. Lomonosov and M. Y. Sinev, *Kinet. Catal.*, 2016, **57**, 647.
48. D. Kwon, I. Yang, J. H. Cho and J. C. Jung, *Mol. Catal.*, 2021, **516**, 111982.
49. Y. Gambo, A. A. Jalil, S. Triwahyono and A. A. Abdulrasheed, *J. Ind. Eng. Chem.*, 2018, **59**, 218.
50. R. Xi, J. Xu, Y. Zhang, Z. Zhang, X. Xu, X. Fang and X. Wang, *Catal. Today*, 2021, **364**, 35.
51. Y. Gong, C. Lu, X. Zhong, J. Xu, R. Ouyang, X. Fang, X. Xu, C. Deng and X. Wang, *Eur. J. Inorg. Chem.*, 2024, **27**, 1.
52. V. J. Ferreira, P. Tavares, J. L. Figueiredo and J. L. Faria, *Catal. Commun.*, 2013, **42**, 50.
53. Y. Zhang, J. Xu, X. Xu, R. Xi, Y. Liu, X. Fang and X. Wang, *Catal. Today*, 2020, **355**, 518.
54. Y. Osada, S. Koike, T. Fukushima, S. Ogasawara, T. Shikada and T. Ikariya, *Appl. Catal.*, 1990, **59**, 59.
55. T. L. Yang, L. B. Feng and S. K. Shen, *J. Catal.*, 1994, **145**, 384.
56. I. Kim, G. Lee, H. B. in Na, J. M. Ha and J. C. Jung, *Mol. Catal.*, 2017, **435**, 13.
57. G. Kumar, S. L. J. Lau, M. D. Krcha and M. J. Janik, *ACS Catal.*, 2016, **6**, 1812.
58. W. Sun, Y. Gao, G. Zhao, J. Si, Y. Liu and Y. Lu, *J. Catal.*, 2021, **400**, 372.

59. J. Xu, R. Xi, Z. Zhang, Y. Zhang, X. Xu, X. Fang and X. Wang, *Catal. Today*, 2021, **374**, 29.
60. Z. Zhang, Y. Gong, J. Xu, Y. Zhang, Q. Xiao, R. Xi, X. Xu, X. Fang and X. Wang, *Catal. Today*, 2022, **400–401**, 73.
61. A. Pandey, G. Jain, D. Vyas, S. Irusta and S. Sharma, *J. Phys. Chem. C*, 2017, **121**, 481.
62. P. Li, X. Chen, Y. Li and J. W. Schwank, *Catal. Today*, 2019, **327**, 90.
63. W. Song, A. S. Poyraz, Y. Meng, Z. Ren, S. Y. Chen and S. L. Suib, *Chem. Mater.*, 2014, **26**, 4629.
64. R. Singh Pal, S. Rana, S. Kumar Sharma, R. Khatun, D. Khurana, T. Suvra Khan, M. Kumar Poddar, R. Sharma and R. Bal, *Chem. Eng. J.*, 2023, **458**, 141379.
65. R. Feng, P. Niu, B. Hou, Q. Wang, L. Jia, M. Lin and D. Li, *J. Energy Chem.*, 2022, **67**, 342.
66. R. Feng, P. Niu, Q. Wang, B. Hou, L. Jia, M. Lin and D. Li, *Fuel*, 2022, **308**, 121848.
67. D. D. Petrolini, F. F. C. Marcos, J. M. Assaf and E. M. Assaf, *Chem. Eng. J. Adv.*, 2021, **7**, 100119.
68. F. Cheng, J. Yang, L. Yan, J. Zhao, H. Zhao, H. Song and L. J. Chou, *J. Rare Earths*, 2020, **38**, 167.
69. H. Idriss, *Surf. Sci.*, 2021, **712**, 2.
70. T. J. Frankcombe and Y. Liu, *Chem. Mater.*, 2023, **35**, 5468.
71. F. Papa, P. Luminita, P. Osiceanu, R. Birjega, M. Akane and I. Balint, *J. Mol. Catal. A: Chem.*, 2011, **346**, 46.
72. P. Huang, Y. Zhao, J. Zhang, Y. Zhu and Y. Sun, *Nanoscale*, 2013, **5**, 10844.
73. J. Xu, Y. Zhang, X. Xu, X. Fang, R. Xi, Y. Liu, R. Zheng and X. Wang, *ACS Catal.*, 2019, **9**, 4030.
74. J. Xu, R. Xi, Q. Xiao, X. Xu, L. Liu, S. Li, Y. Gong, Z. Zhang, X. Fang and X. Wang, *J. Catal.*, 2022, **408**, 465.
75. M. Zhao, S. Ke, H. Wu, W. Xia and H. Wan, *Ind. Eng. Chem. Res.*, 2019, **58**, 22847.
76. Y. Gordienko, T. Usmanov, V. Bychkov, V. Lomonosov, Z. Fattakhova, Y. Tulenin, D. Shashkin and M. Sinev, *Catal. Today*, 2016, **278**, 127.
77. B. Zhang, D. Li and X. Wang, *Catal. Today*, 2010, **158**, 348.
78. G. L. Catuzo, Y. L. De Lima, D. D. Petrolini and E. M. Assaf, *Catal. Today*, 2025, **444**, 114994.
79. J. Xu, R. Ouyang, X. Zhong, X. Fang, X. Xu and X. Wang, *Mol. Catal.*, 2023, **547**, 113386.
80. L. Tang, D. Yamaguchi, L. Wong, N. Burke and K. Chiang, *Catal. Today*, 2011, **178**, 172.
81. F. Papa, P. Luminita, P. Osiceanu, R. Birjega, M. Akane and I. Balint, *J. Mol. Catal. A: Chem.*, 2011, **346**, 46.
82. A. M. Maitra, I. Campbell and R. J. Tyler, *Appl. Catal., A*, 1992, **85**, 27.
83. J. Cho, D. Kwon, I. Yang, S. An and J. C. Jung, *Mol. Catal.*, 2021, **510**, 111677.

84. D. Matras, S. D. M. Jacques, S. Poulston, N. Grosjean, C. Estruch Bosch, B. Rollins, J. Wright, M. Di Michiel, A. Vamvakeros, R. J. Cernik and A. M. Beale, *J. Phys. Chem. C*, 2019, **123**, 1751.
85. V. R. Choudhary, V. H. Rane and S. T. Chaudhari, *Appl. Catal., A*, 1997, **158**, 121.
86. V. H. Rane, S. T. Chaudhari and V. R. Choudhary, *J. Chem. Technol. Biotechnol.*, 2006, **81**, 208.
87. W. Jeon, J. Y. Lee, M. Lee, J.-W. Choi, J.-M. Ha, D. J. Suh and I. W. Kim, *Appl. Catal., A*, 2013, **464–465**, 68.
88. H. Wan, Z. Chao, W. Weng, X. Zhou, J. Cai and K. Tsai, *Catal. Today*, 1996, **30**, 67.
89. D. S. Lima and O. W. Perez-Lopez, *Inorg. Chem. Commun.*, 2020, **116**, 107928.
90. T. Matsumoto, M. Saito, S. Ishikawa, K. Fujii, M. Yashima, W. Ueda and T. Motohashi, *ChemCatChem*, 2020, **12**, 1968.
91. V. R. Choudhary, V. H. Rane and M. Y. Pandit, *J. Chem. Technol. Biotechnol.*, 1997, **68**, 177.
92. X. Peng, D. A. Richards and P. C. Stair, *J. Catal.*, 1990, **121**, 99.
93. K. Qian, R. You, Y. Guan, W. Wen, Y. Tian, Y. Pan and W. Huang, *ACS Catal.*, 2020, **10**, 15142.
94. T. Jiang, J. Song, M. Huo, N. T. Yang, J. Liu, J. Zhang, Y. Sun and Y. Zhu, *RSC Adv.*, 2016, **6**, 34872.
95. Z. Q. Wang, D. Wang and X. Q. Gong, *ACS Catal.*, 2020, **10**, 586.
96. C. Chu, Y. Zhao, S. Li and Y. Sun, *Phys. Chem. Chem. Phys.*, 2016, **18**, 16509.
97. S. Wang, S. Li and D. A. Dixon, *Catal. Sci. Technol.*, 2020, **10**, 2602.
98. R. C. Schucker, K. J. Derrickson, A. K. Ali and N. J. Caton, *Appl. Catal., A*, 2020, **607**, 117827.
99. V. J. Ferreira, P. Tavares, J. L. Figueiredo and J. L. Faria, *Ind. Eng. Chem. Res.*, 2012, **51**, 10535.
100. H. Özdemir, *ChemistrySelect*, 2021, **6**, 7999.
101. A. C. Chien, I. Z. Xie and C. H. Yeh, *Mol. Catal.*, 2023, **538**, 112974.
102. C. Guan, Y. Yang, Y. Pang, Z. Liu, S. Li, E. I. Vovk, X. Zhou, J. P. H. Li, J. Zhang, N. Yu, L. Long, J. Hao and A. P. van Bavel, *J. Catal.*, 2021, **396**, 202.
103. Y. Gong, C. Lu, X. Zhong, J. Xu, R. Ouyang, X. Fang, X. Xu, C. Deng and X. Wang, *Eur. J. Inorg. Chem.*, 2024, **27**, 1.
104. B. M. Sollier, M. Bonne, N. Khenoussi, L. Michelin, E. E. Miró, L. E. Gómez, A. V. Boix and B. Lebeau, *Ind. Eng. Chem. Res.*, 2020, **59**, 11419.
105. D. Noon, A. Seubsai and S. Senkan, *ChemCatChem*, 2013, **5**, 146.
106. D. Noon, B. Zohour and S. Senkan, *J. Nat. Gas Sci. Eng.*, 2014, **18**, 406.
107. J. J. Ternero-Hidalgo, J. Torres-Liñán, M. O. Guerrero-Pérez, J. Rodríguez-Mirasol and T. Cordero, *Catal. Today*, 2019, **325**, 131.
108. E. Reichelt, M. P. Heddrich, M. Jahn and A. Michaelis, *Appl. Catal., A*, 2014, **476**, 78.
109. X. Xu, F. Liu, J. Tian, H. Peng, W. Liu, X. Fang, N. Zhang and X. Wang, *Chem. Phys. Chem.*, 2017, **18**, 1533.

110. C. Wang, Y. Wang, A. Zhang, Y. Cheng, F. Chi and Z. Yu, *J. Mater. Sci.*, 2013, **48**, 8133.
111. M. Lang, F. Zhang, J. Zhang, J. Wang, J. Lian, W. J. Weber, B. Schuster, C. Trautmann, R. Neumann and R. C. Ewing, *Nucl. Instrum. Methods Phys. Res., Sect. B*, 2010, **268**, 2951.
112. J. Xu, L. Peng, X. Fang, Z. Fu, W. Liu, X. Xu, H. Peng, R. Zheng and X. Wang, *Appl. Catal., A*, 2018, **552**, 117.
113. T. Wu, P. Zhang, Y. Wei, J. Xiong, D. Han, T. Li, Y. Yang, Z. Zhao and J. Liu, *ACS Catal.*, 2024, **14**, 1882.
114. D. V. Ivanov, L. A. Isupova, E. Y. Gerasimov, L. S. Dovlitova, T. S. Glazneva and I. P. Prosvirin, *Appl. Catal., A*, 2014, **485**, 10.
115. S. Lim, J. W. Choi, D. J. Suh, K. H. Song, H. C. Ham and J. M. Ha, *J. Catal.*, 2019, **375**, 478.
116. M. A. Peña and J. L. G. Fierro, *Chem. Rev.*, 2001, **101**, 1981.
117. J. Zhu, H. Li, L. Zhong, P. Xiao, X. Xu, X. Yang, Z. Zhao and J. Li, *ACS Catal.*, 2014, **4**, 2917.
118. L. B. Lopes, L. H. Vieira, J. M. Assaf and E. M. Assaf, *Catal. Sci. Technol.*, 2021, **11**, 283.
119. V. Fung, F. Polo-Garzon, Z. Wu and D. Jiang, *Catal. Sci. Technol.*, 2018, **8**, 702.
120. W. Ding, Y. Chen and X. Fu, *Appl. Catal., A*, 1993, **104**, 61.
121. S. Lim, J. W. Choi, D. Jin Suh, U. Lee, K. H. Song and J. M. Ha, *Catal. Today*, 2020, **352**, 127.
122. C. Yu, W. Li, W. Feng, A. Qi and Y. Chen, *Stud. Surf. Sci. Catal.*, 1993, **75**, 1119.
123. Z. Fakhroueian, F. Farzaneh and N. Afrookhteh, *Fuel*, 2008, **87**, 2512.
124. J. Hao, F. Cai, J. Wang, Y. Fu, J. Zhang and Y. Sun, *Chem. Phys. Lett.*, 2021, **771**, 138562.
125. Y. Sim, I. Yang, D. Kwon, J. M. Ha and J. C. Jung, *Catal. Today*, 2020, **352**, 134.
126. Y. Sim, D. Kwon, S. An, J. M. Ha, T. S. Oh and J. C. Jung, *Mol. Catal.*, 2020, **489**, 110925.
127. S. Ji, T. Xiao, S. Li, L. Chou, B. Zhang, C. Xu, R. Hou, A. P. E. York and M. L. H. Green, *J. Catal.*, 2003, **220**, 47.
128. T. Chukeaw, W. Tiyatha, K. Jaroenpanon, T. Witoon, P. Kongkachuichay, M. Chareonpanich, K. Faungnawakij, N. Yigit, G. Rupprechter and A. Seubsai, *Process Saf. Environ. Prot.*, 2021, **148**, 1110.
129. V. Fleischer, U. Simon, S. Parishan, M. G. Colmenares, O. Görke, A. Gurlo, W. Riedel, L. Thum, J. Schmidt, T. Risse, K. P. Dinse and R. Schomäcker, *J. Catal.*, 2018, **360**, 102.
130. M. Yildiz, Y. Aksu, U. Simon, K. Kailasam, O. Goerke, F. Rosowski, R. Schomacker, A. Thomas and S. Arndt, *Chem. Commun.*, 2014, **50**, 14440.
131. L. H. Park, Y. R. Jo, J. W. Choi, D. J. Suh, K. H. Song and J. M. Ha, *RSC Adv.*, 2020, **10**, 37749.
132. D. Dissanayake, J. H. Lunsford and M. P. Rosynek, 1993, **143**, 286.

133. S. Damasceno, F. J. Trindade, F. C. Fonseca, D. Z. de Florio and A. S. Ferlauto, *Fuel Process. Technol.*, 2022, **231**, 107255.
134. A. Palermo, J. P. H. Vazquez, A. F. Lee, M. S. Tikhov and R. M. Lambert, *J. Catal.*, 1998, **177**, 259.
135. A. Palermo, J. P. Holgado Vazquez and R. M. Lambert, *Catal. Lett.*, 2000, **68**, 191.
136. W. Zheng, D. Cheng, F. Chen and X. Zhan, *J. Nat. Gas Chem.*, 2010, **19**, 515.
137. R. Ghose, H. T. Hwang and A. Varma, *Appl. Catal., A*, 2014, **472**, 39.
138. P. Wang, G. Zhao, Y. Liu and Y. Lu, *Appl. Catal., A*, 2017, **544**, 77.
139. P. Wang, G. Zhao, Y. Wang and Y. Lu, *Sci. Adv.*, 2017, **3**, 1.
140. H. Wang, R. Schmack, B. Paul, M. Albrecht, S. Sokolov, S. Rümmler, E. V. Kondratenko and R. Kraehnert, *Appl. Catal., A*, 2017, **537**, 33.
141. H. Liu, D. Yang, R. Gao, L. Chen, S. Zhang and X. Wang, *Catal. Commun.*, 2008, **9**, 1302.
142. M. Yildiz, U. Simon, T. Otremba, Y. Aksu, K. Kailasam, A. Thomas, R. Schomäcker and S. Arndt, *Catal. Today*, 2014, **228**, 5.
143. W. Zheng, D. Cheng, N. Zhu, F. Chen and X. Zhan, *J. Nat. Gas Chem.*, 2010, **19**, 15.
144. S. M. K. Shahri and A. N. Pour, *J. Nat. Gas Chem.*, 2010, **19**, 47.

CHAPTER 4

Exploring Chemical Looping for Methane Utilisation

A. F. B. ABU KASIM AND E. J. MAREK*

Department of Chemical Engineering and Biotechnology, University of Cambridge, Philippa Fawcett Drive, CB3 0AS, Cambridge, UK
*Email: ejm94@cam.ac.uk

4.1 Chemical Looping – Concept, History, and Fundamentals

Chemical looping processes represent an innovative class of reactions in which a solid material undergoes a cyclic process of reduction and oxidation, first donating a component to drive the key reaction and then taking up the component to regenerate to the original structure. The most common chemical looping reactions concern oxidation reactions with metal oxides (M_xO_y) used as the reactive solid in which the lattice oxygen represents the component that is cyclically released and reabsorbed from the material's crystal structure. Because the solid donates oxygen, M_xO_y is commonly termed an "oxygen carrier". Taking hematite, Fe_2O_3, as a potential carrier of oxygen for a reaction with methane, a chemical loop can be presented as a two-step process:

Step 1 – reduction of M_xO_y

$$12Fe_2O_{3(s)} + CH_{4(g)} \rightarrow 8Fe_3O_{4(s)} + CO_{2(g)} + 2H_2O_{(g)} \quad (4.1)$$

Step 2 – oxidation (regeneration) of M_xO_{y-z}

$$8Fe_3O_{4(s)} + 2O_{2(g)} \rightarrow 12Fe_2O_{3(s)} \quad (4.2)$$

where x, y, and z are stoichiometric coefficients.

Catalysis Series No. 49
Catalytic Activation of Small Molecules
Edited by Mustafa Yasin Aslan, Angela Daisley, Justin S. J. Hargreaves and José L. Rico
© The Royal Society of Chemistry 2025
Published by the Royal Society of Chemistry, www.rsc.org

The "looping" term stems from the fact that the sole source of oxygen – the solid oxide – undergoes the two-step process, but returns to the original form, closing the "chemical loop". The regeneration step requires a gaseous source of oxygen, and air is commonly used, but for some oxides, a milder oxidiser can be sufficient, such as water or CO_2.[1]

The overall reaction from eqn (4.1) and (4.2) boils down to the combustion of methane:

$$CH_{4(g)} + 2O_{2(g)} \rightarrow CO_{2(g)} + 2H_2O_{(g)} \quad (4.3)$$

While the oxide undergoes reversible redox steps, generalised as:

$$M_xO_{y(s)} \leftrightarrow M_xO_{y-z(s)} + 0.5O_{2(g)} \quad (4.4)$$

Compared to the flame-wise and air-based combustion of methane, chemical looping offers a range of advantages. The main one is the avoidance of mixing the combustion products with nitrogen from the gaseous oxidiser. When combusting methane in air, methane reacts with oxygen, leading to a flue gas mixture of CO_2, H_2O, and N_2 as the main components. If CO_2 were not to be released, first, the mixture components would need to be separated, which requires a minimum theoretical work input in the range of 8–20 kJ mol_{CO2}^{-1} when using aqueous monoethanolamine.[2] In contrast, the chemical looping scheme produces a concentrated stream of gaseous reaction products because CH_4 in eqn (4.1) never comes into contact with air. Since CO_2 is concentrated without gas separation, chemical looping has been described as capable of circumventing the second law limitations.[3]

Chemical looping linked with carbon capture has been proposed as an attractive route towards low-emissions power production, resembling a more advanced version of oxy-fuel technology.[1,4] Consequently, the majority of chemical looping research has focused on combustion, including possible reactor configurations, achievable process and CO_2 capture efficiencies, and an extensive quest to identify suitable M_xO_y materials. However, the idea of sourcing components, particularly oxygen, from a solid material offers further advantages. Using lattice-sourced oxygen prevents the creation of explosive gaseous mixtures (as in eqn (4.1)) and, consequently, allows for potential process intensification. Additionally, reactions separated in time or space (see Figure 4.1(a)) offer more flexibility for heat integration and overall optimisation with other processes, improving efficiencies.

Because chemical looping offers more control over reactions with benefits applicable beyond combustion, recently, it has been proposed as a method for producing value-added chemicals, with over 40 reactions demonstrated in the literature and more expected.[3] In such processes, the role of the M_xO_y oxide is to provide oxygen at the partial pressure that will promote the creation of selective products while avoiding overoxidation to CO_2 and H_2O. The selective reaction route often requires mediation with a catalyst, which can be combined with particles of M_xO_y or, occasionally, the oxide can play a double role, *e.g.* Mg–Mn–mixed oxides in oxidative coupling of methane.[5]

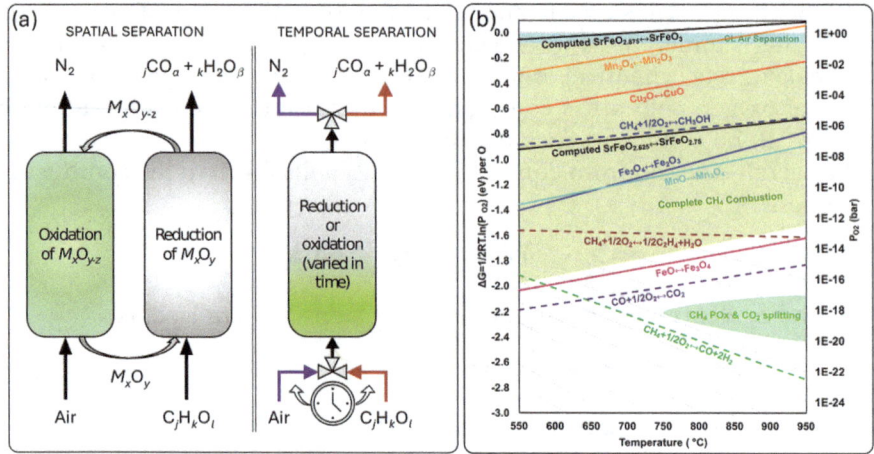

Figure 4.1 (a) Chemical looping schemes with redox steps separated in space or time, (b) Ellingham diagram to guide the selection of an oxygen carrier to a reaction, based on the provided vs. required pO_2 (or equivalently, ΔG). Reproduced from ref. 6 with permission from the Royal Society of Chemistry.

From the economical perspective, a catalytic carrier should comprise inexpensive, abundant, stable, and safe for the environment components.

The theoretical propensity of an oxide to provide oxygen at the chemical potential that matches the reaction requirements can be assessed through eqn (4.4), graphically expressed in the form of phase diagrams, which present the phase transitions of M_xO_y to a lower oxide as a function of temperature and oxygen partial pressure, pO_2. The two oxides represent the thermodynamically stable solids when exposed to gas atmospheres used in the two chemical looping steps. The maximum amount of oxygen available in the reaction, known as the oxygen capacity of the carrier, can then be calculated by comparing the stoichiometry of the starting oxide and the one after its reduction.

Phase diagrams (including Ellingham diagrams, with an example in Figure 4.1(b)) provide a useful starting point when selecting the potential oxygen carriers for a reaction. Depending on the reaction, the required pO_2 from an oxide can differ by up to 20 orders of magnitude.[6] Depending on the oxide, if it can generate high pO_2 (as in eqn (4.4)), it might be a good match for combustion applications, while a lower pO_2 indicates the oxide may be better suited for selective routes. Therefore, the equilibrium oxygen chemical potential is a useful metric for identifying potential oxygen carriers. However, the experimental data and theoretical models used in constructing phase diagrams are only available for relatively simple oxides.

In selective catalytic processes, the final form of the oxide, and thus, the actual amount of oxygen donated, can differ from those expected at equilibrium because of the slow kinetics of oxygen release from the solids, especially at low temperatures needed in some selective reaction routes.[7] Additionally, when oxides are in intimate contact with catalysts, they can interact, affecting both the kinetics and thermodynamics of oxygen release and the final form of the oxide in each redox step.[8] Further manipulation of the reducibility of the main oxide[9] may come from theoretically inactive oxides, such as Al_2O_3 or SiO_2, which are often added for structural support to improve the lifetime of oxygen carriers. Thus, the final activity of such multiphase, multimetallic composites depends on various factors – materials' interactions, porosity, surface morphology, ionic mobilities at reaction temperatures, and structural changes upon cycling. Selectivity for a specific catalytic route then results from the chemical potential of the catalytic composite, kinetics of its redox reactions, energy of formation of oxygen vacancies, capability to adsorb reactive species, and reincorporation of oxygen from the molecular form (O_2, CO_2, H_2O) to the lattice.

Consequently, finding suitable catalytic oxygen carriers for specific reactions remains a non-trivial task with multiple degrees of freedom. In a few pioneering studies, the pO_2 of a theoretical multimetallic oxide was predicted, matched with the reaction requirements, and then experimentally validated,[6] implementing high-throughput material screening with modelling based on density functional theory, accelerated with machine learning. Similar studies for composites comprising oxides, catalysts, and supports are yet to come. Thus, an alternative approach used in most experimental studies is still based on trial and error.

4.2 Chemical Looping with Methane

Methane is a widely available hydrocarbon sourced from natural gas, thus of fossil origin, or from biogas, produced by biomass fermentation or pyrolysis. Since CH_4 is relatively safe to transport and work with, it is utilised not only for power production but also for industrial and residential heating. The use of bio-methane, if paired with carbon capture and storage, can lead to carbon drawdown and net-negative CO_2 emissions; especially attractive as bio-methane can be a drop-in replacement for natural gas, offering a straightforward route to decarbonising a wide range of processes that currently use CH_4.

As a source of carbon, methane is a valuable feedstock, easier to transform than CO_2, and easier to obtain from CO_2 than other hydrocarbons. Overall, CH_4 can be seen as a key participant in the circular carbon economy, with the transformation of CH_4 to high-value-added products becoming increasingly desirable (see Figure 4.2 for value comparison). But in comparison to longer hydrocarbons, methane is a relatively stable molecule with four primary C–H bonds that are difficult to activate. Using CH_4 to produce functionalised products such as alkanes or oxygenates is challenging because the products are less stable than methane – the starting molecule.[10] Even combusting methane requires high activation energies; thus, high reaction rates necessitate high temperatures, making methane a convenient

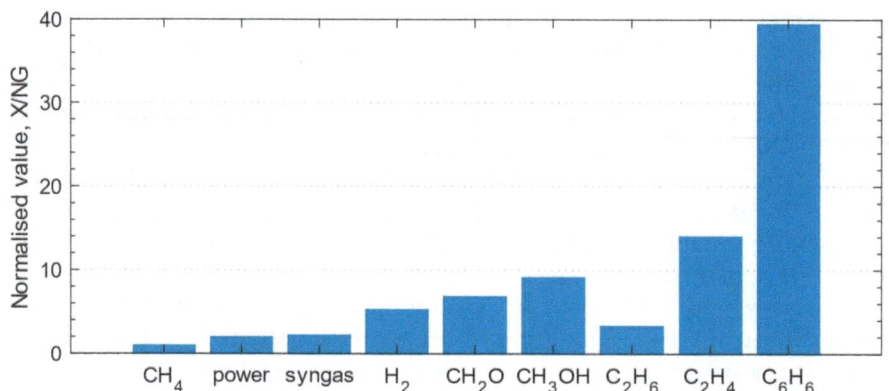

Figure 4.2 Comparison of value creation of CH_4 and other chemicals to assess the economic incentive of using CH_4 as a feedstock for producing value-added chemicals. Adapted from ref. 6 with permission from the Royal Society of Chemistry.

fuel for power generation but also linking methane combustion to a significant production of nitrogen oxides – harmful pollutants.

Utilising chemical looping when transforming methane can help with the listed problems. For combustion applications, chemical looping can help lower the process temperature, thus minimising the creation of pollutants. Additionally, the stream of CO_2 is concentrated (see eqn (4.1)) and ready for capture without the energy penalty for gas separation. Beyond power generation, catalytic combustion of methane to CO_2 and H_2O is of importance in environmental protection, where combustion neutralises methane – a potent greenhouse gas – instead of venting into the atmosphere. In such cases, the concentration of CH_4 is low, beyond the flammability limits, requiring the use of catalysts.

Catalytic chemical looping towards value-added chemicals can offer flexibility in process design towards methane activation (see Figure 4.3). As discussed, an oxygen carrier can be selected to match the chemical potential of the desired reactions, promoting high selectivity to specific products. Schematic representation of chemical looping processes, now mapping the required chemical potential of the oxide, is given in Figure 4.4. Besides the large space for design of new processes, chemical looping subsections can be manipulated to increase product yields, achieve good heat integration,[12] or incorporate CO_2 capture, as in chemical looping super-dry reforming of CH_4[13] (discussed in Section 4.5).

4.3 Chemical Looping Combustion (CLC) of CH_4

Chemical looping commonly utilises fluidisation, with solid oxygen carriers fluidised by a reactive gas. This promotes good contact between the solids and gas and enables the solids to move between reactors to either donate

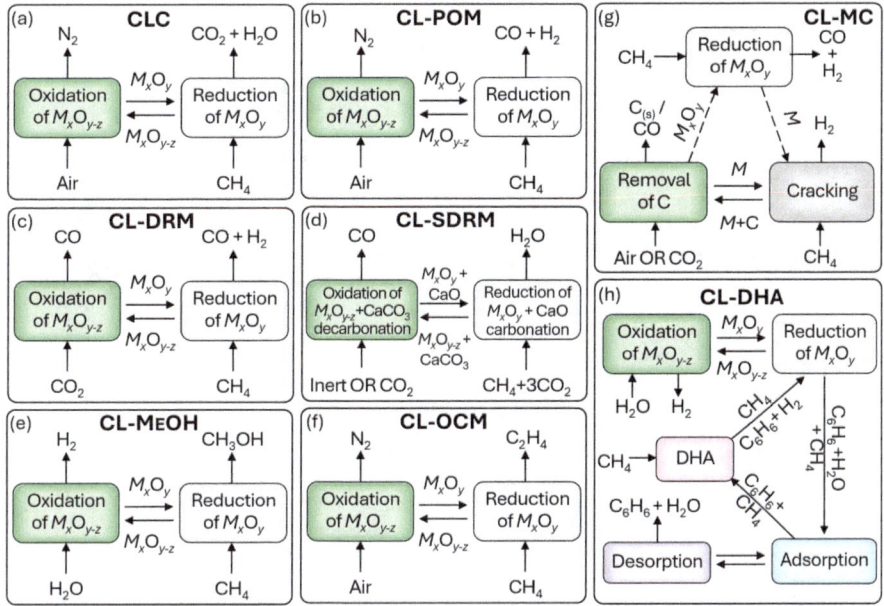

Figure 4.3 Schematics of chemical looping arrangements with CH_4 as feedstock discussed in this chapter. (a) Chemical looping combustion (CLC), (b) chemical looping–partial oxidation of methane (CL–POM), (c) chemical looping–dry reforming of methane (CL–DRM), (d) chemical looping–superdry reforming of methane (CL–SDRM), (e) chemical looping–methanol synthesis (CL–MeOH), (f) chemical looping–oxidative coupling of methane (CL–OCM), (g) chemical looping–cracking of methane (CL–MC), (h) chemical looping–dehydroaromatisation to benzene (CL–DHA). Adapted from ref. 71 with permission from Springer Nature, Copyright 2018.

oxygen or to regenerate, separating the CL steps in space (see Figure 4.1(a)). Fluidised reactors are also commonly used in power generation, but mainly to combust solid fuels. Gaseous fuels are more difficult to combust in fluidised beds because their oxidation is commonly facilitated by reactive radicals, but mobile solids act as very efficient flame arrestors, quenching the radicals.[14] Indeed, the lifetime of OH radicals in fluidised beds has been assessed as shorter than the characteristic time of OH reactions when combusting carbon monoxide in air.[15] Previous studies report that combustion of CH_4 is particularly difficult in fluidised beds, requiring beds at temperatures above 900 °C.[14] Since chemical looping involves heterogeneous gas–solid reactions, the scheme supports oxidative reactions without the presence of radicals, making the fluidised technologies attractive again when using methane as a fuel.[16] Early work on CL for CH_4 combustion (CH_4-CLC) demonstrated that the main quality parameter is the activity of the solid oxide – with iron oxides not interacting with CH_4, whilst nickel oxide supported its complete conversion to CO_2.[17] Marek et al. speculated that oxygen carriers should show some catalytic affinity to activate and then

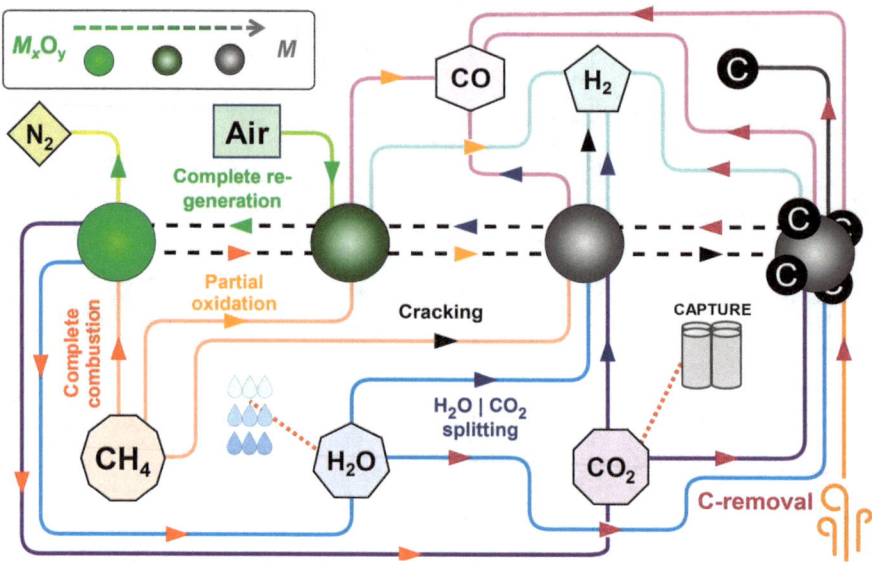

Figure 4.4 Reaction map of methane in chemical looping processes with an oxygen carrier in oxidised (M_xO_y) and gradually reduced form (M), aligned with variations in its chemical potential. Species involved and changes in the oxidation state of the solid are colour-coded.

convert methane.[16] They reported a perovskite oxide, $SrFeO_{3-\delta}$, to assist with methane combustion above 550 °C, discussing two possible catalytic mechanisms, suprafacial – where O_2 and fuel are adsorbed on the surface of the catalytic material where they react – and intrafacial – where lattice oxygen is donated. Because $O_{2(g)}$ released by $SrFeO_{3-\delta}$ was also detected, two types of oxygen forms – molecular and lattice – could have been active, and, as confirmed by Wang *et al.* for the $LaCoO_3$ family of perovskites, the dominating catalytic route changes with process temperature.[18] Compared with catalyst-assisted reactions of CH_4 with O_2 in fluidised beds, $SrFeO_3$'s activity was similar to metallic catalysts: Pd/Al_2O_3 above 450 °C[19] and Cu/Al_2O_3 above 650 °C[20].

For oxide-assisted conversion of methane, the catalysis community widely accepts the dominating mechanism (the Mars–van Krevelen mechanism),[10,21] where the reactants adsorb at the surface of the catalytic solid, with CH_4 reacting directly with the lattice oxygen, while $O_{2(g)}$ – co-fed with CH_4 – replenishes the consumed $O_{lattice}$. Chemical looping is thus akin to such a scheme;[22] if methane activation occurs directly at the surface of the oxide, then the process can still be supported by an oxygen carrier even in the absence of $O_{2(g)}$ (see Figure 4.5). Mechanistic chemical looping studies support the Mars–van Krevelen mechanism as dominating, with CO_2 arising as the primary product of CH_4 oxidation,[23] preferred when the amount of oxygen vacancies is low, *i.e.* when $O_{lattice}$ is freely available. Then, with the time of reaction, unless oxygen is replenished or delivered in a timely

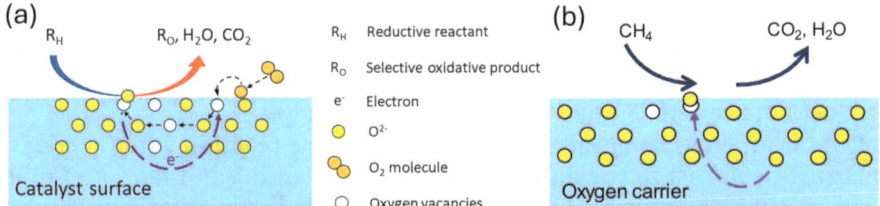

Figure 4.5 (a) Schematic representation of the Mars–van Krevelen mechanism where the reductive component, R_H, is co-fed with an oxygen source – here O_2, (b) methane reacting with oxygen species at the surface of an oxygen carrier. Reproduced from ref. 11, https://doi.org/10.1016/j.cattod.2020.09.023, under the terms of the CC-BY 4.0 license, https://creativecommons.org/licenses/by/4.0/.

fashion from the bulk of the oxide, the selectivity of products shifts towards partial oxidation routes and, with oxygen substantially depleted, towards coking. This explains why in CL combustion of methane, a high inventory of oxides with large oxygen capacity is needed, aligning with the focus on stoichiometric oxides in the early studies on CL combustion,[24] and mainly on Fe_2O_3, NiO, CuO, and Mn_3O_4. Because stoichiometric oxides undergo substantial crystallographic changes upon reduction, which leads to sintering and deactivation, they are commonly synthesised with a support oxide, *e.g.* Al_2O_3, $MgAl_2O_4$, SiO_2, TiO_2, ZrO_2, and CeO_2. Alternatively, the pO_2 of the process can be controlled to limit the reduction to a desired form of the oxide. As recently summarised by Li *et al.*, the most commonly studied oxygen carriers are based on iron oxides, demonstrating selectivities to CO_2 above 60% at $T \sim 650$ °C and between 80–100% at $T \sim 900$ °C, with similarly high conversions for both lean and CH_4-rich mixtures, and at industrially relevant space velocities, both in bubbling beds and circulating fluidised beds.[24]

Overall, the chemical looping combustion of CH_4 is a relatively mature and well-studied area, with installation ranging from kW to a few MWth.[25] Because in CH_4 combustion, large inventories of oxides are needed (4 moles of $O_{lattice}$ per mole of CH_4) and materials deactivate upon cycles, natural ores are commonly suggested. Ores containing oxides of Fe and Mn are promising candidates owing to their good availability, low price, and the potential for reuse in the metal industry if they deactivate during CLC.[24]

4.4 Partial Oxidation of Methane with Oxygen Carriers (CL–POM)

Syngas, a mixture of CO and H_2, is a valuable feedstock for the chemical industry, commonly prepared by steam reforming of methane (SRM): $CH_4 + H_2O \leftrightarrow CO + 3H_2$. The technology is fully mature but energy-intensive because SRM is endothermic; thus, alternative reaction pathways for methane to syngas are still of interest. In the considered processes, methane undergoes partial oxidation, but the form of the oxidiser changes. In the partial oxidation of methane (POM), the oxidiser is $O_{2(g)}$ but of limited availability: $CH_4 + 0.5O_2 \leftrightarrow CO + 2H_2$. Promoting CO and H_2 as the final products instead of CO_2 and H_2O requires a catalyst.

POM is an attractive route to syngas because the reaction with O_2 releases energy (in contrast to the endothermic SRM), resulting in the H_2/CO molar ratio of 2, which is useful for methanol synthesis.[26] But POM is difficult to achieve, requiring high temperatures to promote the selective route over complete combustion or reforming with H_2O, but high temperatures lead to catalyst deactivation through coking.[27]

Looking at the POM reaction, chemical looping can offer a substitution for $O_{2(g)}$ with oxygen sourced from M_xO_y to help eliminate the need for costly air separation and the explosion risks from CH_4–O_2 mixtures,[28] but with potential problems from coking or overoxidation if the oxide is not matched well to POM. An alternative approach has also been proposed where instead of engineering the most suitable oxide, a stoichiometric oxide is employed but partially depleted of oxygen and never fully regenerated in the chemical looping reactors.[29]

The first studies on the chemical looping approach for POM have been attributed to Otsuka et al., who, in the 1990s, proposed to use CeO_2 to source oxygen when producing syngas from CH_4.[27,30,31] While not calling the scheme chemical looping at that time, they also highlighted the need for regeneration, which for CeO_{2-x} is feasible when using mild oxidisers, CO_2 and H_2O, making the regeneration step contribute to the chemical production. Thermodynamically, the partial oxidation of CH_4 with CeO_2 shows high selectivity towards CO and H_2, and CH_4 conversion. However, the slow reaction kinetics necessitate high temperatures or the addition of a catalyst. For the latter, alkaline metal oxides were used as they add vacancies to CeO_2, helping with the ionic transport of oxygen, but only MgO resulted in the actual acceleration of POM. The most effective catalyst was Pt-black, decreasing the activation energy for POM on CeO_2 by a factor of three.[27]

Since Otsuka et al.'s pioneering studies, chemical looping for POM has been studied extensively in the last 30 years. Progress in developing oxygen carriers for POM, either the full catalytic composites or the M_xO_y components, has been summarised by Zhang et al.[32] and Zhang et al.[30] Notably, most CL-POM carriers employ ceria, iron, and nickel, with a few studies using tungsten oxides. The metals can be present as a catalytic additive (commonly Ni as it is selective for CO), nominal monometallic oxides (mainly Fe_2O_3, while NiO is avoided because it is unselective for CO) with or without dopants and supports, or multimetallic oxides, including spinels and perovskites (e.g. $LaFeO_3$, $La_{0.95}Ce_{0.05}Ni_xFe_{1-x}O_3$). The reported conversions of CH_4 commonly range between 50–100%, with selectivities to CO between 70–100%, at 600–950 °C.

The mechanistic insight into the involved reaction pathways is not yet fully established. Most often, the ability of an oxygen carrier to participate effectively in POM has been ascribed to the availability of lattice oxygen,[30] its activity and mobility,[28] fast desorption rates of the products,[27] the ability of the material to activate methane or resist the cracking of CH_4, and coke deposition.[31] The need for such relatively complex functionality from oxygen carriers, on top of the mechanical requirements and their overall stability in cycles, explains why composite oxides are the mainstream trend in searching for POM-suitable carriers.[33]

Alternatively, improvements to CL–POM can also be introduced through manipulating process configurations. Zheng *et al.* investigated the production of H_2 using dielectric-barrier discharge for plasma-assisted chemical looping where non-thermal plasma enabled the partial oxidation of CH_4 to occur at moderate temperatures (300–500 °C).[34]

4.5 Dry Reforming of Methane with Chemical Looping (CL–DRM)

The dry reforming of methane (DRM) produces syngas, CO, and H_2, using CO_2 as an oxidising agent. Compared to POM, DRM is also a sink of CO_2, offering a potential route for carbon capture and utilisation, for example, within the circular carbon economy.

$$CH_4 + CO_2 \rightarrow 2CO + 2H_2 \tag{4.5}$$

This endothermic reaction is spontaneous above 650 °C but kinetically slow. Similar to POM, nickel is the most commonly used catalyst, with the same problems of coking and deactivation at higher temperatures.[35] Additionally, the presence of CO_2 leads to a competitive reverse water gas shift reaction (rWGSR), consuming some of the created H_2 and undesirably reducing the final $CO:H_2$ ratio.[36,37]

$$CO_2 + H_2 \Leftrightarrow CO + H_2O \tag{4.6}$$

With high energy requirements, DRM remains challenging. Chemical looping dry reforming of methane (CL–DRM) introduces a potentially attractive solution, splitting the reaction into two steps. First, the oxygen carrier (M_xO_y) supplies oxygen for the partial oxidation of CH_4 (the same as in CL–POM):

$$CH_4 + M_xO_y \rightarrow CO + 2H_2 + M_xO_{y-1} \tag{4.7}$$

Then, in a separate step, the (M_xO_{y-1}) is reoxidised by CO_2:

$$CO_2 + M_xO_{y-1} \rightarrow CO + M_xO_y \tag{4.8}$$

resulting in the same net reaction as eqn (4.5). The benefit of CO_2 being introduced only in the second step is the potential process intensification and minimised impact from rWGSR. Additionally, because oxygen is sourced from M_xO_y, catalysts in CL–DRM are less affected by coking than in eqn (4.5). The reason for this is that coke can be destabilised by the oxide, reacting with the lattice oxygen and producing CO. The reoxidation step in CO_2 can also gasify the coke to CO through the reverse Boudouard reaction[38] ($CO_2 + C \leftrightarrow 2CO$) when the oxide regenerates as in eqn (4.8). The overall scheme of CL–DR is presented graphically in Figure 4.6.

The reaction scheme in CL–DRM is similar to CL–POM, with methane being oxidised by the solid oxide, and differs in the second chemical looping step, where the solid regenerates. In CL–DRM, the use of CO_2 as the oxidiser of the solid means that the material needs to be capable of splitting CO_2 – the characteristic that adds additional design requirements when looking for oxygen carriers for CL–DRM vs. CL–POM. The ability to split CO_2, or H_2O, as both offer a similar oxidative potential, results in the production of CO, or H_2; thus, the role of the second CL step switches from mainly heat production to chemical production.

Materials capable of splitting water or CO_2 are also the focus of another technology – solar thermochemical splitting cycles, where the oxygen carrier M_xO_y is reduced by applying heat from the concentrated solar energy, leading to the spontaneous release of $O_{2(g)}$, after which the solid replenishes lattice oxygen by splitting CO_2 or water. Experimental investigations showed promising results for ZnO, Fe_3O_4, CeO_2, La, Sr-perovskites, and composite materials. Thermodynamic studies indicated that binary oxides require very high temperatures for the two CL steps, and finding suitable materials is challenging.[40] The theoretical framework indicates the importance of the solids' enthalpy and entropy change to allow CO_2 splitting to fall in the ranges 0–80 J (mol K)$^{-1}$ for Δs_{solid} and 280–460 kJ mol^{-1} for Δh_{solid}.[41] The thermodynamic data are usually limited for more complex oxides and composites; thus, thermodynamic screening of materials is an ongoing process, with the missing input predicted from first-principles calculations, e.g. density functional theory.

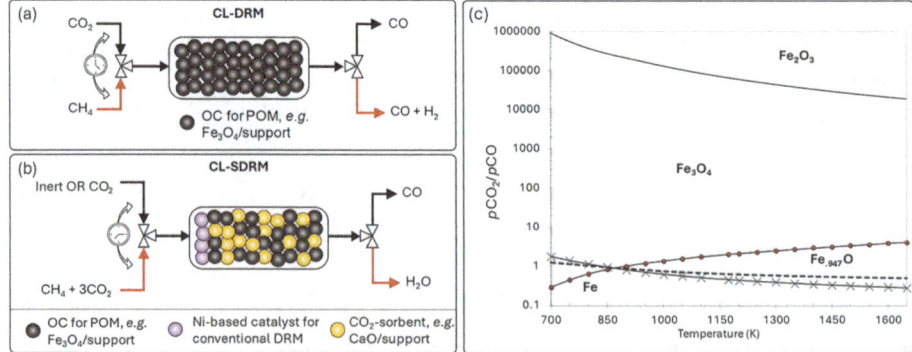

Figure 4.6 Schematic comparison of (a) chemical looping–dry reforming of methane (CL–DR), with (b) chemical looping–super-dry reforming of methane (CL–SDRM). Also shown (c) is a phase diagram for iron oxides (data from ref. 39).

Similar to solar-powered cycles, research in CL–DRM has focused on theoretical screening and experimental development of oxygen carriers. Most studied solids that can react with CO_2 – a mild-oxidiser – are iron-based oxides. Figure 4.6(c) shows the main transitions between Fe_3O_4, $Fe_{0.947}O$, and the metallic Fe,[42] indicating the operability range for CO_2 splitting. Other materials commonly involve Ce and Zr. Reviews by Li et al.[43] and Zheng et al.[44] summarise progress on new materials for CL–DR from 25 separate studies, and most notable new materials, not covered by the two reviews, are summarised in Table 4.1. Only the best-performing candidates are reported for studies that analysed more than one OC.

Variations of the CL–DR schemes are also an active field of research. Buelens et al.[45] adapted the technique for CO_2 capture from calcium-looping technology (CaL)[46] into CL–DR, resulting in a process called super-dry reforming of methane (CL–SDRM). As shown in Figure 4.6(b), they used a multilayered packed bed containing an upstream layer of particles of 10 wt% NiO supported on $MgAl_2O_4$, followed by a layer of mixed particles of 50 wt% Fe_2O_3 on $MgAl_2O_4$ and 90 wt% CaO on Al_2O_3 – the three materials at a 1:3:6 weight ratio. The role of CaO was to absorb CO_2 during the reduction step between Fe_2O_3 and CH_4, then desorb the gas and react it with the Fe-based oxide during the regeneration step. The overall reaction is:

$$CH_4 + 3CO_2 \rightarrow 4CO + 2H_2O \tag{4.9}$$

with H_2O and CO being produced separately in the reduction and reoxidation stages, respectively. Up to 25 consecutive redox cycles were conducted when an optimised CO_2 to CH_4 ratio – 3:1 by mole was achieved.

In reaching eqn (4.9), the syngas produced from co-feeding CH_4 and CO_2 over the nickel catalyst (DR) primarily comprises CO since the high CO_2 content in the feed shifts the equilibrium towards the product of eqn (4.6), i.e. CO and H_2O. The Fe_3O_4 reduces down to Fe, as opposed to mainly $Fe_{0.947}O$ expected in $CO_2:CH_4$ mixtures, because of the simultaneous absorption of CO_2 on CaO (CaL), which pulls the equilibrium towards a high pCO/pCO_2 ratio (see Figure 4.6(c)), promoting the creation of CO. Thus, in the first step, Buelens et al.[45] were able to push the oxidation of CH_4 beyond dry reforming, indirectly leading to complete oxidation to CO_2, captured by CaO, with H_2O released. Then, instead of feeding CO_2 in the reoxidation step, an inert gas, He, with pCO_2 that induces the decarbonation of $CaCO_3$ back to CaO, was used. The CO_2 released immediately reoxidised Fe to Fe_3O_4, closing the looping cycle. The overall process excels as a CO_2 sink, consuming 3 moles of CO_2 per mole of CH_4 to produce 4 moles of CO rather than syngas – see eqn (4.9).

Zheng et al.[47] also integrated a simultaneous CO_2 capture with CaL but maintained the conventional CL–DR scheme with a composite oxygen carrier, $NiO/Ca_2Fe_2O_5/CaO$.[48] Performing reduction at 700 °C, now with all of the produced CO_2 absorbed by CaO, resulted in syngas rich in H_2. Rather than having excess CO_2 in chemical looping superdry reforming, the

Table 4.1 Overview of new oxygen carriers developed for chemical looping dry reforming of methane.

Oxygen carrier	Conversion of CH_4[a] (%)	Selectivity to CO[a,b] (%)	Conversion of CO_2[c] (%)	Reaction temperature (°C)	No. of cycles	Ref.
1 wt% Ru/$La_2Ce_2O_7$	65	~100	95	650	10	38
0.5 wt% Ni/10 wt%WO_3/ZrO_2	70	97	65	750	180	105
1 wt% Pt-20%SiO_2/V_2O_3	70	99.5	~100	850	8	106
$LaFe_{0.8}Al_{0.2}O_3$	80	~100	90	900	20	107
Ni–(α-MoC)/Al_2O_3	63	95	60	500	1140	108
70 mol%($Mg_{0.2}Co_{0.2}Ni_{0.2}Mn_{0.2}Fe_{0.2}$)$O_x$/$ZrO_2$	43	97	(~67)[d]	800	100	109

[a] In the reduction step, averaged over all performed cycles.
[b] Moles of CO produced over moles of CH_4 consumed
[c] In the re-oxidation step, averaged over all performed cycles.
[d] Undisclosed, the value reported in brackets was calculated taking the reported CO concentration at the reactor outlet and then the corresponding pCO/pCO_2 at equilibrium.

CO_2-lean gas shifted the equilibrium towards the forward WGSR (reversed eqn (4.6)), enriching the product stream in H_2. The reoxidation step, however, required temperatures of up to 900 °C to decompose $CaCO_3$ in the presence of CO_2, recombining the CaO and Fe back to $Ca_2Fe_2O_5$, producing CO. Stable performance was achieved over 30 redox cycles showing no substantial degradation even under temperature cycling (700–900 °C).

Chemical looping permits the separation of the product and introduces an approach to rectify the detrimental effects of coking, but challenges remain with the high energy requirement. Efforts made in lowering the operating temperature come at the expense of CH_4 conversion and the need for OCs doped with precious materials (see Table 4.1). Nevertheless, chemical looping can offer new avenues to perform DR as with the super-dry reforming scheme[45] and DR enhanced with CO_2 sorption,[47] thus paving the way for innovative approaches for efficient CH_4 reforming.

4.6 Methane Cracking with Chemical Looping (CL–MC)

Methane cracking produces H_2 and solid carbonaceous products (see eqn (4.10)), with thermal cracking above 1200 °C, but the temperature can be lowered to 650–950 °C when applying catalysts.[49] The first step in methane cracking is its adsorption on the catalyst, followed by sequential dehydrogenation until $C_{(s)}$ is formed and dissolved. Which step is the rate-limiting one remains debatable.[50]

$$CH_{4(g)} \rightarrow C_{(s)} + 2H_{2(g)} \qquad (4.10)$$

Methane cracking has been proposed as a method for producing H_2 without CO_2 release, and instead with carbon trapped in the solid form, making it easier to store than CO_2. A new focus on CH_4 cracking also concerns the production of high-value carbon products.[51] Amorphous carbon is a relatively reactive, disordered solid, often a product of coking, which can be readily gasified. Graphitic and filamentous forms are less reactive and more attractive as a commodity.[50] The catalyst applied dictates the type of carbon formed, with nickel being the popular choice for filamentous carbons.[52]

In the chemical looping community, methane cracking is more often discussed as the unintentional process accompanying dry reforming (CL–DRM) and less as a standalone process. For example, Zheng et al.[47] ascribed the production of hydrogen-rich syngas to CH_4 cracking on Ni/$Ca_2Fe_2O_5$, finding both whisker-like (filamentous) and less-structured carbon growing on the segregated Ni and Fi nanoparticles. Although initially, the regeneration step in CO_2 gasified all the deposited carbon to CO, in the follow-up study, Zheng et al. demonstrated the possibility of separating the carbon structures from the oxygen carrier when using a fluidised bed reactor.[53] A combination of (1) a regeneration step in air to selectively remove the amorphous carbon, together with (2) high gas velocity, helped dislodge the carbon filaments from the oxide. The carbonaceous solids, collected on a filter above the fluidised bed, comprise multi-walled carbon nanotubes and carbon onions. Further work, *e.g.* integrating the swirling flow separator designed by Hatanaka *et al.*,[54] could aid in full separation.

Zheng *et al.* also investigated a variant of CH_4 cracking at mild temperatures (~400 °C) supported by non-thermal plasma.[34,55] Heavy hydrocarbons, mainly C2–C4, tend to form when cracking methane using plasma,[50,56] lowering the selectivity to H_2. The presence of chemical looping materials, *i.e.* active oxides and plasma, resulted in a mix of products from cracking and partial oxidation of CH_4, again promoting the production of H_2-rich syngas.

Studies dedicated to methane cracking *via* chemical looping (CL–MC) share a common theme, *i.e.* applying a catalyst supported on a metal oxide to crack CH_4, followed by the reverse Boudouard reaction gasifying the carbon with CO_2 to produce CO,[57–61] or with steam to produce H_2.[62] Unlike CL–DRM, H_2 and CO can be produced in separate process steps, allowing more flexibility in producing the desired H_2/CO blends for downstream processes. However, the purity of H_2 in the cracking stage depends on the level of conversion of CH_4. Keller *et al.*[57–59,63] pioneered the studies in valorising CH_4 cracking through chemical looping. The team focused on Fe-oxides as the active materials because of the abundance of iron, being also environmentally benign compared to the toxic Ni. The process is proposed to regenerate the Fe-oxide up to Fe_2O_3, which supports CH_4 combustion rather than cracking, but the extra extent of the oxidation releases heat – needed to drive the energy-intensive, endothermic cracking. Besides the overall conceptual design[57] and the assessment of limits for carbon deposition,[58] the use of $BaZr_{0.9-x}Fe_{-x}Y_{0.1}O_{3-\delta}$ (BZFY), with $x=0.4$ being the most optimal,[57,59] was proposed as an active material for accelerating the kinetics of the slow reduction step, attributing the rapid decomposition of CH_4 to the high proton conductivity of BZFY *via* a proposed H_2-abstraction mechanism.[57] Besides the works by Keller *et al.*, Table 4.2 summarises the independent studies in CL–MC.

Table 4.2 Overview of new oxygen carriers developed for chemical looping cracking of methane.

Looping material	Conversion of CH_4 (%)	Conversion of CO_2^a (%)	Conversion of H_2O^b (%)	Reaction temperature (°C)	No. of cycles	Ref.
62.5 mol% Ni/CeO_2–Al_2O_3	90	~100	—	600	20	60
$Ni_yFe_{3-y}O_4$–$Ca_2Fe_xAl_{2-x}O_5$	95	~100	—	800	20	61
Fe/CaO–$Ca_{12}Al_{14}O_{33}$	72	—	~100	850	8	62

a Peak conversion of CO_2 during the regeneration step.
b Peak conversion of H_2O during the regeneration step, achieved with simultaneous CO_2 capture from CaL.

Further innovation is needed before CL–MC can distinguish itself as a bespoke process in chemical looping. The most prevalent focus on producing syngas from CH_4 puts CL–MC forward as an alternative to CL–POM and CL–DR processes. Considering CL–MC as an opportunity for producing value-added carbonaceous solids, as proposed by Zheng et al.[53] and Keller et al.,[63] could be the way forward.

4.7 Chemical Looping for Oxidative Coupling of Methane (CL–OCM)

One of the first reactions proposed for transforming methane to higher value chemicals was the oxidative coupling of methane (OCM) producing longer-chain hydrocarbons, but in practice, mainly its dimers,[64] ethane in eqn (4.11), and ethylene in subsequent eqn (4.12) and in eqn (4.13).

$$4CH_4 + O_2 \rightarrow 2C_2H_6 + 2H_2O \quad (4.11)$$

$$2C_2H_6 + O_2 \rightarrow 2C_2H_4 + 2H_2O \quad (4.12)$$

$$2CH_4 + O_2 \rightarrow C_2H_4 + 2H_2O \quad (4.13)$$

While the addition of oxygen aids in activating methane at lower temperatures than non-oxidative dehydrogenation, it also competes with the full oxidation of methane to CO and CO_2. These challenges underscore the complexity of the OCM process, where high selectivity is typically achieved at low CH_4 conversions, but decreases with further process intensification. Despite intensive research, including in industry, the OCM process is yet to make a significant breakthrough to compete with hydrocarbon production from crude oil refining.[65]

The oxidative processes can be carried out using purified oxygen (direct OCM) or providing oxygen from a solid oxide (indirect OCM). Interestingly, the chemical looping approach can support both routes. In the classical approach, the solids participate in a heterogeneous reaction, donating lattice oxygen and often activating methane, having a dual function of oxygen carrier and catalyst. But some oxides can easily release oxygen to the gas phase, without interacting directly with CH_4. The released $O_{2(g)}$ replaces the need to supply O_2 with CH_4, offering an *in situ* $O_{2(g)}$ delivery.[66] Both methods eliminate the need for air separation units, but only the direct one helps mitigate the safety considerations around flammable hydrocarbon and $O_{2(g)}$ mixtures.

In the last 40 years, a broad range of catalysts have been studied for OCM – the recent review by Deng *et al.* mentions thousands of combinations,[67] including materials supporting the chemical looping routes. The most notable results were reported for composite oxides comprising 4–5 different alkali and rare earth metals. In the chemical looping and direct approach with $O_{2(g)}$, oxides of Mn, Mg, Na, Li, and W have been demonstrated to be particularly effective, owing to their ability to promote the formation and coupling of methyl radicals.[66] Promising catalytic behaviour has been linked with high electrical conductivity,[68] which correlates with fast rates of oxygen delivery to reactions. Additionally, the performance in OCM is closely related to the surface defects,[69] which facilitate methane adsorption and activation.

Those oxides, and especially perovskite oxides, can incorporate significant amounts of defects in the structure, promoting further the activation of CH_4.

Overall, the best catalysts for OCM seem to be those that affect the oxygen availability close to the active sites for methane activation. Thus, chemical looping seems especially promising for OCM when recalling the fundamental principle that in chemical looping, the chemical potential of oxides should match the one needed in the selective reaction.

An example of manipulating the chemical potential and the rate of oxygen delivery was demonstrated by Luongo *et al.*, who covered a perovskite-based oxygen carrier with a molten salt of Na_2CO_3.[70] The carbonate layer on top of the perovskite facilitated oxygen transport, albeit at a significantly lower rate than observed with the bare perovskite. This manipulation of the chemical potential and the kinetics of oxygen delivery improved the selectivity of ethane ODH to ethylene (eqn (4.13)).

Further engineering efforts for the design and delivery of well-fitted oxygen carriers are underway. In their review on metal oxide redox chemistry, Zeng *et al.* highlighted that the rates of delivery of lattice oxygen can already be manipulated quite well,[71] besides tailoring the chemistry of the solid materials but also using the reactor design and engineering to distribute materials with different oxidative potentials. Alternatively, the gaseous components can be injected in specific locations, regulating the redox capability of the reactive system from the gas side. Such an approach is applicable both to the processes with $O_{2(g)}$ and using chemical looping.

Further advances were also proposed introducing plasma to drive reactions at lower temperatures.[72,73] Here, however, the interaction of plasma with catalysts and the lifetime of reactive plasma species remains underexplored. Thus, the potential benefits of plasma-supported chemical looping for OCM are still unclear.

4.8 Methane to Other Products *via* Chemical Looping Routes

Chemical looping with oxides can be a straightforward replacement in catalytic oxidation reactions, providing the oxide donates $O_{lattice}$ at the right chemical potential. Thus, further efforts often focus on identifying the starting CL arrangements for a given new reaction, demonstrating the feasibility of the chemical looping scheme. For transforming methane, successful attempts were presented for the synthesis of methanol and benzene, and both were met with significant interest as both exemplify single-step conversion of methane to liquid chemicals.

4.8.1 Chemical Looping Methanol Synthesis

In biological systems, enzymatic catalysts can drive methane to methanol conversion very efficiently, reaching up to 30% yield and dwarfing the results achieved so far with heterogeneous catalysis. Direct methanol synthesis from CH_4 in a mixture with a gaseous oxidiser was first studied with V, Fe, and Mo oxides as catalysts, using O_2 or nitrogen oxides as the source of oxygen.[74] Liu et al. noticed that the reaction mechanism involves surface oxygen species O^-, produced when N_2O decomposed on Mo(v) sites, following the Mars–van Krevelen mechanism.[75] Adding H_2O to the process hindered the total combustion pathways. More recent research focused on embedding catalytic sites into zeolites, creating metallic centres, similar to those found in enzymes. Here, the reaction is usually carried out sequentially, i.e. similarly to in the chemical looping approach, first heating up the catalytic solid and exposing it to oxygen or NO for activation, then introducing methane, followed by the addition of steam – needed to extract the methanol from the zeolite surface. The successful attempts demonstrated methanol synthesis when using Cu, Fe, and Co for the catalytic centres embedded in ZSM-5 and MOR zeolites. Of the three steps needed, the last two overlap with the commonly understood chemical looping:

$$CH_4 + M_xO_y \rightarrow CH_3OH + M_xO_{y-1} \qquad (4.14)$$

$$H_2O + M_xO_{y-1} \rightarrow H_2 + M_xO_y \qquad (4.15)$$

while the exposure to higher temperature, oxygen, or NO in step 1 differs from the usual chemical looping pathways. Here, step 1 is needed to reactivate the surface of the zeolite catalyst, which loses activity upon exposure to water in eqn (4.15).[76] In most published studies, the activation step required elevating the temperature to ~400 °C – significantly higher than the ~200 °C used in the reaction steps.[74] This was first circumvented by Narsimhan et al. when using Cu-SSZ-13 but operating the process in a continuous mode, i.e. with $CH_4/O_2/H_2O$ mixtures,[77] thus losing the chemical looping benefits of producing undiluted products. Tomkins et al. demonstrated that all three chemical looping steps can be performed at 200 °C when using Cu-MOR zeolite,[78] with further improvements achieved after swapping MOR with an erionite zeolite and carrying out both the activation and eqn (4.14) steps at 300 °C, but changing the H_2O extraction step to room temperature.[79]

Chemical looping methanol synthesis has been under development for the last 20 years and represents one of the first selective reactions demonstrated in the chemical looping mode. For the transition-metal–zeolite composites, the process translates to stoichiometric reactions, i.e. the active sites participate in one reactive event only per cycle, giving a turnover of 1.[80] Thus, the commonly proposed reactive solids are not considered strictly catalytic. Other selective chemical looping schemes commonly source $O_{lattice}$ from a large reservoir of the bulk oxide, with reactions taking place at the surface of the catalyst located on top of the oxide.[65] Such a scheme has not been demonstrated for methane to methanol synthesis yet. Additionally, while most chemical looping studies provide information on the cycling behaviour of oxygen carriers, such results are commonly missing in the studies on methanol synthesis, with a few exceptions reporting improved yields for CH_3OH with cycling.[81] Thus, while chemical looping has been particularly promising for methanol synthesis, multiple research questions remain unaddressed.

4.8.2 Chemical Looping Benzene Synthesis

The dehydroaromatisation (DHA) of methane to produce benzene C_6H_6 is a one-step gas-to-liquid process, most often discussed in relation to the non-oxidative route:

$$6CH_4 \rightarrow C_6H_6 + 9H_2 \tag{4.16}$$

Thermodynamically, the yield of benzene is limited,[82–84] competing against the prevalent cracking to solid carbon, as illustrated in Figure 4.7(a), and dehydrogenation to lower-molecular-weight hydrocarbons and higher-molecular-weight substituted aromatics (not shown in Figure 4.7a).[85,86] The pathway to C_6H_6 can be promoted with a catalyst, though. For example, at 700 °C, between 60–80% selectivity to C_6H_6 can be achieved over Mo/ZSM-5 at 10–12% conversion of CH_4, reaching yields of ~10% – close to the thermodynamic limits.[87]

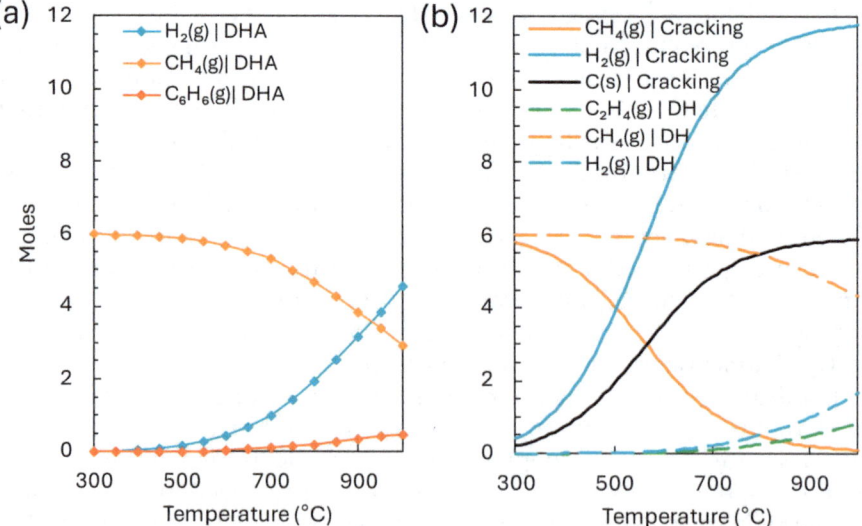

Figure 4.7 Thermodynamic limitations of dehydroaromatisation (DHA) of methane when (a) only constituents of the DHA reaction are considered, and (b) cracking and dehydrogenation (DH) are also considered. Calculated using FactSage.[88]

Spivey and Hutchings[84] outlined the spontaneity of performing DHA oxidatively – which might give rise to the potential of using chemical looping materials as an active support but at the additional risk of complete or partial oxidation. Brady and Lobo et al.[89–91] incorporated chemical looping schemes into the DHA process to circumvent the thermodynamic limits. The primary goal was to selectively oxidise the H_2 from eqn (4.16), using Fe_3O_4, then recirculate the unreacted CH_4 and the produced C_6H_6. The reduced oxide was regenerated with H_2O, recreating the H_2. At 700 °C, the highest conversion of CH_4 achieved after 32 passes of recirculation was ~55% at ~70% selectivity to C_6H_6.[90] Other researchers followed with the same principle (see ref. 92–94), albeit not explicitly denoting the process as chemical looping. Compared to other CL processes, the main focus of CL–DHA is to use the oxidative step to shift the equilibrium towards C_6H_6. Further development can involve integrating the DHA step with selective oxidation, employing a dual-function catalyst, as used, for example, in CL–OCM.

4.9 Chemical Looping for CH_4 Conversion – Perspectives

Over the last two decades, the chemical looping processes have been extensively researched with a variety of applications developed from conceptual studies to pilot-scale reactors, and for CLC with a few examples of scaled-up experimental facilities, with the recently commissioned largest one at 4 MWth in Deyang City, China.[95] This versatility of the chemical looping concept is one of the strongest points of working with solid materials, attracting interest beyond combustion. Extensive reviews by Li et al.[3] and Liu[96] summarised chemical looping for catalytic and selective reactions, emphasising the thermodynamic fundamentals and the principles of designing materials with desired chemical potential.

New reactions that are considered feasible to proceed using solid oxides are those that follow the Mars–van Krevelen mechanism, thus those that may already involve metal oxides but rather as catalysts than the source of oxygen. For methane conversion, the simplest C1 oxygen-containing derivatives are formic acid (HCOOH), methanol (CH_3OH), and formaldehyde (HCHO). Methanol synthesis from CH_4 can be performed via chemical looping using transition-metal-exchanged zeolites, but – as discussed in Section 4.8 – the earlier studies demonstrated that direct oxidation with a gaseous source of oxygen over Mo, V, Co-catalytic oxides also involves the Mars–van Krevelen mechanism. Interestingly, in such processes (with zeolites and oxides), the main competitive organic product to methanol is formaldehyde,[75,97] with selectivity to either oxygenate varying with the content of water in the reactive system. Similarly, high selectivity towards formaldehyde has been demonstrated when directly reacting CH_4 with $O_{2(g)}$ over B_2O_3.[98] Despite the potential promise, deliberate oxidation of methane to formaldehyde in the chemical looping mode has not yet been reported.

Following the leads from reports in direct oxidation, the use of Pt for selective methane oxidation has been proposed as promising, if accompanied by the presence of water, as H_2O enables high coverage of Pt surfaces with oxygen-containing species.[99] Nanoparticles of Pt support the conversion of CH_4 to methanol and formaldehyde, and, further, methanediol.[99] At the same time, Pt has been shown to actively interact with perovskite oxides, which can pass lattice oxygen onto Pt to facilitate catalytic reactions.[100] Because of the high cost, Pt is rarely proposed in chemical looping studies, but its use would be justified when synthesising value-added chemicals. Similarly, other noble metals show promise in chemical looping reactions. Silver was so far mainly explored in the C2 chemistry *via* looping, with ethylene oxide[101] and acetaldehyde[102] as the main products, again supporting the oxygen donation from a perovskite oxide.[103] With methane, silver catalyses the oxidative coupling pathway *via* direct reaction with O_2, demonstrating the ability to selectively activate methane.[104] This indicates opportunities for Ag-supported CH_4 conversion if implemented *via* chemical looping. Looking for new avenues for chemical looping will most likely involve exploring possible catalyst–oxide systems, thus searching through wide design areas. Automated material screening and machine learning can also accelerate further discoveries in chemical looping, opening new design space for chemical looping processes in methane conversion.

References

1. L. Zeng, Z. Cheng, J. A. Fan, L.-S. Fan and J. Gong, *Nat. Rev. Chem.*, 2018, **2**, 349.
2. M. Karimi, M. Shirzad, J. A. C. Silva and A. E. Rodrigues, *Environ. Chem. Lett.*, 2023, **21**, 2041.
3. X. Zhu, Q. Imtiaz, F. Donat, C. R. Müller and F. Li, *Energy Environ. Sci.*, 2020, **13**, 772.
4. H. Zheng, X. Jiang, Y. Gao, A. Tong and L. Zeng, *Discover Chem. Eng.*, 2022, **2**, 5.
5. S. Damasceno, F. J. Trindade, F. C. Fonseca, D. Z. de Florio and A. S. Ferlauto, *Fuel Process. Technol.*, 2022, **231**, 107255.
6. X. Wang, Y. Gao, E. Krzystowczyk, S. Iftikhar, J. Dou, R. Cai, H. Wang, C. Ruan, S. Ye and F. Li, *Energy Environ. Sci.*, 2022, **15**, 1512.
7. E. J. Marek and E. García-Calvo Conde, *Chem. Eng. J.*, 2021, **417**, 127981.
8. A. R. P. Harrison, K. Y. Kwong, Y. Zheng, A. Balkrishna, A. Dyson and E. J. Marek, *Energy Fuels*, 2023, **37**, 9487.
9. N. L. Galinsky, A. Shafiefarhood, Y. Chen, L. Neal and F. Li, *Appl. Catal., B*, 2015, **164**, 371.
10. Y. Wang, P. Hu, J. Yang, Y.-A. Zhu and D. Chen, *Chem. Soc. Rev.*, 2021, **50**, 4299.
11. Z.-Z. Pan, Y. Li, Y. Zhao, C. Zhang and H. Chen, *Catal. Today*, 2021, **364**, 2.
12. V. V. Galvita, H. Poelman and G. B. Marin, *Energy Fuels*, 2017, **31**, 11509.

13. L. C. Buelens, V. V. Galvita, H. Poelman, C. Detavernier and G. B. Marin, *Science*, 2016, **354**, 449.
14. R. P. Hesketh and J. F. Davidson, *Combust. Flame*, 1991, **85**, 449–467.
15. A. N. Hayhurst, *Combust. Flame*, 1991, **85**, 155.
16. E. Marek, W. Hu, M. W. Gaultois, C. P. Grey and S. A. Scott, *Appl. Energy*, 2018, **223**, 369.
17. M. Johansson, T. Mattisson and A. Lyngfelt, *Energy Fuels*, 2006, **20**, 2399.
18. T. Wang, C. Zhang, J. Wang, H. Li, Y. Duan, Z. Liu, J. Y. Lee, X. Hu, S. Xi, Y. Du, S. Sun, X. Liu, J.-M. Lee, C. Wang and Z. J. Xu, *J. Catal.*, 2020, **390**, 1.
19. Z. Yang, P. Yang, L. Zhang, M. Guo and Y. Yan, *RSC Adv.*, 2014, **4**, 59418.
20. M. Iamarino, P. Ammendola, R. Chirone, R. Pirone, G. Ruoppolo and G. Russo, *Ind. Eng. Chem. Res.*, 2006, **45**, 1009.
21. L. He, Y. Fan, J. Bellettre, J. Yue and L. Luo, *Renewable Sustainable Energy Rev.*, 2020, **119**, 109589.
22. *Chemical Looping Partial Oxidation: Gasification, Reforming, and Chemical Syntheses*, ed. L.-S. Fan, Cambridge University Press, Cambridge, 2017, pp. 307–369.
23. J. Yang, E. Bjørgum, H. Chang, K.-K. Zhu, Z.-J. Sui, X.-G. Zhou, A. Holmen, Y.-A. Zhu and D. Chen, *Appl. Catal., B*, 2022, **301**, 120788.
24. D. Li, R. Xu, Z. Gu, X. Zhu, S. Qing and K. Li, *Energy Technol.*, 2020, **8**, 1900925.
25. N. E. L. Haugen, Z. Li, V. Gouraud, S. Bertholin, W. Li, Y. Larring, K. Luo, A. Szlęk, T. A. Flach, Ø. Langørgen, X. Liu and M. Yazdanpanah, *Int. J. Greenhouse Gas Control*, 2023, **129**, 103975.
26. T. J. Siang, A. A. Jalil, A. A. Abdulrasheed, H. U. Hambali and W. Nabgan, *Energy*, 2020, **198**, 117394.
27. K. Otsuka, Y. Wang, E. Sunada and I. Yamanaka, *J. Catal.*, 1998, **175**, 152.
28. C. Wang, S. Xu, Y. Li, M. Long, D. Chen and H. Duan, *Sep. Purif. Technol.*, 2024, **343**, 127066.
29. S. Bhavsar, M. Najera, A. More and G. Veser, in *Reactor and Process Design in Sustainable Energy Technology*, ed. F. Shi, Elsevier, Amsterdam, 2014, pp. 233–280.
30. J. Zhang, Y. Cui, W. Si-ma, Y. Zhang, Y. Gao, P. Wang and Q. Zhang, *Catalysts*, 2024, **14**, 246.
31. *Chemical Looping Partial Oxidation: Gasification, Reforming, and Chemical Syntheses*, ed. L.-S. Fan, Cambridge University Press, Cambridge, 2017, pp. 55–171.
32. X. Zhang, F. Zhang, Z. Song, L. Lin, X. Zhao, J. Sun, Y. Mao and W. Wang, *Fuel*, 2022, **325**, 124964.
33. H. Zheng, X. Jiang, Y. Gao, A. Tong and L. Zeng, *Discover Chem. Eng.*, 2022, **2**, 5.
34. Y. Zheng, R. Grant, W. Hu, E. Marek and S. A. Scott, *Proc. Combust. Inst.*, 2019, **37**, 5481.
35. K. Wittich, M. Krämer, N. Bottke and S. A. Schunk, *ChemCatChem*, 2020, **12**, 2130.

36. A. G. S. Hussien and K. Polychronopoulou, *Nanomaterials*, 2022, **12**, 3400.
37. H. C. Mantripragada and G. Veser, *J. CO2 Util.*, 2021, **49**, 101555.
38. K. Kang, N. Kayama, T. Higo, C. Sampson and Y. Sekine, *Catal. Sci. Technol.*, 2024, **14**, 3609–3617.
39. R. Davies, A. Dinsdale, J. Gisby, J. Robinson and S. Martin, *Calphad*, 2002, **26**, 229.
40. B. Meredig and C. Wolverton, *Phys. Rev. B*, 2009, **80**, 245119.
41. B. Chen, H. Yang, Q. Dong, L. Tong, Y. Ding and L. Wang, *Int. J. Hydrogen Energy*, 2024, **84**, 1058.
42. C. R. Müller, C. D. Bohn, Q. Song, S. A. Scott and J. S. Dennis, *Chem. Eng. J.*, 2011, **166**, 1052.
43. Y. Li, M. Chen, L. Jiang, D. Tian and K. Li, *Phys. Chem. Chem. Phys.*, 2024, **26**, 1516.
44. H. Zheng, X. Jiang, Y. Gao, A. Tong and L. Zeng, *Discover Chem. Eng.*, 2022, **2**, 5.
45. L. C. Buelens, V. V. Galvita, H. Poelman, C. Detavernier and G. B. Marin, *Science*, 2016, **354**, 449.
46. R. Han, Y. Wang, S. Xing, C. Pang, Y. Hao, C. Song and Q. Liu, *Chem. Eng. J.*, 2022, **450**, 137952.
47. Y. Zheng, M. S. Sukma and S. A. Scott, *Chem. Eng. J.*, 2023, **465**, 142779.
48. M. S. Sukma, Y. Zheng, P. Hodgson and S. A. Scott, *Energy Fuels*, 2022, **36**, 9410.
49. M. Pudukudy, Z. Yaakob, Q. Jia and M. S. Takriff, *New J. Chem.*, 2018, **42**, 14843.
50. M. Hamdan, L. Halawy, N. Abdel Karim Aramouni, M. N. Ahmad and J. Zeaiter, *Fuel*, 2022, **324**, 124455.
51. J. Prabowo, L. Lai, B. Chivers, D. Burke, A. H. Dinh, L. Ye, Y. Wang, Y. Wang, L. Wei and Y. Chen, *Carbon*, 2024, **216**, 118507.
52. I. R. Hamdani, A. Ahmad, H. M. Chulliyil, C. Srinivasakannan, A. A. Shoaibi and M. M. Hossain, *ACS Omega*, 2023, **8**, 28945.
53. K. Alkhatib, M. Santihayu Sukma, S. A. Scott and Y. Zheng, *Fuel*, 2024, **377**, 132816.
54. T. Hatanaka and Y. Yoda, *Int. J. Hydrogen Energy*, 2022, **47**, 20176.
55. Y. Zheng, E. J. Marek and S. A. Scott, *Chem. Eng. J.*, 2020, **379**, 122197.
56. S. Wang, J. Wang, D. Feng, F. Wang, Y. Zhao and S. Sun, *Int. J. Hydrogen Energy*, 2024, **63**, 284.
57. M. Keller, Y. Matsuzaki and J. Otomo, *Chem. Eng. J.*, 2018, **349**, 249.
58. M. Keller, A. Matsumura and A. Sharma, *Chem. Eng. J.*, 2020, **398**, 125612.
59. S. Hikima, M. Keller, H. Matsuo, Y. Matsuzaki and J. Otomo, *Chem. Eng. J.*, 2021, **417**, 128012.
60. Y.-C. Zeng, Z. X. Law and D.-H. Tsai, *ACS Sustainable Chem. Eng.*, 2024, **12**, 12200.
61. Z. Sun, T. Cai, C. K. Russell, J. K. Johnson, R.-P. Ye, W. Xiang, X. Chen, M. Fan and Z. Sun, *Appl. Catal., B*, 2020, **271**, 118938.

62. Z. Chu, J. Zhang, W. Zhao, Y. Yang, J. Zhao and Y. Li, *Chem. Eng. J.*, 2024, **497**, 154599.
63. M. Keller and A. Sharma, *Energy Fuels*, 2021, **35**, 847.
64. S. Damasceno, F. J. Trindade, F. C. Fonseca, D. Z. de Florio and A. S. Ferlauto, *Fuel Process. Technol.*, 2022, **231**, 107255.
65. S. Bhavsar and G. Veser, *RSC Adv.*, 2014, **4**, 47254.
66. *Chemical Looping Partial Oxidation: Gasification, Reforming, and Chemical Syntheses*, ed. L.-S. Fan, Cambridge University Press, Cambridge, 2017, p. 172.
67. J. Deng, P. Chen, S. Xia, M. Zheng, D. Song, Y. Lin, A. Liu, X. Wang, K. Zhao and A. Zheng, *Atmosphere*, 2023, **14**, 1538.
68. A. Malekzadeh, A. Khodadadi, M. Abedini, M. Amini, A. Bahramian and A. K. Dalai, *Catal. Commun.*, 2001, **2**, 241.
69. H. Borchert and M. Baerns, *J. Catal.*, 1997, **168**, 315.
70. G. Luongo, F. Donat, A. H. Bork, E. Willinger, A. Landuyt and C. R. Müller, *Adv. Energy Mater.*, 2022, **12**, 2200405.
71. L. Zeng, Z. Cheng, J. A. Fan, L.-S. Fan and J. Gong, *Nat. Rev. Chem.*, 2018, **2**, 349.
72. T. Liu, C. Wang, Y. Song, W. Ou, R. Xiao and D. Zeng, *Sustainable Energy Fuels*, 2023, **7**, 2455.
73. S. Kang, J. Deng, X. Wang, K. Zhao, M. Zheng, D. Song, Z. Huang, Y. Lin, A. Liu, A. Zheng and Z. Zhao, *Catalysts*, 2023, **13**, 557.
74. M. Ravi, M. Ranocchiari and J. A. van Bokhoven, *Angew. Chem., Int. Ed.*, 2017, **56**, 16464.
75. R.-S. Liu, M. Iwamoto and J. H. Lunsford, *J. Chem. Soc., Chem. Commun.*, 1982, **0**, 78.
76. V. L. Sushkevich, D. Palagin, M. Ranocchiari and J. A. van Bokhoven, *Science*, 2017, **356**, 523.
77. K. Narsimhan, K. Iyoki, K. Dinh and Y. Román-Leshkov, *ACS Cent. Sci.*, 2016, **2**, 424.
78. P. Tomkins, A. Mansouri, S. E. Bozbag, F. Krumeich, M. B. Park, E. M. C. Alayon, M. Ranocchiari and J. A. van Bokhoven, *Angew. Chem., Int. Ed.*, 2016, **55**, 5467.
79. J. Zhu, V. L. Sushkevich, A. J. Knorpp, M. A. Newton, S. C. M. Mizuno, T. Wakihara, T. Okubo, Z. Liu and J. A. van Bokhoven, *Chem. Mater.*, 2020, **32**, 1448.
80. M. Ravi, V. L. Sushkevich, A. J. Knorpp, M. A. Newton, D. Palagin, A. B. Pinar, M. Ranocchiari and J. A. van Bokhoven, *Nat. Catal*, 2019, **2**, 485.
81. S. E. Bozbag, E. M. C. Alayon, J. Pecháček, M. Nachtegaal, M. Ranocchiari and J. A. van Bokhoven, *Catal. Sci. Technol.*, 2016, **6**, 5011.
82. N. Kosinov and E. J. M. Hensen, *Adv. Mater.*, 2020, **32**, 2002565.
83. U. Menon, M. Rahman and S. J. Khatib, *Appl. Catal., A*, 2020, **608**, 117870.
84. J. J. Spivey and G. Hutchings, *Chem. Soc. Rev.*, 2014, **43**, 792.
85. Y. Liu, M. Ćoza, V. Drozhzhin, Y. van den Bosch, L. Meng, R. van de Poll, E. J. M. Hensen and N. Kosinov, *ACS Catal.*, 2023, **13**, 1.

86. Z. R. Ismagilov, E. V. Matus and L. T. Tsikoza, *Energy Environ. Sci.*, 2008, **1**, 526.
87. J. J. Spivey and G. Hutchings, *Chem. Soc. Rev.*, 2014, **43**, 792.
88. C. W. Bale, E. Bélisle, P. Chartrand, S. A. Decterov, G. Eriksson, A. E. Gheribi, K. Hack, I.-H. Jung, Y.-B. Kang, J. Melançon, A. D. Pelton, S. Petersen, C. Robelin, J. Sangster, P. Spencer and M.-A. Van Ende, *Calphad*, 2016, **54**, 35.
89. C. Brady, B. Murphy and B. Xu, *ACS Catal.*, 2017, **7**, 3924.
90. C. Brady, Q. Debruyne, A. Majumder, B. Goodfellow, R. Lobo, T. Calverley and B. Xu, *Chem. Eng. J.*, 2021, **406**, 127168.
91. S. Kim, L. Annamalai and R. F. Lobo, *Chem. Eng. J.*, 2023, **455**, 140919.
92. X. Ji, Y. Liu, J. Liu and J. Zhang, *Appl. Catal., B*, 2022, **307**, 121194.
93. M. Sakbodin, E. Schulman, Y. Pan, E. D. Wachsman and D. Liu, *Catal. Today*, 2021, **365**, 80.
94. N. K. Razdan, A. Kumar and A. Bhan, *J. Catal.*, 2019, **372**, 370.
95. N. E. L. Haugen, Z. Li, V. Gouraud, S. Bertholin, W. Li, Y. Larring, K. Luo, A. Szlęk, T. A. Flach, Ø. Langørgen, X. Liu and M. Yazdanpanah, *Int. J. Greenhouse Gas Control*, 2023, **129**, 103975.
96. W. Liu, *React. Chem. Eng.*, 2021, **6**, 1527.
97. J. Ohyama, A. Hirayama, Y. Tsuchimura, N. Kondou, H. Yoshida, M. Machida, S. Nishimura, K. Kato, I. Miyazato and K. Takahashi, *Catal. Sci. Technol.*, 2021, **11**, 3437.
98. J. Tian, J. Tan, Z. Zhang, P. Han, M. Yin, S. Wan, J. Lin, S. Wang and Y. Wang, *Nat. Commun.*, 2020, **11**, 5693.
99. E. van Steen, J. Guo, N. Hytoolakhan Lal Mahomed, G. M. Leteba and S. V. L. Mahlaba, *ChemCatChem*, 2023, **15**, e202201238.
100. M. Kothari, Y. Jeon, D. N. Miller, A. E. Pascui, J. Kilmartin, D. Wails, S. Ramos, A. Chadwick and J. T. S. Irvine, *Nat. Chem.*, 2021, **13**, 677.
101. E. J. Marek, S. Gabra, J. S. Dennis and S. A. Scott, *Appl. Catal., B*, 2020, **262**, 118216.
102. J. C. Gebers, A. F. B. Abu Kasim, G. J. Fulham, K. Y. Kwong and E. J. Marek, *ACS Eng. Au*, 2023, **3**, 184.
103. A. R. P. Harrison, K. Y. Kwong, Y. Zheng, A. Balkrishna, A. Dyson and E. J. Marek, *Energy Fuels*, 2023, **37**, 9487.
104. X. Bao, M. Muhler, R. Schlögl and G. Ertl, *Catal. Lett.*, 1995, **32**, 185–194.
105. S. Miyazaki, Z. Li, L. Li, T. Toyao, Y. Nakasaka, Y. Nakajima, K. Shimizu and Z. Maeno, *Energy Fuels*, 2023, **37**, 7945.
106. Y. Ge, T. He, Z. Wang, D. Han, J. Li, J. Wu and J. Wu, *AIChE J.*, 2020, **66**, e16772.
107. X. Xia, W. Chang, S. Cheng, C. Huang, Y. Hu, W. Xu, L. Zhang, B. Jiang, Z. Sun, Y. Zhu and X. Wang, *ACS Catal.*, 2022, **12**, 7326.
108. X. Zhang, Y. Xu, Y. Liu, L. Niu, Y. Diao, Z. Gao, B. Chen, J. Xie, M. Bi, M. Wang, D. Xiao, D. Ma and C. Shi, *Chem*, 2023, **9**, 102.
109. Y. Shao, C. Wu, S. Xi, P. Tan, X. Wu, S. Saqline and W. Liu, *Appl. Catal., B*, 2024, **355**, 124191.

CHAPTER 5

Catalytic Dehydroaromatisation of Methane

J. S. J. HARGREAVES

School of Chemistry, Joseph Black Building, University of Glasgow, Glasgow G12 8QQ, UK
Email: Justin.Hargreaves@glasgow.ac.uk

5.1 Introduction

The direct conversion of methane, the major component of natural gas, is a topic of enduring interest. This interest is primarily driven through the perspectives of the production of products of greater value and/or added functionality in a single step, the utilisation of stranded natural gas reserves and methane produced as a by-product or waste where it is often simply combusted or flared. With reference to the transformation of stranded reserves, there is the additional benefit of target products often being more easily transportable. One area which has attracted particular interest over the past few decades in particular has been the dehydroaromatisation of methane for which the target reaction is:

$$6CH_4 \rightarrow C_6H_6 + 9H_2 \quad (5.1)$$

where the product benzene is of greater value and higher functionality than the methane from which it is derived as well as being a liquid under ambient conditions (and therefore more easily transportable than methane). Whilst much of the literature in the area concentrates on the hydrocarbon product, it is self-evident that, in molar terms, hydrogen is the major product.

Catalysis Series No. 49
Catalytic Activation of Small Molecules
Edited by Mustafa Yasin Aslan, Angela Daisley, Justin S. J. Hargreaves and José L. Rico
© The Royal Society of Chemistry 2025
Published by the Royal Society of Chemistry, www.rsc.org

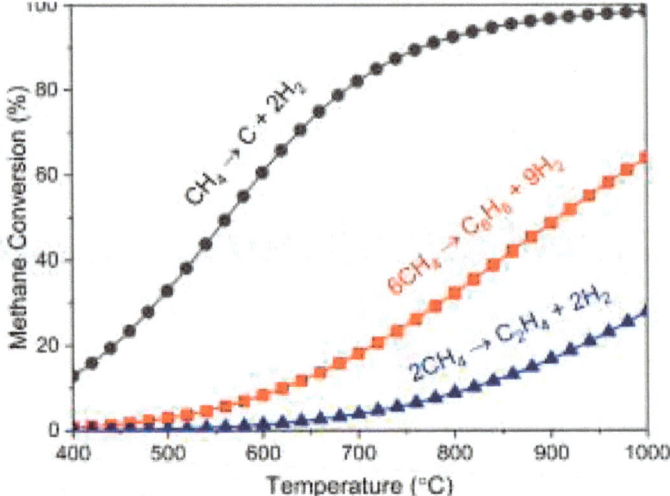

Figure 5.1 Thermodynamic equilibrium limits of methane conversion processes at 1 bar pressure. Reproduced from ref. 2 with permission from Elsevier, Copyright 2020.

The reaction itself has an obvious relationship with the direct cracking of methane:

$$CH_4 \rightarrow C + 2H_2 \quad (5.2)$$

as well as its direct dehydrogenation to produce C_2 products such as:

$$2CH_4 \rightarrow C_2H_4 + 2H_2 \quad (5.3)$$

Similarly to the above two processes, the dehydroaromatisation reaction is highly endothermic with an enthalpy change of $+531$ kJ mol^{-1} being reported for $6CH_{4(g)} \rightarrow C_6H_{6(g)} + 9H_{2(g)}$,[1] necessitating high temperatures to be of potential interest due to equilibrium limitations. The thermodynamic limits of methane conversion corresponding to the above processes at 1 bar pressure are presented in Figure 5.1.[2]

5.2 Molybdenum-containing Catalysts

Whilst the direct conversion of methane to benzene had been reported in earlier studies, *e.g.* ref. 3, it is arguable that much of the current interest in the area derives from a landmark publication by Wang *et al.* dating from 1993.[4] In this publication, it was shown that the intrinsic low activity of HZSM-5 catalysts for methane conversion to benzene at 700 °C could be enhanced by the incorporation of either Mo or Zn components, with Mo being particularly effective. Ethane and hydrogen were reported to be co-products of the reaction. This system, in which the molybdenum component is generally introduced in the form of ammonium molybdate or heptamolybdate by impregnation, has been a particular focus of research ever since. Catalytic performance has been shown to be a strong function of the zeolitic properties and molybdenum loading with maximum performance for HZSM-5 systems being associated with a SiO_2/Al_2O_3 ratio of *ca.* 40.[5] Bronsted acidity is also important in relation to activity. Whilst Bao and coworkers have reported that Mo/HMCM-22 catalysts prepared by impregnation perform better than their Mo/HZSM-5 counterparts,[6] it is the latter which has been the dominant system studied in the research literature. In a comparison of different zeolite-, zeotype- and mesoporous-based materials, Zhang *et al.* concluded that silica–alumina zeolites of two-dimensional structure and pore size close to the kinetic diameter of benzene (5.9 Å) were suitable, and they reported the following order of activities:[7]

Mo/HZSM-11 > Mo/HZSM-5 > Mo/HZSM-8 > Mo/Hβ

> Mo/HMCM-41 > Mo/HSAPO-34 > Mo/HMOR ~ Mo/HX

> Mo/HY > Mo/HSAPO-5 > Mo/HSAPO-11

Typical molybdenum loadings employed for HZSM-5-based catalysts are in the range of 2–6 wt% MoO_3, with greater loadings leading to loss of performance due to pore blockage and dealumination.[8] Within the literature, there has been extensive discussion relating to the distribution of the MoO_x species formed following the calcination of the materials employed as catalysts,[9] with no overall consensus being apparent. Some species are believed to be located within the host zeolite channel structure, for example Iglesia *et al.*'s proposal of $(Mo_2O_5)^{2+}$ bridging two framework Al sites,[10] Ha *et al.*'s suggestion of $(Mo_2O_5)^{2+}$ or MoO_2^+,[11] Li *et al.*'s proposal of $Mo_5O_{12}^{6+}$ located at the channel intersections (with molybdenum oxide on the external surface)[12] and Zhou *et al.*'s proposal of $MoO_2(OH)_2$ within the pores.[13] Improper catalyst pre-treatment (in terms of calcination temperature, *etc.*) can be deleterious for catalytic behaviour leading to extraction of framework Al, the formation of aluminium molybdate and ultimately the collapse of the zeolite structure. In this context it is interesting to note that Al XAS studies relating to the activation process have indicated a distortion of the local Al environment which can be related to changes in Bronsted acid site strength for the pre-catalytic material.[14] NMR-based studies have indicated that the aluminium local environment changes as a function of reaction time on stream.[15]

The typical reaction conditions employed for dehydroaromatisation of methane are 700 °C and atmospheric pressure with fixed-bed microreactors operating at low space velocity being most commonly employed. These conditions correspond to a thermodynamically limited methane conversion of around 11.5%[16]; the catalyst performance profile is as shown in Figure 5.2[17] in which it can be seen that there is an induction period, followed by the attainment of maximum activity and then the occurrence of deactivation.

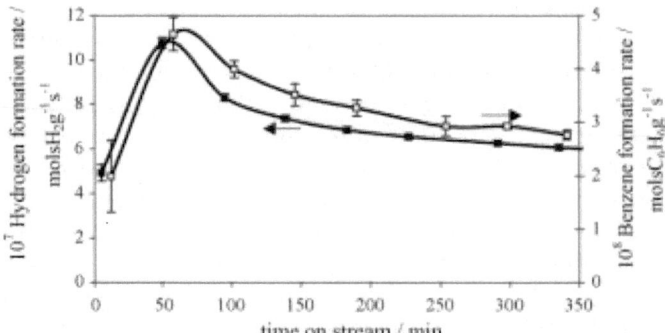

Figure 5.2 Benzene and hydrogen formation rates as a function of time on stream for 3 wt% MoO_3/ZSM-5 at ambient pressure and 700 °C. Reproduced from ref. 17 with permission from Elsevier, Copyright 2006.

In addition to hydrogen, the stoichiometric co-product of the reaction, ethylene is frequently reported as a significant co-product for which variable space velocity experiments indicate that it may be primary in nature. As shown in Figure 5.3, Ichikawa and co-workers have identified that higher aromatics, especially naphthalene, are also formed, although at much lower levels than benzene.[18]

In relation to the induction period, it is believed that the transformation of the oxidic molybdenum species occurs such that molybdenum oxycarbide or molybdenum carbide clusters are formed generating the active form of the catalyst, with those residing within the zeolite channel structure generally being considered more important. Different proposals for the active phase of molybdenum have been reported. Beale and co-workers have recently proposed that $Mo_{1.6}C_3$ is formed,[19] Ichikawa and co-workers suggest Mo_2C clusters are produced[5] and Iglesia and co-workers report that MoC_x clusters 6 Å in diameter containing ca. 10 Mo species are formed.[10] A bifunctional reaction mechanism has been proposed to be operative wherein methane is initially dehydrogenated to form a C_2 intermediate (with ethylene most commonly being proposed e.g. ref. 20, but acetylene also in other studies[21]), which then undergoes further dehydrogenation and cyclisation catalysed by the Bronsted acid sites. The process can be summarised by the schematic presented in Figure 5.4.[18]

As stated earlier, there have been indications, through the effect of varying space velocity upon product distribution, that ethylene is a primary product of the reaction. Following the landmark publication of Boudart and Levy, where platinum-like catalytic properties of tungsten carbide were claimed,[22] molybdenum carbide is also believed to have precious metal-like properties[23] and is thus expected to possess strong dehydrogenation functionality, generating the C_2 intermediate. As stated elsewhere,[24] caution should be exercised in drawing the parallels between such binary carbides too strongly. In this context, it is notable that Pd/HZSM-5 has been found to be

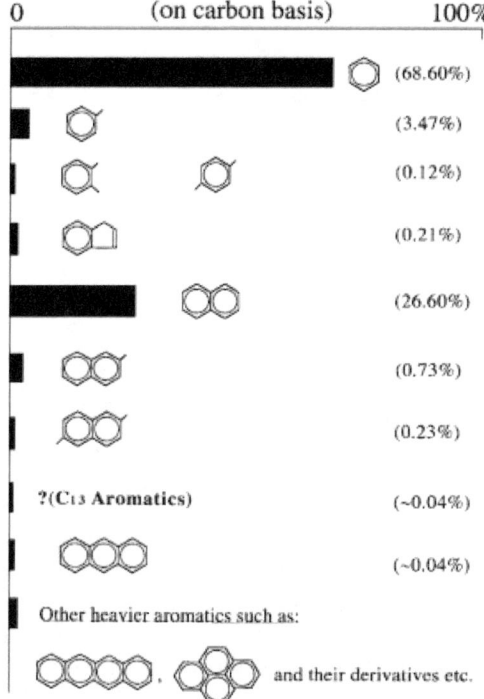

Figure 5.3 Average distribution of aromatic compound product on carbon basis formed in methane dehydroaromatisation over 3 wt% Mo/HZSM-5 after 4 hours on stream at 700 °C. Reproduced from ref. 18 with permission from Elsevier, Copyright 1999.

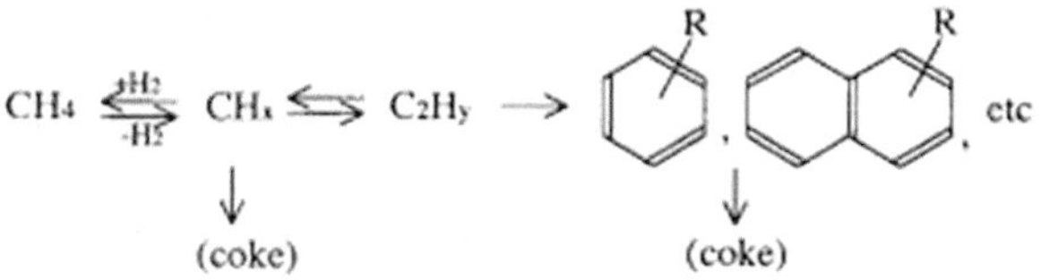

Figure 5.4 Reaction pathway for the dehydroaromatisation of methane. Reproduced from ref. 18 with permission from Elsevier, Copyright 1999.

an effective methane cracking catalyst (generating C and H_2) as opposed to a dehydroaromatisation catalyst.[25] These observations, in conjunction with the pathway shown in Figure 5.4, suggest that a future direction towards the development of higher activity methane dehydroaromatisation catalysts may lie in the partial poisoning of high activity methane cracking catalysts as discussed in the literature.[25] It is also pertinent to note that not all aromatic products comprise an even number of carbon atoms (*e.g.* methyl naphthalene), meaning that the mechanism cannot just be a simple matter of C_2 oligomerisation steps. In more recent years, proposals for the formation of

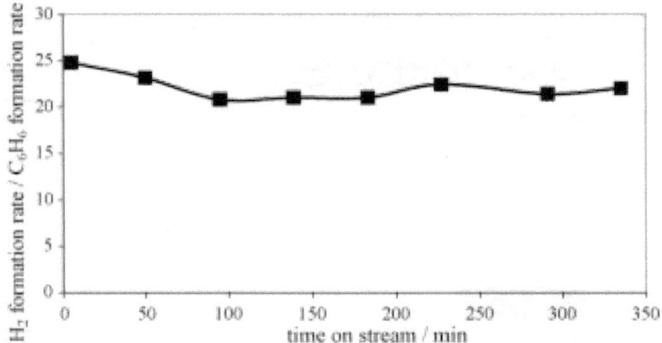

Figure 5.5 Benzene and hydrogen formation rate ratio as a function of time on stream for 3 wt% MoO_3/ZSM-5 at ambient pressure and 700 °C. Reproduced from ref. 17 with permission from Elsevier, Copyright 2006.

an active hydrocarbon pool, as has been reported for methanol to hydrocarbon zeolite catalysts, *e.g.* ref. 26, have gained more traction in the literature.[27]

In view of the fact, that in molar terms, hydrogen is the major reaction product of dehydroaromatisation, it is somewhat surprising that so many studies have failed to quantify its production, with some notable exceptions to this, including ref. 28. This possibly relates to the difficulty in its quantification which by gas chromatography necessitates the use of a thermal conductivity detector operating with a suitable carrier gas such as argon. Ideally, in the absence of side-reactions, it is anticipated that the hydrogen to benzene formation ratio would be 9 : 1. However, as shown in Figure 5.5, in a study which has investigated this, a much larger ratio has been found.[17] This ratio, which is essentially independent of the activation and deactivation phases of the reaction, is indicative of the significant occurrence of side reactions with post-reaction analysis showing methane cracking to be important in this regard.

Catalyst deactivation is generally attributed to (excessive) coke formation, although there have been suggestions of the full carbidisation of molybdenum species causing their detachment from the zeolite support leading to their sintering in this respect.[19] In an XPS study, Lunsford and co-workers documented three different types of near-surface coke – with a hydrogen-poor sp hybridised or pre-graphitic coke gradually covering the surface of the zeolite and the molybdenum carbide phase leading to deactivation.[29] Applying ^{13}C MAS NMR, Jiang *et al.* reported the formation of two types of carbonaceous species – one leading to a reduction in the number of Bronsted acid sites and the other covering the active molybdenum-containing phase.[30] In some studies, the formation of carbon nanotubes has been reported, for example ref. 25 and 31.

The inclusion of low levels of CO or CO_2 in the reaction feed has been reported to have a promotional effect upon benzene production and catalyst stability.[18] In terms of the addition of CO, the proposal has been made that CO_2 and C formation occur *via* the Boudouard reaction and that the resultant C formed is preferentially incorporated into benzene and naphthalene and that the CO_2 formed further reacts with the deposited coke to regenerate CO, resulting in a reduction in its formation.

In relation to the improvement of activity and stability, it is notable that a wide range of promoters have been added to the Mo/HZSM-5 system in this respect. Initially these included Fe and Co,[18,32] but studies have extended far beyond this[2,33] and even now include dual promotion such as by Pt–Bi.[34] As noted,[2] the reported effects of dopants are often contradictory. Figure 5.6 summarises some of the effects reported for different dopants and Figure 5.7 shows how the dopants can affect the nature and quantity of the carbonaceous species formed upon reaction. Figure 5.8 shows the open-ended multiwalled carbon nanotubes formed.

Regeneration of deactivated catalysts can be accomplished *via* the removal of deactivating coke by controlled oxidation (for which caution has to be exercised due to the potential loss of volatile MoO_3 and also zeolite framework destruction), reduction[35] or in cyclic operation.[36]

In practical application, consideration would need to be given to the use of different reactor configurations such as fluidised bed and also the application of catalyst pellets which are formed employing the use of a binder. The use of novel reactor configurations employing hydrogen permeable membranes may also be of interest for driving the reaction in a forward direction, as might the application of a two-zone reactor as reported in the review by Hutchings and Spivey.[1]

Whilst the literature has demonstrated a positive effect upon performance upon some zeolite-catalysed reactions related to binder effects,[37] some studies have shown binders to have negative effects for methane dehydroaromatisation.[38,39]

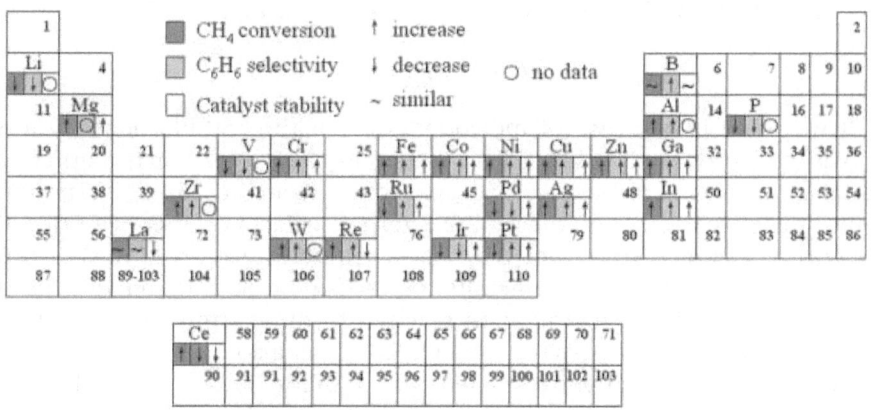

Figure 5.6 Effect of additives on the performance of Mo/zeolite catalysts for methane dehydroaromatisation. Reproduced from ref. 33 with permission from Elsevier, Copyright 2013.

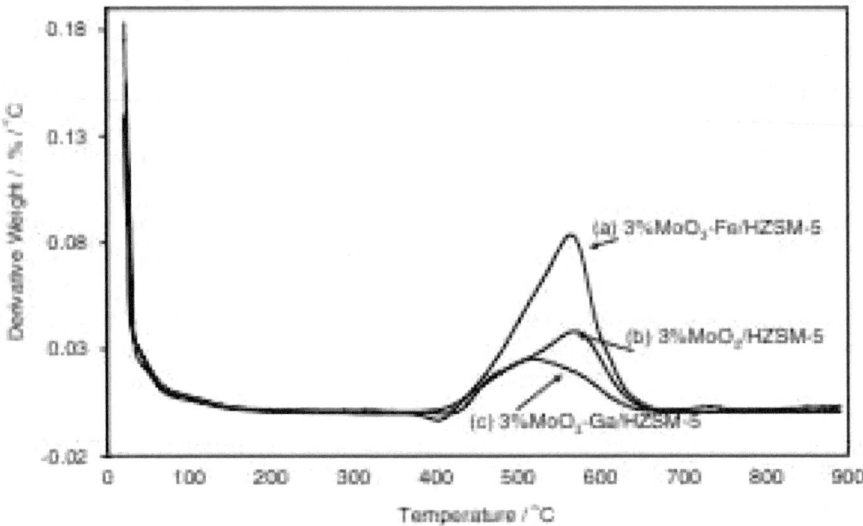

Figure 5.7 First derivative thermogravimetric analysis plots (5 °C min^{-1} in flowing air) of 3 wt% MoO$_3$/HZSM-5 derived catalysts after 6.5 hours on stream at 700 °C from methane dehydroaromatisation. Reproduced from ref. 25 with permission from Springer Nature, Copyright 2007.

5.3 Other Catalytic Systems

A number of other catalytic systems have been found to be effective for methane dehydroaromatisation. As stated previously, in their landmark publication of 1993,[4] Wang et al. reported Zn/HZSM-5 to be active. Lunsford and co-workers made a comparison of the performance of Mo-, W-, Fe-, V- and Cr-containing catalysts.[40] After 3 hours reaction at 750 °C, following CO pre-reduction, the order of reaction was found to be:

$$Mo > W > Fe > V > Cr$$

It was only in the case of molybdenum that the carbide phase was observed and proposed to be of importance for activity.[41] For the other materials, the active phases were proposed to relate to the corresponding suboxides dispersed upon the zeolite external surface, i.e. WO$_2$, Fe$_3$O$_4$, V$_2$O$_3$, Cr$_2$O$_3$.[41,42] Detailed studies of W/HZSM-5 reported recently have shown the importance of tuning zeolite acidity and W-zeolite interaction with partial carburisation of W occurring upon reaction forming a mixture of oxidation states between W(IV) and W(II).[43] Re/HZSM-5 has also been shown to be an active catalyst[44,45] with encapsulated ~8.2 Å diameter Re clusters being reported to be of importance for performance.[46] In a study comparing Mo/ZSM-5, Re/ZSM-5 and Fe/ZSM-5, metallic Re and Fe(II) oxycarbides were stated to be of importance for the latter two materials.[27] The length of the induction period, during which an active hydrocarbon pool species formed,

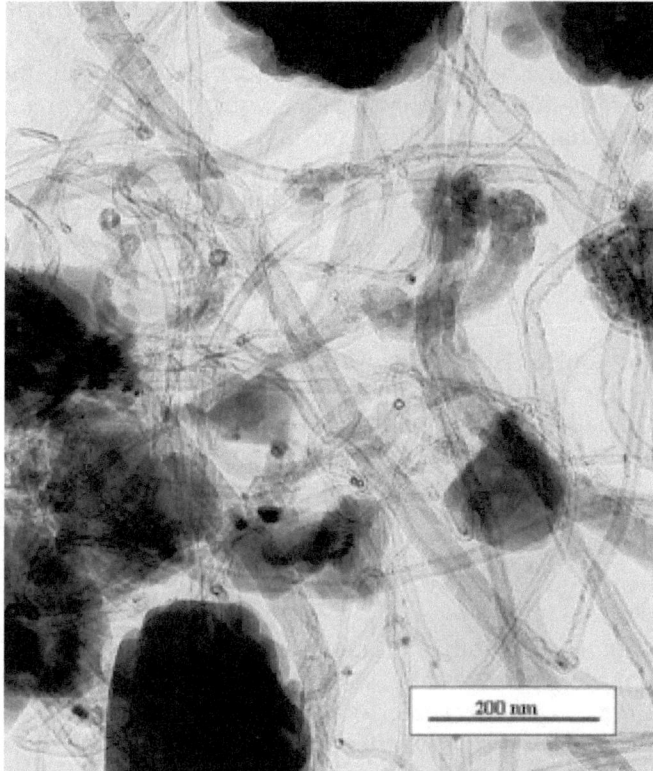

Figure 5.8 Transmission electron micrograph of post-reaction Fe^{3+}-MoO_3/H-ZSM-5 following 6.5 hours on stream for methane dehydroaromatisation at 700 °C. Reproduced from ref. 25 with permission from Springer Nature, Copyright 2007.

was reported to be a good descriptor in relation to catalytic behaviour with there being an activity–stability trade-off. The performance of Fe/ZSM-5, which displayed a long induction time, was found to be improved by increasing the iron loading, the reaction pressure and the space velocity.

5.4 Conclusions

As shown in this short review, methane dehydroaromatisation continues to be a topic of interest. Whilst it is arguable that much of the attention has been focused upon Mo/HZSM-5, it is clear that this is a complex system and that there is still a lack of consensus relating to some of the key points. Additional reviews relating to the topic have been published by others over the years, *e.g.* ref. 1, 2, 33 and 47–50. Thermodynamic considerations dictate that reactions are conducted at high temperatures, with 700 °C being commonly employed, in order to achieve appreciable levels of aromatic product. Tests are generally employed at low space velocity and there is significant room for the development of more active catalysts such that higher levels of productivity such as benzene space time yield can be achieved. It is also apparent that despite being the major product of reaction in molar terms, only a minority of studies employ hydrogen quantification, presumably due to the inherent analytical challenges involved. However, hydrogen in itself is a very valuable product of reaction and consideration of its productivity sheds additional insight on the occurrence of side-reactions and de-activation pathways.

References

1. J. J. Spivey and G. Hutchings, *Chem. Soc. Rev.*, 2014, **43**, 792.
2. U. Menon, M. Rahman and S. J. Khatib, *Appl. Catal., A*, 2020, **608**, 117870.
3. H. L. Mitchell and R. H. Waghorn, *US Pat.*, 4239658, 1980.
4. L. Wang, L. Tao, M. Xie, G. Xu, J. Huang and Y. Xu, *Catal. Lett.*, 1993, **21**, 35.
5. S. T. Liu, L. Wang, R. Ohnishi and M. Ichikawa, *J. Catal.*, 1999, **181**, 175.
6. D. Ma, Y. Y. Shu, M. J. Cheng, Y. D. Xu and X. H. Bao, *J. Catal.*, 2000, **194**, 105.
7. C. L. Zhang, S. A. Li, Y. Yuan, W. X. Zhang, T. H. Wu and W. Lin, *Catal. Lett.*, 1998, **56**, 207.
8. Y. D. Xu, X. H. Bao and L. W. Lin, *J. Catal.*, 2003, **216**, 386.
9. P. Cong, I. Lezcano-Gonzalez, A. Longo, W. Bras and A. M. Beale, *Phys. Chem. Chem. Phys.*, 2024, **26**, 30055.
10. R. W. Borry, Y. H. Kim, A. Huffsmith, J. A. Reimer and E. Iglesia, *J. Phys. Chem. B*, 1999, **103**, 5787.
11. V. T. T. Ha, L. V. Tiep, P. Meriaudeau and C. Naccache, *J. Mol. Catal. A: Chem.*, 2002, **181**, 283.
12. B. Li, S. Li, N. Li, H. Chen, W. Zhang, X. Bao and B. Lin, *Microporous Mesoporous Mater.*, 2006, **88**, 244.
13. D. Zhou, D. Ma, X. Liu and X. Bao, *J. Mol. Catal. A: Chem.*, 2001, **168**, 225.
14. S. Burns, J. S. J. Hargreaves, M. Stockenhuber and R. P. K. Wells, *Microporous Mesoporous Mater.*, 2007, **104**, 97.
15. H. Zheng, D. Ma, X. Liu, W. Zhang, X. Han, Y. Xu and X. Bao, *Catal. Lett.*, 2006, **111**, 111.
16. Y. Y. Shu, Y. D. Xu, S. T. Wong, L. S. Wang and X. X. Guo, *J. Catal.*, 1997, **170**, 11.
17. S. Burns, J. S. J. Hargreaves, P. Pal, K. M. Parida and S. Parija, *Catal. Today*, 2006, **114**, 383.
18. R. Ohnishi, S. Liu, Q. Dong, L. Wang and M. Ichikawa, *J. Catal.*, 1999, **182**, 92.
19. M. Agote-Arán, A. B. Kroner, H. U. Islam, W. J. Sławiński, D. S. Wragg, I. Lezcano-González and A. M. Beale, *ChemCatChem*, 2019, **11**, 473.
20. D. J. Wang, J. H. Lunsford and M. P. Rosynek, *J. Catal.*, 1997, **169**, 347.
21. P. Meriaudeau, V. T. T. Ha and L. V. Tiep, *Catal. Lett.*, 2000, **64**, 49.
22. R. B. Ley and M. Boudart, *Science*, 1973, **181**, 547.
23. S. T. Oyama, *Catal. Today*, 1992, **15**, 179.

24. A.-M. Alexander and J. S. J. Hargreaves, *Chem. Soc. Rev.*, 2010, **39**, 4388.
25. S. Burns, J. G. Gallagher, J. S. J. Hargreaves and P. J. F. Harris, *Catal. Lett.*, 2007, **116**, 122.
26. U. Olsbye, S. Svelle, M. Bjørgen, P. Beato, T. V. W. Janssens, F. Joensen, S. Bordiga and K. P. Lillerud, *Angew. Chem., Int. Ed.*, 2012, **51**, 5810.
27. Y. Liu, M. Ćoza, V. Drozhzhin, Y. van den Bosch, L. Meng, R. van de Poll, E. J. M. Hensen and N. Kosninov, *ACS Catal.*, 2023, **13**(1), 8128.
28. A. Beuque, H. Hao, E. Berrier, N. Batalha, A. Sachse, J.-F. Paul and L. Pinard, *Appl. Catal., B*, 2022, **309**, 121274.
29. B. M. Weckhuysen, M. P. Rosynek and J. H. Lunsford, *Catal. Lett.*, 1998, **52**, 31.
30. H. Jiang, L. S. Wang, W. Cui and Y. D. Xu, *Catal. Lett.*, 1999, **57**, 95.
31. S. Qi and B. Yang, *Catal. Today*, 2004, **98**, 639.
32. S. Liu, Q. Dong, R. Ohnishi and M. Ichikawa, *Chem. Commun.*, 1997, 1455.
33. S. Ma, X. Guo, L. Zhao, S. Scott and X. Bao, *J. Energy Chem.*, 2013, **22**, 1.
34. P. Zhu, W. Bian, B. Liu, H. Deng, L. Wang, X. Huang, S. L. Spence, F. Lin, C. Duan, D. Ding, P. Dong and H. Ding, *Nat. Commun.*, 2024, **15**, 3280.
35. Y. Liu, H. Zhang, A. S. G. Wijpkema, F. J. A. G. Coumans, L. Meng, E. A. Uslamin, A. Longo, E. J. M. Hensen and N. Kosinov, *Chem. - Eur. J.*, 2022, **28**, e202103894.
36. M. T. Portilla, F. J. Llopis and C. Martínez, *Catal. Sci. Technol.*, 2015, **5**, 3086.
37. J. S. J. Hargreaves and A. L. Munnoch, *Catal. Sci. Technol.*, 2013, **3**, 1165.
38. K. Honda, X. Chen and Z.-G. Zhang, *Appl. Catal., A*, 2008, **351**, 122.
39. Y. Xu, H. Ma, Y. Yamamoto, Y. Suzuki and Z. Zhang, *J. Nat. Gas Chem.*, 2012, **21**, 729.
40. B. M. Weckhuysen, D. Wang, M. P. Rosynek and J. H. Lunsford, *J. Catal.*, 1998, **175**, 338.
41. B. M. Weckhuysen, D. Wang, M. P. Rosynek and J. H. Lunsford, *J. Catal.*, 1998, **175**, 347.
42. B. M. Weckhuysen, D. Wang, M. P. Rosynek and J. H. Lunsford, *Angew. Chem., Int. Ed.*, 1997, **36**, 2374.
43. M. Çağlayan, A. Nasserewddine, S.-A. F. Natase, A. Aguilar-Tapia, A. Dikhtiarenko, S.-H. Chung, G. Shterk, T. Shoinkhorova, J.-L. Hazemann, J. Ruiz-Martinez, L. Cavallo, S. Ould-Chikh and J. Gascon, *Catal. Sci. Technol.*, 2023, **13**, 2748.
44. L. S. Wang, R. Ohnishi and M. Ichikawa, *Catal. Lett.*, 1999, **62**, 29.
45. L. S. Wang, R. Ohnishi and M. Ichikawa, *J. Catal.*, 2000, **190**, 276.
46. H. S. Lacheen, P. J. Cordeiro and E. Iglesia, *Chem. – Eur. J.*, 2007, **13**, 3048.
47. T. V. Choudhary, E. Aksoylu and D. W. Goodman, *Catal. Rev.:Sci. Eng.*, 2003, **45**, 151.
48. K. D. Sun, D. M. Ginosar, T. He, Y. L. Zhang, M. H. Fan and R. P. Chen, *Ind. Eng. Chem. Res.*, 2018, **57**, 1768.
49. S. V. Konnov, *Pet. Chem.*, 2022, **62**, 280.
50. Y. Ogawaa, Y. Xub, Z. Zhang, H. Ma and Y. Yamamoto, *Resour. Chem. Mater.*, 2022, **1**, 80.

CHAPTER 6

Activation of CO, CO_2, and H_2 Toward Synthetic Fuel Manufacture by Fischer–Tropsch Synthesis

A. N. AKIN,*[a] O. OZCAN,[a] M. DOGAN-OZCAN[b] AND D. UNER*[b]

[a] Chemical Engineering, Kocaeli University, Baki Komsuoglu Bulvari, Kocaeli 41001, Turkiye; [b] Chemical Engineering, Middle East Technical University, Dumlupinar Bulvari no 1, Ankara 06800, Turkiye
*Emails: akinn@kocaeli.edu.tr; uner@metu.edu.tr

6.1 Overview

Catalytic hydrogenation of carbon monoxide to hydrocarbons over supported transition metals (group VIII of the periodic table) has been studied for many years. The history of the process has been adequately described in numerous articles, including review papers by Storch,[1] Pichler,[2] Biloen and Sachtler,[3] Dry,[4] Janardanarao,[5] and Hindermann et al.[6] The following highlights must be noted: The synthesis of hydrocarbons from carbon monoxide hydrogenation was first recorded by Sabatier and Senderens in 1902 when they found that CO could be hydrogenated to form methane using cobalt and nickel catalysts. However, it was the pioneering research of Franz Fischer and Hans Tropsch that was to be of immense commercial significance. Their work revealed that iron, cobalt, and nickel were the best metals for the industrial hydrogenation of carbon monoxide to hydrocarbons. Cobalt had the greatest tendency to produce aliphatic hydrocarbons with more than one carbon atom per molecule. Iron was less active and rapidly deactivated, while nickel showed too high a hydrogenation activity favoring the formation of methane. After considerable research a cobalt/thorium oxide catalyst was selected for the first commercial plant for which a license was granted to Ruhrchemie A.G. in 1934. In 1954, Sasol commercialized the use of an alkali-promoted iron catalyst in South Africa which has successfully been in continuous production since that date.

Hydrogenation of carbon monoxide remains a cornerstone technology for converting syngas into a range of products, including gasoline, diesel, kerosene, wax, alcohols, ethers, and lower olefins, using catalysts such as cobalt and iron. Precious metals such as ruthenium also exhibit high activity, but their commercialization potential is limited due to their high cost and limited availability. The selectivity of the resulting products depends on the catalyst used and the specific process conditions such as temperature, pressure, and space velocity.[7,8] The last quarter of the 20th century witnessed rapid progress in surface science and successful implementation of computational tools such as density functional theory (DFT) for catalysis research. Enabled by such tools, it was possible to develop sophisticated surface reaction models, and critically evaluate the surface intermediates and determine the bottlenecks of the reaction. The surface science of catalysis has led to realistic surface reaction models, which could be justified through *in situ* and *operando* studies. Better selection of catalytic components could lead to catalyst design or "catalysis by design" optimizing objective functions such as activity, and in the case of Fischer–Tropsch synthesis (FTS), most importantly, selectivity. Nevertheless, challenges remain in the efficient activation of C–O and C–C bonds and the control of carbon chain growth, which are essential for the selective synthesis of long-chain hydrocarbons and other high-value products.[9]

The driving force for the increased interest in Fischer–Tropsch technology in the 21st century primarily stems from the global need to produce energy from renewable resources. The bottleneck in FTS as a viable option for sustainable fuel production resides in the high energy demands of syngas manufacture. If renewable resources are used for syngas production, the exothermic FTS process bears significant potential to produce sustainable fuels. FTS fuels can be envisioned as fully renewable as long as waste and renewable carbon resources are used to produce syngas. Furthermore, renewable energy has to be used to drive the endothermic reforming reactions. FTS can close the carbon cycle by incorporating CO_2 in the reforming process as well as utilizing rapid-growing biomass such as algae and switchgrass, with their already existing capacity to capture and convert CO_2 from the atmosphere. In addition, renewable hydrogen sources can be used for converting CO_2 through the reverse water–gas shift (RWGS) reaction to produce CO, which can then be used in subsequent hydrogenation processes.[10] Examples of current investments in FTS range from the 140 000 bbl per day plant of Shell in Qatar, operating using methane as the raw material,[11] to the 90 000 bbl per day capacity coal to liquid plants in China† and a high-temperature FTS pilot plant of Synfuels in Mongolia with 4000 bpd capacity can be included here.[11]

6.2 Production of Hydrocarbons from Carbon Monoxide Hydrogenation

Product formation in the FTS can be presented by the following general equations:[12,13]

Paraffins:

$$(2n+1)H_2 + nCO = C_nH_{2n+2} + nH_2O \tag{6.1}$$

or

$$(n+1)H_2 + 2nCO = C_nH_{2n+2} + nCO_2 \tag{6.2}$$

Olefins:

$$2nH_2 + nCO = C_nH_{2n} + nH_2O \tag{6.3}$$

or

$$nH_2 + 2nCO = C_nH_{2n} + nCO_2 \tag{6.4}$$

Alcohols:

$$2nH_2 + nCO = C_nH_{2n+1}OH + (n-1)H_2O \tag{6.5}$$

or

$$(n+1)H_2 + (2n-1)CO = C_nH_{2n+1}OH + (n-1)CO_2 \tag{6.6}$$

Aromatics, aldehydes, ketones, and acids can also be produced by similar reactions. A number of complicating reactions can occur concurrently with the synthesis reaction, and some of the most important ones are as follows:[14]

Water–gas shift reaction:

$$CO + H_2O = CO_2 + H_2 \tag{6.7}$$

Boudouard reaction:

$$2CO = C + CO_2 \tag{6.8}$$

Coke deposition:

$$H_2 + CO = C + H_2O \tag{6.9}$$

Carbide formation:

$$xM + C = M_xC \tag{6.10}$$

---

† Accessed from http://www.news.cn/english/2021-12/21/c_1310386110.htm.

All of the above reactions can find a window of operational parameters to be thermodynamically favorable during FTS reactions. Therefore it is theoretically possible to produce an enormous variety of compounds at reasonable reaction temperatures.[14] However, one or more of these reactions can be favored by choosing appropriate catalysts and suitable reaction conditions, such as temperature, pressure, space velocity, and H_2/CO ratio. As shown in Table 6.1, prepared by considering review articles[12,15,16] and books,[17,18] different catalysts and promoters under a wide range of temperature and pressure conditions produce various products from the carbon monoxide hydrogenation reaction. It must be noted here that carbide formation is one of the critical steps during Fischer–Tropsch synthesis: Hägg carbide is an active form of carbon involved in the chain growth process, while free carbon formation leads to catalyst deactivation. The addition of chromium oxide to precipitated iron catalysts has been shown to inhibit the free carbon formation and to favor branched hydrocarbons and oxygenates.[19]

Figure 6.1 gives a general idea of the influence of reaction conditions on the resultant product pattern. The product distribution in FTS is significantly influenced by process parameters such as temperature, pressure, ratio of hydrogen to carbon monoxide (H_2/CO), space velocity, and water presence. Temperature is a critical factor; higher temperatures favor the production of lower carbon number species like methane, while lower temperatures increase the selectivity toward heavier hydrocarbons (C_{5+}). Reaction thermochemistry dictates that lower temperatures and higher pressures suppress methane selectivity and favor C_{5+} selectivity. At higher pressures, in addition to heavier hydrocarbons, production of oxygenates is also observed.[20] Nevertheless, very high pressures can accelerate catalyst deactivation. H_2/CO ratio is another key parameter. Higher ratios tend to favor the formation of lighter hydrocarbons, whereas lower ratios encourage chain growth and higher-molecular-weight products. For example, an increase in the H_2/CO ratio typically results in a higher methane selectivity and a lower C_{5+} selectivity.[16,21] Furthermore, space velocity affects product selectivity as well; lower space velocities tend to reduce methane and olefin selectivity without significantly impacting paraffin selectivity. Additionally, the presence of water, a by-product of the FTS reaction, also influences the product distribution. In iron-based catalysts, water can cause re-oxidation and consequently affect the water–gas shift reaction. In the case of cobalt-based catalysts, the presence of water has been observed to influence the CO conversion and the selectivity toward methane and C_{5+} hydrocarbons. The effect of water varies depending on the catalyst support. In the case of silica-supported cobalt catalysts, increased CO conversion has been observed in the presence of water. Conversely, alumina-supported catalysts may experience a negative effect in the presence of water. Furthermore, smaller cobalt particles are more susceptible to oxidation by water. The interplay of these parameters determines the efficiency and selectivity of the FTS process,

Table 6.1 Products of reaction between carbon monoxide and hydrogen.

Reaction	Catalysts	Supports and promoters	Temp. (°C)	Pressure (atm)	Products
Methanation	Ru, Ni	ThO_2, MgO	220–500	1	Mainly methane
Fischer–Tropsch synthesis	Fe, Co	TiO_2, SiO_2, MgO, Al_2O_3, K_2O, ZrO_2, CeO_2, hydroxyapatite (HAP), carbon nanotube (CNT), K, Mn, Bi, Pb, Na, La	200–360	1–40	Lighter olefins and oxygenates and long-chained hydrocarbons
Methanol synthesis	ZnO, Cu, Cr_2O_3, MnO, Al_2O_3		200–450	10–400	Methanol
Higher alcohol synthesis	Rh, Mo, ZnO, Cu, Cr_2O_3, MnO, CuCo, CuFe	Alkalis, Mn, Li, Fe, Al_2O_3, TiO_2, SiO_2, K_2CO_3, Na_2CO_3, NaOH, KOH, Rb_2CO_3, Cs_2CO_3	200–450	10–400	Methanol and higher alcohol
Iso-synthesis	Zn-Cr, ThO_2, ZrO_2, CeO_2, ZnO + Al_2O_3	K_2O, Li, Na, Cs	350–500	70–100	Saturated branched hydrocarbons

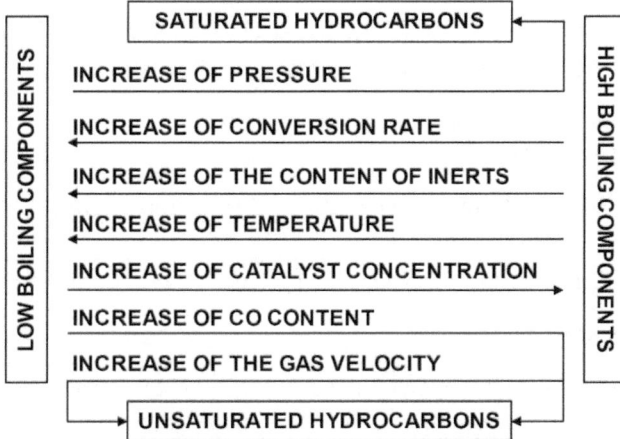

Figure 6.1 Influence of the process parameters on the product distribution.[5] Reproduced from ref. 5 with permission from American Chemical Society, Copyright 1990.

necessitating careful optimization to achieve the desired product distribution.[16]

6.3 Thermodynamics of the CO Hydrogenation Reaction

The thermodynamics of synthesis gas reactions has been discussed in detail in several review papers such as those by Storch *et al.*[22] and Mills and Steffgen.[23] Although thermodynamic analysis of FTS is very complex, and conclusions drawn from thermodynamics have limited applicability since most synthesis gas reactions are not operated under equilibrium conditions, it is still pertinent to highlight important thermodynamic consequences of this reaction system in order to obtain useful information concerning competing reactions.

6.3.1 Heats of Reaction

FTS is an exothermic reaction system. Information on the heats of reaction is of great practical importance, since removal of the heat is one of the most difficult engineering aspects in FTS. This high heat release makes it difficult to prevent overheating and deactivation of the catalyst and also it can cause temperature rises to such a degree that desirable product yield becomes limited due to thermodynamic equilibrium. The reaction energetics summarized in Box 6.1 reveal the interplay of the endothermic and exothermic reactions. Generally, the temperature dependencies of these reactions are not very strong and standard values can be practically used to infer conclusions about the synthesis.

> **Box 6.1 Enthalpy changes of major reactions leading to Fischer–Tropsch synthesis.**
>
> **FTS reactions and their energetics at a glance**
> *Reforming reactions*
>
> $$CH_4 + H_2O \Leftrightarrow CO + 3H_2 \quad \Delta H^0 = 206 \text{ kJ mol}^{-1}$$
>
> $$CH_4 + CO_2 \Leftrightarrow 2CO + 2H_2 \quad \Delta H^0 = 247 \text{ kJ mol}^{-1}$$
>
> Water gas shift
>
> $$CO + H_2O \Leftrightarrow CO_2 + H_2 \quad \Delta H^0 = -41 \text{ kJ mol}^{-1}$$
>
> Fischer Tropsch Synthesis
>
> $$CO + 2H_2 \Leftrightarrow -CH_2- + H_2O \quad \Delta H^0 = -165 \text{ kJ mol}^{-1}$$

6.3.2 Free Energy of Reaction

The free energy characteristics of the FTS were described by Anderson,[24] Shah and Perrotra,[12] Newsome,[25] and Hindermann et al.[6] The major conclusions that can be drawn are as follows:

i. Methane production is always thermodynamically preferred over reactions producing alcohols, higher alkanes, and alkenes since the standard free energy changes per carbon atom for reactions producing methane are more negative.
ii. The order of preference for product type is paraffin > mono-olefin > di-olefins = alcohols.
iii. Among olefin products, ethylene is the thermodynamically least preferred product below 430 °C. It is therefore clear that if C_2–C_4 olefins or alcohols are the desired products, then the reaction requires a catalyst that exerts kinetic control.
iv. Reactions forming carbon dioxide rather than water have more negative values for the Gibbs free energy change.
v. The equilibrium constant of the water–gas shift reaction is large (>20) at the FTS temperatures (180–350 °C). Hence if equilibrium were attained for the water–gas shift, almost all of the water produced by the primary synthesis reaction should be converted to CO_2, but, in fact, for cobalt catalysts the kinetics rather than the thermodynamics control product distribution, since water–gas shift has a relatively low rate of reaction within this temperature range.
vi. The equilibrium conversion of synthesis increases with increasing pressure. For a given conversion, the higher the operating pressure the higher the synthesis temperature needed.

vii. Carbon formation is favored at high temperatures. Therefore, the practical upper limits of synthesis temperature and pressure for iron are about 400 °C and 30–40 atms. For cobalt and nickel catalysts, they are considerably lower.
viii. In this range of temperature and pressure, significant yields of all hydrocarbons with the exception of acetylene are thermodynamically possible.
ix. Hydrogenation of carbon dioxide to hydrocarbons or alcohols, except for acetylene and methanol is thermodynamically possible under most synthesis conditions. Hydrogenation of olefins and dehydration of alcohols are also thermodynamically possible.
x. Reactions of water and carbon to give hydrocarbons have positive Gibbs free energy values. Therefore, the production of synthetic fuels directly from coal is thermodynamically limited under all practical conditions.

6.4 Product Distribution

The synthesis of hydrocarbons from carbon monoxide and hydrogen on Fischer–Tropsch catalysts can be viewed as starting from a C_1 unit which gradually builds up in molecular size to produce higher hydrocarbons.[26] On the basis of extensive experimental data, mathematical formulae have been derived which can very adequately fit the product distribution. The models are based on the assumptions that (a) chain growth in the carbon monoxide hydrogenation occurs *via* a polymerization process with the addition of one C_1 unit at a time to the growing chain and (b) that the probability of chain growth is independent of chain length. This type of molecular weight distribution was demonstrated by Schulz for polycondensation and by Flory for free radical polymerization,[6] and the combination of both approaches has become known as the Schulz–Flory distribution function. If the chain growth probability α is considered to be independent of the chain length, the reaction scheme is as shown in Box 6.2.

Box 6.2 Basic mechanism for FTS if the chain growth probability α is independent of chain length.[11] Adapted from ref. 11 with permission from Elsevier, Copyright 2013.

$$C* \rightarrow C_1* \xrightarrow{1-\alpha} C_1 + *$$

$$2\,C_1* \xrightarrow{\alpha} C_2* + * \xrightarrow{1-\alpha} C_2 + *$$

$$C_2* + C_1* \xrightarrow{\alpha} C_3* + * \xrightarrow{1-\alpha} C_3 + *$$

When the prevailing conditions lead to a chain growth mechanism as shown in Box 6.2, the total amount of carbon contained in products with n carbon atoms, C_n, can be inferred through the following relationships:

$$C_1 = 1(1-\alpha) \tag{6.11}$$

$$C_2 = 2(1-\alpha)\alpha \tag{6.12}$$

$$C_3 = 3(1-\alpha)\alpha^2 \tag{6.13}$$

$$C_n = n(1-\alpha)\alpha^{(n-1)} \tag{6.14}$$

Once the sum is evaluated, the total amount of the carbon is related to the chain growth probability as follows:[11]

$$\sum_1^\infty C_n = \sum_1^\infty n(1-\alpha)\alpha^{(n-1)} = \frac{1}{1-\alpha} \tag{6.15}$$

This equation can be used to obtain a relationship for the selectivity toward the n carbon containing product as

$$S_n = \frac{C_n}{\sum_1^\infty C_n} = n(1-\alpha)^2 \alpha^{n-1} \tag{6.16}$$

or in the logarithmic domain

$$\ln\left(\frac{S_n}{n}\right) = n \ln \alpha + \ln \frac{(1-\alpha)^2}{\alpha} \tag{6.17}$$

a is the chain growth probability factor, which can be defined as

$$a = \frac{r_p}{(r_p + r_t)} \tag{6.18}$$

where r_p and r_t are the rates of propagation and chain termination, respectively. Friedel and Anderson developed this approach with specific application to the Fischer–Tropsch synthesis and the resulting equations were equivalent to those developed by Schulz and Flory.[6] If this distribution function applies, a plot of $\ln(S_n/n)$ versus n should yield a straight line where both the slope and intercept give the same calculated value of a. A consequence of such a polymerization mechanism is that the theoretical maximum yield possible for any product rather than methane is relatively low, and only methane can be obtained with 100% selectivity. Figure 6.2 shows how the product distribution changes with the increasing chain growth probability factor which demonstrates that as this approaches unity, the product mainly comprises heavy waxes, and the synthesis of light hydrocarbons is limited. Hydrocarbons in the proper boiling range for gasoline can be produced at a maximum selectivity of slightly less than 50%, the maximum selectivity to a C_2 species is

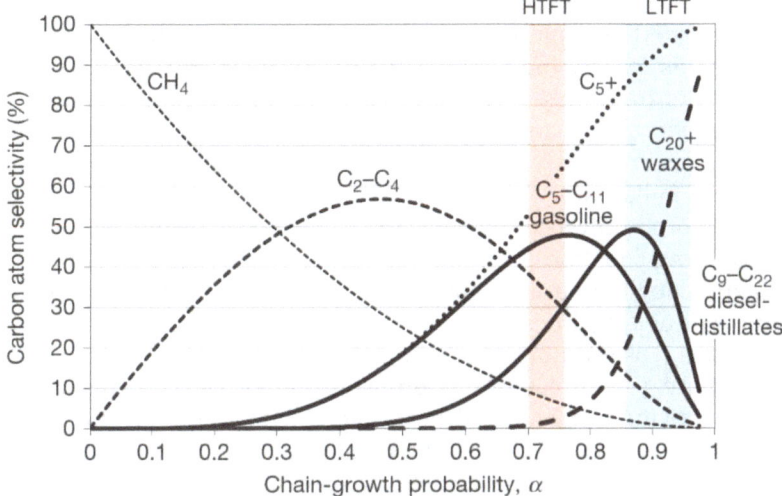

Figure 6.2 Product distribution for CO hydrogenation as a function of increasing chain growth probability factor (α).[11] Reproduced from ref. 11 with permission from Elsevier, Copyright 2013.

approximately 30%. Thus, to use the Anderson–Schulz–Flory product distribution to the advantage of lower hydrocarbons, a choice must be made between two approaches: (a) operation at high α values (>0.9) and develop catalysts that facilitate this at mild conditions, or (b) development of catalysts that do not yield the Anderson–Schulz–Flory (ASF) product distribution.

Deviations from the ASF distribution are common due to secondary reactions of the olefinic products, such as the incorporation of C_2–C_4 olefins into the chain and hydrogenolytic cleavage of olefins, leading to excessive methane formation and altered molecular weight distributions.[27] Experimental studies have shown that product distributions vary with the type of catalyst employed. Catalyst composition and interaction play significant roles in product distribution. In a study conducted by Zhang and Lin, different iron-based catalysts combined with ZSM5 zeolites exhibited varied product distributions by adjusting the Si/Al ratio to control the C_5–C_{11} hydrocarbon selectivity, highlighting the impact of catalyst composition on ASF deviations.[28]

6.5 Carbon Monoxide Hydrogenation Catalysts

Basically, the hydrogenation of carbon monoxide to hydrocarbons can occur on almost any group VIII transition metal. They contain partially occupied d-orbitals, and, under reaction conditions, may be converted by carbon monoxide or hydrogen to lower, mostly metallic or carbidic, oxidation states. In these electronic states, the catalytically active metals are able to interact with the syngas and adsorb or chemisorb the syngas components. It is generally conceded that iron, cobalt, and ruthenium yield high-molecular-weight compounds while nickel favors an almost exclusive production of methane, platinum and iridium show very little activity, and palladium yields largely methane and very minimal ethane and ethylene, rhodium results in mainly methane, and osmium produces primarily oxygenated products.[29] Ethylene selectivity declines in the order iridium > cobalt > ruthenium > iron > rhodium and no ethylene is produced on nickel, platinum, and palladium.[29]

The most commonly used metals for carbon monoxide hydrogenation catalysts are iron, cobalt, ruthenium, and nickel. From a practical viewpoint, only iron and cobalt are useful commercially.[5] Cobalt catalysts are preferred for producing linear hydrocarbons and for their high selectivity toward long-chain paraffins. Disadvantages are the high costs of cobalt and low water–gas shift activity. Cobalt catalysts are not inhibited by water, resulting in a higher productivity at a high synthesis-gas conversion.[30] Furthermore, industry uses cobalt catalysts for their resistance to the carbon deposition problem.[31] The effect of metal particle size on the activity and selectivity of the supported cobalt catalysts has established new directions in catalyst preparation and utilization toward product selectivity. Joint research from Shell Global Solutions, and Utrecht and Delft universities revealed that for cobalt nanoparticles smaller than 8 nm, the turnover frequency (TOF) increases with catalyst particle size reaching a plateau at which further change in particle size does not alter the TOF for the reaction.[32] The ability to observe the structure sensitivity at this scale further enabled the design of cobalt catalyst for tuned activity and selectivity.

Iron catalysts are more cost-effective and have been applied industrially for Fischer–Tropsch synthesis for many years. These catalysts have a high water–gas shift activity and high selectivity to olefins.[30] They are particularly advantageous for processing synthesis gas with a low H_2/CO ratio, typical of coal gasification outputs, without requiring an additional WGS reactor.[33] A recent study revealed the role of the pure χ-iron carbide (χFe_5C_2) on the selective production of linear α-olefins while decreasing CO_2 selectivity.[34] Lower CO_2 selectivity is especially important not only for addressing global emissions concerns but also for maximizing carbon utilization in carbon-based raw material economies. The choice between iron and cobalt catalysts involves a balance between desired resistance to deactivation, selectivity, price, and operating conditions. Additionally, nickel and ruthenium are also used in CO hydrogenation, with nickel producing lower-molecular-weight hydrocarbons in comparison to cobalt and iron, with ruthenium being less common due to its high cost.[20,35]

Some metals other than the group VIII metals can also be used as Fischer–Tropsch catalysts, for instance, molybdenum and tungsten. It is believed that molybdenum itself is sulfur resistant, and molybdenum carbide is the form active for carbon monoxide hydrogenation and displays a high resistance to carbon and sulfur poisoning.[36,37] Also, they have high olefin selectivity with the promotion of potassium.[38]

According to Krylov[39] certain oxides are also capable of yielding high conversions of carbon monoxide into hydrocarbons. Most active among typical oxide catalysts are MoO_3, WO_3, V_2O_5, and MnO. Less active are SiO_2, MgO, ThO_2, Al_2O_3, and Cr_2O_3. This reaction is also catalyzed by nitrides, carbides, and borides of transition metals.[39] Among borides, the order of decreasing activity is: $Ni_2B > Co_2B$ and Fe_2B. Cobalt carbide is less active than cobalt metal. In contrast, the carbides, nitrides, and carbonitrides of iron are more catalytically active than metallic iron. Prokheronko et al.[40] studied the effect of the chemical nature of the support on the activity of nickel and cobalt metals in the hydrogenation of carbon monoxide. The dispersion of nickel and cobalt changes as a function of the type of oxide support used, and a change is observed in the activity and selectivity. The activity of the cobalt catalysts at 540 K decreases in the order $SiO_2 > TiO_2 > ZrO_2 > BeO > Cr_2O_3 > Al_2O_3 > MgO \gg WO_3$, while selectivity for higher hydrocarbons decreases in the order $ZrO_2 > MgO > BeO > Al_2O_3$, $Cr_2O_3 > SiO_2 > TiO_2 > WO_3$. Selectivity is also a function of the acid–base properties of the carrier. Selectivity for higher hydrocarbons in Fischer–Tropsch synthesis increases with an increase in the acidity of the support, attaining the maximum value in the presence of ZrO_2. A further increase in the acidity results in a decrease in selectivity, and the samples with the highest acidity (WO_3) exhibit 100% selectivity for methane.

To obtain lower olefins via carbon monoxide hydrogenation, the oxides of titanium, vanadium, molybdenum, tungsten, manganese, zinc, and potassium are suitable as activators.[5,41] Chen[42] studied the synthesis and characterization of a series of methyl carbonyl cluster-derived, carbon-supported Fe–Co catalysts. There is a high selectivity to olefins and a lack of C_6 or higher-molecular-weight hydrocarbons produced in these catalysts.

Recent studies have investigated molybdenum carbides as catalysts for CO hydrogenation, revealing their notable resistance to carbon and sulfur poisoning. It has been observed that alkaline earth metal promoters enhance both catalytic activity and carbon chain growth, while alkali promoters, such as potassium, allow for tuning of selectivity toward alcohols and olefins. Likewise, mixed potassium–molybdenum sulfides have been suggested to improve olefin and alcohol selectivity. Research by Santos et al.[43] proposes that alkoxy species are critical to alcohol formation, with potassium promotion likely stabilizing these intermediates and altering the electronic structure of MoS_2. Although initial results for these Mo-based catalyst systems are promising, further work is needed to elucidate the molecular view of active sites and of the reaction mechanism.[36]

6.5.1 Active Metals

Iron is the first transition metal on which Fischer and Tropsch produced hydrocarbons which marked the discovery of the Fischer–Tropsch synthesis. The inexpensiveness and ready availability of iron have stimulated extensive studies on iron catalysts since the 1930s when Fischer and Pichler discovered medium-pressure synthesis on iron.[3] Iron requires higher pressure and temperature for Fischer–Tropsch synthesis than cobalt and nickel.

Water is the principal primary oxygenated product, but a subsequent water–gas shift reaction proceeds at about the same rate as the primary process. Iron has perhaps the greatest tendency to produce elemental carbon among the usual metals for carbon monoxide hydrogenation.[37] According to Dry,[4] silica is the best support for iron in terms of both activity and wax production. Yang et al.[44] studied the catalytic synthesis of light olefinic hydrocarbons from carbon monoxide and hydrogen over some iron catalysts. In these catalysts the main component is iron and the list includes the synthetic ammonia catalyst, two native iron ores (hematite and magnetite), and potassium, zinc, copper, magnesium, manganese, and calcium promoted iron catalysts.

Promoters significantly enhance the activity and selectivity of iron catalysts for carbon monoxide (CO) hydrogenation in FTS. In particular, zinc, copper, manganese, potassium, cerium, and molybdenum have been found to be highly effective. Copper outperforms Zn, K, and Ce in boosting the performance of iron–manganese catalysts for converting synthesis gas to light olefins and C_5–C_{12} hydrocarbon products.[45] The synergistic interaction between alkali and structural promoters (e.g., silica, alumina, etc.) helps maintain an adequate iron carbide layer, crucial for the active phase of the catalyst.[46] It was reported in the literature that iron carbides are crucial for high FTS activity, while magnetite (Fe_3O_4) is the most active phase for the WGS reaction.[33] Koelbel et al.,[47] Barrault et al.,[48] and Malessa et al.[49] investigated the catalytic performance of iron/manganese oxides having different manganese contents varying from 0% to 80%. Small amounts of manganese (3–15 wt%) enhanced the formation of olefins and lowered simultaneously the isomerization activity. Das et al.[50,51] also improved the olefin selectivity of iron–manganese oxide catalysts by adding potassium and/or silicalite. On the other hand, the addition of vanadium separately or together with zinc has greatly increased the olefin selectivity of iron catalysts.[52] The amount of olefin hydrocarbon fractions reached over 80%, although the addition of zinc separately has been less effective in terms of the olefin selectivity of the catalyst. Cornils et al.[53] studied on the one hand how to maximize the yield of C_2–C_4 olefins and on the other hand how to optimize the production of long-chain hydrocarbons. Iron catalysts with suitable promoters, such as oxides of titanium, vanadium, molybdenum, or manganese were used. They found that the selectivity of the formation of C_2–C_4 olefins decreases with an increase in syngas conversion, catalyst packing height, tube diameter, temperature gradients in the catalyst packing, and use of a gas circulation system. Adding Mo improves catalytic stability and reduces deactivation rates by mitigating carbon deposition, which is often a common issue in FTS reactions, and increases sulfur poisoning tolerance, maintaining catalyst activity under harsh conditions.[54] Feimer et al.[55] used a copper and potassium promoted iron catalyst to find out whether forced cyclical changes in feed composition could be used to improve selectivity of the carbon monoxide hydrogenation reaction in favor of gasoline range products. Except for methane, the product distribution followed the steady-state ASF model. Nonetheless, these experiments demonstrated that cyclic operation can affect product distribution, in this case cycling between a synthesis mixture and pure hydrogen substantially increased the rate of methane formation. Kikuchi et al.[56] observed that graphite-supported iron was the most active catalyst for carbon monoxide hydrogenation at atmospheric pressure and 400 °C. Methane was the main product. The addition of potassium promoted the activity of the catalyst but decreased the selectivity for hydrocarbon production. In another study it was shown that alkali promotion of iron catalysts did not affect their activity.[12] Although the activity was unaffected by alkali content, the selectivity of iron catalysts was changed. A substantial increase in the wax concentration occurs with a decrease in both the LPG and the liquid hydrocarbons.

The catalytic performance of iron catalysts can be significantly enhanced by incorporating other metals such as palladium, nickel, cobalt, and copper. The study of Gustafson and Wihner[57] reports the use of Pd–Fe bimetallics supported on ZnO for the selective production of olefins. The addition of palladium to iron resulted in a significant increase in catalytic activity and at the same time selectivity to olefins remained high. Incorporating cobalt into iron catalysts alters the crystal structure and reduction behavior, forming bimetallic CoFe phases that enhance catalytic performance in terms of alcohol formation and short-chain hydrocarbons.[58,59]

The first standard Fischer–Tropsch catalyst in commercial synthesis in Germany during World War II was a cobalt catalyst with a composition of 100 Co : 5 ThO_2 : 8 MgO : 200 kieselguhr.[24] It is generally considered that cobalt has one of the highest activities for carbon monoxide hydrogenation and has the greatest tendency to produce aliphatic hydrocarbons with more than one carbon atom per molecule.[60] Olefins, particularly α-olefins, seem to be a primary product of the synthesis on cobalt, as the fraction of olefin increases with increasing space velocity, which limits the secondary hydrogenation of olefins. The percentage of olefin in carbon number fractions has a maximum at about C_3–C_4 and then diminishes. Alcohols are hardly produced on cobalt catalysts.

The performance of cobalt catalysts is influenced by the support material, cobalt particle size, promoters, dispersion, metal loading, and the preparation method used.[61–63] Prohorenko et al.,[40] Wang et al.,[64] Lapidus et al.,[65] and Bessel[66] investigated the influence of the nature of the support; SiO_2, TiO_2, ZrO_2, BeO, Cr_2O_3, Al_2O_3, MgO, WO_3, NaX, kieselguhr, bentonite, zeolite Y, mordenite, and ZSM-5 were used as support materials. It was found that the support material is an important factor in influencing the specific activity of the catalyst. It is noted that in carbon monoxide hydrogenation, reaction selectivity for higher hydrocarbons increases with an increase in the acidity of the support, attaining the maximum value in the presence of zirconia. A further increase in the acidity results in a decrease in selectivity, and the catalysts with strong acidity, such as WO_3 and NaX, exhibit high selectivity for low-molecular-weight hydrocarbons. Also, the nature of the hydrocarbons produced depends very much on the acidity of the support. While straight-chained hydrocarbons are observed with low acidity supports, such as kieselguhr, silica, alumina, bentonite, more highly branched and aromatic products are obtained with acidic zeolite supports. Bessell[66] pointed out that if a high octane gasoline range product is desired, ZSM-5 is an excellent support for cobalt.

On the other hand, Varma et al.[67] reported the performance of MnO-supported nickel and cobalt catalysts of varying compositions for producing lower hydrocarbons, particularly olefins. 2% Ni–5% Co/MnO was found to be superior to other metal compositions in terms of the highest olefin/hydrocarbon selectivity with a corresponding high hydrocarbon yield. Van der Riet et al.[68] and Colley et al.[69] reported stable Co/Mn catalysts with high selectivity for C_3 hydrocarbons. The results indicate that total C_3 selectivity increases with increasing conversion, whether this is achieved by increasing temperature or pressure. Copperthwaite et al.[70] investigated the reaction conditions on alkene selectivity of the manganese oxide matrix catalyst containing comparable amounts of iron and cobalt. The reaction was carried out in a multi-fixed-bed laboratory reactor over coprecipitated Co/MnO and Fe/MnO catalysts. At comparable reaction conditions the iron catalyst was much less active than the cobalt catalyst. Inui et al.[71] developed a composite catalyst system of Co–Mn–Ru supported on alumina to synthesize a gaseous fuel of high calorific value from syngas. The catalyst showed high activity with high stability for syngas conversion into methane containing large amounts of C_2–C_4 hydrocarbons. The reaction was carried out in a continuous flow reactor under 10 kg cm^{-2} in the temperature range from 180° to 300 °C. Colley et al.[72] enhanced the selectivity toward light alkenes by adding potassium promoters to the Co/MnO catalyst, however, they reported also that adding chromium as a promoter increases the selectivity toward wax fractions.

The most commonly used supports include alumina, silica, and titania, which affect cobalt dispersion and reducibility differently, impacting the catalyst's activity and selectivity. Support materials, such as alumina, offer mechanical strength; however, strong metal–support interactions limit cobalt reducibility, which can be mitigated by promoters like Pt or Re.[73] Silica provides higher surface area and stability with weaker metal–support interactions than aluminum, enhancing cobalt reducibility but risking agglomeration and reduced dispersion.[74] To address these issues, carbon allotropes (e.g., carbon nanotubes and graphene) have been explored for their chemical inertness and high surface area, though they may suffer from rapid activity decline due to weak metal–support interactions leading to nanoparticle sintering.[75] Metal–organic frameworks (MOFs) have emerged as promising supports, transforming into porous carbon matrices upon pyrolysis, preventing nanoparticle sintering and controlling product selectivity in FTS. For example, a Co-MOF-71-derived catalyst (Co@C-500 catalyst) demonstrated high CO conversion and selectivity toward gasoline-range hydrocarbons, maintaining stability over extended periods.[75] Mesoporous silica supports, such as MCM-41, SBA-15, and SHS, improve cobalt dispersion and active metal surface area, leading to higher catalytic performance and selectivity for C_{5+} hydrocarbons. The physical structure of the support also plays a role, with periodic mesoporous silicas facilitating better access to active sites and transport of higher hydrocarbon products.[76] Studies in the literature reported that nitrogen-containing groups in Co/SBA-15 catalysts could suppress methane selectivity and enhance lower olefin selectivity, including ethylene, due to the electron-donating properties of N atoms. These properties accelerate CO dissociation and increase the concentration of CH_2 monomers.[77]

There is an enormous number of studies on cobalt catalysts supported on alumina to improve the selectivity toward light hydrocarbons for carbon monoxide hydrogenation. The major interest in Co/alumina catalysts has centered on interactions between the metal and support. Since most of the studies in the literature[64,78–83] indicate that cobalt oxide supported on alumina is not completely reduced to the metallic state due to a strong metal oxide–support interaction, the extent of reduction and cobalt dispersion are the major properties that affect the catalytic properties and product selectivity of cobalt catalysts. Specifically, cobalt can form unreducible cobalt aluminate spinels on alumina, which limit the reducibility of cobalt oxides and decrease the number of active sites available for FTS.[61,83] Moon et al.[80] and Wang et al.[81] found that the selectivity of unsaturated hydrocarbons increases as the catalyst becomes poorly reduced, and this is explained as the result of hydrogen suppression on poorly reduced catalysts. It is proposed that the unreduced cobalt oxide interacts electronically with the surface cobalt metal and consequently modifies the overall catalytic properties. Bartholomew and co-workers[63,84,85] studied the role of surface structure and dispersion on the product distribution. They found that although product selectivity is a function of support, dispersion, metal loading, and preparation, it is best correlated with dispersion and extent of reduction. In other words, lower-molecular-weight hydrocarbon products are observed for catalysts having higher dispersions and lower extents of reduction. Lapidus et al.[65] pointed out that pretreatment by calcination in air decreases the cobalt reducibility and cobalt dispersion for the 10% Co/alumina catalyst. Akin and Onsan[86] also found that calcination prior to reduction has a detrimental effect on metal surface area and hence on catalytic activity for the 35% Co/alumina catalyst prepared by the coprecipitation method. Lee et al.[87] studied the effect of carbon deposits on the hydrogenation of carbon monoxide over alumina-supported cobalt catalysts and they observed that hydrocarbon product distributions are not affected by carbon deposits.

The catalytic properties of Co/Al_2O_3 catalysts are notably influenced by preparation conditions, including the reduction temperature, which plays a crucial role in determining the extent of cobalt species reduction and the formation of active Co^0 species.[88] Furthermore, the pore structure and surface area of the support also influence cobalt dispersion, with narrow-pore alumina supports generally leading to higher dispersion and smaller cobalt particle sizes compared to wide-pore supports.[61] The interaction between cobalt and alumina can also result in the diffusion of cobalt ions into the alumina structure, forming cobalt aluminates that hinder reduction and adversely affect catalytic performance.[89,90] Despite these challenges, alumina continues to be a popular support due to its mechanical robustness and the potential to modify its surface properties to optimize catalyst performance. The formation of smaller cobalt particles on alumina-supported catalysts enhances dispersion and catalytic activity, highlighting the significance of support interactions in determining the overall efficacy of cobalt-based FTS catalysts.[90]

Other studies on cobalt catalysts to improve lower-molecular-weight hydrocarbons are either based on different promoters or different reactor types and preparation conditions of catalysts. Yang et al.[91] studied the reaction over Co–Cu–Al_2O_3 catalyst to obtain light hydrocarbon products. The best formulation for maximum C_2–C_4 selectivity was about 0.7 wt% Cu. It was also found that in order to maximize C_2–C_4 production, the reaction should be carried out in the low temperature range; however, since conversion decreases with temperature, a lower space velocity would be required to compensate for the loss in conversion. Alkali addition to this catalyst was ineffective as a promoting agent. Chen et al.[42] studied the synthesis on methyl carbonyl cluster–derived, carbon-supported iron and cobalt catalysts. A high selectivity to olefins is observed, and there is a lack of C_6 or higher hydrocarbons produced. These observations could be the result of small particle size in these catalysts. Rosynek et al.[92] investigated the effect of cobalt source on the reduction properties of silica-supported cobalt catalysts. They prepared the catalysts from nitrate, chloride, and acetate salts of cobalt. At comparable levels of carbon monoxide conversion, reaction selectivities for all of the catalysts are virtually identical, and methane accounts for ca. 80% of the hydrocarbon product in every case, with the alkene/alkane ratio in the C_2–C_3 product fraction being low, typically <0.4. Dalai and Bakhshi[93] performed the carbon monoxide hydrogenation reaction in a tube-wall reactor over a plasma sprayed cobalt catalyst. They studied the effect of temperature, pressure, and feed gas exposure velocity on the activity and selectivity of the catalyst. The results showed that the formation of olefins, especially ethylene and propylene, was favored at low pressures, low temperatures, and high exposure velocities. The promoting effects of magnesium, vanadium, and cerium oxides on cobalt and ruthenium supported on carbon have been compared by Guerrero-Ruiz et al.[94] In all cases the added oxide promoter enhanced the activity and selectivity for alkenes and long-chain hydrocarbons. Chen and Adesina[95] investigated the Fischer–Tropsch synthesis performance of a Co–Mo bimetallic catalyst containing 6 Co : 1 Mo : 4 K : 100 SiO_2 and compared it with a similar monometallic cobalt catalyst. The bimetallic system showed nearly 100% improvement in the alkene/alkane ratio at the experimental conditions of 101 kPa and 280 °C and CO/H_2 ratio of 1/19 to 19/1. The increase in chain growth probability and alkene content was attributed to substantial methane suppression caused by the introduction of molybdenum to the catalyst.

A significant fraction of the carbon monoxide hydrogenation studies has centered on nickel, the metal which has been accepted as the best methanation catalyst. The use of nickel catalysts for methanation has been discussed extensively by Mills and Steffgen,[23] Shah and Perotta,[12] and Vannice.[60] Several studies have been directed toward determining the influence of the support and of promoters on the catalytic properties of nickel. Vannice[14] examined nickel supported on a variety of typical support materials and observed only a small variation in specific activities. There was some evidence which implied that the crystallite size of nickel may influence the specific activity, and the existence of a broad methanation activity maximum for 10–20 nm nickel particles was suggested. This conclusion seems to have been verified by Mathews and co-workers,[96] who claim to have observed a maximum in 15 nm nickel particles, and also by Bartholomew et al.[97] Prokhorenko et al.[40] studied the chemical nature of the support on the activity of the nickel catalysts. They found that the activity of the nickel at 540 K decreases in the order $ZrO_2 > TiO_2 > BeO > MgO > Al_2O_3 > SiO_2 > Cr_2O_3 \gg WO_3$, while selectivity for higher hydrocarbons decreases in the order $ZrO_2 > Al_2O_3 > Cr_2O_3 > MgO > BeO$, $SiO_2 > TiO_2 > WO_3$. Huang and Richardson[98] have studied the effect of the addition of an alkali metal, in this case sodium, on the catalytic behavior of a series of Ni/silica–alumina catalysts. The authors concluded that small amounts of sodium poison acid sites on the support and reduce carbon build-up, while larger quantities result in poisoning of the metal sites. The Agency of Industrial Sciences and Technology in Japan[99] manufactured high calorific value gases containing methane (40%) and $C_{\geq 2}$ (6%) hydrocarbons by reacting hydrogen and carbon monoxide over a supported catalyst containing $MoNi_2$, $MoNi_3$, and/or $MoNi_4$. Thus a catalyst containing $MoNi_4$ was prepared from a mixture containing NiO 20, MoO_3 25, calcined alumina 55 weight parts and water by pelletizing, calcining at 500 °C and reducing with hydrogen at 700 °C.

6.5.2 Supports

Selection of support is a very important step in the preparation of supported catalysts. As Snel[100] pointed out in his review article, a support can influence the catalytic behavior of a supported metal catalyst in four different ways, namely, support basicity effect, support dispersion effect, electronic modification effect, and strong metal–support interaction effect.

In FTS, each support material offers distinct advantages and influences the catalytic properties differently. The most widely used supports in carbon monoxide hydrogenation are Al_2O_3 and SiO_2. Other metal oxides such as TiO_2, MnO, Cr_2O_3, Nb_2O_5, V_2O_5, ZrO_2, ZnO, and MgO[101] are also employed as supports in carbon monoxide hydrogenation catalysts for improved activity and selectivity.

As stated earlier, alumina interacts with cobalt to form cobalt aluminate spinels, which impact the reducibility and dispersion of cobalt particles, that are crucial for catalytic activity and selectivity.[61,89] Alkaline earth metals such as magnesium, calcium, strontium, and barium can effectively modify alumina supports, leading to improved cobalt reducibility and dispersion.

These modifications enhance selectivity for C_{5+} hydrocarbons while reducing methane formation. The presence of these metals facilitates electron transfer to cobalt, increasing its electron density, which in turn promotes CO adsorption and dissociation, beneficial for the Fischer–Tropsch reaction.[102] Silica supports, particularly those with mesoporous structures, promote high dispersion of cobalt nanoparticles, enhancing catalytic performance and increasing selectivity for long-chain hydrocarbons.[76,103] Titania (TiO_2) is another commonly used support material in FTS due to its low cost, safety, and chemical stability. The most widely encountered polymorphs of TiO_2 are anatase, rutile, and brookite, with anatase and rutile being particularly significant. Beyond these three structures, the Ti–O phase diagram is rich with intermediate oxidation states that are stable. As a result of these partial oxidation states of titanium, the material has a very diverse chemical potential range. Anatase is often preferred for its larger surface area, while rutile is known for its greater mechanical strength and thermal stability.[16] However, the reducibility of TiO_2 can lead to what has been referred to strong metal–support interactions, which complicate the reduction of cobalt species. This is likely due to a strong Co–O interaction with the support, leading to very high temperatures for reduction.[104]

As catalyst supports other metal oxides such as zirconia, magnesia, chromia, thorium oxide, urania, niobia, tin oxide, *etc.* have been discussed in detail by Anderson[105] and Stiles.[106] Ishihara *et al.*[101] studied 10 kinds of single metal oxide and 21 kinds of mixed oxide as possible support materials for cobalt–nickel bimetallic catalysts in the hydrogenation of carbon monoxide and found that the activity and selectivity is strongly affected by the type of oxide support. The best gasoline selectivity, for example, was obtained with MnO–ZrO_2 mixed oxide and the best selectivity for C_2–C_4 olefins with Al_2O_3–TiO_2. Magnesia provides a basic environment that promotes the adsorption and activation of CO molecules, enhancing the hydrogenation process. The modification of γ-alumina with magnesia has been demonstrated to facilitate the reduction of metal oxides and improve catalytic activity in FTS. The incorporation of a small amount of magnesia in γ-Al_2O_3 has been shown to increase the activity of cobalt catalysts. However, an excess of magnesia (>0.8 wt%) can lead to the formation of a difficult-to-reduce intermediate (MgO–CoO), which diminishes the catalyst's activity.[16]

Carbon supports, on the other hand, offer high surface areas and unique electronic properties that enhance the dispersion of active metal particles. Materials such as activated carbon (AC), carbon black (CB), graphite, carbon nanotubes (CNTs), carbon nanofibers (CNFs), graphene, and mesoporous carbons provide distinct advantages as catalyst supports. These benefits include customizable porous structures, ease of metal phase reduction, resistance to both acidic and basic environments, structural stability at high temperatures, and cost-effectiveness compared to traditional oxide supports like Al_2O_3 and SiO_2. However, carbon supports also present challenges, including weak metal–support interactions that can lead to metal sintering, susceptibility to gasification at high temperatures, poor reproducibility, and low bulk density in powder form, which can result in high reactor volumes and pressure drops. The interaction between cobalt and carbon supports is generally weaker than with oxide supports, which can be advantageous. This weaker interaction facilitates easier reduction of the metal phase and increases the specific surface area, enhancing catalytic performance. Studies have shown that applying carbon coatings on oxide supports, such as silica, can significantly improve cobalt dispersion and reduce particle size, leading to higher CO conversion and increased selectivity toward light hydrocarbons in FTS.[107]

Zeolite supports are critical in determining the performance and stability of catalysts used in FTS. The incorporation of mesopores in zeolite supports, such as ZSM-5, has been shown to significantly enhance selectivity toward higher hydrocarbons (C_{5+}) while reducing the formation of methane and light hydrocarbons (C_2–C_4).[108]

6.5.3 Promoters

Promoters play a crucial role in significantly enhancing the activity and selectivity of catalysts used in carbon monoxide hydrogenation. Alkali metals have been widely used as promoters to improve the activity and selectivity in the carbon monoxide hydrogenation reaction. Presence of alkali metals on transition metal surfaces changes the binding energy as well as the sticking probability of reactive molecules.[109,110] In general, the presence of potassium or an increase in the potassium content of an iron catalyst increases the carbon monoxide dissociation and olefin/paraffin ratio while decreasing methane selectivity. However, the distribution of hydrocarbons shifts toward high molecular weights. Benziger and Madix[111] reported that the heats of adsorption of both hydrogen and carbon monoxide increase by the addition of potassium to nickel surfaces. Reactivity studies indicate that the poisoning of iron oxide during the hydrogenation reaction can be inhibited by the addition of potassium.[112] It appears that alkali metals change markedly the binding characteristics of adsorbates such as carbon monoxide and hydrogen. As a result, their relative surface concentration and their dissociation probability are significantly altered, leading to a change in catalyst activity and selectivity.[110] A cautionary note is in order here, to mention that studies performed under ultra high vacuum conditions over single crystals may bear little validity for the industrial processes as they are limited to the metallic form of the alkali promoters. Alkali metals do not survive in their metallic form under FTS conditions, where oxygen is inherent in the reaction chemistry.

^{1}H NMR spectroscopy was used under *operando* conditions to monitor hydrogen adsorption dynamics over Ru/SiO_2, and the effect of alkali promoters.[113] The dynamic studies were only amenable to NMR spectroscopy as the technique enables the researcher to perturb the nuclear spins without disturbing chemical equilibria. A hole-burning experiment revealed that in the presence of alkali promoters, the adsorption desorption equilibrium of hydrogen over Ru surface was significantly perturbed: this was due to alkali promoters blocking the defect-like sites which were responsible for the dissociative adsorption of hydrogen. In the presence of alkali promoters, not only the adsorption equilibrium was blocked, but also, the migration of hydrogen on the surface of the metal particles and their spillover to the support was also inhibited.[114]

Manganese compounds have also been used as promoters in small quantities relative to the catalytically active component such as cobalt, nickel, and iron. It has been found that the addition of small amounts of manganese to iron enhances the formation of olefins and lowers simultaneously the isomerization activity.[5,45] Similar effects for manganese oxide as a catalytic component were observed for cobalt[115] and ruthenium.[116] On the other hand, catalysts with high manganese to iron or cobalt ratios selectively yield C_2–C_4 olefins with low methane selectivity, as mentioned in Section 6.5.1.

Mn and Zn can modify the electronic structure and facilitate the formation of active phases, while alkaline earth metals (Ca, Mg, Sr, and Ba) improve the dispersion, reduction, and carburization of iron species, enhancing the stability and activity of the catalysts. The transfer of electrons from promoters to iron species weakens carbon–oxygen bonds and suppresses hydrogen dissociation, ultimately promoting the production of olefins and hydrocarbons with longer carbon chains.[117] Furthermore, the Na–Zn–Fe_5C_2 catalyst, modified with Zn and Na, has demonstrated high CO conversion rates of up to 74.9%, with a C_{5+} selectivity of 58.1% and a CO_2 selectivity of 27.3%. The addition of Na alters the electronic structure of Fe_5C_2, influencing intermediate adsorption and hydrogenation behaviors, which leads to increased alkene production.[118] Adding rhenium to cobalt catalysts has been shown to increase cobalt reducibility, resulting in more active sites and improved synthesis gas conversion, although it also leads to a higher rate of deactivation over time.[119] Lanthanum, a rare earth oxide promoter, helps suppress the formation of cobalt-support species, increases cobalt dispersion, and improves reducibility, resulting in higher CO conversion rates and lower methane selectivity in cobalt-based FTS catalysts.[120]

Lee and Ponec[121] have summarized general promoter effects as follows:

i. An alkali additive enhances the mobility of supports such as silica or titania and similar ones. This can influence the spreading of the support over the metal and enhances the promoting effect of the support.
ii. Alkali additives undoubtedly decrease the accessible active metal surface area by blocking it; site blocking is more prominent for hydrogen than for CO and as such, chain growth and selectivity toward CO_2 production is favored.
iii. Alkali additives are mostly found to increase the rate of CO dissociation and consequently alter catalyst properties,
iv. Alkali additives, as well as some other basic additives such as MgO, decrease the rate of M^{n+} (active metal oxide) reduction and result in more M^{n+} ions surviving a standard reduction; additives can also influence directly or indirectly the reduction of transition metal

Activation of CO, CO₂, and H₂ Toward Synthetic Fuel Manufacture

oxides such as vanadium and titanium oxides to the lower oxides, which can adhere better to metal surfaces.

v. Oxides of transition metals such as titanium, vanadium, and niobium enhance the rate of carbon monoxide dissociation but stabilize, at the same time, oxygen-containing intermediates like formates or acetates; unstable formates seem to react easily to form methane.

vi. It is very likely that promoters activate the carbon monoxide molecule by influencing the carbon–oxygen bond strength and by increasing the migratory ("insertion") properties of carbon monoxide.

6.6 CO and H₂ Adsorption on Fischer–Tropsch Catalysts

It is well known that catalysis involves the adsorption of one or more reactants, the reaction of adsorbed reactants or an adsorbed reactant with another in the gas phase, followed by the desorption of products. Therefore, it is clear that the direction of the reaction will depend upon the nature of the adsorbed complex, and a study of the adsorption of carbon monoxide and hydrogen on FT catalysts, namely transition metals, provides a better understanding of the mechanism of the synthesis reaction, since the interaction of the adsorbed carbon monoxide and hydrogen produces the intermediates which are precursors of the hydrocarbons formed.

6.6.1 Electronic Configuration of CO

The carbon monoxide molecule is itself, and by its possibility of interaction with metals, very interesting. Although the CO molecule is isoelectronic with the nitrogen molecule, significant changes occur in the relative energy levels of the molecular orbitals because of the difference in electronegativity between the carbon atom and the oxygen atom. Localization occurs with the orbital of lower energy existing on the more electronegative atom, oxygen, and since the average electronegativity of carbon monoxide is similar to that of nitrogen, the lowering of one orbital results in the raising of another, as shown schematically in Figure 6.3.

In its ground state the electronic configuration of carbon monoxide is:[122]

$$1\sigma^2\ 2\sigma^2\ 3\sigma^2\ 4\sigma^2\ 1\pi^4\ 5\sigma^2$$

1σ and 2σ orbitals have a nonbonding character. The 3σ orbital has a substantial bonding character and the center of its charge lies nearer to the oxygen than to the carbon atom but inside of the C–O interval. The 1π orbital is essentially bonding and, together with the 3σ orbital, contributes to the formation of the triple bond on the CO molecule.[6]

The energy of the 5σ orbitals is higher than that of the 1π orbital at the equilibrium internuclear distance. Thus the 5σ orbital is the higher occupied molecular orbital and the $2\pi^*$ antibonding molecular orbital is

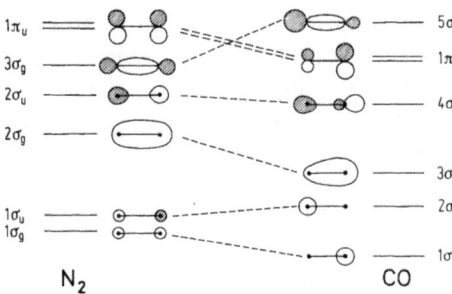

Figure 6.3 A comparison of N_2 and CO molecular orbital energy levels.[60] Reproduced from ref. 60 with permission from Springer Nature, Copyright 1982.

the lower unoccupied molecular orbital. However, an increase of the internuclear spacing can cause a change in orbital energy levels and 1π orbital will then become the higher occupied molecular orbital of CO.[6] It is the higher occupied molecular orbital (5σ) that results in the small dipole moment of CO, and which is responsible for allowing a strong interaction between CO and metal surfaces since it has the high energy and requisite directional character to donate electrons into empty metal orbitals. The $2\pi^*$ antibonding orbitals which represent the next higher level and tend to be localized on the carbon atom facilitate back donation from the metal to the CO molecule.[60] Figure 6.4 provides a perspective of the shape of these various molecular orbitals of CO.

6.6.2 Carbon Monoxide Adsorption

Adsorption of carbon monoxide on supported metal catalysts is well studied because of its simple molecular structure and its technological importance for surface chemistry. Extensive evidence from the literature[6,60,123] indicates that with few exceptions carbon monoxide is bonded perpendicularly to the surface of transition metals through the carbon end of the molecule. The metal carbon bond may be formed with either a single metal atom giving rise to linear bonding or may be shared between two or more metal atoms to produce bridge bonding. The metal carbon bond is comprised of two components. The first arises from an overlap of the occupied 5σ orbital of carbon monoxide and unoccupied metal orbitals. This results in the donation of electrons from the molecule to the metal. The second component of the metal carbon bond is formed by back donation of electrons from occupied metal orbitals to the unoccupied $2\pi^*$ orbitals of carbon monoxide. Figure 6.5 provides a visualization of the configuration of adsorbed carbon monoxide.

The 4σ and 1π orbitals are not thought to participate appreciably in bond formation, and the 5σ orbital in gaseous carbon monoxide is considered to be essentially non-bonding with respect to carbon and oxygen atoms.[60] Consequently, it is assumed that transfer of electrons from the 5σ orbital will not

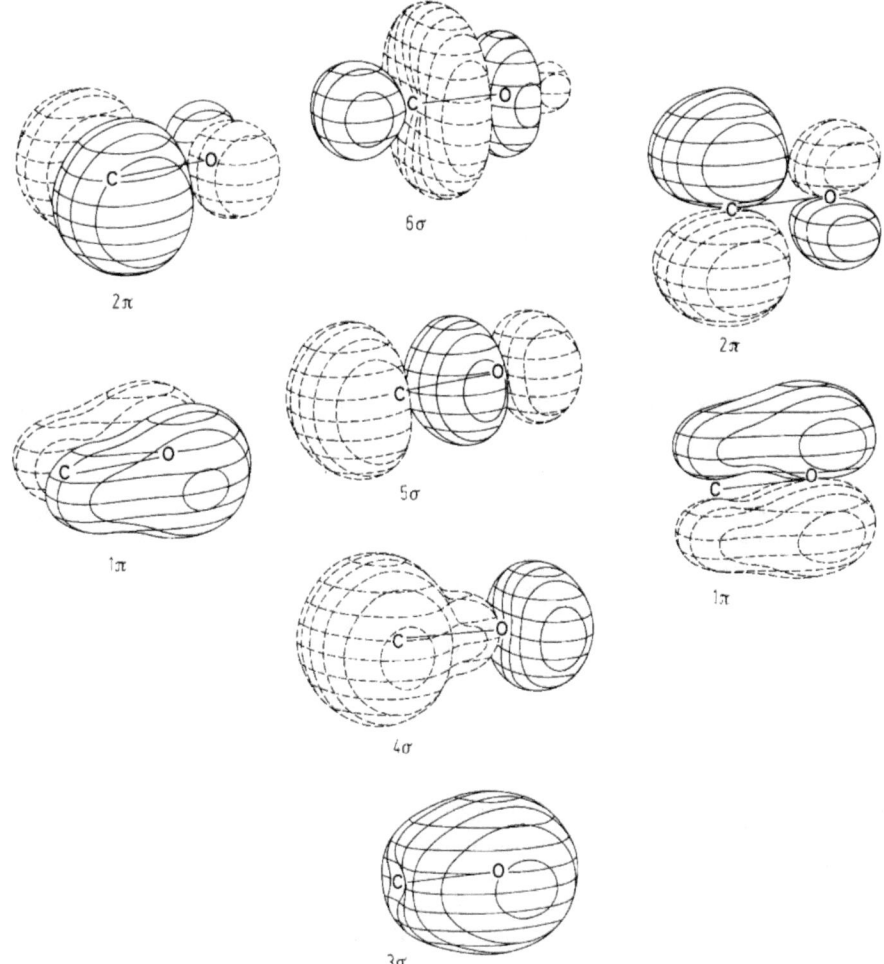

Figure 6.4 Configuration of different molecular orbitals of CO.[60] Animated versions of the orbital structures of CO can be found at https://www.chemtube3d.com/orbitalsco/. Reproduced from ref. 60 with permission from Springer Nature, Copyright 1982.

markedly affect the carbon–oxygen bond strength in the chemisorbed state. This conclusion has been verified by the work of Doyen and Ertl.[124] However, the $2\pi^*$ orbitals are anti-bonding, and back donation into these orbitals will tend to weaken the carbon–oxygen bond. Therefore, the net result of the chemisorption process is expected to be a weakened carbon–oxygen bond.

Dissociative chemisorption of carbon monoxide is also known to occur on the surface of many transition metals, and it is most likely that it proceeds *via* the molecularly adsorbed state. Differences in catalytic activity among group VIII metals are usually attributed to the different strengths of the metal–carbon bond, and ultimately the strength of the metal–carbon bond

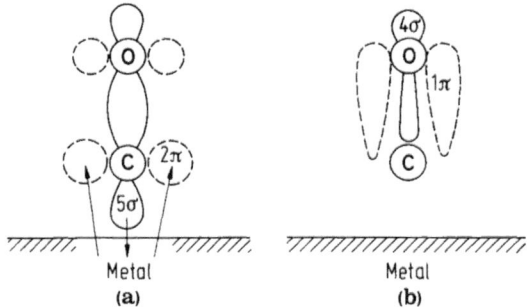

Figure 6.5 CO chemisorption configuration on a metal surface (a) 5σ and 2π* orbitals are involved, (b) 4σ and 1π orbitals are involved insignificantly.[60] Reproduced from ref. 60 with permission from Springer Nature, Copyright 1982.

Table 6.2 Trend of nature of carbon monoxide adsorption on transition metals.[123] Adapted from ref. 123 with permission from Taylor & Francis, Copyright 1981.

III B	IV B	V B	VI B	VII B	VIII	VIII	VIII	I B
Sc	Ti	V	Cr	Mn	Fe	Co	Ni	Cu
Y	Zr	Nb	Mo	Tc	**Ru**	**Rh**	Pd	Ag
La	Hf	Ta	W	**Re**	**Os**	Ir	Pt	Au

determines whether carbon monoxide is molecularly or dissociatively adsorbed. As a result, the easier any transition metal donates electrons into the 2π* orbital of the carbon monoxide, the higher will be the probability of a bond rupture and hence dissociative chemisorption. It is well known that transition elements located at the left of the periodic table can adsorb carbon monoxide dissociatively at room temperature. The tendency for dissociative adsorption increases moving from right to left across a row and upwards in each group and, as may be seen from Table 6.2, iron and tungsten mark the borderline of dissociative and non-dissociative adsorption at room temperature. However, at temperatures of 200–300 °C (temperatures of usual carbon monoxide hydrogenation) this borderline is shifted to the right in the periodic table and is now situated at the vicinity of osmium, rhodium, and nickel.

The structure of the transition state for carbon monoxide dissociation is not known but may be assumed to involve simultaneous interactions of both the carbon and the oxygen portions of the molecule with two or more sites on the catalyst. Since the initial bonding of chemisorbed carbon monoxide is through the carbon end of the molecule, interaction of the surface with the oxygen end of the molecule very likely proceeds *via* vibrational deformation of the metal–carbon–oxygen bond angle.[123] As expected, this process occurs more readily for bridge-bonded carbon monoxide.

Consistent microkinetic analyses coupled to DFT can reveal significant insights about the surface species, their concentration, and their proximity to the equilibrium during the reaction.[125] CO adsorption on metal surfaces is known to go beyond CO : M 1 : 1 stoichiometry. Multicarbonyls can be observed on the surfaces and their presence is strongly linked to how CO molecules are activated on the metal surfaces.[126] The attractive forces between CO ad-molecules on the surface can lead to the island formation and subsequent periodic oscillations are observed during CO oxidation studies.[127,128] These fluctuations were noted and also reported during Fischer–Tropsch synthesis also.[129]

The multicarbonyls on the surface bear potential to form stable multicarbonyl groups that can be volatile in the gas phase or soluble in the reaction wax. Many of the active metals have stable multicarbonyls, such as iron pentacarbonyl, dicobalt octacarbonyl nickel tetracarbonyl, and triruthenium dodecacarbonyl. The next challenge of heterogeneous catalysis research is to develop strategies to reveal whether these mobile and active carbonyls play any role during the complex process of Fischer–Tropsch catalysis.

Cobalt-based catalysts, which are renowned for their stability and selectivity toward long-chain hydrocarbons, generally favor dissociative adsorption, resulting in higher hydrocarbon yields and reduced water–gas shift activity.[103,130] The specific cobalt phases, such as metallic Co, cobalt oxides, and cobalt carbides, also influence adsorption behavior and FTS activity. Cobalt carbides, particularly Co_3C, exhibit increased activity toward CO activation, enhancing selectivity toward light olefins and alcohols.[130] The interaction between the active metal and the support plays a critical role in determining the adsorption mechanism. Supports such as Al_2O_3, SiO_2, and other mesoporous materials facilitate the dispersion of the active metal, improving catalyst reduction and dispersion, which in turn affects CO adsorption.[16,103] Additionally, operational conditions including temperature and pressure, and influence the adsorption mechanism by affecting heat transfer and reaction kinetics, thereby controlling product selectivity.[16] Cobalt catalysts supported on alumina with low acidity tend to exhibit higher reducibility, promoting the formation of stronger bridged-type CO and providing more active sites for the adsorption of hydrogen and carbon monoxide. In contrast, cobalt catalysts supported on high-acidity alumina suppress catalyst reduction, leading to the formation of linear-type CO.[131]

6.6.3 Hydrogen Adsorption

Hydrogen may adsorb dissociatively[132] on transition metals by the withdrawal of electron density from the vacant d-atomic orbitals of the transition metal[24] to form a bond that is partly ionic and partly covalent. The strength of the hydrogen–metal interaction is expected to increase with the number of such orbitals per atom. It has been unambiguously established that hydrogen chemisorption on the FT catalysts yields one hydrogen atom per metal atom. The hydridic character of the metal–hydrogen bond is evidenced by the ease with which many transition metal hydride complexes are formed.[133]

IR spectroscopy studies[134] have indicated that there are two types of chemisorption mechanisms, both resulting in hydrogen adatoms as identified by IR bonds at 4.86 µm and 4.76 µm for the strong and weak chemisorption, respectively. A similar effect was also observed for deuterium adatoms. If the weak adsorption leads to molecular adspecies, a bond should be seen at 4.4 µm when HD is adsorbed. However, when HD was contacted with the catalyst, no IR adsorption was found at this wavelength. It has been suggested that the concentration of weakly chemisorbed hydrogen adatoms has a slight positive dependence on pressure,[135] which explains why the hydrogen adsorption isotherm does not level out completely at higher hydrogen pressures as expected for the Langmuir adsorption isotherm.

Hydrogen adsorption on cobalt is strongly activated, and the extent of activation depends upon the type of support and level of metal loading. Also, it is highly reversible at room temperature with the degree of reversibility being affected by metal loading, type of support, and method of preparation.[63] Zowtiak and Bartholomew[136] reported that the activation energy of adsorption is a function of catalyst support and metal loading. They found that the activation energy of adsorption ranged from 5.8 kJ mol^{-1} for unsupported cobalt to 43 kJ mol^{-1} for 3% Co/silica, it ranges from 145 kJ mol^{-1} on 10% Co/silica to 105 kJ mol^{-1} on 10% Co/alumina. The interaction between the Co precursor and oxide supports such as Al_2O_3, SiO_2, and TiO_2 significantly affects H_2 and/or CO adsorption and activation, influencing Fischer–Tropsch (FT) performance.[8,137] For example, SiO_2-supported Co catalysts exhibit high activity at 250 °C due to the formation of a unique CoO–Co interface that enhances secondary hydrogenation reactions, leading to increased production of paraffinic products. In contrast, TiO_2 supports show less activity at 250 °C because their high porosity and low surface area result in the formation of agglomerates with a broad cluster size distribution.[8]

Because of the electron withdrawal character of hydrogen adsorption, any electron donor additives like alkali metals, notably potassium, will enhance hydrogen adsorption while electron acceptor additives such as nitrogen,[138] oxygen, sulfur, phosphorus,[111,139] chlorine, and bromine [140] will suppress hydrogen chemisorption. Additionally, promoting FT catalysts with alkali metals such as K, Na, and Rb can alter hydrogen mobility on the catalyst surface, reducing methane production and increasing the olefin-to-paraffin ratio.[141]

6.6.4 Co-adsorption and Interaction of Carbon Monoxide and Hydrogen

The information gained from studies of the interaction of carbon monoxide and hydrogen on metal surfaces allows much insight into the chemical nature of the surface complex. This knowledge can clarify and substantiate the reaction paths proposed for both the methanation and Fischer–Tropsch synthesis reactions.

Carbon monoxide is adsorbed more strongly than hydrogen (the maximum heat of carbon monoxide adsorption is about 50% higher than that of hydrogen) and therefore at temperatures below 77 °C, carbon monoxide displaces hydrogen from the surface of metal atoms,[13] and, pre-adsorbed carbon monoxide also inhibits chemisorption of hydrogen, whereas carbon monoxide is not influenced by hydrogen pre-adsorption.[138] This chemisorption behavior may be explained on the basis of competition for the withdrawal of electrons from the surface. Because of its higher electron affinity and heat of adsorption, carbon monoxide is able to displace hydrogen. At higher temperatures carbon monoxide dissociates to a clearly detectable extent; in this range, the interaction of hydrogen and carbon monoxide in the adsorbed layer also becomes observable, and therefore, this competition decreases and hydrogen is not displaced by carbon monoxide under synthesis conditions.[142]

A further effect associated with co-adsorption is an increase in the number of adsorption sites. It has been observed that simultaneous adsorption of hydrogen and carbon monoxide gave higher heats of adsorption than the individual adsorption of the gases.[60] A similar observation was made by Sastri *et al.*[143] over a cobalt FT catalyst. The reason for the increase in uptake compared to either gas alone is not clear, but it could be speculated that hydrogen coordinates with the lone-pair electrons of the electron-rich oxygen end of the carbonyl ligand, thereby decreasing the electron density in the metal–carbon bond. The possibility that the oxygen atom of carbonyl ligands coordinates with electron acceptors was demonstrated by Wade.[144] The decrease in the metal–carbon bond order may then result in lower metal coordination and hence an increase in the number of sites available for carbon monoxide adsorption.

Since enhancement in total adsorption indicates an interaction on the metal surface, in a number of cases a stoichiometric surface complex appears to be formed. Although the hydrogen/carbon monoxide ratio in the adsorbed layer can be dependent upon temperature and pressure, ample evidence is present to show that this ratio is frequently close to one on a variety of metal surfaces. The exact chemical nature of this surface species is not clear, but it is frequently represented as either a keto or enol structure with at least one carbon–metal bond.

Sastri *et al.*[143] have investigated the co-adsorption of carbon monoxide and hydrogen on a Co/ThO_2/kieselguhr catalyst. In the temperature range around 100 °C, both kinetic and volumetric adsorption measurements indicated that carbon monoxide was molecularly adsorbed. It is, however, likely that since their experiments were carried out at low temperatures, these conclusions may be inappropriate at practical FT temperatures. A series of unsupported cobalt catalysts as well as cobalt supported on silica, alumina, carbon, titania, and magnesia have been studied by Reuel and Bartholomew,[63] who found that the stoichiometry for irreversible carbon monoxide adsorption varies from 0.4 to 2.3 molecules of carbon monoxide per metal atom. This observation indicates that there may be a mixture of both linearly bonded and bridge-bonded carbon monoxide on the same catalyst.

6.7 Mechanism of Carbon Monoxide Hydrogenation

Due to the complex nature of the synthesis that involves multiple monomers that compete in chain growth reactions, a basic problem of the carbon monoxide hydrogenation reaction is the control of product selectivity. The problem of selectivity is closely related to the reaction mechanism and behavior of the reaction intermediates. The kinetics of the process are influenced by the type of catalyst, typically iron or cobalt, as well as the catalyst's physicochemical properties, including particle size, metal dispersion, and support material.[8,16] Additionally, the interaction between CO and H_2 is affected by the partial pressures of these gases; higher H_2/CO ratios favor the formation of lighter hydrocarbons, while increased total pressure shifts selectivity toward heavier products and more oxygenates.[16]

The discussion on the FTS reaction mechanism has been contentious since the early work of Fischer and Tropsch. The mechanism of the hydrogenation of carbon monoxide has been the subject of intensive research extensively reviewed for many years.[16,27,31,36,123,145–148]

The overall reaction has been interpreted as being a sequence of the following steps.

i. chemisorption of carbon monoxide and hydrogen on the catalyst surface,
ii. formation of a primary complex, which leads to the weakening of a C–O and the formation of a C–H bond (chain initiation),
iii. chain growth by interference of the primary complex with reactant gases or synthesis products,
iv. chain termination, and
v. re-adsorption and further reaction.

While the products are in agreement with the Anderson–Schulz–Flory polymerization model, the process is clearly not a simple polymerization reaction, since "the monomer" has to be produced *in situ* from carbon monoxide and hydrogen on the surface of the catalyst. It is the nature of the monomer that has to be resolved, and mechanistic studies focus on steps (ii), (iii), and (iv) of the above sequence. The various mechanisms reported can be grouped into four types:[6,146]

1. The carbide mechanism,
2. The enolic mechanism,
3. The CO insertion mechanism, and
4. The alkoxy mechanism.

6.7.1 Carbide Mechanism

One of the earliest proposals for hydrocarbon synthesis was the carbide theory proposed by Fischer and Tropsch (in 1926). They hypothesized that

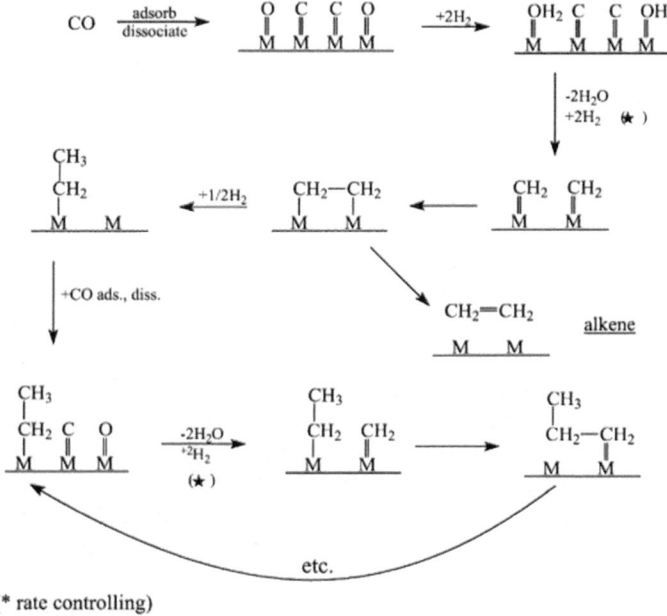

Figure 6.6 Carbide mechanism.[147] Reproduced from ref. 147, with permission from Elsevier, Copyright 2001.

upon chemisorption the carbon monoxide molecule dissociates completely to form a bulk carbide, which subsequently undergoes hydrogenation to form methylene (CH_x) groups. The methylene species were assumed to polymerize to form hydrocarbon chains then desorb from the surface as saturated and/or unsaturated hydrocarbons. Figure 6.6 shows schematically the carbide mechanism.

Craxford and Rideal[149] gave a more detailed version of this mechanism, especially with respect to chain growth. From studies of the para to ortho hydrogen conversion under various conditions, they concluded that under normal reaction conditions only molecular hydrogen was involved in the synthesis. Atomic hydrogen was only present when large amounts of CH_4 were produced. It was postulated that adsorbed CO reacted with hydrogen to form single carbon atoms. These were then converted to CH_2 entities which linked up to form long chains. The chains were cracked by hydrogen to yield a wide range of hydrocarbons.

There is a considerable amount of experimental data that support the carbide mechanism in principle.[3,4,6,13,146,150–152] However, there are several objections. Based on the results of thermodynamics, ^{14}C tracer and synthesis studies performed with iron and cobalt catalysts, Kummer et al.[153] and Browning et al.[154] concluded that the bulk phase carbide participates in the synthesis to a negligible extent and that hydrogenation of surface carbides could account for only a small fraction (10–20%) of the hydrocarbons formed during steady-state synthesis. However, these results do not refute the carbide mechanism. One of the key objections was that the postulated mechanism does not predict the formation of oxygen-containing products, e.g. alcohols. This objection has been addressed to some extent by the proposal that CO insertion into the metal–carbon bond of the growing hydrocarbon chain could lead to chain termination via oxygenate formation. However, although CO insertion/migration are well documented for homogeneous systems, they have not been shown to be important in heterogeneous systems. Another limitation of carbide mechanism about branched hydrocarbons has been overcome by modified carbide mechanisms such as alkyl, alkenyl, and alkylidene-hydride-methylidyne. The alkyl mechanism proposes that chemisorbed CH_2 serves as the monomer, while CH_3 initiates chain growth, leading to the formation of alkanes and alkenes. The alkenyl mechanism suggests that an adsorbed vinyl species (CH_2=CH) initiates chain growth, with chemisorbed CH_2 acting as the monomer. In contrast, the alkylidene-hydride-methylidyne mechanism proposes that adsorbed CH and H are the chain-growth monomers, with an isomeric vinyl compound (CH–CH_2) initiating chain growth. This mechanism is supported by ab initio calculations for Fischer–Tropsch synthesis over ruthenium.[16,155]

It thus seems that there are no serious objections in the literature against a "dissociative" mechanism of carbon monoxide hydrogenation, and a number of observations can easily be rationalized by a mechanism which starts with carbon monoxide dissociation. In conclusion, the surface carbide mechanism still remains the most plausible mechanism, although it cannot fully explain the entire product spectrum. It also seems, however, that it is unlikely to be the unique mechanism for this process. The current understanding of the carbide mechanism also includes a structure with a core of magnetite (Fe_3O_4) in a shell of iron carbide, balancing the local carbon to hydrogen ratio, through the water–gas shift reaction catalyzed by the magnetite.[156] Within one catalyst particle, co-existence of an oxide and a carbide layer can tune the selectivity and ensure oxygen removal.

6.7.2 Enolic Mechanism

Introduced in the 1950s, this mechanism was proposed as an alternative to the carbide mechanism, which had limitations in explaining the formation of branched hydrocarbons and oxygenated products such as alcohols and acids.[16] In this mechanism it is believed that the intermediate (HCOH) is formed via the partial hydrogenation of associatively adsorbed carbon monoxide. Chain growth occurs by water elimination between two enolic complexes. Any oxygenated product is assumed to be the result of desorption of the hydroxy complex. Furthermore, re-adsorption of these alcohols, aldehydes, or esters produces hydrocarbons upon dehydrogenation or cleavage of the adsorbed complex. This mechanism readily explains the formation of both oxygenated and hydrocarbon products and can also be

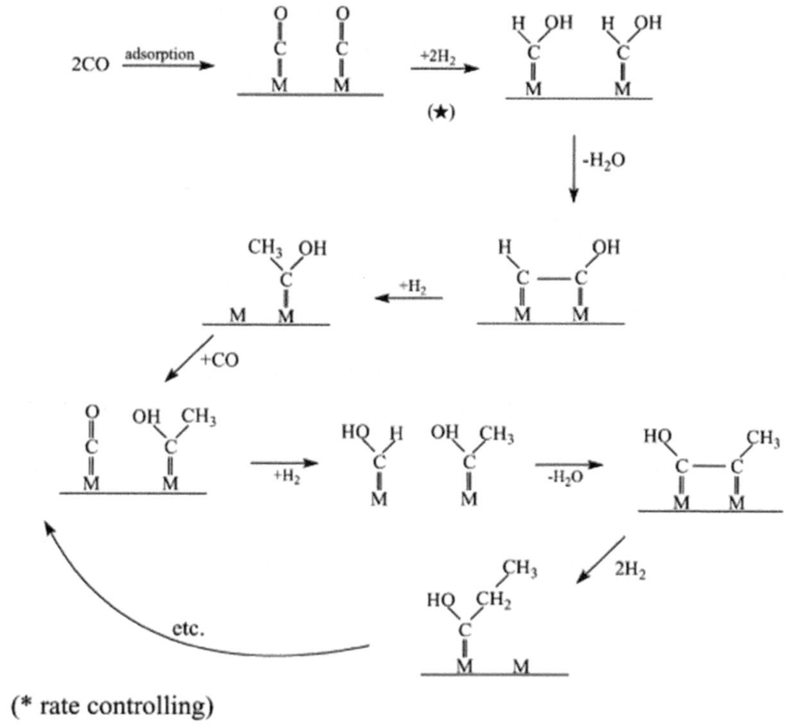

Figure 6.7 Enolic mechanism.[147] Reproduced from ref. 147, with permission from Elsevier, Copyright 2001.

modeled with the Anderson–Schulz–Flory distribution. Figure 6.7 illustrates the mechanism schematically.

Elvins and Nash[3] were the first to suggest the enolic mechanism. Storch et al.[3] presented a detailed reaction scheme in which HCOH entities were the basic building blocks. Emmett and co-workers[3,16,157] studied the incorporation of a series of radioactively labeled alcohols in the carbon monoxide hydrogenation reaction. Their results appear to support the concept that oxygenated surface complexes were involved in the chain growth mechanism. Additional evidence has been provided by Balaji et al.[158] for a cobalt catalyst and Golodets et al.[148] for several transition metals supported on γ-alumina. The results from Ruhrchemie AG also agreed with this mechanism.[159] Further research on Fe and Co catalysts has shown that alcohols and aldehydes can form alkoxide intermediates, which act as nucleophilic initiators for chain growth rather than CHOH.[155]

This mechanism can account for the formation of oxygenated compounds and many other aspects of synthesis. However, it is not free of objections. It has never been explained why alcohol can initiate the synthesis but evidently cannot contribute to the chain growth by dehydrocondensation.[13] On the other hand, olefins can start the synthesis as well, and it is then a question of whether the initiating species must contain oxygen. So, the major problem is how the two enolic intermediates, which are electrophilic at carbon, interact to form a carbon–carbon bond, while the self-reaction of methylene groups is a known reaction and, hence, lends support to the carbide mechanism.

6.7.3 CO Insertion Mechanism

Since the surface carbide and enolic mechanisms involve difficulties mainly in the propagation step, Pichler and Schulz[160] suggested a mechanism by which the synthesis is started by CO insertion into a metal–hydrogen bond and propagated by CO insertion into a metal–alkyl bond (Figure 6.8). Intermediates such as Me(CO)R are subsequently converted into metal–alkyl species by hydrogenolysis.

The CO insertion mechanism has explained quite well the experimental evidence that when CO is removed from the reaction mixture methane formation not only continues but in fact increases, since the adsorption of hydrogen is less hindered due to the absence of carbon monoxide.

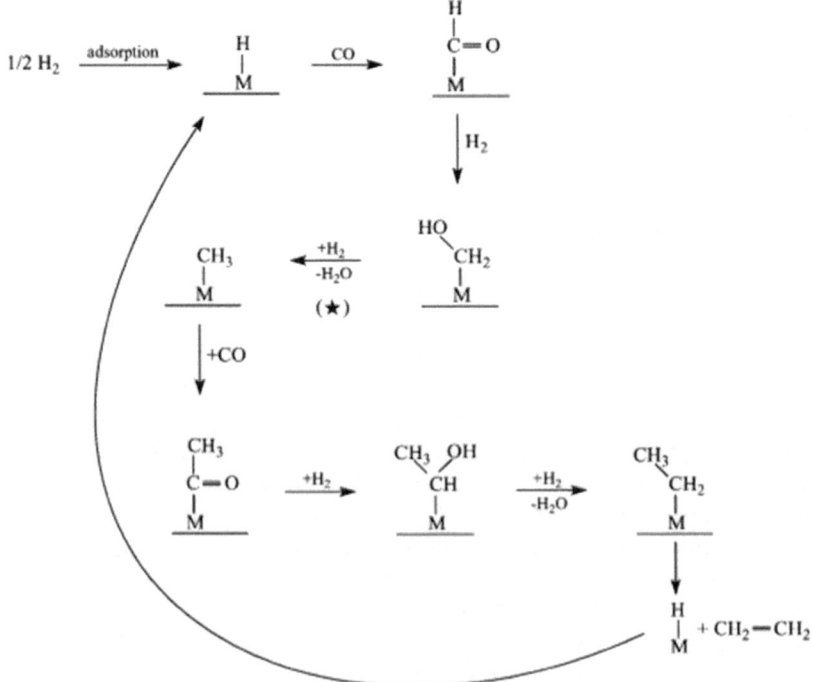

(* rate controlling)

Figure 6.8 CO insertion mechanism.[147] Reproduced from ref. 147, with permission from Elsevier, Copyright 2001.

In contrast, the production of higher hydrocarbons simply drops, which indicates that carbon monoxide is necessary for chain growth. The isotopic labeling experiments conducted by Blyholder and Emmett[161,162] supported this mechanism. No experimental evidence has disputed this mechanism except for its initiation step which involves CO insertion into a metal–hydrogen bond. Clearly, CO insertion as a propagation step can overcome the difficulties encountered in carbide and enolic proposals.

Supported by micro-kinetic models, isotope tracer studies, and kinetic investigations, this mechanism highlights the importance of C–O scission after C–C coupling.[16] However, its role as the primary pathway in Fischer–Tropsch synthesis (FTS) remains debated. Some studies suggest that minimal molecularly adsorbed CO is present on Fe surfaces under certain conditions (*e.g.*, 700 mbar), indicating that CO insertion may not always be the predominant route.[163]

6.7.4 Alkoxy Mechanism

Sapienza *et al.*[164] advanced a version of the alkoxy mechanism which they called the oxide mechanism and claimed that the oxide mechanism encompasses the observations of earlier theorists and is useful in predicting catalyst behavior which was evidenced by the work done by Sapienza's group. The alkoxy mechanism is based on the idea that carbon monoxide chemisorbs on a metal surface and reacts with hydrogen to yield an oxygen rather than a carbon coordinated species. Thus, the bond strength of metal oxides should correspond directly to the specific activity of methanation but should be related inversely to methane selectivity. The metal–oxide bond strength will also determine the probable product distribution. Weak oxide formers, such as rhodium, should produce ethylene glycol, whereas a strong oxide former, like iron or manganese, should yield ethylene. A mixed system should give a good yield of ethanol. The same logic also applies to the effects of promoters. The addition of a weak oxide former should decrease methane and olefin formation and *vice versa*. A schematic representation of the alkoxy mechanism is given in Figure 6.9.

A major concern with this mechanism is that it involves structures that are unusual and even unprecedented in the literature. Examples are those involving tetravalent oxygen coordinated to the metal and traditionally it is also difficult to accept the production of such species as M–O=C, M=O=CH$_2$, M–O=CH$_2$, and M–O–CH$_3$.[159]

6.7.5 Other Alternative Mechanisms

The mechanisms discussed so far all share the common feature that they involve only one active chain growth intermediate. As such, these proposals are restrictive and may not readily explain the full product distribution that is obtained during reaction. Variations have therefore been proposed to overcome these difficulties.

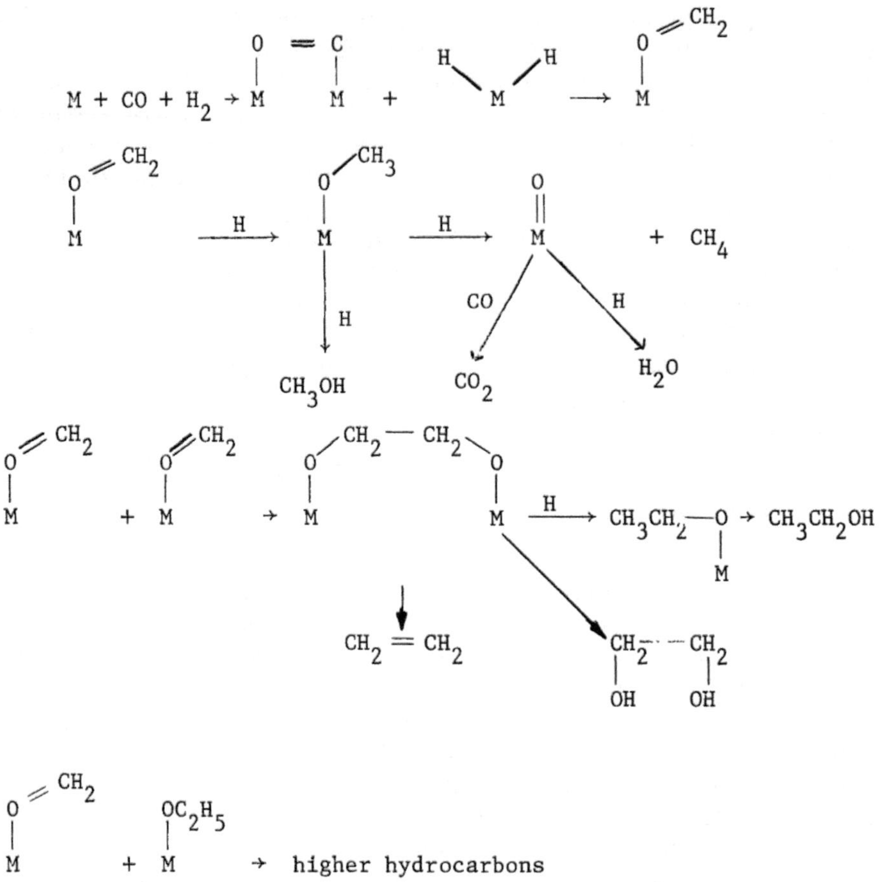

Figure 6.9 Alkoxy mechanism.[164] Reproduced from ref. 164 with permission from Springer Nature, Copyright 1979.

Based on the four mechanisms discussed above, Bell[123] postulated a rather comprehensive scheme shown in eqn (6.19)–(6.31). In this mechanism it is suggested that the synthesis of hydrocarbons is initiated by the stepwise hydrogenation of single carbon atoms formed from carbon monoxide dissociation on the catalyst surface in such a way that the methyne (CH), methylene (CH_2), and methyl (CH_3) groups are in equilibrium. The addition of a hydrogen atom to the methyl group leads to methane production, while the insertion of a methylene group into the metal–carbon bond of a metal–methyl complex results in chain growth. Thus one may visualize the formation of C_{2+} hydrocarbons as a polymerization process in which the methylene groups act as the monomer and the alkyl groups are active centers for chain growth.

$$CO + M \leftrightarrow CO\text{–}M \qquad (6.19)$$

$$CO\text{–}M + M \leftrightarrow C\text{–}M + O\text{–}M \qquad (6.20)$$

$$H_2 + 2M \leftrightarrow 2H-M \tag{6.21}$$

$$O-M + H_2 \text{(or 2H-M)} \leftrightarrow H_2O + 2M \tag{6.22}$$

$$O-M + CO \text{(or CO-M)} \leftrightarrow CO_2 + 2M \tag{6.23}$$

$$C-M + H-M \leftrightarrow CH-M + M \tag{6.24}$$

$$CH-M + H-M \leftrightarrow CH_2-M + M \tag{6.25}$$

$$CH_2-M + H-M \leftrightarrow CH_3-M + M \tag{6.26}$$

$$CH_3-M + H-M \leftrightarrow CH_4 + M \tag{6.27}$$

$$CH_3-M + CH_2-M \leftrightarrow CH_3CH_2-M + M \tag{6.28}$$

$$CH_3CH_2-M + CH_2-M \leftrightarrow CH_3CH_2CH_2-M + M \tag{6.29}$$

$$CH_3CH_2-M + M \leftrightarrow CH_2CH_2 + H-M + M \tag{6.30}$$

$$CH_3CH_2-M + H-M \leftrightarrow CH_3CH_3 + 2M \tag{6.31}$$

For the formation of oxygenated products, Bell[123] postulated the mechanism shown in eqn (6.32)–(6.42) In this scheme, direct hydrogenation of a carbon monoxide molecule bonded through both its carbon and oxygen ends could lead to the formation of methoxide species which might serve as a precursor to methanol. The insertion of a CO molecule into a metal alkyl bond and subsequent addition of hydrogen could provide the pathway for the formation of aldehydes and higher alcohols.

Methanol

$$M=C=O + M \leftrightarrow M=C=O \cdots M \tag{6.32}$$

$$M=C=O \cdots M + H-M \leftrightarrow M-CH=O \cdots M \tag{6.33}$$

$$M-CH=O \cdots M + H-M \leftrightarrow M-CH_2-O-M \tag{6.34}$$

$$M-CH_2-O-M + H-M \rightarrow CH_3-O-M + 2M \tag{6.35}$$

$$CH_3-O-M + H-M \rightarrow CH_3OH + M + M \tag{6.36}$$

Aldehydes

$$M-R + CO-M \rightarrow M-CR=O \tag{6.37}$$

$$M-CR=O + H-M \rightarrow RCHO + 2M \tag{6.38}$$

Higher alcohols

$$M-CR=O + M \leftrightarrow M-CR=O \cdots M \tag{6.39}$$

$$M-CR=O\cdots M + H-M \leftrightarrow M-CRH-O-M \tag{6.40}$$

$$M-CRH-O-M + H-M \rightarrow RCH_2-O-M + 2M \tag{6.41}$$

$$RCH_2-O-M + H-M \rightarrow RCH_2OH + M + M \tag{6.42}$$

Based on a study of the kinetics and product distribution of carbon monoxide hydrogenation over a cobalt catalyst supported on kieselguhr, Wojciechowski[146] argued the following mechanistic statements:

i. Adsorption of all species proceeds on the catalyst surface onto one set of sites. The hydrogen decomposes into adsorbed atoms while carbon monoxide decomposes to adsorbed carbon and adsorbed oxygen. A sequence of hydrogenations by adsorbed hydrogen atoms leads to various surface species: CH_x, OH, *etc.*
ii. The monomeric species for oligomerization is CH_2 and the formation of this monomer is the rate-limiting step.
iii. The growing radical on the surface is immobile, except perhaps for species up to C_4. In order to grow, it must have a monomer appearing next to it. This monomer either is formed next to the growing chain or migrates there by surface diffusion among an appropriate set of sites.
iv. During the growth process the surface radical can undergo 1–2 spontaneous shifts of attachment to the surface. Such events lead to side-chain methyls and, if two or more shifts occur in a row, to side-chain ethyls, *etc.*
v. Termination is the step-determining product distribution. Each termination event is due to a site near to a growing radical and containing the appropriate termination function.
vi. Changes in temperature, P_{H_2}, and P_{CO} are the fundamental governing factors which affect both the kinetics and product distribution. Total pressure and H_2/CO ratio are compound variables with complex influences on the reaction.
vii. Re-adsorption of products is kinetically unimportant, although certain catalysts can hydrogenate crack and aromatize olefins. However, these are secondary reactions.

Dry[165] proposed another mechanism which involves both CH_2 and CO as active surface intermediates. In this proposal, which is given in Figure 6.10, chain growth occurs *via* non-oxygenated surface intermediates, and a parallel mechanism of CO insertion occurs to give oxygenated products. The overall concept of these parallel pathways has been proposed and supported in a number of other studies.[123,166–168] Hence, there appears to be a general consensus that more than one active surface intermediate present on the catalyst surface may be responsible for the product distribution observed in the hydrogenation of carbon monoxide.

Activation of CO, CO₂, and H₂ Toward Synthetic Fuel Manufacture

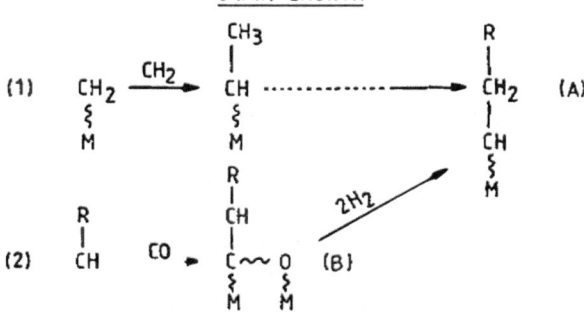

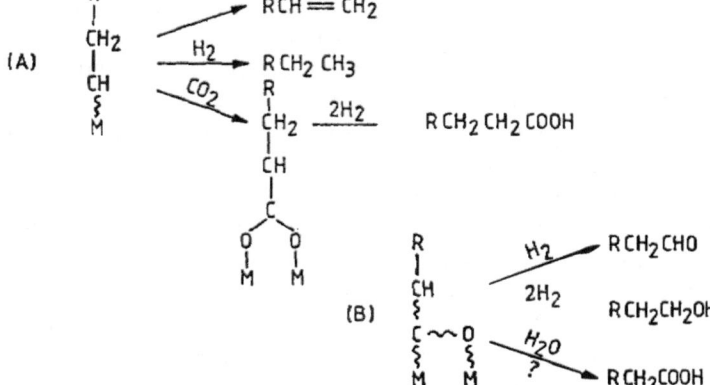

Figure 6.10 CO hydrogenation mechanism proposed by Dry.[165] Reproduced from ref. 165 with permission from Elsevier, Copyright 1990.

The CO insertion-carbide mechanism and the H-assisted CO dissociation mechanism are other prominent mechanisms proposed for FTS. The CO insertion-carbide mechanism, proposed by Gaube and Klein, suggests that CO insertion and alkyl mechanisms occur simultaneously. In contrast, the H-assisted CO dissociation mechanism involves the hydrogenation of surface carbon formed after the dissociation of adsorbed CO, leading to the formation of CH_x intermediates that serve as monomers for chain growth. Quantum-chemical density functional theory (DFT) studies support this mechanism, indicating that direct C–O bond dissociation is dominant on the stepped sites of Fe carbide, while H-assisted CO dissociation occurs on the surface. These studies reveal that the energy barrier for CO dissociation via HCO intermediate formation (123 kJ mol^{-1}) is lower than for direct CO dissociation (142 kJ mol^{-1}). This mechanism has gained significant attention for Fe and Co catalysts, with strong experimental evidence supporting its relevance.[16,155,169]

6.8 Kinetics of Carbon Monoxide Hydrogenation

The kinetics of carbon monoxide (CO) hydrogenation is complex, influenced by catalyst composition, particle size, reaction conditions, and surface phenomena. Kinetic modeling of CO hydrogenation typically employs the Langmuir–Hinshelwood–Hougen–Watson (LHHW) approach, incorporating parameters such as activation energies and surface coverage effects.[170,171] Empirical reaction rates for CO consumption are dependent on factors including partial pressures of reactants and temperature.[172] Therefore the kinetic equations for carbon monoxide hydrogenation published in the literature do not present a uniform picture. This subject has been reviewed by several authors. Vannice[14] summarized almost all known rate expressions for the reaction proposed before 1974 concerning catalysts with a variety of active materials. Later, Wojciechowski[146] and Sarup and Wojciechowski[173] presented a rather comprehensive kinetic study on carbon monoxide hydrogenation over an impregnated cobalt catalyst supported on kieselguhr. It is obvious from the literature how the derived kinetic expressions were influenced by the type of catalyst and the operating conditions employed by the various investigators. The kinetic expressions vary from simple to complex. The dependence of the rate equations on the partial pressure of hydrogen varies from zero to two, and that of carbon monoxide varies from one to minus one. Comparisons of different equations are complicated by the fact that the various catalysts were prepared in different ways, which inevitably result in differences in metallic area and in the pore structures of the final catalyst particles. The types of products also vary from predominantly wax to pure methane. These differences should influence the rate of diffusion of reactants to and within the catalyst particles to varying degrees and this could change the form of the equations derived.[4]

Generally speaking, most syngas conversion kinetics studies are focused on iron and cobalt catalysts. For iron-based catalysts it is well known that the partial pressure of water has a negative effect on the rate of reaction while the partial pressure of hydrogen has a positive influence.[174] The Anderson[24] equation is widely used for fitting experimental data from micro- to commercial-scale reactors. Eqn (6.43) is the Anderson rate equation which can be derived from the enolic mechanism by assuming that the hydrogenation of chemisorbed carbon monoxide is the rate-determining step:[175]

$$r = \frac{k_0 P_{CO} P_{H_2}}{P_{CO} + \alpha P_{H_2O}} \quad (6.43)$$

To handle the observed hydrogen dependence, Huff and Satterfield[176] derived an alternate rate form of the equation:

$$r = \frac{k_0 P_{CO} P_{H_2}^2}{P_{CO} P_{H_2} + b P_{H_2O}} \quad (6.44)$$

using two different mechanisms, the carbide mechanism taking the hydrogenation of surface carbon as the rate-determining step, and an enol/carbide mechanism with hydrogenation of surface enol as the rate-limiting step.

In order to accommodate situations with high water–gas shift activity and/or low H_2/CO ratios, Ledakowicz et al.[177] developed the following relationship:

$$r = \frac{k_0 P_{CO} P_{H_2}}{P_{CO} + P_{CO_2}} \quad (6.45)$$

based on the enolic mechanism with the hydrogenation of surface carbon monoxide as the rate-determining step assuming that carbon monoxide and dioxide are the only gaseous species which adsorb significantly on the catalyst surface. They also presented the following rate equation:

$$r = \frac{k_0 P_{CO} P_{H_2}}{P_{CO} + \alpha P_{H_2O} + c P_{CO_2}} \quad (6.46)$$

to account for inhibition by both water and carbon dioxide. This generalized rate expression may be used for catalysts with low water–gas shift activity, where water concentrations are high, as well as for catalysts with high shift activity which show inhibition by carbon dioxide. The activation energy reported by the authors for iron catalysts is about 80–103 kJ mol^{-1}, regardless of catalyst type. It is worthwhile to note that, at low CO conversions, eqn (6.43)–(6.46) can all be reduced to a power function form which is first order in the hydrogen partial pressure.

For cobalt-based catalysts the literature also proposes several rate equations. In contrast to most iron-based catalysts, cobalt catalysts are not very active for the water–gas shift reaction, thus the carbon dioxide inhibition effect is usually not a significant term in the kinetics of CO hydrogenation over cobalt catalysts.

Golodets et al.[148] presented a fairly complicated kinetic equation for the carbon monoxide hydrogenation reaction based on the following mechanism:

Step 1

$$CO + * \leftrightarrow *CO \quad (6.47)$$

Step 2
$$H_2 + 2* \leftrightarrow 2*H \tag{6.48}$$

Step 3
$$*CO + *H \leftrightarrow *X + * \tag{6.49}$$

Step 4
$$*X + *H \rightarrow *X_0 + * \tag{6.50}$$

Step 5
$$*X_0 + *H \rightarrow *X_0^{*H} \; CH_4 + H_2O + * \tag{6.51}$$

Step 6
$$*X_0 + *CO \rightarrow *X_1^{*H} \; C_2H_6 + H_2O + * \tag{6.52}$$

Step 7
$$*X_1 + *CO \rightarrow X_2 + *\ldots \tag{6.53}$$

where * represents active sites and X enolic intermediates. They assumed that equilibrium is attained in steps 1–3, step 4 is the rate-limiting step, and selectivity is controlled by steps 5–7. Thus they derived the rate equation for the overall conversion of carbon monoxide as:

$$r = k_4 \frac{K_3 b_{CO} P_{CO} b_{H_2} P_{H_2}}{\left[1 + b_{CO}P_{CO} + (b_{H_2}P_{H_2})^{\frac{1}{2}}(1 + K_3 b_{CO} P_{CO})\right]^2} \tag{6.54}$$

Rhodium catalysts agree well with the equation above. They also developed the following equations for methane selectivity and selectivity for higher hydrocarbons respectively:

$$S_{CH_4} = \frac{r5}{r5 + r6} = \frac{1}{1 + \dfrac{k_6 b_{CO} P_{CO}}{k_5 (b_{H_2} P_{H_2})^{1/2}}} \tag{6.55}$$

$$S_{C_{1+}} = 1 - S_{CH_4} = \frac{1}{1 + \dfrac{k_5 (b_{H_2} P_{H_2})^{1/2}}{k_6 b_{CO} P_{CO}}} \tag{6.56}$$

where k_5 and k_6 are the rate constants for steps 5 and 6.

Sarup and Wojciechowski[173] described six different possible mechanisms for the Fischer–Tropsch reaction on cobalt catalysts. The rate of reaction was measured by both the rate of carbon disappearance and the rate of oxygen appearance as water plus carbon dioxide. Four of their proposed expressions presume that dissociated carbon monoxide participates in the reaction, while two postulate that carbon monoxide is adsorbed but not dissociated. The general form of rate expression they proposed is:

$$-r = \frac{kP_{CO}^a P_{H_2}^b}{\left(1 + \sum_{i=1}^{n} K_i P_{CO}^{ci} P_{H_2}^{di}\right)^2} \quad (6.57)$$

In eqn (6.57) k is a kinetic parameter, a and b are the reaction orders of the rate-controlling step, K_i represents the adsorption parameter for the ith adsorption term, and c_i and d_i describe the dependence of surface coverage of the ith adsorption term on the reactant partial pressure.

The activation energy reported for carbon monoxide hydrogenation over cobalt catalysts is reported to be in the range of 93–103 kJ mol^{-1}.[178]

6.9 Fischer–Tropsch Synthesis Through Carbon Dioxide Hydrogenation

Due to the acknowledged long-term climate impact of carbon dioxide, its incorporation in synthetic chemistry attracts significant attention. Carbon dioxide hydrogenation is a critical process with substantial environmental and industrial implications. It presents a promising approach to mitigating anthropogenic CO_2 emissions by converting this greenhouse gas into valuable fuels and chemicals, thereby contributing to carbon recycling and energy storage.[179,180] This process is central to the "power-to-gas" concept, where CO_2 is converted into methane (CH_4) through methanation, enabling large-scale storage of hydrogen produced *via* sustainable methods, such as solar energy and hydropower. The resulting synthetic natural gas can be easily stored, transported, or integrated into existing natural gas infrastructure.[181] Moreover, CO_2 hydrogenation to methanol plays a key role in the methanol economy, which aims to incorporate CO_2 into a sustainable carbon cycle, provided that hydrogen is derived from green technologies and efficient CO_2 capture methods are used. Copper-based catalysts, such as Cu/ZnO/Al$_2$O$_3$, have shown promise in this conversion. However, challenges persist due to competing reactions like the reverse water–gas shift (RWGS), which converts CO_2 into CO.[182] Additionally, CO_2 hydrogenation can produce formic acid, an important feedstock and hydrogen source for fuel cells, highlighting its role in C_1 chemistry and the broader chemical industry.[179] The process also has the potential to synthesize higher hydrocarbons and olefins, which are essential building blocks in the chemical industry, through methods such as the methanol-to-gasoline (MTG) process or modified Fischer–Tropsch synthesis.[109] Despite its potential, CO_2 hydrogenation faces challenges including low activity and selectivity due to the chemical inertness of CO_2. This necessitates the development of efficient catalysts that can operate under mild conditions. Recent advances in single atom catalysts (SACs) have shown promise in improving the selectivity and stability of CO_2 hydrogenation, offering new pathways for producing high value-added chemicals like methanol and formate.[180] CO_2 hydrogenation can proceed through various catalytic methods, including thermal hydrogenation, photocatalysis, and photothermal catalysis, each of which has distinct advantages and challenges. Thermal catalysis is widely studied due to its high efficiency and scalability, as demonstrated by commercial operations like the e-gas plant in Germany, which efficiently converts CO_2 to hydrocarbons. Photothermal catalysis, which combines the benefits of both thermal and photocatalysis, utilizes solar energy to enhance reaction rates, though it faces challenges related to catalyst and reactor design.[183]

6.10 Products of Carbon Dioxide Hydrogenation

Carbon dioxide (CO_2) hydrogenation is a versatile chemical reaction capable of producing valuable products, depending on the catalysts and reaction conditions employed. One of the most attractive products is methanol (CH_3OH), a clean, biodegradable, and high-energy fuel. Methanol can be further converted into other valuable chemicals such as dimethyl ether, olefins, and longer-chain alcohols and hydrocarbons, making it a highly versatile product.[184] The reaction typically involves the conversion of CO_2 and hydrogen (H_2) over catalysts like copper (Cu), with methanol being a primary product due to its ease of storage and transport as a hydrogen carrier.[184] In addition to methanol, CO_2 hydrogenation can yield by-products such as carbon monoxide (CO), methane (CH_4), and water (H_2O), particularly when using Cu-based catalysts.[185] The process can be optimized further by employing bifunctional catalysts, which facilitate the conversion of methanol into higher-value C_{2+} compounds, such as dimethyl ether, hydrocarbons like gasoline, and light olefins, through dehydration or coupling reactions over zeolites or alumina.[186] Another pathway in CO_2 hydrogenation is the reverse water–gas shift (RWGS) reaction, which leads to the production of CO with high selectivity, especially when using bifunctional catalysts that combine methanol synthesis catalysts with zeolites.[187]

Methanation, a key reaction in this process, can occur either through the hydrogenation of CO to methane or the direct methanation of CO_2, known as the Sabatier reaction. Both are exothermic reactions that are favored at moderately high temperatures (200–500 °C).[183] CO_2 hydrogenation to methanol is particularly promising due to its role as an intermediate in producing olefins and aromatics through methanol-to-olefins (MTO) or methanol-to-aromatics (MTA) reactions. Methanol synthesis may proceed *via* CO-mediated or formate-mediated routes, with CO_2 hydrogenated through various intermediates.[188] Additionally, the conversion of CO_2 to formic acid represents a significant step in reducing CO_2 to other chemicals, often catalyzed by advanced systems such as Schiff-base-modified gold nanocatalysts.[179] The catalytic hydrogenation of CO_2 to hydrocarbons, including paraffins (alkanes) and olefins (alkenes), can also be achieved through the Fischer–Tropsch process. This process, favored by elevated pressures, involves the conversion of CO and hydrogen into long-chain hydrocarbons.[183] The choice of catalyst, reaction conditions, and reactor type, such as slurry or fluidized-bed reactors, plays a crucial role in influencing the product distribution, conversion rates, and selectivity toward various hydrocarbons.[189–191]

6.11 Thermodynamics of the Carbon Dioxide Hydrogenation Reactions

The thermodynamics of carbon dioxide hydrogenation involves a complex interaction of reaction conditions, catalyst properties, and the inherent stability of CO_2. As a highly stable molecule with low Gibbs free energy, CO_2 requires significant activation energy, which can be reduced when paired with a higher-energy reactant like hydrogen (H_2).[192] The hydrogenation of CO_2 can produce a range of products, including methane (CH_4) and methanol, each characterized by distinct thermodynamic profiles. Methanol synthesis from CO_2 is an exothermic reaction as shown in eqn (6.58) and it favors low temperatures. However, this process is thermodynamically limited due to the chemically inert nature of CO_2. The CO_2 molecule is difficult to be activated due to its thermodynamic stability ($\Delta G_f^0 = -394.38$ kJ mol^{-1}) and kinetic inertness.[193] It is reported that CO_2 activation and methanol formation are promoted at the increased reaction temperature, *i.e.* around 240 °C, although a higher temperature is completely debatable from the point of view of reaction thermodynamics.[194] At higher temperatures, the endothermic reverse water–gas shift (RWGS) reaction as shown in eqn (6.59) becomes more prominent, complicating the process.[195] Undesirable products such as higher alcohols and hydrocarbons are also produced when the reaction is performed at higher temperatures. Therefore, the production of methanol is most favorable at high pressures to maximize conversion due to the reduction in the number of gas molecules.

$$CO_2 + 3H_2 \Leftrightarrow CH_3OH + H_2O \quad \Delta H^0 = -49.5 \text{ kJ mol}^{-1} \quad (6.58)$$

$$CO_2 + H_2 \Leftrightarrow CO + H_2O \quad \Delta H^0 = +41.1 \text{ kJ mol}^{-1} \quad (6.59)$$

CO_2 methanation, which is highly exothermic, is kinetically limited and requires catalysts that function efficiently at mild thermal conditions to achieve high conversion rates. The Sabatier reaction given in eqn (6.60) is the process by which CO_2 reacts with H_2 to produce CH_4 and water. The reaction is exothermic—CO_2 conversion and CH_4 yield both are increased with the pressure and decreased with temperature. To overcome the kinetic barrier at low temperatures, catalysts consisting mostly of group VIIB metals supported by various porous materials are used. Ni-based catalysts supported on Al_2O_3, SiO_2, MgO, CeO_2, and ZrO_2, and zeolites have been widely studied in CO_2 methanation.[196] During the methanation process, the gas molecules decrease. Therefore, the CO_2 methanation is favored at high pressure (Le Chatelier principle).

$$CO_2 + 4H_2 \Leftrightarrow CH_4 + 2H_2O \quad \Delta H^0 = -164.7 \text{ kJ mol}^{-1} \quad (6.60)$$

6.12 Product Distribution

The products of CO_2 hydrogenation can be divided into two groups: C_1 compounds and C_{1+} compounds.[196] C_1 compounds include methane, methanol, formaldehyde, and formic acid, while C_{1+} compounds include hydrocarbons and oxygenates. To improve process efficiency toward target products, factors such as thermodynamics (temperature and pressure) and H_2/CO_2 ratio as well as optimizing catalyst properties like surface area, porosity, and metal dispersion are imperative. Achieving the desired transformation is difficult due to the lack of favorable thermodynamic conditions, increasing the need for catalyst development to address this limitation.

Various catalysts have been investigated to optimize selectivity and yield in carbon dioxide (CO_2) hydrogenation that converts CO_2 into valuable hydrocarbons and oxygenates. For example, Fe_3O_4-based nanocatalysts have demonstrated promise in producing light olefins with a high olefin/paraffin ratio and selectivity toward C_2–C_4 olefins, while minimizing the formation of CO and CH_4.[197] In contrast, Cu/γ-Al_2O_3 catalysts, when combined with nonthermal plasma, significantly improve CO_2 conversion (10%) and methanol (CH_3OH) selectivity (65%), overcoming kinetic limitations at room temperature and atmospheric pressure.[198]

Pt/In_2O_3 catalysts, particularly those with low Pt loading (e.g., 0.13 wt% Pt), have also been shown to enhance methanol selectivity (58.4%), although increasing Pt loadings tend to reduce this effect.[199] Similarly, bimetallic In–Pd catalysts influence product distribution, with the In:Pd(2:1)/SiO_2 catalyst achieving the highest methanol activity and selectivity (61%), however, increasing Pd content leads to a decrease in methanol selectivity.[200] Additionally, iron-based catalysts derived from delafossite–$CuFeO_2$ have been shown to produce heavier hydrocarbons, similar to those obtained from Fischer–Tropsch synthesis, with a high olefin-to-paraffin ratio and low methane selectivity, highlighting the critical role of the catalyst precursor in determining product distribution.[201]

6.13 Carbon Dioxide Hydrogenation Catalysts

The products of CO_2 hydrogenation, such as methane (CH_4), methanol (CH_3OH), and long-chain hydrocarbons, are determined by the catalysts and reaction conditions used. Catalysts play a critical role in enhancing the efficiency and selectivity of these reactions. For example, the catalytic hydrogenation of CO_2 to methane, known as CO_2 methanation, is facilitated by catalysts like Ru/MgAl. Under non-thermal plasma (NTP) conditions, these catalysts significantly improve CO_2 conversion and CH_4 yield by promoting alternative reaction pathways and reducing activation barriers.[181]

Methanol synthesis from CO_2, often catalyzed by copper-based systems such as $Cu/ZnO/Al_2O_3$ and Cu/ZrO_2, is effective due to the formation of formate and methoxy intermediates at the metal–support interface.[182,202] The hydrogenation of CO_2 to long-chain hydrocarbons can be achieved through modified Fischer–Tropsch synthesis or methanol-mediated pathways. Catalysts like $Na-Fe_3O_4$/HMCM-22 exhibit high selectivity for isoparaffins, which are ideal components of clean fuels, by effectively matching tandem reactions to improve both selectivity and yield.[203]

The development of transition metal-based and main group metal-based catalysts, including novel nanostructured catalysts like metal–organic frameworks (MOFs) and zeolitic imidazolate frameworks (ZIFs) derived catalysts, shows potential for improving activity and selectivity in the conversion of CO_2 to methanol.[204] Single atom catalysts (SACs) have also gained considerable attention due to their high atom utilization efficiency and their ability to produce high value-added chemicals. For instance, Cu-N4 single atom catalysts have exhibited remarkable methanol selectivity (95.5%) and productivity under mild conditions (150 °C), out-performing conventional catalysts like $Cu-ZnO/Al_2O_3$.[180] These advancements highlight ongoing efforts to address the challenges of CO_2 hydrogenation, including low activity and selectivity, by leveraging innovative catalyst designs, sustainable hydrogen sources, and advanced reactor technologies.

Fe-based catalysts are complex due to the co-existence of several iron carbide and iron oxide phases along with metallic iron under FT conditions. Fe-based catalysts are moreover often heavily promoted and the selectivity of Fe-based FT catalysts can be tuned. Their water–gas shift (WGS) activity makes Fe-based catalysts a good choice for CO_2-based FT processes.[36]

6.13.1 Active Metals

In carbon dioxide hydrogenation, various active metals serve as effective catalysts, each exhibiting unique properties and mechanisms. Copper is extensively studied for its catalytic activity and stability in CO_2 hydrogenation to methanol. Copper-based catalysts utilize mixed valence states, including Cu^0, Cu^+, and $Cu^{\delta+}$, which contribute to the hydrogenation process.[205] The addition of palladium to copper, forming Pd–Cu bimetallic catalysts, enhances activity by promoting electron transfer, facilitating the formation of intermediates like COOH, and lowering activation energy, thus increasing methanol productivity.[206]

Single-atom catalysts like Cu_1/ZrO_2 have demonstrated high turnover frequency and selectivity for methanol, emphasizing the importance of isolated active sites.[205] While traditional $Cu-ZnO-Al_2O_3$ catalysts are effective for converting synthesis gas to methanol, they are less efficient with CO_2-rich feeds, especially at lower space velocities. However, supports like ZrO_2 enhance activity and selectivity due to the strong $Cu-ZrO_2$ interaction and the presence of Lewis acidic sites, which promote the formation of formate and methoxy intermediates.[182,207,208] The Cu–ZrO$_2$ interface is particularly effective in promoting methanol synthesis while suppressing the formation of by-products like CO and methane.[209] Additionally, transition-metal carbides, such as Cu/Mo$_2$CT$_x$ (MXene) materials, have shown promise in enhancing the catalytic performance of copper *via* electronic metal–support interactions.[207]

Cobalt is another active metal used in CO$_2$ hydrogenation, particularly for hydrocarbon production. Cobalt oxide (CoO) on titania support is highly active, following a hydrogen-assisted pathway, while metallic cobalt typically follows a direct dissociation pathway.[210] Cobalt catalysts supported on silica (Co@Si$_x$) have exhibited enhanced methanol selectivity and CO$_2$ conversion rates, outperforming conventional cobalt catalysts and some noble-metal catalysts.[211] Cobalt-based catalysts supported on other metal oxides, such as CeO$_2$, ZrO$_2$, Gd$_2$O$_3$, and ZnO have also been explored for CO$_2$ methanation, with Co/CeO$_2$ showing superior performance.[212]

Iron-based catalysts play a crucial role in CO$_2$ hydrogenation, particularly in the reverse water–gas shift (RWGS) reaction and subsequent Fischer–Tropsch synthesis (FTS) for hydrocarbon production. The process typically involves converting CO$_2$ to CO *via* the RWGS reaction, catalyzed by the Fe$_3$O$_4$ phase, followed by hydrogenation of CO to hydrocarbons, facilitated by iron carbides such as Hägg iron carbide (χ-Fe$_5$C$_2$).[213,214] The introduction of other metals, such as cobalt, can enhance the catalytic activity of iron-based catalysts. The formation of Fe–Co alloys, for example, improves hydrogen dissociation and promotes chain growth, though it may also increase methane selectivity due to cobalt's higher hydrogenation ability.[215]

Bimetallic catalysts, such as Na-promoted CoFe$_2$O$_4$/CNT, have demonstrated superior performance in CO$_2$ hydrogenation, achieving higher CO$_2$ conversion and selectivity toward light olefins compared to single-metal catalysts like Na-promoted Fe$_3$O$_4$/CNT.[213] The interaction between iron and other metals, like copper, also influences hydrocarbon distribution. Cu-promoted Fe catalysts show improved selectivity for C$_{5+}$ hydrocarbons due to enhanced olefin adsorption and secondary conversion.[216]

6.13.2 Supports

Support materials play a crucial role in CO$_2$ hydrogenation, significantly impacting the activity, selectivity, and stability of catalysts. Various supports have been investigated for different catalytic systems, including γ-Al$_2$O$_3$, SiO$_2$, TiO$_2$, ZrO$_2$, and metal–organic frameworks (MOFs), showing considerable potential for enhancing iron-based catalysts. Among these, γ-Al$_2$O$_3$ is often preferred due to its strong metal–support interaction, which help prevent the sintering of active particles and promote the formation of C$_{2+}$ hydrocarbons.[217,218]

ZrO$_2$ and TiO$_2$ are widely studied for Cu-based catalysts, with ZrO$_2$ exhibiting higher CO$_2$ conversion rates and methanol selectivity compared to

TiO_2. This advantage is attributed to the ability of ZrO_2 to more effectively bind key reaction intermediates, facilitating methanol formation via the reverse water–gas-shift (RWGS) pathway.[219] The crystalline phase of ZrO_2, whether monoclinic or tetragonal, also affects methanol selectivity and yield during CO_2 hydrogenation, with monoclinic ZrO_2 demonstrating better electron transfer capabilities that enhance the hydrogenation of formate intermediates to methanol.[220] Additionally, ZrO_2 supports create oxygen vacancies, which are essential for CO_2 dissociation, thereby improving selectivity toward light olefins and reducing methane formation.[191] The incorporation of Zr into ceria supports further increases the number of weak and medium basic sites required for CO_2 methanation, resulting in higher methane selectivity.[191]

For Cu-based catalysts in methanol synthesis, supports such as ZnO play a crucial role by enhancing the dispersion and stabilization of copper. The lattice oxygen vacancies in ZnO are active in methanol synthesis.[217] Pd–Cu nanoparticles supported on P25 (a form of TiO_2) demonstrate high selectivity in CO_2 hydrogenation to ethanol, with the oxygen vacancies on P25 facilitating the process.[221] Reducible supports like TiO_2 and CeO_2, when combined with Cu catalysts, tend to promote CO formation via the reverse water–gas shift (RWGS) reaction, highlighting the importance of support materials in controlling catalyst performance and product selectivity.[207]

Cu/CeO_2 catalysts exhibit even higher methanol selectivity than Cu/ZrO_2, attributed to carbonate intermediates formed by oxygen vacancies on CeO_2. These intermediates enhance the dispersion of Cu and the formation of active sites.[222] The interaction between Cu and the support plays a critical role, as demonstrated in Cu/SiO_2 catalysts, where the presence of Cu^+ species leads to high catalytic activity and stability. This is due to the strong metal–support interaction and high Cu dispersion.[223] Novel supports like MXene materials, particularly Mo_2CT_x, have also shown increased methanol formation rates. This improvement is linked to the high dispersion of Cu and the presence of Lewis acidic Cu^+ sites at the Cu/Mo_2CT_x interface.[207]

Cobalt/ceria-based catalysts demonstrate that varying the particle size of the ceria–zirconia support can effectively tune metal–support interactions, significantly enhancing the CO_2 hydrogenation rate to methane by stabilizing cobalt particles and generating oxygen vacancies.[224] Cobalt catalysts supported on different crystal phases of TiO_2 exhibit varied selectivity and conversion rates, with Co/rutile-TiO_2 producing methane and Co/anatase-TiO_2 yielding CO. The addition of promoters such as Zr, K, and Cs can further improve selectivity toward C_{2+} hydrocarbons by adjusting the surface C/H ratio.[225] Silica-supported cobalt catalysts, particularly those with an optimized cobalt-to-silica ratio, demonstrate enhanced methanol selectivity and CO_2 conversion, outperforming conventional Co/SiO_2 catalysts.[211]

6.13.3 Promoters

Catalyst promoters play a critical role in improving the efficiency and selectivity of catalysts used in carbon dioxide (CO_2) hydrogenation. These promoters modify the electronic and structural properties of the catalyst to improve the yield of desired hydrocarbons. For example, potassium is a well-known promoter for iron-based catalysts, enhancing the production of olefinic hydrocarbons, particularly light olefins.[191,226] Potassium-modified iron catalysts increase CO_2 adsorption and chain growth probability, resulting in higher selectivity and conversion rates, while also suppressing secondary hydrogenation of olefins, which minimizes the formation of unwanted by-products like methane.[109,226] Additionally, potassium also influences the surface properties of catalysts, such as Al_2O_3, improving the reducibility of metal oxides like Cu, which is beneficial for methanol synthesis.[204]

Manganese (Mn) is an important promoter, particularly in enhancing catalyst selectivity for light olefin production. Mn promotes the formation of Fe_5C_2 species in Na/Fe catalysts, which is crucial for improving selectivity without altering the overall CO_2 conversion rate. This enhancement results from the increased Fe_5C_2 content and the interaction between Mn and Fe, which reduces CO adsorption, thereby suppressing excessive chain-growth reactions.[227]

Zinc also plays a vital role as a promoter in CO_2 hydrogenation to methanol, especially in copper-based catalysts. Zinc oxide (ZnO) improves the dispersion of metallic copper particles and facilitates electron transfer from ZnO to copper, which is essential for CO_2 activation and subsequent hydrogenation to methanol. Additionally, zinc induces defects in the copper structure, creating additional active sites that enhance the catalytic activity.[228] The presence of zinc also generates oxygen vacancies on the catalyst surface, further promoting the hydrogenation process.[229] Furthermore, zinc alters the reaction mechanism by enhancing the hydrogenation of CO_2 over CO, making CO_2 the primary carbon source in the reaction.[230] Studies indicate that ZnO-promoted Cu catalysts exhibit weaker CO binding, shifting the reaction pathway toward CO_2 hydrogenation.[230]

Magnesium oxide (MgO) is another important promoter, particularly for enhancing the selectivity and activity of catalysts used in methanol synthesis from CO_2 hydrogenation. Adding MgO to catalysts such as $Cu–ZnO–ZrO_2/Al_2O_3$ increases the number of alkali sites on the catalyst surface, preventing methane formation and promoting methanol production.[191]

6.14 Electronic Configuration of CO_2

Carbon dioxide (CO_2) is a linear non-polar molecule consisting of one carbon atom double-bonded to two oxygen atoms, resulting in a symmetrical O=C=O structure. However, during CO_2 adsorption and activation on catalyst surfaces, its linear structure can change. CO_2 may either physisorb in its linear form or chemisorb in a bent configuration, which is essential for its activation and subsequent reactions that lead to various final products.[231] Linear adsorption involves physisorption, where CO_2 retains its structure, while bent adsorption occurs through chemisorption, in which CO_2 bends due to interactions with the catalyst surface. In the bent mode, CO_2 can form various chemisorption configurations by coordinating with metal atoms serving as Lewis acid sites or oxygen atoms acting as Lewis base sites. This coordination may involve the oxygen atoms in CO_2 donating electrons, or the carbon atom accepting electrons, resulting in a mixed coordination where CO_2 simultaneously donates and accepts electrons.[231]

The bent configuration plays a critical role in CO_2 activation, as it reduces the energy barrier for further reactions, thereby influencing reaction pathways and product distribution. For instance, the formaldehyde, carbene, and glyoxal pathways for CO_2 reduction are influenced by the adsorption mode, with the bent configuration facilitating the formation of intermediates such as CO and bidentate formate. These intermediates are essential for the production of methanol, methane, and other C_2 products.[231]

6.15 Carbon Dioxide Adsorption

Understanding the interaction between CO_2 molecules and catalyst surfaces is essential, as it is influenced by the nature of the catalyst and the adsorption conditions. Typically, CO_2 is activated into carbonate species, which are subsequently hydrogenated into formate and methoxy species, as observed in studies involving Ni/CeO_2–ZrO_2 catalysts.[218] Surface frustrated Lewis pairs (FLPs) in heterogeneous catalysis, which consist of proximal Lewis acid–base sites, highlight the significance of heterolytic H_2 dissociation and CO_2 activation on metal oxide surfaces. The interaction between surface acidity and basicity, which is affected by the crystal face and surface reconstruction, determines the adsorption strength and overall catalytic activity.[232]

CO_2 can adsorb on both stoichiometric and non-stoichiometric metal oxide surfaces, coordinating either with metal sites or surface oxygen sites *via* its terminal oxygen atoms. In non-stoichiometric metal oxides, surface oxygen vacancies act as additional adsorption sites, further enhancing CO_2 interaction.[233] Catalysts such as Cu/ZnO have been extensively studied for CO_2 hydrogenation, but recent density functional theory (DFT) studies emphasize the high activity of In_2O_3 in this process, indicating the importance of catalyst development in improving CO_2 activation.[234] The stabilization of key intermediates, such as CO, is essential for controlling the selectivity of CO_2 hydrogenation pathways. Stronger CO binding tends to favor methanol synthesis, whereas weaker binding promotes the reverse water–gas shift (RWGS) reaction.[218]

6.16 Mechanism of Carbon Dioxide Hydrogenation

One significant pathway in carbon dioxide (CO_2) hydrogenation involves converting CO_2 to methanol, where formate (HCOO*) serves as a key intermediate. Oxygen vacancies on the catalyst surface are crucial for activating CO_2 and H_2, thus facilitating the hydrogenation process.[235] The use of single-atom catalysts (SACs) introduces mechanisms like the formate pathway, in which CO_2 is activated on isolated metal atoms, leading to methanol formation through a series of hydrogenation steps.[180] Density functional theory (DFT) calculations have revealed the formation of intermediates like *HCOO and *HOCO, which play a critical role in conversion processes on various catalyst surfaces.[180]

Another pathway includes hydrogenating CO_2 to methane and other hydrocarbons *via* the reverse water–gas shift (RWGS) reaction, followed by further hydrogenation of CO or direct methanation *via* the Sabatier reaction.[183] The RWGS reaction, which converts CO_2 into CO, serves as a precursor for further processes such as Fischer–Tropsch synthesis (FTS), leading to the production of olefins and aromatics. While the RWGS reaction is slightly endothermic, the subsequent FTS is exothermic, making the overall process thermodynamically favorable for hydrocarbon production.[188]

Non-thermal plasma (NTP) activation offers an alternative route by converting CO_2 into bicarbonate and formate species, which then react with hydrogen to produce methane and water at lower temperatures. This method can alter reaction pathways depending on the catalyst, such as Ru/MgAl layered double hydroxide.[181] Additionally, a modified Fischer–Tropsch mechanism involves CO_2 reduction by iron, followed by hydrogen radical abstraction, resulting in hydrocarbons with higher olefin selectivity due to limited H_2 uptake.[236]

6.16.1 Conversion of CO_2 to Methanol: Formate and CO Pathway

The conversion of CO_2 to methanol is a promising strategy for reducing greenhouse gas emissions and supporting a sustainable methanol-based economy. This process involves the hydrogenation of CO_2 using hydrogen derived from renewable sources, offering an alternative to traditional fossil-fuel-based methods.[237,238] Methanol is a versatile chemical that serves as a feedstock for producing aromatics, lower olefins, and as a fuel, highlighting its industrial significance.[239] Catalytic conversion of CO_2 to methanol typically utilizes supported copper catalysts such as Cu–ZnO, Cu–ZrO_2, and Cu–ZnO–ZrO_2 (CZZ) due to their high performance and effective operation at relatively low temperatures (180–240 °C).[240] The enhanced catalytic activity of CZZ catalysts is attributed to the weak hydrophilic character of the ZrO_2 support, which mitigates the poisoning effect of water on active sites during methanol synthesis.[240]

This process involves two main steps: CO_2 adsorption and activation, followed by H_2 dissociation, which facilitates reactions with activated intermediates to produce methanol rather than by-products like methane.[239] The most widely accepted mechanism involves CO_2 adsorption on metal oxide surfaces, such as ZnO and ZrO_2, forming bicarbonate species. Dissociatively adsorbed hydrogen on copper surfaces then transfers to these bicarbonate species, leading to the formation of formate intermediates, which further hydrogenate to form dioxomethylene and eventually methoxide species, crucial for methanol production.[241] The formate pathway, where CO_2 initially forms *HCOO, which hydrogenates to *H_2COO and then to *H_2CO via HCOOH, eventually results in methanol formation.[242] In_2O_3-based catalysts also follow this pathway, with oxygen vacancies playing a crucial role in facilitating formate formation and conversion.[235] The reverse water–gas shift reaction provides an additional pathway, where CO_2 is first converted to CO and then hydrogenated to methanol.[241] Density functional theory studies have identified the conversion of HCOO to H_2COO as a rate-determining step in this process.[242]

The CO pathway, involving the RWGS reaction where CO_2 is initially reduced to CO and then hydrogenated to methanol, is significant in industrial applications, particularly when using syngas (a mixture of CO, H_2, and CO_2) for methanol production. This method allows for the utilization of CO_2, providing a more sustainable approach to methanol synthesis.[242–244] Despite its potential, challenges remain in achieving high conversion rates and selectivity, which are critical for commercial viability. Current research focuses on developing more effective catalysts and reactor configurations to optimize CO_2 hydrogenation technology, addressing issues such as competitive reactions and thermodynamic limitations.[242]

6.16.2 Hydrogenation of CO_2 to Methane

The hydrogenation of CO_2 to methane, also known as CO_2 methanation, is a significant process for converting carbon dioxide into valuable fuels, primarily methane. This highly exothermic reaction is a key component of "power-to-gas" technology, which stores hydrogen produced from renewable sources by converting it into synthetic natural gas, thereby integrating seamlessly into existing gas infrastructures. The reaction can proceed through various catalytic methods, including thermal catalysis, photocatalysis, and photothermal catalysis, with thermal catalysis being the most researched due to its high efficiency and scalability for industrial applications.[181,183]

CO_2 methanation typically follows two primary pathways: the reverse water–gas shift reaction, where CO_2 is first converted to CO and then hydrogenated to methane, or the direct Sabatier reaction.[183,216] Catalysts, such as iron- and cobalt-based materials, play a crucial role. Iron catalysts are widely used in Fischer–Tropsch synthesis to convert CO_2 into hydrocarbons, while cobalt catalysts exhibit high activity in methane production.[189,216] Recent advancements in catalyst development, such as cobalt-based systems, have significantly improved methanation performance. For example, Co/CeO_2 catalysts have demonstrated methane yields of approximately 96% at 300 °C, attributed to enhanced reducibility due to Co–ceria interactions.[212] Additionally, non-thermal plasma (NTP)-assisted CO_2 hydrogenation offers a potentially more energy-efficient pathway by activating the reaction through plasma.[181]

The Sabatier reaction, which involves the exothermic conversion of CO_2 and hydrogen into methane and water, is a key process for closed-loop CO_2 cycling, making it essential for sustainable energy and environmental mitigation efforts.[191] Despite challenges like high hydrogen consumption and a lower energy density per unit volume compared to other CO_2 utilization routes, the Sabatier reaction remains promising, particularly when integrated with renewable energy sources for hydrogen production.[191,212] The reaction typically occurs at moderately high temperatures (200–500 °C) and elevated pressures to favor methane formation.[183] Efficient catalyst systems and reactor configurations are critical for optimizing performance, as cooling equipment is often required to manage the exothermic nature of the reaction. This helps prevent hot spots and maintain steady-state operating conditions, ensuring both efficiency and safety in the process.[191]

6.16.3 Fischer–Tropsch Synthesis Mechanism

The Fischer–Tropsch synthesis mechanism is typically integrated with the reverse water–gas shift reaction, where CO_2 is first converted to CO and subsequently hydrogenated to form hydrocarbons. This process follows a stepwise chain growth polymerization reaction, where the feed is initially converted into an initiator and monomer, which then polymerize to produce hydrocarbons. Iron-based catalysts are particularly effective in this process due to their dual activity in RWGS and FTS reactions, resulting in the formation of olefins and paraffins.[245]

CO_2 hydrogenation produces a variety of hydrocarbons, ranging from light hydrocarbons like methane to heavier hydrocarbons (C_{5+}), though the latter are less commonly formed as the process tends to favor the production of lower-molecular-weight hydrocarbons, with CO serving as the chain growth agent.[201] Multifunctional catalysts, such as iron-based catalysts combined with zeolites, can alter the product distribution, shifting it toward high-octane gasoline-range hydrocarbons by promoting oligomerization and isomerization reactions.[245] The structure and composition of catalysts, particularly metal–oxide interfaces, play a crucial role in determining catalytic performance.[246]

The methanol-mediated mechanism, where CO_2 is first converted to methanol and then to olefins, faces challenges due to the high temperatures required and low methanol selectivity.[227,247] However, the direct conversion of CO_2 into iso-paraffins, ideal clean hydrocarbon fuel components, can be achieved using multifunctional catalysts like Na-Fe_3O_4/HMCM-22. This catalyst facilitates tandem reactions, including reverse water–gas shift, C–C coupling, and isomerization, enabling high selectivity for C_{4+} hydrocarbons.[203] Zeolites in these catalysts provide unique pore structures and Bronsted acid sites, which help suppress unwanted side reactions and promote isomerization, a key process for producing iso-paraffins with high octane numbers.[203,245] Additionally, alkali-promoted iron catalysts enhance CO_2 adsorption while reducing H_2 adsorption, beneficial for increasing light olefin selectivity.[245,248]

6.17 Kinetics of Carbon Dioxide Hydrogenation

Kinetic modeling of carbon dioxide hydrogenation to methanol is essential for understanding the reaction mechanisms and rate-determining steps, which are crucial for optimizing catalyst performance and reaction conditions. The CO_2 hydrogenation reaction is commonly modeled using the Langmuir–Hinshelwood (LH) mechanism, which accounts for the adsorption of reactants on catalyst surfaces followed by surface reactions. For example, the hydrogenation of formate is often proposed as the rate-determining step in producing methanol in these models.[249] The kinetic equations for CO_2 hydrogenation depend on factors such as catalyst type and reaction conditions. For example, the apparent activation energy for methanol production over $CuCeTiO_x$ catalysts is significantly lower than that for traditional Cu-based catalysts, emphasizing the role of catalyst composition in influencing reaction kinetics. Oxide supports like ZrO_2 and CeO_2 enhance CO_2 adsorption and lower activation energy, thereby improving methanol selectivity and yield.[250]

Kinetic parameters, including reaction rate and equilibrium constants, are important for modeling CO_2 hydrogenation. Specific values for rate constants (k) and equilibrium constants (K) are provided for different reaction steps.[251] Bifunctional catalysts have been investigated to enhance CO_2 conversion rates by coupling methanol synthesis with methanol-to-olefins (MTO) reactions. Although kinetic modeling studies for this approach are limited, experimental data suggest that combined catalysts can outperform individual catalysts, indicating potential synergies.[249] Furthermore, the formation and decomposition of intermediates such as formate and carbonyl species also influence reaction kinetics, with transient kinetic experiments offering valuable insights into these processes.[246]

The Langmuir–Hinshelwood (LH) mechanism, commonly applied to CO_2 hydrogenation, suggests that the reaction rate depends on the dual-site mechanism for H_2 and CO_2 adsorption and activation for methanol synthesis.[204] Competitive adsorption models indicate that the presence of CO can interfere with hydrogen adsorption, affecting the rate of methanol formation.[204] Additionally, the partial pressure of CO produced *via* the reverse water–gas shift reaction influences the chain-growth probability of hydrocarbons, with lower H_2/CO_2 ratios slightly increasing the probability of forming C_{2+} hydrocarbons.[215] Methanol synthesis involves several hydrogenation steps, with the hydrogenation of H_2CO* to CH_3O* identified as the rate-determining step due to its high energy barrier.[204]

6.18 Outlook and Summary

Fischer–Tropsch synthesis technology has been proved to offer a viable synthetic fuel alternative to oil-based fuel. Already mature and energy-efficient Fischer–Tropsch synthesis has its economic barriers in the syngas manufacture. When the syngas production is based on fossil-based fuels such as coal or natural gas, fuel production through Fischer–Tropsch synthesis will not be much different from the petroleum-based fuel technologies. The technology areas open to development reside in the waste-to-fuel domains. The waste products generated by urban settings range from solid household waste, sewage, to CO_2 produced during heating and energy conversion processes, including transportation. Conversion of these wastes to synthetic fuels with our existing understanding of technology passes through highly endothermic syngas manufacture. If renewable energy can be used to supply the energy needed for the syngas manufacture from waste, the thermodynamic as well as economic advantages can be immediately gained. FT synthesis technology can be applied in small, medium, and large scales with not much difference in process economics. Synthetic fuels manufactured as such will be free from compounds containing sulfur, nitrogen, and other contaminants that are inherent to petroleum, giving the process significant technological, economical, and energy advantages. Incorporation of CO_2 to the Fischer–Tropsch process will further assist a major global challenge. CO_2 is an FTS product, managing the oxygen chemical potential of the catalyst along with H_2O. Addition of CO_2 to the FTS feed stream has two potential influences. (i) In the presence of excess hydrogen, a mildly endothermic reverse water–gas shift reaction can take place. CO is synthesized, while removing some of the reaction heat and helping local thermal energy management near the catalytic site. When local heat is controlled as such, the product selectivity can move forward to longer chains. (ii) Since the product is already present in the feed stream, that limits the forward rates, such as by adjusting CO_2 partial pressure in the gas phase, enabling product selectivity tuning.

Acknowledgements

Financial support from TUBITAK under Grant no: 120C150 under BIDEB National Leading Researcher Program is kindly acknowledged.

References

1. H. H. Storch, *Adv. Catal.*, 1948, **1**, 115.
2. H. Pichler, *Adv. Catal.*, 1952, **4**, 271.
3. P. Biloen and W. M. H. Sachtler, *Adv. Catal.*, 1981, **30**, 165.
4. M. E. Dry, in *Catalysis Science and Technology*, ed. J. R. Anderson, De Gruyter, Berlin, Boston, 1982, ch. 4, vol. 1, pp. 159–256.
5. M. Janardanarao, *Ind. Eng. Chem. Res.*, 1990, **29**, 1735.
6. J. P. Hindermann, G. J. Hutchings and A. Kiennemann, *Catal. Rev.*, 1993, **35**, 1.
7. Z. Liu, H. Liu, Y. Gao and Y. Xing, *React. Kinet., Mech. Catal.*, 2024, **137**, 879.
8. N. C. Shiba, X. Liu and Y. Yao, *Reactions*, 2023, **4**, 420.
9. C. Du, P. Lu and N. Tsubaki, *ACS Omega*, 2020, **5**, 49.

10. H. Tang, T. Qiu, X. Wang, C. Zhang and Z. Zhang, *Molecules*, 2024, **29**, 1194.
11. J. van de Loosdrecht, F. G. Botes, I. M. Ciobica, A. Ferreira, P. Gibson, D. J. Moodley, A. M. Saib, J. L. Visagie, C. J. Weststrate and J. W. Niemantsverdriet, in *Comprehensive Inorganic Chemistry II*, ed. J. Reedijk and K. Poeppelmeier, Elsevier, 2nd edn, 2013, vol. 7, pp. 525–557.
12. Y. T. Shah and A. J. Perrotta, *Product R&D*, 1976, **15**, 123.
13. V. Ponec, *Catal. Rev.*, 1978, **18**, 151.
14. M. A. Vannice, *Catal. Rev.*, 1976, **14**, 153.
15. A. Yahyazadeh, A. K. Dalai, W. Ma and L. Zhang, *Reactions*, 2021, **2**, 227.
16. Z. Teimouri, N. Abatzoglou and A. K. Dalai, *Catalysts*, 2021, **11**, 1.
17. B. H. Davis and M. L. Occelli, *Fischer–Tropsch synthesis, catalysts and catalysis: advances and applications*, CRC Press, Boca Raton, 1st edn, 2016.
18. J. Hu, W. Zhou, K. Cheng, Q. Zhang and Y. Wang, in *The Chemical Transformations of C1 Compounds*, ed. X. F. Wu, B. Han, K. Ding and Z. Liu, Wiley-VCH GmbH, 2022, ch. 20, pp. 861–907.
19. T. C. Bromfield and R. Visagie, WIPO (PCT), WO2005049765A1, 2005.
20. M. Arsalanfar, A. Nouri, M. Abdouss and E. Rezazadeh, *J. Chem. Technol. Biotechnol.*, 2024, **99**, 133.
21. D. Dhamo, J. Kühn, S. Lüttin, M. Rubin and R. Dittmeyer, *Sustainable Energy Fuels*, 2024, **8**, 2094.
22. H. H. Storch, *The Fischer–Tropsch and related syntheses, including a summary of theoretical and applied contact catalysis*, Wiley, New York, 1951.
23. G. A. Mills and F. W. Steffgen, *Catal. Rev.*, 1974, **8**, 159.
24. R. B. Anderson, *Catalysis*, 1956, **4**, 1.
25. D. S. Newsome, *Catal. Rev.*, 1980, **21**, 275.
26. D. L. King, J. A. Cusumano and R. L. Garten, *Catal. Rev.*, 1981, **23**, 233.
27. I. Puskas and R. S. Hurlbut, *Catal. Today*, 2003, **84**, 99.
28. Y. Zhang and X. Lin, *New J. Chem.*, 2024, **48**, 7875.
29. M. Vannice, *J. Catal.*, 1975, **37**, 462.
30. G. P. Van Der Laan and A. A. C. M. Beenackers, *Catal. Rev.*, 1999, **41**, 255.
31. M. Jamaati, M. Torkashvand, S. Sarabadani Tafreshi and N. H. de Leeuw, *Molecules*, 2023, **28**, 6525.
32. G. L. Bezemer, J. H. Bitter, H. P. C. E. Kuipers, H. Oosterbeek, J. E. Holewijn, X. Xu, F. Kapteijn, A. J. Van Diilen and K. P. De Jong, *J. Am. Chem. Soc.*, 2006, **128**, 3956.
33. A. A. Mirzaei, R. M. Kiai, H. Atashi, M. Arsalanfar and S. Shahriari, *J. Ind. Eng. Chem.*, 2012, **18**, 1242.
34. P. Wang, F.-K. Chiang, J. Chai, A. I. Dugulan, J. Dong, W. Chen, R. J. P. Broos, B. Feng, Y. Song, Y. Lv, Q. Lin, R. Wang, I. A. W. Filot, Z. Men and E. J. M. Hensen, *Nature*, 2024, **635**, 102.
35. K. Xu, B. Sun, J. Lin, W. Wen, Y. Pei, S. Yan, M. Qiao, X. Zhang and B. Zong, *Nat. Commun.*, 2014, **5**, 5783.

36. K. T. Rommens and M. Saeys, *Chem. Rev.*, 2023, **123**, 5798.
37. R. B. Anderson, *The Fischer–Tropsch Synthesis*, Academic Press Inc, Orlando, Florida, 1984.
38. K. Y. Park, W. K. Seo and J. S. Lee, *Catal. Lett.*, 1991, **11**, 349.
39. O. V. Krylov, *Catalysis by nonmetals: rules for catalyst selection*, Academic Press, 1970.
40. E. V. Prokhorenko, N. V. Pavlenko and G. I. Golodets, *Kinet. Katal.*, 1988, **29**, 820.
41. B. Büssemeier, C. D. Frohning and B. Cornils, *Hydrocarbon Process.*, 1976, 105.
42. A. Chen, M. A. Vannice, M. Kaminsky and G. Geoffroy, *Am. Chem. Soc., Div. Pet. Chem., Prepr.; (United States)*, New York, NY, USA, 1986.
43. V. P. Santos, B. van der Linden, A. Chojecki, G. Budroni, S. Corthals, H. Shibata, G. R. Meima, F. Kapteijn, M. Makkee and J. Gascon, *ACS Catal.*, 2013, **3**, 1634.
44. C. Yang, F. E. Massoth and A. G. Oblad, *Adv. Chem.*, 1979, **178**, 35.
45. M. Feyzi, N. Yaghobi and V. Eslamimanesh, *Mater. Res. Bull.*, 2015, **72**, 143.
46. B. H. Davis, *Catal. Today*, 2009, **141**, 25.
47. H. Kölbel, K. D. Tillmetz, *US Pat.*, US4177203A, 1979.
48. J. Barrault, C. Renard, L. T. Yu and J. Gal, Proceedings of 8th International Congress on Catalysis, Berlin (West), 1984.
49. R. Malessa and M. Baerns, *Ind. Eng. Chem. Res.*, 1988, **27**, 279.
50. D. Das, G. Ravichandran, D. K. Chakrabarty, S. N. Piramanayagam and S. N. Shringi, *Appl. Catal., A*, 1993, **107**, 73.
51. C. K. Das, N. S. Das, D. P. Choudhury, G. Ravichandran and D. K. Chakrabarty, *Appl. Catal., A*, 1994, **111**, 119.
52. M. Saglam, *Ind. Eng. Chem. Res.*, 1989, **28**, 150.
53. B. Cornils, D. Frohning and K. Moraw, Proceedings of 8th International Congress on Catalysis, Berlin (West), 1984.
54. S. Qin, C. Zhang, B. Wu, J. Xu, H. Xiang and Y. Li, *Catal. Lett.*, 2010, **139**, 123.
55. J. L. Feimer, P. L. Silveston and R. R. Hudgins, *Can. J. Chem. Eng.*, 1984, **62**, 241.
56. E. Kikuchi, T. Ino and Y. Morita, *J. Catal.*, 1979, **57**, 27.
57. B. L. Gustafson and P. S. Wehner, *Prepr. Pap., Am. Chem. Soc., Div. Fuel Chem.*, Santa Clara, CA, USA, 1986.
58. V. A. De la Peña O'Shea, N. N. Menéndez, J. D. Tornero and J. L. G. Fierro, *Catal. Lett.*, 2003, **88**, 123.
59. H. Du, H. Zhu, T. Liu, Z. Zhao, X. Chen, W. Dong, W. Lu, W. Luo and Y. Ding, *Catal. Today*, 2017, **281**, 549.
60. M. A. Vannice, in *Catalysis Science and Technology*, ed. J. R. Anderson and M. Boudart, Springer, Berlin, Heidelberg, 1982, ch. 3, pp. 139–198.
61. E. Rytter, S. Eri, T. H. Skagseth, D. Schanke, E. Bergene, R. Myrstad and A. Lindvåg, *Ind. Eng. Chem. Res.*, 2007, **46**, 9032.

62. A. Jean-Marie, A. Griboval-Constant, A. Y. Khodakov and F. Diehl, *C. R. Chim.*, 2009, **12**, 660.
63. R. C. Reuel and C. H. Bartholomew, *J. Catal.*, 1984, **85**, 78.
64. W.-J. Wang and Y.-W. Chen, *Appl. Catal.*, 1991, **77**, 21.
65. A. Lapidus, A. Krylova, J. Rathousky, A. Zukal and M. Jancalkova, *Appl. Catal., A*, 1992, **80**, 1.
66. S. Bessell, *Appl. Catal., A*, 1993, **96**, 253.
67. R. L. Varma, L. Dan-Chu, J. F. Mathews and N. N. Bakhshi, *Can. J. Chem. Eng.*, 1985, **63**, 72.
68. M. van der Riet, G. J. Hutchings and R. G. Copperthwaite, *J. Chem. Soc., Chem. Commun.*, 1986, **1**, 798.
69. S. Colley, R. G. Copperthwaite, G. J. Hutchings and M. Van der Riet, *Ind. Eng. Chem. Res.*, 1988, **27**, 1339.
70. R. G. Copperthwaite, G. J. Hutchings, M. Van der Riet and J. Woodhouse, *Ind. Eng. Chem. Res.*, 1987, **26**, 869.
71. T. Inui, A. Sakamoto, T. Takeguchi and Y. Ishigaki, *Ind. Eng. Chem. Res.*, 1989, **28**, 427.
72. S. E. Colley, R. G. Copperthwaite, G. J. Hutchings, G. A. Foulds and N. J. Coville, *Appl. Catal., A*, 1992, **84**, 1.
73. F. Rohr, O. A. Lindvåg, A. Holmen and E. A. Blekkan, *Catal. Today*, 2000, **58**, 247.
74. Y. Zhang, Y. Liu, G. Yang, S. Sun and N. Tsubaki, *Appl. Catal., A*, 2007, **321**, 79.
75. L. M. Kabir, M. K. Albolkany, M. M. Mohamed and A. A. El-Moneim, *Catal. Lett.*, 2024, **154**, 3372.
76. J. S. Jung, S. W. Kim and D. J. Moon, *Catal. Today*, 2012, **185**, 168.
77. Y. Zhao, S. Huang, C. Liu, Y. Zhang, L. Wang, A. lin and J. Li, *J. Taiwan Inst. Chem. Eng.*, 2024, **156**, 105328.
78. R. B. Greegor, F. W. Lytle, R. L. Chin and D. M. Hercules, *J. Phys. Chem.*, 1981, **85**, 1232.
79. R. L. Chin and D. M. Hercules, *J. Phys. Chem.*, 1982, **86**, 360.
80. S. H. Moon and K. E. Yoon, *Appl. Catal.*, 1985, **16**, 289.
81. W.-J. Wang and Y.-W. Chen, *Appl. Catal.*, 1991, **77**, 223.
82. P. G. Dimitrova and D. R. Mehandjiev, *J. Catal.*, 1994, **145**, 356.
83. A. N. Akin, A. E. Aksoylu and Z. I. Önsan, *React. Kinet. Catal. Lett.*, 1999, **66**, 393.
84. L. Fu and C. H. Bartholomew, *J. Catal.*, 1985, **92**, 376.
85. B. G. Johnson, C. H. Bartholomew and D. W. Goodman, *J. Catal.*, 1991, **128**, 231.
86. A. N. Akin and Z. İ Önsan, *J. Chem. Technol. Biotechnol.*, 1997, **69**, 337.
87. D. K. Lee, J. H. Lee and S. K. Ihm, *Appl. Catal.*, 1988, **36**, 199.
88. A. A. Khassin, T. M. Yurieva, G. N. Kustova, I. S. Itenberg, M. P. Demeshkina, T. A. Krieger, L. M. Plyasova, G. K. Chermashentseva and V. N. Parmon, *J. Mol. Catal. A: Chem.*, 2001, **168**, 193.
89. R. Bechara, D. Balloy and D. Vanhove, *Appl. Catal., A*, 2001, **207**, 343.

90. Y. Zhang, S. Nagamori, S. Hinchiranan, T. Vitidsant and N. Tsubaki, *Energy Fuels*, 2006, **20**, 417.
91. C. Yang, F. E. Massoth and A. G. Oblad, *Am. Chem. Soc., Div. Pet. Chem., Prepr.*; United States, Anaheim, CA, USA, 1978.
92. M. P. Rosynek and C. A. Polansky, *Appl. Catal.*, 1991, **73**, 97.
93. A. K. Dalai, M. N. Esmai and N. N. Bakhshi, *Can. J. Chem. Eng.*, 1992, **70**, 278.
94. A. Guerrero-Ruiz, A. Sepúlveda-Escribano and I. Rodríguez-Ramos, *Appl. Catal., A*, 1994, **120**, 71.
95. H. Chen and A. A. Adesina, *Appl. Catal., A*, 1994, **112**, 87.
96. S. Bhatia, N. N. Bakhshi and J. F. Mathews, *Can. J. Chem. Eng.*, 1978, **56**, 575.
97. C. H. Bartholomew, R. B. Pannell and J. L. Butler, *J. Catal.*, 1980, **65**, 335.
98. C. P. Huang and J. T. Richardson, *J. Catal.*, 1978, **51**, 1.
99. A. of I. S. and T. of Japan, Chemical Abstracts, 1981, 96, 55261.
100. R. Snel, *Catal. Rev.*, 1987, **29**, 361.
101. T. Ishihara, N. Horiuchi, K. Eguchi and H. Arai, *Appl. Catal.*, 1990, **66**, 267.
102. S. Guo, Z. Ma, Q. Wang, J. Wang, H. Guo, C. Chen, B. Hou, L. Jia and D. Li, *Mol. Catal.*, 2024, **557**, 113962.
103. C. Ahn and J. W. Bae, *Catal. Today*, 2016, **265**, 27.
104. H. Atashi, F. Siami, A. A. Mirzaei and M. Sarkari, *J. Ind. Eng. Chem.*, 2010, **16**, 952.
105. J. R. Anderson, *Characterization of catalysts: structure of metallic catalysts*, Academic Press, New York, 1975.
106. A. B. Stiles, *Catalyst supports and supported catalysts*, Butterworth Publishers, Stoneham, MA, United States, 1987.
107. A. C. Ghogia, A. Nzihou, P. Serp, K. Soulantica and D. Pham Minh, *Appl. Catal., A*, 2021, **609**, 117906.
108. J. Horáček, *Monatsh. Chem.*, 2020, **151**, 649.
109. C. G. Visconti, M. Martinelli, L. Falbo, A. Infantes-Molina, L. Lietti, P. Forzatti, G. Iaquaniello, E. Palo, B. Picutti and F. Brignoli, *Appl. Catal., B*, 2017, **200**, 530.
110. G. A. Somorjai, *Catal. Rev.*, 1981, **23**, 189.
111. J. Benziger and R. J. Madix, *Surf. Sci.*, 1980, **94**, 119.
112. G. A. Somorjai, *Catal. Rev.*, 1978, **18**, 173.
113. D. O. Uner, N. Savargoankar, M. Pruski and T. S. King, *Stud. Surf. Sci. Catal.*, 1997, 315.
114. D. O. Uner, M. Pruski and T. S. King, *Top. Catal.*, 1995, **2**, 59.
115. J. Barrault and C. Renard, *Chem. Abstr.*, 1986, **105**, 155071.
116. E. L. Kugler, Am. Chem. Soc., Div. Pet. Chem., *Prepr.;* United States, San Francisco, CA, USA, 1980.
117. K. Liu, D. Xu, H. Fan, G. Hou, Y. Li, S. Huang and M. Ding, *ACS Sustainable Chem. Eng.*, 2024, **12**, 2070.
118. B. Zhao, P. Zhai, P. Wang, J. Li, T. Li, M. Peng, M. Zhao, G. Hu, Y. Yang, Y.-W. Li, Q. Zhang, W. Fan and D. Ma, *Chem*, 2017, **3**, 323.

119. T. K. Das, G. Jacobs, P. M. Patterson, W. A. Conner, J. Li and B. H. Davis, *Fuel*, 2003, **82**, 805.
120. Y. Zhang, K. Liew, J. Li and X. Zhan, *Catal. Lett.*, 2010, **139**, 1.
121. G. V. D. Lee and V. Ponec, *Catal. Rev.*, 1987, **29**, 183.
122. S. Ichi Ishi, Y. Ohno and B. Viswanathan, *Surf. Sci.*, 1985, **161**, 349.
123. A. T. Bell, *Catal. Rev.*, 1981, **23**, 203.
124. G. Doyen and G. Ertl, *Surf. Sci.*, 1974, **43**, 197.
125. A. A. Gokhale, J. A. Dumesic and M. Mavrikakis, *J. Am. Chem. Soc.*, 2008, **130**, 1402.
126. E. Iglesia and D. Hibbitts, *J. Catal.*, 2022, **405**, 614.
127. G. Ertl, *Angew. Chem., Int. Ed.*, 2008, **47**, 3524.
128. G. Ertl, *Science*, 1991, **254**, 1750.
129. R. Zhang, Y. Wang, P. Gaspard and N. Kruse, *Science*, 2023, **382**, 99.
130. X. Shen, X. Han, T. Zhang, H. Y. Suo, L. Yan, Y. Li and Y. Yang, *Mol. Catal.*, 2024, **555**, 113889.
131. J. Zhang, J. Chen, J. Ren, Y. Li and Y. Sun, *Fuel*, 2003, **82**, 581.
132. H. P. Bonzel and H. J. Krebs, *Surf. Sci.*, 1982, **117**, 639.
133. E. L. Muetterties, *Transition Metal Hydrides*, Marcel Dekker, New York, 1971.
134. C. Neue, W. A. Pliskin, R. P. Eischens, R. P. Eischens and A. Francis, *Z. Phys. Chem.*, 1960, **24**, 11.
135. J. J. F. Scholten, A. P. Pijpers and A. M. L. Hustings, *Catal. Rev.*, 1985, **27**, 151.
136. J. M. Zowtiak and C. H. Bartholomew, *J. Catal.*, 1983, **83**, 107.
137. Y. Jiang, K. Wang, Y. Wang, X. Gao, J. Zhang, T. S. Zhao and M. Yao, *Mol. Catal.*, 2024, **556**, 113950.
138. Y. Amenomiya and G. Pleizier, *J. Catal.*, 1973, **28**, 442.
139. T. N. Rhodin and C. F. Brucker, *Solid State Commun.*, 1977, **23**, 275.
140. R. Queau, D. Labroue and R. Poilblanc, *J. Catal.*, 1981, **69**, 249.
141. A. Cosultchi, M. Pérez-Luna, J. A. Morales-Serna and M. Salmón, *Catal. Lett.*, 2012, **142**, 368.
142. K. Subramanyam and M. R. A. Rao, *J. Res. Inst. Catalysis, Hokkaido Univ.*, 1970, **18**, 124.
143. M. V. C. Sastri, R. B. Gupta and B. Viswanathan, *J. Catal.*, 1974, **32**, 325.
144. B. F. G. Johnson, *Transition Metal Clusters*, J. Wiley, New York, 1980.
145. R. C. Baetzold and J. R. Monnier, *J. Phys. Chem.*, 1986, **90**, 2944.
146. B. W. Wojciechowski, *Catal. Rev.*, 1988, **30**, 629.
147. B. H. Davis, *Fuel Process. Technol.*, 2001, **71**, 157.
148. G. I. Golodets, N. V. Pavlenko and A. I. Tripolskii, *React. Kinet. Catal. Lett.*, 1986, **32**, 481.
149. S. R. Craxford and E. K. Rideal, *J. Chem. Soc.*, 1939, 1604.
150. J. G. Ekerdt and A. T. Bell, *J. Catal.*, 1979, **58**, 170.
151. P. K. Agrawal, J. R. Katzer and W. H. Manogue, *Ind. Eng. Chem. Fundam.*, 1982, **21**, 385.
152. R. S. Dixit and L. L. Tavlarldee, *Ind. Eng. Chem. Process Des. Dev.*, 1983, **22**, 1.

153. J. T. Kummer, T. W. DeWitt and P. H. Emmett, *J. Am. Chem. Soc.*, 1948, **70**, 3632.
154. L. C. Browning, T. W. DeWitt and P. H. Emmett, *J. Am. Chem. Soc.*, 1950, **72**, 4211.
155. S. Mousavi, A. Zamaniyan, M. Irani and M. Rashidzadeh, *Appl. Catal., A*, 2015, **506**, 57.
156. B. Qian, S. Yang, J. Zhang, S. Zhou, B. Etschmann, C. Liu, B. Dai, J. Cashion, Y. Wang, H. Wang and L. Zhang, *Fuel Process. Technol.*, 2022, **232**, 107265.
157. E. D. e Smit and B. M. Weckhuysen, *Chem. Soc. Rev.*, 2008, **37**, 2758.
158. R. Balaji, M. V. C. Sastri and B. Viswanathan, *J. Catal.*, 1972, **217**, 212.
159. J. Haggin, *Chem. Eng. News*, 1981, **59**, 22.
160. H. Pichler and H. Schulz, *Chem. Ing. Tech.*, 1970, **42**, 1162.
161. G. Blyholder and P. H. Emmett, *J. Phys. Chem.*, 1959, **63**, 962.
162. G. Blyholder and P. H. Emmett, *J. Phys. Chem.*, 1960, **64**, 470.
163. J. Chai, J. Jiang, Y. Gong, P. Wu, A. Wang, X. Zhang, T. Wang, X. Meng, Q. Lin, Y. Lv, Z. Men and P. Wang, *Catalysts*, 2023, **13**, 1052.
164. R. S. Sapienza, M. J. Sansone, L. D. Spaulding and J. F. Lynch, in *Fundamental Research in Homogeneous Catalysis*, ed. M. Tsutsui, Springer US, Boston, MA, 1979, pp. 179–197.
165. M. E. Dry, *Catal. Today*, 1990, **6**, 183.
166. M. Pijolat and V. Perrichon, *Appl. Catal.*, 1985, **13**, 321.
167. K. G. Anderson and J. G. Ekerdt, *J. Catal.*, 1985, **95**, 602.
168. N. O. Egiebor, W. C. Cooper and B. W. Wojciechowski, *Can. J. Chem. Eng.*, 1985, **63**, 826.
169. M. Mansouri, H. Atashi, F. F. Tabrizi, A. A. Mirzaei and G. mansouri, *J. Ind. Eng. Chem.*, 2013, **19**, 1177.
170. M. Shiva, H. Atashi, F. F. Tabrizi and A. A. Mirzaei, *J. Ind. Eng. Chem.*, 2012, **18**, 1112.
171. A. A. Mirzaei, B. Shirzadi, H. Atashi and M. Mansouri, *J. Ind. Eng. Chem.*, 2012, **18**, 1515.
172. M. Sarkari, F. Fazlollahi, H. Atashi, A. A. Mirzaei and V. Hosseinpour, *Fuel Process. Technol.*, 2012, **97**, 130.
173. B. Sarup and B. W. Wojciechowski, *Can. J. Chem. Eng.*, 1989, **67**, 620.
174. B. Jager and R. Espinoza, *Catal. Today*, 1995, **23**, 17.
175. M. E. Dry, *Product R&D*, 1976, **15**, 282.
176. G. A. Huff and C. N. Satterfield, *Ind. Eng. Chem. Process Des. Dev.*, 1984, **23**, 696.
177. S. Ledakowicz, H. Nettelhoff, R. Kokuun and W. D. Deckwer, *Ind. Eng. Chem. Process Des. Dev.*, 1985, **24**, 1043.
178. I. C. Yates and C. N. Satterfield, *Ind. Eng. Chem. Res.*, 1989, **28**, 9.
179. Q. Liu, X. Yang, L. Li, S. Miao, Y. Li, Y. Li, X. Wang, Y. Huang and T. Zhang, *Nat. Commun.*, 2017, **8**, 1407.
180. T. Yang, X. Mao, Y. Zhang, X. Wu, L. Wang, M. Chu, C.-W. Pao, S. Yang, Y. Xu and X. Huang, *Nat. Commun.*, 2021, **12**, 6022.

181. S. Xu, S. Chansai, Y. Shao, S. Xu, Y. chi Wang, S. Haigh, Y. Mu, Y. Jiao, C. E. Stere, H. Chen, X. Fan and C. Hardacre, *Appl. Catal., B*, 2020, **268**, 118752.
182. E. Lam, J. J. Corral-Pérez, K. Larmier, G. Noh, P. Wolf, A. Comas-Vives, A. Urakawa and C. Copéret, *Angew. Chem., Int. Ed.*, 2019, **58**, 13989.
183. W. K. Fan and M. Tahir, *Chem. Eng. J.*, 2022, **427**, 131617.
184. P. S. Murthy, W. Liang, Y. Jiang and J. Huang, *Energy Fuels*, 2021, **35**, 8558.
185. Y. Zhang, J. Fei, Y. Yu and X. Zheng, *Energy Convers. Manage.*, 2006, **47**, 3360.
186. R.-P. Ye, J. Ding, W. Gong, M. D. Argyle, Q. Zhong, Y. Wang, C. K. Russell, Z. Xu, A. G. Russell, Q. Li, M. Fan and Y.-G. Yao, *Nat. Commun.*, 2019, **10**, 5698.
187. Z. Li, J. Wang, Y. Qu, H. Liu, C. Tang, S. Miao, Z. Feng, H. An and C. Li, *ACS Catal.*, 2017, **7**, 8544.
188. D. Wang, Z. Xie, M. D. Porosoff and J. G. Chen, *Chem*, 2021, **7**, 2277.
189. J. Liu, A. Zhang, M. Liu, S. Hu, F. Ding, C. Song and X. Guo, *J. CO2 Util.*, 2017, **21**, 100.
190. H. Wang, G. Zhang, G. Fan, L. Yang and F. Li, *Ind. Eng. Chem. Res.*, 2021, **60**, 16188.
191. S. Saeidi, S. Najari, V. Hessel, K. Wilson, F. J. Keil, P. Concepción, S. L. Suib and A. E. Rodrigues, *Prog. Energy Combust. Sci.*, 2021, **85**, 100905.
192. F. Sha, Z. Han, S. Tang, J. Wang and C. Li, *ChemSusChem*, 2020, **13**, 6160.
193. R. Guil-López, N. Mota, J. Llorente, E. Millán, B. Pawelec, J. L. G. Fierro and R. M. Navarro, *Materials*, 2019, **12**, 3902.
194. S. Kanuri, S. Roy, C. Chakraborty, S. P. Datta, S. A. Singh and S. Dinda, *Int. J. Energy Res.*, 2022, **46**, 5503.
195. M. Huš, V. D. B. C. Dasireddy, N. Strah Štefančič and B. Likozar, *Appl. Catal., B*, 2017, **207**, 267.
196. S. Ebrahimian and S. Bhattacharya, *Energies*, 2024, **17**, 3701.
197. J. Wei, J. Sun, Z. Wen, C. Fang, Q. Ge and H. Xu, *Catal.: Sci. Technol.*, 2016, **6**, 4786.
198. Z. Cui, S. Meng, Y. Yi, A. Jafarzadeh, S. Li, E. C. Neyts, Y. Hao, L. Li, X. Zhang, X. Wang and A. Bogaerts, *ACS Catal.*, 2022, **12**, 1326.
199. Z. Han, C. Tang, J. Wang, L. Li and C. Li, *J. Catal.*, 2021, **394**, 236.
200. J. L. Snider, V. Streibel, M. A. Hubert, T. S. Choksi, E. Valle, D. C. Upham, J. Schumann, M. S. Duyar, A. Gallo, F. Abild-Pedersen and T. F. Jaramillo, *ACS Catal.*, 2019, **9**, 3399.
201. Y. H. Choi, Y. J. Jang, H. Park, W. Y. Kim, Y. H. Lee, S. H. Choi and J. S. Lee, *Appl. Catal., B*, 2017, **202**, 605.
202. C. Huang, J. Wen, Y. Sun, M. Zhang, Y. Bao, Y. Zhang, L. Liang, M. Fu, J. Wu, D. Ye and L. Chen, *Chem. Eng. J.*, 2019, **374**, 221.
203. J. Wei, R. Yao, Q. Ge, Z. Wen, X. Ji, C. Fang, J. Zhang, H. Xu and J. Sun, *ACS Catal.*, 2018, **8**, 9958.

204. X. Jiang, X. Nie, X. Guo, C. Song and J. G. Chen, *Chem. Rev.*, 2020, **120**, 7984.
205. H. Zhao, R. Yu, S. Ma, K. Xu, Y. Chen, K. Jiang, Y. Fang, C. Zhu, X. Liu, Y. Tang, L. Wu, Y. Wu, Q. Jiang, P. He, Z. Liu and L. Tan, *Nat. Catal.*, 2022, **5**, 818.
206. E. J. Choi, Y. H. Lee, D. W. Lee, D. J. Moon and K. Y. Lee, *Mol. Catal.*, 2017, **434**, 146.
207. H. Zhou, Z. Chen, A. V. López, E. D. López, E. Lam, A. Tsoukalou, E. Willinger, D. A. Kuznetsov, D. Mance, A. Kierzkowska, F. Donat, P. M. Abdala, A. Comas-Vives, C. Copéret, A. Fedorov and C. R. Müller, *Nat. Catal.*, 2021, **4**, 860.
208. T. Witoon, J. Chalorngtham, P. Dumrongbunditkul, M. Chareonpanich and J. Limtrakul, *Chem. Eng. J.*, 2016, **293**, 327.
209. C. Wu, L. Lin, J. Liu, J. Zhang, F. Zhang, T. Zhou, N. Rui, S. Yao, Y. Deng, F. Yang, W. Xu, J. Luo, Y. Zhao, B. Yan, X. D. Wen, J. A. Rodriguez and D. Ma, *Nat. Commun.*, 2020, **11**, 5767.
210. I. C. Te Have, J. J. G. Kromwijk, M. Monai, D. Ferri, E. B. Sterk, F. Meirer and B. M. Weckhuysen, *Nat. Commun.*, 2022, **13**, 324.
211. L. Wang, E. Guan, Y. Wang, L. Wang, Z. Gong, Y. Cui, X. Meng, B. C. Gates and F.-S. Xiao, *Nat. Commun.*, 2020, **11**, 1033.
212. J. Díez-Ramírez, P. Sánchez, V. Kyriakou, S. Zafeiratos, G. E. Marnellos, M. Konsolakis and F. Dorado, *J. CO2 Util.*, 2017, **21**, 562.
213. K. Y. Kim, H. Lee, W. Y. Noh, J. Shin, S. J. Han, S. K. Kim, K. An and J. S. Lee, *ACS Catal.*, 2020, **10**, 8660.
214. M. Ronda-Lloret, G. Rothenberg and N. R. Shiju, *ChemSusChem*, 2019, **12**, 3896.
215. L. Guo, J. Sun, Q. Ge and N. Tsubaki, *J. Mater. Chem. A*, 2018, **6**, 23244.
216. J. Liu, A. Zhang, X. Jiang, M. Liu, Y. Sun, C. Song and X. Guo, *ACS Sustainable Chem. Eng.*, 2018, **6**, 10182.
217. H. Yang, C. Zhang, P. Gao, H. Wang, X. Li, L. Zhong, W. Wei and Y. Sun, *Catal.: Sci. Technol.*, 2017, **7**, 4580.
218. M. D. Porosoff, B. Yan and J. G. Chen, *Energy Environ. Sci.*, 2016, **9**, 62.
219. S. Kattel, B. Yan, Y. Yang, J. G. Chen and P. Liu, *J. Am. Chem. Soc.*, 2016, **138**, 12440.
220. C. Yang, C. Pei, R. Luo, S. Liu, Y. Wang, Z. Wang, Z.-J. Zhao and J. Gong, *J. Am. Chem. Soc.*, 2020, **142**, 19523.
221. S. Bai, Q. Shao, P. Wang, Q. Dai, X. Wang and X. Huang, *J. Am. Chem. Soc.*, 2017, **139**, 6827.
222. W. Wang, Z. Qu, L. Song and Q. Fu, *J. Energy Chem.*, 2020, **40**, 22.
223. Z.-Q. Wang, Z.-N. Xu, S.-Y. Peng, M.-J. Zhang, G. Lu, Q.-S. Chen, Y. Chen and G.-C. Guo, *ACS Catal.*, 2015, **5**, 4255.
224. A. Parastaev, V. Muravev, E. Huertas Osta, A. J. F. van Hoof, T. F. Kimpel, N. Kosinov and E. J. M. Hensen, *Nat. Catal.*, 2020, **3**, 526.
225. W. Li, G. Zhang, X. Jiang, Y. Liu, J. Zhu, F. Ding, Z. Liu, X. Guo and C. Song, *ACS Catal.*, 2019, **9**, 2739.

226. J. Zhang, S. Lu, X. Su, S. Fan, Q. Ma and T. Zhao, *J. CO2 Util.*, 2015, **12**, 95.
227. B. Liang, T. Sun, J. Ma, H. Duan, L. Li, X. Yang, Y. Zhang, X. Su, Y. Huang and T. Zhang, *Catal.: Sci. Technol.*, 2019, **9**, 456.
228. M. M.-J. Li, C. Chen, T. Ayvalı, H. Suo, J. Zheng, I. F. Teixeira, L. Ye, H. Zou, D. O'Hare and S. C. E. Tsang, *ACS Catal.*, 2018, **8**, 4390.
229. X. Liu, M. Wang, H. Yin, J. Hu, K. Cheng, J. Kang, Q. Zhang and Y. Wang, *ACS Catal.*, 2020, **10**, 8303.
230. F. Studt, M. Behrens, E. L. Kunkes, N. Thomas, S. Zander, A. Tarasov, J. Schumann, E. Frei, J. B. Varley, F. Abild-Pedersen, J. K. Nørskov and R. Schlögl, *ChemCatChem*, 2015, **7**, 1105.
231. T. Kong, Y. Jiang and Y. Xiong, *Chem. Soc. Rev.*, 2020, **49**, 6579.
232. K. K. Ghuman, L. B. Hoch, T. E. Wood, C. Mims, C. V. Singh and G. A. Ozin, *ACS Catal.*, 2016, **6**, 5764.
233. J. Jia, C. Qian, Y. Dong, Y. F. Li, H. Wang, M. Ghoussoub, K. T. Butler, A. Walsh and G. A. Ozin, *Chem. Soc. Rev.*, 2017, **46**, 4631.
234. K. Sun, Z. Fan, J. Ye, J. Yan, Q. Ge, Y. Li, W. He, W. Yang and C. J. Liu, *J. CO2 Util.*, 2015, **12**, 1.
235. J. Wang, G. Zhang, J. Zhu, X. Zhang, F. Ding, A. Zhang, X. Guo and C. Song, *ACS Catal.*, 2021, **11**, 1406.
236. S. Saeidi, S. Najari, F. Fazlollahi, M. K. Nikoo, F. Sefidkon, J. J. Klemeš and L. L. Baxter, *Renewable Sustainable Energy Rev.*, 2017, **80**, 1292.
237. P. Borisut and A. Nuchitprasittichai, *Front. Energy Res.*, 2019, **7**, 1.
238. A. García-Trenco, E. R. White, A. Regoutz, D. J. Payne, M. S. P. Shaffer and C. K. Williams, *ACS Catal.*, 2017, **7**, 1186.
239. F. Jiang, S. Wang, B. Liu, J. Liu, L. Wang, Y. Xiao, Y. Xu and X. Liu, *ACS Catal.*, 2020, **10**, 11493.
240. Y. Wang, W. Gao, K. Li, Y. Zheng, Z. Xie, W. Na, J. G. Chen and H. Wang, *Chem*, 2020, **6**, 419.
241. T. Phongamwong, U. Chantaprasertporn, T. Witoon, T. Numpilai, Y. Poo-arporn, W. Limphirat, W. Donphai, P. Dittanet, M. Chareonpanich and J. Limtrakul, *Chem. Eng. J.*, 2017, **316**, 692.
242. S. Roy, A. Cherevotan and S. C. Peter, *ACS Energy Lett.*, 2018, **3**, 1938.
243. Y. Slotboom, M. J. Bos, J. Pieper, V. Vrieswijk, B. Likozar, S. R. A. Kersten and D. W. F. Brilman, *Chem. Eng. J.*, 2020, **389**, 124181.
244. A. A. Kiss, J. J. Pragt, H. J. Vos, G. Bargeman and M. T. de Groot, *Chem. Eng. J.*, 2016, **284**, 260.
245. J. Wei, Q. Ge, R. Yao, Z. Wen, C. Fang, L. Guo, H. Xu and J. Sun, *Nat. Commun.*, 2017, **8**, 1.
246. A. Parastaev, V. Muravev, E. H. Osta, T. F. Kimpel, J. F. M. Simons, A. J. F. van Hoof, E. Uslamin, L. Zhang, J. J. C. Struijs, D. B. Burueva, E. V. Pokochueva, K. V. Kovtunov, I. V. Koptyug, I. J. Villar-Garcia, C. Escudero, T. Altantzis, P. Liu, A. Béché, S. Bals, N. Kosinov and E. J. M. Hensen, *Nat. Catal.*, 2022, **5**, 1051.

247. Y. Han, C. Fang, X. Ji, J. Wei, Q. Ge and J. Sun, *ACS Catal.*, 2020, **10**, 12098.
248. L. Guo, J. Sun, X. Ji, J. Wei, Z. Wen, R. Yao, H. Xu and Q. Ge, *Commun. Chem.*, 2018, **1**, 1.
249. P. Sharma, J. Sebastian, S. Ghosh, D. Creaser and L. Olsson, *Catal.: Sci. Technol.*, 2021, **11**, 1665.
250. K. Chang, T. Wang and J. G. Chen, *Appl. Catal., B*, 2017, **206**, 704.
251. T. N. Do and J. Kim, *J. CO2 Util.*, 2019, **33**, 461.

CHAPTER 7
Electrocatalytic Carbon Dioxide Activation

J. L. RICO

Laboratorio de Catálisis, Facultad de Ingeniería Química, Universidad Michoacana de San Nicolás de Hidalgo, Edificio V1, C.U. Morelia Mich., C.P.58060, México
Email: jose.rico@umich.mx

7.1 Introduction

During the combustion of fuels, with the aim to obtain energy, carbon dioxide (CO_2) and water are produced and commonly released into the environment. As a consequence, this activity by humans has resulted in the continuous increase in the concentration of CO_2 and other pollutants in the atmosphere. The actual CO_2 concentration in the atmosphere is about 422.38 ppm, which is higher than the 248 ppm of the pre-industrial period.[1] It is therefore urgent to reduce this pollutant, if we want to preserve life on Earth. The transformation of CO_2 to other useful products such as methane, methanol, ethanol, formaldehyde, formic acid, *etc.*, is an endothermic process and requires energy. It could be argued that this transformation is not economically attractive; however, it is not a question of profits; the problem needs to be addressed in order to preserve the habitable life conditions on Earth. Natural fixation of CO_2 by the vegetal world is a green transformation in which CO_2 is first captured and then slowly transformed using the energy from the sun to produce thousands of different organic compounds. Although natural fixation of CO_2 occurs, the natural vegetal world is not capable of fixing the huge CO_2 quantities eliminated to the atmosphere by humans. As a consequence of this and other pollutants in the atmosphere, the temperature of the planet is increasing and this problem needs to be urgently addressed, otherwise it might pose fatal consequences to life.

Zero CO_2 emission is needed by 2050 if a cap of 1.5 °C above the pre-industrial temperature is to be achieved.[2] In addition to CO_2 fixation by natural photosynthesis, three strategies are envisaged to reduce CO_2 emissions: decarbonization, CO_2 recycling, and CO_2 sequestration. Continuous recycling of CO_2 is necessary since the reduction of CO_2 produces fuels and hydrocarbons that after combustion form CO_2 again. There are currently about 18 processes for carbon capture and storage (CCS) worldwide; however, it represents only about 0.1% of the global CO_2 emissions.[3] Unfortunately, most of these CCS projects are aimed to enhance oil recovery (EOR) by injection of CO_2 into wells to improve oil extraction.

There are some well-developed catalytic processes related to the usage of CO_2, for instance, the reverse water–gas shift reaction, methane and methanol synthesis, and Fischer–Tropsch synthesis.[4–7] The former reaction generates CO and H_2, which are subsequently transformed to methane, methanol, and hydrocarbons by other catalytic processes. For instance, Ni, Fe–Co, and Cu/ZnO, are some common catalysts used for methane, hydrocarbon, and methanol synthesis, respectively. The reverse water–gas shift reaction occurs at about 700 °C and at pressures from 1 up to 6 bars and in the presence of a catalyst.[8] In addition, all of these processes take place at high temperature and pressure and the products commonly end up as CO_2 and water, which are released into the atmosphere. It is again therefore important to close the carbon cycle.

7.2 Theoretical Aspects of CO_2

Carbon dioxide is a chemically inert and thermodynamically stable molecule. In addition, it is a linear compound presenting two C=O bonds, each with a dissociation energy of 750 kJ mol^{-1}.

As previously mentioned, the reduction of CO_2 independently of the desired product is an endothermic process and the addition of energy is therefore required. The use of a catalyst decreases this energy, compared to that needed in its absence. The attention of the present review will focus on the catalytic transformation of CO_2 in an electrochemical cell. The required energy for this reaction could be obtained from solar cells, but it is beyond the scope of this chapter. An electrochemical cell is a chemical reactor which uses electrical current to perform a chemical transformation. For the electrocatalytic CO_2 reaction many research studies are reported in the literature, with methane, methanol, ethylene, ethanol, formic acid, and carbon monoxide as the main reaction products. The electrocatalytic CO_2 transformation to produce CO and formic acid, which are two-electron transfer products, can be attained with low overpotential and high faradaic efficiency (FE), whereas the formation of the other compounds, which are multiple electron transfer products, requires higher overpotentials and the transformation shows low selectivity.[9] Several selective metals, non-metals, and hybrid electrocatalysts, which are suitable for CO_2 reduction, are reported in the literature.

7.3 Electrocatalytic Reduction of CO_2

The highest oxidation state of carbon, +4, is observed in CO_2 and carbonate minerals. The transfer of electrons to CO_2 is required to reduce this molecule. Since CO_2 is generated during the oxidation of carbon and hydrocarbons through an exothermic process, the reverse transformation, the addition of electrons into CO_2, is endothermic as previously mentioned.

As expected, the reduction of CO_2 is non-spontaneous and when water is used as an electrolyte, dissolution of CO_2 in this medium occurs at about 0.33 mol of CO_2/L at 25 °C and 1 bar. Commonly during the electrocatalytic process, the reduction of CO_2 takes place simultaneously with the hydrogen evolution reaction (HER). It is therefore desired to design a catalyst with high activity and high selectivity.[10,11] For the electrocatalytic reduction of CO_2 some reaction steps are envisaged: (1) diffusion of CO_2 through the electrolyte to the electrode, (2) adsorption of CO_2 on the catalytic surface, (3) catalytic transformation of CO_2 at the electrode by transferring some electrons and/or protons to break the C–O bonds and form other bonds, (4) desorption of the products from the catalytic surface, and (5) diffusion of the reaction products through the electrolyte. During operation of the electrocatalytic cell, it is therefore very important to increase as much as possible the mass transport of CO_2 and products through the electrolyte. The electrocatalytic splitting of water generates H_2 that can further reduce CO_2:

$$H_2O \rightarrow \frac{1}{2}O_2 + 2H^+ + 2e^- \quad (7.1)$$

$$2H^+ + 2e^- \rightarrow H_2 \quad (7.2)$$

The former reaction occurs at the anode and is called the oxygen evolution reaction (OER) and the latter, the HER, takes place simultaneously with the reduction of CO_2 at the cathode. These reactions occur at ambient temperature and pressure, producing hydrocarbons and oxygenated compounds. The reduction of CO_2 is represented as follows:

$$CO_2 + nH^+ + ne^- \rightarrow p + mH_2O \quad (7.3)$$

Accordingly, depending on the catalyst and the electrochemical experimental conditions, p could be some of the following compounds: CO, HCOOH, CH_4, CH_3OH, C_2H_6, C_2H_5OH, C_2H_4, CH_3COOH, CH_3CHO, $(COOH)_2$, C_3H_7OH, and C_2H_5CHO, which can be produced at the cathode. As expected, the H_2 generation is competing with the reduction of CO_2 at this electrode and the selected catalyst plays an important role in the selectivity. Concerning selectivity of the electrocatalytic reduction of CO_2, the electrodes such as Cd, Hg, In, Pd, Sn, and Tl produce mainly formate; Ag, Au, Ga, Pb, and Zn favor carbon monoxide; Fe, Ni, Pt, and Ti mainly produce H_2 and oxygen from the water-splitting reaction; and copper which is dependent upon the experimental conditions, favors the production of various hydrocarbons, aldehydes, alcohols, CO, H_2, and O_2.[12–15] Among the previous metals, Cu is the only pure metal that reduces CO_2 to products which require more than two electrons.

It is worth commenting that CO is an important intermediate in the reduction of CO_2 to produce hydrocarbons, alcohols, and aldehydes, and it is also separately studied.[16] The equilibrium potentials of the electrochemical reactions are listed in Table 7.1.

The thermodynamic equilibrium cell potentials reported in Table 7.1 indicate the minimum values to perform the CO_2 reduction and the OER.[17] The experiments are commonly performed at larger potentials than those reported in this table, however, the information can be used to perform economic viability of a certain product.[18,19] It is worth commenting that the activity in the reduction of CO_2 to a specific product is described in terms of the overpotential, which is the difference between the actual potential in the cell and the thermodynamic equilibrium potential to obtain that specific product, whereas the selectivity for a specific product is described in terms of the FE, which is defined as the percentage of the total electrical current

Table 7.1 Electrochemical reactions with equilibrium potentials. Adapted from ref. 17 with permission from American Chemical Society, Copyright 2019.[a]

Reaction	E^0/[V vs. RHE]	Product, name
$2H_2O \rightarrow O_2 + 4H^+ + 4e^-$	1.23	OER
$2H^+ + 2e^- \rightarrow H_2$	0	HER
$xCO_2 + nH^+ + ne^- \rightarrow \text{product} + yH_2O$		CO_2 reduction
$CO_2 + 2H^+ + 2e^- \rightarrow HCOOH_{(aq)}$	−0.12	Formic acid
$CO_2 + 2H^+ + 2e^- \rightarrow CO + H_2O$	−0.10	Carbon monoxide
$CO_2 + 6H^+ + 6e^- \rightarrow CH_3OH + H_2O$	0.03	Methanol
$CO_2 + 4H^+ + 4e^- \rightarrow C_{(s)} + 2H_2O$	0.21	Graphite
$CO_2 + 8H^+ + 8e^- \rightarrow CH_{4(g)} + 2H_2O$	0.17	Methane
$2CO_2 + 2H^+ + 2e^- \rightarrow (COOH)_{2(s)}$	−0.47	Oxalic acid
$2CO_2 + 8H^+ + 8e^- \rightarrow CH_3COOH_{(aq)} + 2H_2O$	0.11	Acetic acid
$2CO_2 + 10H^+ + 10e^- \rightarrow CH_3CHO_{(aq)} + 3H_2O$	0.06	Acetaldehyde
$2CO_2 + 12H^+ + 12e^- \rightarrow C_2H_5OH + 3H_2O$	0.09	Ethanol
$2CO_2 + 12H^+ + 12e^- \rightarrow C_2H_{4(g)} + 4H_2O$	0.08	Ethylene
$2CO_2 + 14H^+ + 14e^- \rightarrow C_2H_6 + 4H_2O$	0.14	Ethane
$3CO_2 + 16H^+ + 16e^- \rightarrow C_2H_5CHO_{(aq)} + 5H_2O$	0.09	Propionaldehyde
$3CO_2 + 18H^+ + 18e^- \rightarrow C_3H_7OH + 5H_2O$	0.10	Propanol
$xCO + nH^+ + ne^- \rightarrow \text{product} + yH_2O$		CO reduction
$CO + 6H^+ + 6e^- \rightarrow CH_{4(g)} + H_2O$	0.26	Methane
$2CO + 8H^+ + 8e^- \rightarrow CH_3CH_2OH_{(aq)} + H_2O$	0.19	Ethanol
$2CO + 8H^+ + 8e^- \rightarrow C_2H_{4(g)} + 2H_2O$	0.17	Ethylene

[a]The standard potentials were determined using Gibbs free energy of reaction and gas thermochemistry data, whereas Henry's law data were used for aqueous products.

used for the production of that specific product under steady-state electrolysis.

Once the standard potential for a certain reaction is known at pH = 0, the standard potential at other pH values under identical conditions can be determined by the following expression.[20]

$$E^o_{(pH)} = E^o_{(pH=0)} - 2.303(RT/F) \times pH \qquad (7.4)$$

where F is the Faraday constant, 96 485 C mol^{-1} of electrons; T is temperature in K; R is the ideal gas constant; and $E^o_{(pH=0)}$ is the standard potential evaluated at pH = 0.

A reaction mechanism proposed[20] for the ECR of CO_2 in the presence of Cu catalyst surface is presented below (see Figure 7.1):

Before commenting on the catalysts used for the reduction of CO_2, it is worth drawing attention to various reviews on the subject.

Bevilacqua et al. presented the technological progress[22] on the ECR of CO_2.

Endrődi et al. commented on various aspects of the ECR of CO_2 such as catalysts, reactor design, ion-exchange membranes, the effect of temperature and pressure, and the applied potential.[23]

Q. Fan and L. Fan separately published reviews focused on catalysts which selectively favor the production of C_{2+} compounds[10,24] during the ECR of CO_2. L. Fan et al. commented that the electrocatalytic C–C formation is still a challenge due to low catalytic activity, selectivity, and catalyst stability and highlighted the importance of catalyst and reactor design. Recent progress on this subject, the current challenges, and future opportunities are addressed in the review.

Rainer published a review[25] and commented on the state-of-the-art low-temperature, molten carbonate, and solid oxide electrolyzers for the production of CO. In addition, direct comparisons of the three technologies using some of the most common figures of merit from each field are also addressed. Owing to the high efficiency and proven durability, the high-temperature electrolysis of CO_2 in solid oxide is highlighted.

Pappijn et al. commented about the challenges and opportunities for carbon capture and utilization and, specifically, the ECR of CO_2 to ethylene.[26] From economical and practical points of view, the authors concluded that the electrochemical production of ethylene from CO_2 is not feasible under the current market conditions, even in the case that the renewable electricity price would be zero.

Zhang et al. presented a review on the subject and highlighted the fundamental concepts, the electrical double layer, the physical and chemical properties of CO_2, the interaction of CO_2 with the electrode, along with a survey of progress in the rational design of new electrocatalysts.[27]

Quan et al. also published a review on the subject. The authors discussed the current development and then focused on the electrocatalytic CO_2 transformation in the presence of small molecules.[28] In addition, they also commented on the perspectives and challenges in this field.

228 Chapter 7

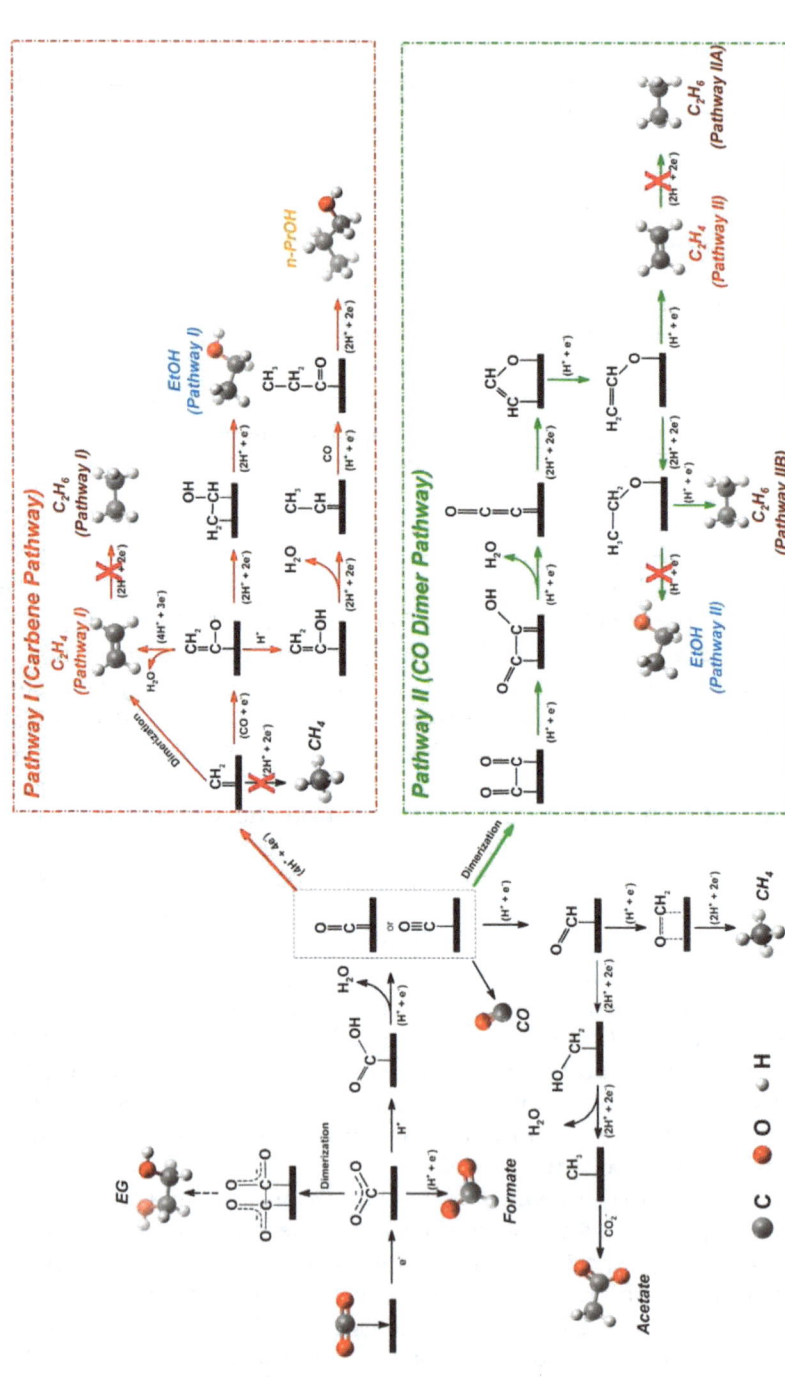

Figure 7.1 Proposed reaction mechanisms for the ECR of CO$_2$ on Cu catalysts. Unfeasible pathways are marked by red crosses. Reproduced from ref. 21, https://doi.org/10.1002/anie.202009498, under the terms of the CC BY 4.0 license, https://creativecommons.org/licenses/by/4.0/.

Brewis et al. reviewed the current challenges and recent developments of catalysts for the production of C3 and liquid products.[29] The authors commented that carbon–carbon bond formation is a key step in the reduction of CO_2 to energy-dense and high-value fuels and highlighted the importance of the catalytic surface morphology and reaction kinetics on the formation of multi-carbon products.

Kou et al. published a review on the subject and commented on the recent advances, basic experimental principles, electrocatalysts, reaction mechanisms, and challenges.[30]

Li et al. published an energetic analysis of the sequential and integrated CO_2 capture and electrochemical transformation.[31] Although the authors present a case study on the coupled amine scrub-to-CO_2 electrochemical conversion, this approach is anticipated to help researchers quickly understand the upper energy limits and targeted performance metrics for different integrated CO_2 capture and electrolysis processes.

Qu et al. published a review on the selective production of ethylene[32] from the ECR of CO_2. Ethylene is of great interest due to it having versatile industrial applications, however, the production of this compound from the electrochemical reduction of CO_2 is still challenging due to the low selectivity to ethylene, the large energy needed for the carbon-double bond formation, and the required large overpotential. The mechanism of the production of ethylene through the electrochemical reduction of CO_2 is addressed. Finally, the major challenges and perspectives on this subject are also discussed.

7.3.1 Electrocatalytic Cells

A short description on the electrochemical cells used for the ECR of CO_2, the electrolytes, and some other aspects will be addressed in the following paragraphs. A review related to these aspects is presented by Tufa et al.[33]

7.3.1.1 H-cells

One of the most frequently utilized electrochemical cells for the reduction of CO_2 is the H-cell, which is useful for studying catalysts and electrolytes. The cell is basically a reactor provided by two electrodes, a cathode and an anode, and an electrolyte placed between the electrodes where CO_2 is dissolved. The reduction of CO_2 is achieved by applying an overpotential in the electrochemical cell. Under operation, the CO_2 enters the cell, dissolves in the electrolyte, and diffuses to reach the cathode surface which contains the catalyst. The CO_2 is reduced on the catalytic surface and the products are desorbed. In addition, the electrochemical cell is provided with a membrane which separates the two electrodes avoiding the transport of products, their re-oxidation at the anode, and the decrease in the efficiency of the cell. The most commonly used membrane is monopolar, but it could be bipolar (BPM). The former includes anion or cation exchange membranes (also called proton exchange membranes [PEM]), AEM, and CEM, respectively, whereas bipolar membrane includes a cation exchange layer on one side and an anion exchange layer on the other.[34–37]

7.3.1.2 Membrane-based Flow Reactors

Some other electrochemical devices are membrane-based flow reactors (MFR), microfluidic reactors (MR), among other designs. In the MFR, the electrodes are separated by a channel of <0.1 mm where a liquid electrolyte flows. The CO_2 gas is supplied directly to the cathode and, as a consequence, the mass transport problems are avoided, compared to the H-cell. This device is suitable for testing catalysts and gas-diffusion electrodes (GDEs). The products in this cell are susceptible to re-oxidation due to the lack of a membrane. Various electrochemical cells are presented here (see Figure 7.2).

7.3.1.3 Microfluidic Reactor

The MR cell could be provided with a channel similarly to in the MFR electrolyzer, or with no channel. In addition, CO_2 could be supplied to the

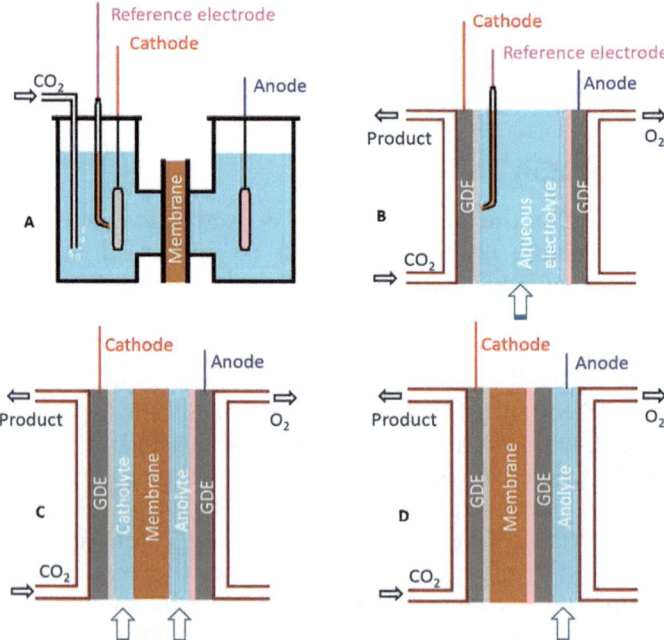

Figure 7.2 Various electrochemical cells used in the reduction of CO_2. (A) H-cell and (B) microfluidic cell, and cell designs of electrolyzers with (C) liquid-phase CO_2 supply and (D) gas-phase CO_2 supply. A diaphragm can also replace the membranes. Reproduced from ref. 33 with permission from Elsevier. Copyright 2020.

Electrocatalytic Carbon Dioxide Activation 231

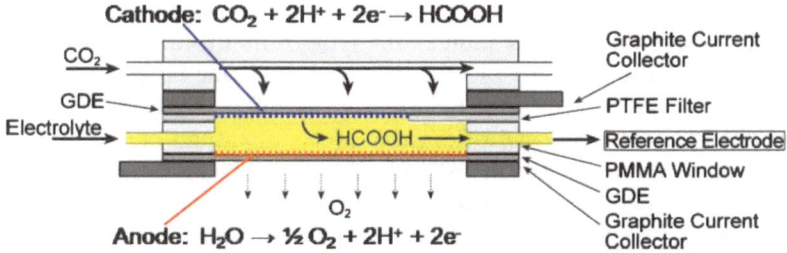

Figure 7.3 Schematic diagram of the microfluidic reactor for CO_2 conversion. Reproduced from ref. 38 with permission from IOP Publishing, Ltd. Copyright © The Electrochemical Society 2010.

cell as a gas, pure or humidified, or dissolved in a liquid electrolyte. The cell also contains an ion exchange membrane which prevents the products from reaching the anode. A microfluidic reactor is illustrated in Figure 7.3.

Delacourt *et al.* reported and compared several cell configurations for the reduction of H_2O and CO_2 on silver to produce syngas.[39] It was found that a configuration derived from PEMFC favors the production of H_2, while a modified configuration based on the insertion of a pH-buffer layer (aqueous $KHCO_3$) between the cathode catalyst layer and the Nafion membrane greatly enhanced the production of CO. The buffer layer is likely to prevent an excessive amount of protons from reaching the cathode.

In another publication presented by Han *et al.*, it was observed that by using GDEs, higher reaction rates for the ECR of CO to ethylene could be achieved when compared to traditional aqueous configurations.[40] The GDE contain Cu nanoparticles which were synthesized and utilized as electrocatalysts for this transformation. With the aim to evaluate the effect of reactant gas transport on the catalytic activity and selectivity of the ECR of CO, two distinctive DGE configurations were compared. In a flow-through configuration, where CO gas flows directly to the catalyst, a high partial current density of 50.8 mA cm^{-2} for the ECR of CO to ethylene was achieved at −0.85 V *vs.* HER in 10 M KOH at −15C, whereas in the flow-by configuration, where CO is directly bubbled into the aqueous electrolyte, and in the presence of the same catalyst, the partial current density for the C_2H_4 production was limited to <1 mA cm^{-2}.

Weekes *et al.* also demonstrated the advantages of directly contacting CO_2 with the electrocatalyst instead of dissolving this gas into the electrolyte before the ECR of CO_2 at the cathode. In the former configuration, the mass transport is enhanced and, as a consequence, higher current densities are achieved, compared to those obtained in the latter configuration.[41] The authors commented that although the use of H-cells for the ECR of CO_2 is useful for screening various catalysts, they do not provide information on how these electrocatalysts behave in scalable flow reactors. The flow reactors used for the ECR of CO_2 are basically classified as membrane-based and microfluid reactors, according to the reactor design. In the case of membrane-based reactors and in addition to the importance of the electrocatalysts, other parameters such as the type of membrane (CEM, AEM, or BPM) can affect the kinetics of ion transport and the range of applicable electrolyte conditions. The authors highlighted the importance of maintaining adequate flow cell hydration to achieve sustained electrolysis.

Higgins et al. commented on the importance of using vapor-phase reactants during the ECR of CO_2, thereby avoiding the intrinsic limitation of aqueous-based systems.[42] Higher reaction rates using vapor-fed reactors compared to the aqueous-based reactors are expected. The authors provided a perspective on the challenges and opportunities for generating fundamental knowledge and technological progress toward the development of practical devices for the ECR of CO_2.

Gabardo et al. published a study related to the electroreduction of CO_2 to C_{2+} products.[43] A membrane electrode assembly (MEA) electrolyzer that converts CO_2 to C_{2+} products is proposed. In comparison to other electrolytic systems, the MEA showed the most stable cell voltage and product selectivity. The authors observed FEs of about 50% and 80% for ethylene and C_{2+} products, respectively. In addition, the cathode-outlet concentrations of about 30 wt% of ethylene and 4 wt% ethanol, were noticed. The reaction system was operated during 100 h for ethylene production at current densities >100 mA cm^{-2}.

Yang and Li presented a review on the reactor design for the electroconversion of CO_2 for large-scale applications.[44] The review is focused on the configurations and operating principles of the used electrolyzers and summarizes the advantages and drawbacks. The authors also commented on the recent progress on the MEA electrolyzer, which is promising for large-scale electroreduction of CO_2.

7.3.2 Electrolytes

7.3.2.1 Pure Water

The electroreduction of CO_2 at the cathode can involve protons and hydroxyls depending on the local pH at the catalyst–electrolyte interface:

$$CO_2 + 2H^+ + 2e^- \leftrightarrow CO + H_2O \quad (7.5)$$

$$CO_2 + H_2O + 2e^- \leftrightarrow CO + 2OH^- \quad (7.6)$$

Simultaneously with the reactions at the cathode, the consumption of hydroxyls occurs at the anode by the OER:

$$2OH^- + 2e^- \leftrightarrow \frac{1}{2}O_2 + H_2O \quad (7.7)$$

The previous reactions indicate the importance of the pH of the electrolytic solution. Moreover, the great significance of the concentration and nature of the anions and cations present in the electrolyte during the transport of CO_2 through the medium is envisaged.

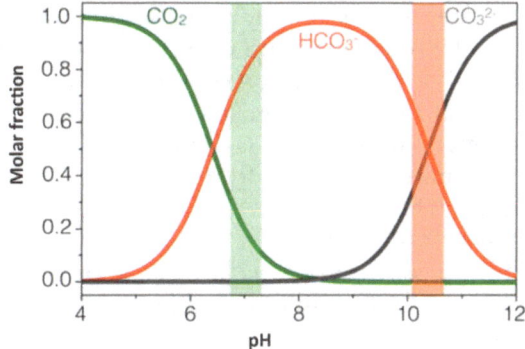

Figure 7.4 Bjerrum plot of carbonate species in a CO_2–H_2O mixture as a function of pH. Reproduced from ref. 45 with permission from American Chemical Society. Copyright 2015.

One step before the CO_2 reduction at the cathode is the hydration of the dissolved CO_2 in water to form carbonic acid:

$$CO_2 + H_2O \leftrightarrow H_2CO_3 \quad K_H = [H_2CO_3]/[CO_2] = 1.7 \times 10^{-3} \quad (7.8)$$

where K_H is the equilibrium constant, $K_H = k_H/k_D$. In addition, k_H and k_D represent the hydration and dehydration rate coefficients, respectively.

It is worth noticing that the small magnitude of K_H indicates the low dissolution of CO_2 in water. Once the H_2CO_3 is formed, a further dissociation occurs:

$$H_2CO_3 \leftrightarrow HCO_3^- + H^+ \quad (7.9)$$

$$HCO_3^- \leftrightarrow CO_3^{2-} + H^+ \quad (7.10)$$

The variation of the equilibrium concentrations of CO_2, HCO_3^-, and CO_3^{2-} species as a function of pH is presented in the Bjerrum plot (see Figure 7.4).

In this figure, it is worth noting the dominance of the species as a function of the pH. The effect of the pressure of CO_2 is also important. An increase in the pressure of CO_2 augments the dissolution of this gas in water, the pH of the electrolyte is altered and, as a consequence, a change in the dominance of the carbon species, according to Figure 7.4, is expected. The solubility of CO_2 is negatively affected by the presence of other ions, for instance in seawater. Contrarily, an increase in the solubility of CO_2 can be achieved by using non-aqueous electrolytes.

7.3.2.2 Aqueous

Salts dissolved in water are frequently used as aqueous-based electrolytes where the CO_2 is in equilibrium. Bicarbonates, carbonates, halides, hydroxides, and sulfates are commonly used for this purpose. Water-based electrolytes prepared by the dissolution of bicarbonates[46] or phosphate[47–49] are frequently used as buffering solution since an alteration of the pH during operation of the ECR of CO_2 is compensated by dissolution of the previous salts. As a consequence, the ECR of CO_2 can occur in a pH range between 6 and 8. When bicarbonate is used, the formation of HCO_3^- serves as a buffer and as a CO_2 source. The dissolved CO_2 is considered as the active species[14,50] rather than HCO_3^- and CO_3^{2-}. However, Dunwell et al. suggested that HCO_3^- species are active in the ECR of CO_2, probably due to the formation of a bicarbonate–CO_2 complex which augments the concentration of CO_2 around the cathode.[51]

7.3.2.3 Non-aqueous

There are some advantages and drawbacks in using non-aqueous electrolytes. The dissolution of CO_2 in these electrolytes is much greater than that in aqueous media.[52] In addition, performing the ECR of CO_2 using non-aqueous electrolytes avoids the competing side reaction, the HER, since many solvents are aprotic; however, the use of this type of electrolyte could pose safety, toxicity, and cost issues. Ionic liquids (IL) in aqueous and non-aqueous solvents are also attractive as electrolytes for the ECR of CO_2 due to the high solubility of CO_2 in IL and high thermal stability.[53–56] In addition, IL can stabilize the $CO_2^{\bullet-}$ radicals. Other solvents such as acetonitrile,[57,58] acetonitrile–water,[59–61] dimethylformamide[62,63], and methanol salts[64–69] are frequently reported as electrolytes for the ECR of CO_2.

7.3.2.4 Solid

We have previously seen the advantages and drawbacks of using aqueous electrolytes. One of the disadvantages is commonly related to the limitation of the mass transport of CO_2 through this phase. In addition, by using aqueous/alkaline electrolytes, the formation of carbonates from the consumption of CO_2 constitutes a challenge due to the decrease in the overall carbon efficiency of the electrocatalytic cell. Conversely, the use of acid electrolytes poses other problems such as the enhancement of the HER. With the aim to increase the CO_2 utilization avoiding the consumption of this reactant through an undesirable side reaction, and to generate high-purity liquid hydrocarbons, various solid electrolytes could be implemented in the ECR of CO_2. In addition, it is commonly possible to operate the electrocatalytic cell at high temperatures when a solid electrolyte is utilized. The following research publications reported the use of solid electrolytes.

Cook et al. commented that by using Cu as an electrocatalyst in the reduction of CO_2 in saturated $KHCO_3$ aqueous solution (pH ~7.5) to produce hydrocarbons, one would observe a progressive dominance of the HER over the CO_2 reduction reaction.[70] The authors suggested that one strategy to overcome this limitation is the use of a solid polymer electrolyte (SPE) cell. In this respect, the authors commented that their work performed in the laboratory related to the following cell configurations: CO_2/Cu/Nafion 117/aq. K_2SO_4, Pt, and CO_2/Cu/Nafion 117/Pt, N_2 (90%) H_2 (10%).

In another investigation, Dewulf and Bard[71] studied the ECR of CO_2 to produce CH_4, C_2H_4, and other hydrocarbons. The system is provided with copper/Nafion electrodes. The FEs of the electrodes under ambient conditions with a counter solution of 1 mM H_2SO_4 at a potential of −2.00 V *versus* saturated calomel reference electrode (SCE) reached a steady-state value of about 20% after 30 min of electrolysis, which corresponded to a rate of total hydrocarbon production of approximately 9.8×10^{-7} mol h^{-1} cm^{-2}. It was found that selecting a more negative potential of the electrode, or increasing the proton concentration of the counter solution, resulted in a decrease in the FEs, due to a relative enhancement in the rate of proton reduction, compared to that for CO_2. Furthermore, when the proton concentration was decreased to an alkaline pH, the production of hydrocarbons quickly ceased.

Komatsu et al.[72] tested Cu–solid polymer electrodes in the ECR of CO_2 in the gas phase. Cu–Rochelle salt and 10% $NaBH_4$ gave the most efficient electrode. With the cation-exchange (Nafion) and anion-exchange (Selemion) electrodes the total current efficiencies for the ECR of CO_2 had maximum values of 19% and 27%, respectively. The use of the former gave C_2H_4 as the major product, whereas HCOOH and CO were produced with the latter. In the case of using simulated exhaust gas from thermal-power plants, NO has no influence on the ECR of CO_2, whereas SO needs to be removed in order to obtain C_2H_4 in higher current densities. It was also found that the CO_2 content in the exhaust gas needs to be *ca.* 30%.

Nishimura et al.[73] applied a solid polymer electrolyte for the ECR of CO_2. In this zero-gap system, SF-17@ (TOSOH, EW = 1100, thickness = 200 pm) was used as an anion-type solid polymer electrolyte. Au was used as the catalyst and a porous $K_2Ti_6O_{13}$, PTFF layer form for the cathode. A Pt-plated Ti porous sheet was used as the anode. A current efficiency of up to 90% for CO production was obtained. The authors commented that this system is expected to be a clean and compact on-site type CO_2 recycling system.

In another research report, Dufek et al.[74] studied the ECR of CO_2 in the presence of an Ag-based cathode in a continuous system operated at high pressure. The authors found that the production of CO was five times higher than that observed at low pressure, with FEs as high as 92% observed at 350 mA cm^{-2} cathode. The advantages of an electrocatalytic cell operated at high pressure are three-fold: it increases the solubility of CO_2; allows the operation of the electrocatalytic cell at high temperature; and the pressurized CO could be suitable as a feed to the Fischer–Tropsch process which occurs at elevated pressure.

Aeshala et al.[75] using Cu_2O as the catalyst in the ECR of CO_2 reported that using suitable functional groups attached to the solid polymer electrolyte can have a profound effect on the efficiency and selectivity of the reaction products. Two different solid electrolytes were synthesized and tested, one labeled as PEI/PVA/KOH (polyethylenimine/polyvenyl alcohol/potassium hydroxide) and the other was obtained by blending quaternized PEI with PVA and doped with KOH(QPEI/PVA/KOH). The authors observed that at certain voltage and similar experimental conditions, methane was suppressed in one electrolyte, whereas ethylene was formed in the other.

Ebbesen et al.[76] presented a review on the ECR of CO_2 at high temperature in alkaline electrolysis cells (AECs), solid proton conducting electrolysis cells (SPCECs), and solid oxide electrolysis cells (SOECs). Concerning AECs, the authors commented that increasing the operation temperature from a conventional cell temperature of 60–90 °C to 250 °C improves the electrolysis performance and efficiency, but decreases the stability of the materials, mainly at high current densities. An advantage of SPCECs and SOECs is the possibility to perform the co-electrolysis of CO_2 and H_2O to produce synthesis gas (a mixture of $CO + H_2$) which can then be catalytically converted to various hydrocarbons by the well-established Fischer–Tropsch process. Pure H_2 can be obtained using SPCECs, which is an advantage compared to SOECs. More research is required concerning electrode materials in the SPCECs. Due to the high temperature of operation of the SOEC at about 800 °C, degradation of the cell is significant at high current densities. To decrease degradation, new development of electrocatalytically cells operated at lower overpotential is required. For the illustration of the principles of the high-temperature AECs, SPCECs, and SOECs, see Figure 7.5.

Aeshala et al.[77] studied the ECR of CO_2 using different solid polymer electrolytes. Cast Nafion and SPEEK membranes were built up and used as cationic solid polymer electrolytes, whereas alkali-doped PVA and (Amberlyst)/SPEEK composites were developed and used as anionic solid polymer electrolytes. In addition, electro-deposited copper on porous carbon paper was used as the cathode, whereas Pt/C on carbon paper was used as the anode. Formic acid, methanol, formaldehyde, carbon monoxide, methane, and H_2 (from the HER) were produced.

In another publication, Zhang et al.[78] presented the recent advances in cathodes used in the co-electrolysis of CO_2 and H_2O performed at high temperatures using solid oxide electrolytes. The catalyst presented on the cathode and described in this reference are: Ni-cermet, perovskite oxides, lanthanum strontium chromium manganites, lanthanum-doped strontium titanate, lanthanum strontium vanadate, and lanthanum strontium vanadate ferrum. The present interesting alternative is aimed to produce H_2 and CO, which can further be used as a feed for the Fischer–Tropsch process to produce hydrocarbons.

It is worth commenting that Xi et al.[7] proposed a novel system to transform CO_2/H_2O directly to hydrocarbons in a single solid oxide electrolytic cell incorporating the Fischer–Tropsch catalyst layer onto the electrochemical catalyst. By this arrangement, a CO_2 conversion rate of 11.5% and a methane flow rate of 0.0047 mL min^{-1} were achieved.

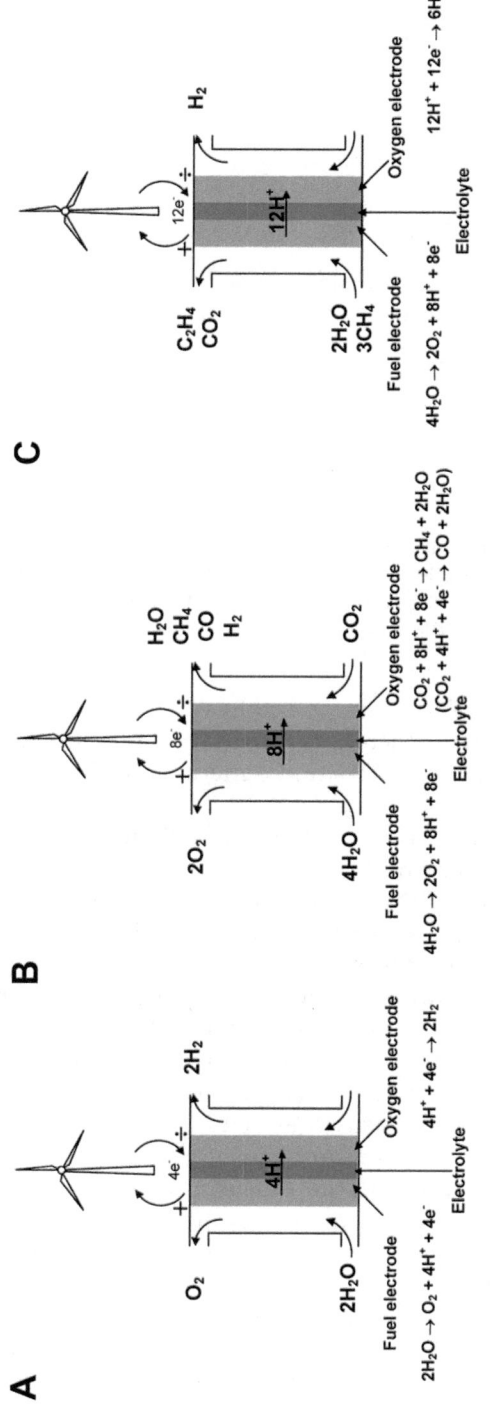

Figure 7.5 Working principle of high temperature solid proton conducting electrolysis cells (SPCECs) for (A) steam electrolysis, (B) electrolysis of water combined with electrochemical conversion of CO_2, and (C) production of hydrogen with simultaneous upgrading of methane to higher hydrocarbons. In this figure an oxygen electrode is the cathode and the fuel electrode is the anode. Adapted from ref. 76 with permission from American Chemical Society, Copyright 2014.

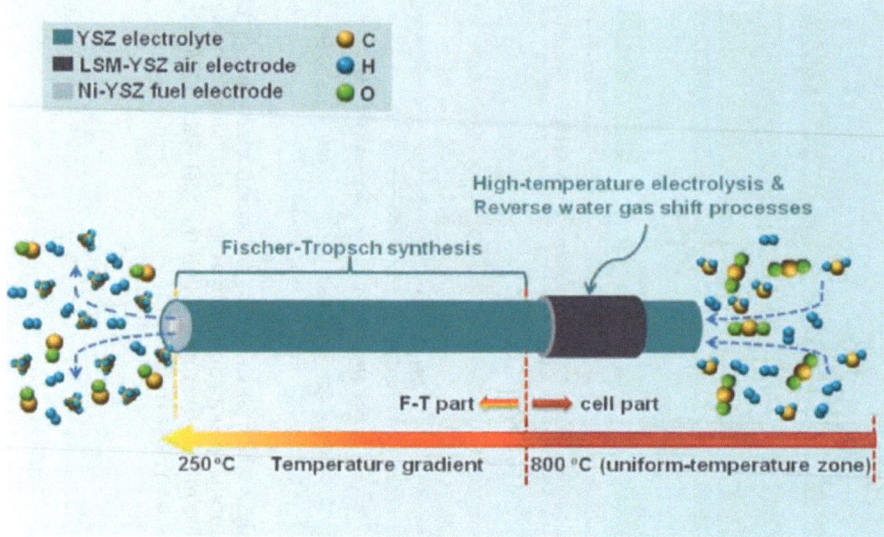

Figure 7.6 Illustration of the direct methane synthesis from CO_2/H_2O co-electrolysis in a single tubular unit combining a high-temperature SOEC and a reduced temperature Fischer–Tropsch reactor. Reproduced from ref. 80 with permission from the Royal Society of Chemistry.

The reactor design was further improved by Chen et al.[80] performing both transformations in a single reactor, as illustrated in Figure 7.6. In this system a CO_2 conversion rate of 64.1% and a methane yield of 0.84 mL min^{-1} were attained.

Xu et al.[81] commented that in the conventional electrolyzers, to produce various hydrocarbons from the ECR of CO_2, 70–95% of the supplied CO_2 forms carbonates, reducing the carbon efficiency. In the present research work, the authors utilized a microchanneled solid electrolyte provided with fixed quaternary ammonium cations. By this arrangement, the (bi)carbonate ions are recaptured and reutilized before reaching the anode, reducing the loss of this gas to about 3%. In addition, the reduction of CO_2 to hydrocarbons showed a selectivity of 77% without the use of metal alkaline cations. Furthermore, the system achieved steady-state operation at an industrially relevant current density over 200 h.

Kim et al.[82] also addressed the problem of CO_2 losses present in traditional electrocatalytic cells during the ECR of CO_2. By placing a permeable and ion-conducting sulfonated polymer electrolyte between the cathode and anode as a buffer layer, the crossover carbonate can combine with protons generated at the anode to re-form CO_2. Using a silver nanowire catalyst for the CO_2 reduction to CO, the authors obtained up to 90% recovery of the crossover CO_2 (purity >99%), while delivering over 90% CO FE under a 200 mA cm^{-2} current. A high continuous CO_2 conversion efficiency of over 90% was achieved by recycling the recovered CO_2.

In another study, Zhang et al.[83] presented a review on solid oxide electrolytes used in the ECR of CO_2 such as zirconia-based, ceria-based, and lanthanum gallates-based oxides. In addition, electrode materials such as metal-ceramics (cermet) and mixed ionic and electronic conductors (MIECs) are also discussed. The authors also commented that at high temperatures only CO is formed, whereas developing other electrolytes and electrode materials, the temperature of operation could be reduced and the production of various reaction products could be achieved.

Vennekötter et al.[84] commented that the long-term stability of the electrolytes under electrolysis is still a challenge. The authors studied the stability of different electrolyte combinations for two electrochemical membrane reactor (ecMR) designs. Gas diffusion electrodes (GDEs) of different composition, synthesized following different procedures, were subsequently tested in a stable electrolyte. Electrolytes in cation exchange membrane-based ecMRs are only stable if the anolyte is an acid with only protons as cations. Pure water usage is possible if a zero-gap assembly is used on the anode side. The supporting catholyte needs to be electrochemically inactive and unreactive to CO_2. In a stable electrolyte system, pressed and non-pressed Nafion- and PTFE-bonded GDEs with copper and silver as catalysts are compared for current densities up to -300 mA cm^{-2}. A C_2H_4 current efficiency of 51% for a copper GDE and a $CO:H_2$ ratio of $2:1$ with a silver GDE were measured for -300 mA cm^{-2}. The authors highlighted the need for a deeper understanding of interfacial phenomena to maximize energetic efficiency.

Xia et al.[85] reported the production of a continuous solution of fuel from the ECR of CO_2 using a solid electrolyte cell and bismuth as catalyst in the cathode. The cell is flexible and can be used for the production of C_{2+} compounds such as ethanol and acetic acid, using a Cu catalyst. By using a solid electrolyte, the generated H^+ and anions such as $HCOO^-$ are combined to produce pure products. The authors demonstrated the production of pure HCOOH solutions with concentrations up to 12 M. They also commented that in a typical cell, $KHCO_3$ in solution is used as an electrolyte to allow the transport of ions between the electrodes; however, energy is required for the separation of the produced hydrocarbons from the electrolyte in downstream operations, which is a disadvantage.

In another research publication, Wu et al.[86] commented on the carbonate formation issue during the ECR of CO_2 in conventional electrolyzers and reviewed the subject in relation to acidic electrolyte. As previously commented, the usage of acidic electrolytes also favors the HER. The authors presented the progress and challenges related to the use of acidic electrolytes on the ECR of CO_2 highlighting the need to improve the catalytic activity, selectivity, stability, and scalability.

Wu et al.[87] commented that the ECR of CO_2 in acidic media provides a pathway to curtail the CO_2 losses by suppressing the formation of (bi)carbonates; however, the HER is enhanced. Contrarily, in an alkaline medium, the HER is suppressed but salt precipitation within the diffusion layer occurs, resulting in poor system durability. To avoid this, the authors replaced the liquid catholyte with a solid-state proton conductor to regulate H^+ transport. This is postulated to allow for a locally alkaline environment at the cathode, enabling selective reduction of CO_2 even in the absence of alkali metal cations. The authors show that this strategy is effective over a broad range of catalyst systems. Using a composite nanoporous Au and single-atom Ni catalyst, with 0.25 M H_2SO_4 as the anolyte, an 87% CO_2 FE at 300 mA cm^{-2}, was observed. In addition, a stable operation over 110 h and a high single-pass carbon efficiency of 82.8% were also successfully achieved. Importantly, the system was suitable for processing feedstocks with low concentration of CO_2 (5%) with a FE of 47.7%, which was much greater than that observed for a conventional system. Figure 7.7 illustrates the proposal.

Chu et al.[88] recently published a review on solid electrolytes for low-temperature operation, applied for the ECR of CO_2. The authors presented fundamental aspects and highlighted the importance of solid polymer electrolytes and a solid electrolyte layer with multiple functions. In addition, they presented recent advances and explored future research, challenges, and opportunities on this subject.

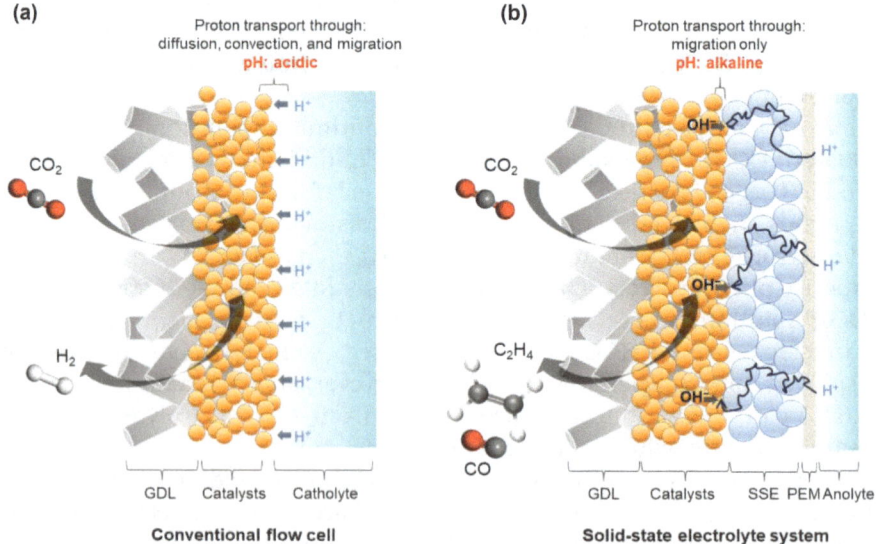

Figure 7.7 Comparison of proton transport during the ECR of CO_2. (a) Conventional flow cell system, where proton transport can occur through diffusion, convection, and migration. This leads to a proton-rich environment and hydrogen evolution dominates. (b) SSE system, where proton transport occurs only through migration. This creates a locally alkaline environment at the catalyst surface, enabling selective CO_2 to occur. GDL stands for gas-diffusion layer and PEM for proton-exchange membrane. Adapted from ref. 87 with permission from American Chemical Society, Copyright 2024.

7.4 Catalysts

7.4.1 Metal-based Catalysts

The first report, concerning the quantification of gaseous and aqueous products in the electrochemical reduction of CO_2 using a current density of 5 mA cm^{-2} and in the presence of various metals as the electrodes, was published by Hori et al.[89] They reported that HCOOH was preferentially formed in the presence of cathodes such as Cd, In, Sn, and Pb, whereas CO was produced when using Au and Ag. They also reported that the formation of CH_4 in significant quantities was observed when using Cu as the cathode.

Some years later, Cook et al.[90,91] reported that CO_2 can be electrochemically reduced to CH_4 and C_2H_4 at high FEs and current densities at a glassy carbon substrate on which Cu was *in situ* deposited. The authors proposed a reaction mechanism for the transformation of CO_2 to CH_4 and C_2H_4.

Nishimura et al.[92] studied the ECR of CO_2 in a zero-gap system provided with an Au electrode and a Pt-plated Ti porous sheet as an anodic electrode. In addition, the cathodic electrode was assembled with a porous PTFE sheet decorated with a solid electrolyte and anion-type solid polymeric electrolyte. The current efficiency for CO production of 90% was achieved. The system is a compact on-site cell that could reduce CO_2 directly from the exhaust gas.

In another research report, Hori et al.[93] noticed during the electrochemical reduction of CO_2 in aqueous media, and depending upon the combination of the modifier atom and substrate electrode, that foreign atoms modified the electrodes such as Ag, Au, Cu, Ga, Ni, Pd, and Zn, with a metal coverage of virtually unity, favoring the formation of CO, whereas others, such as Cd, Hg, In, Pb, Sn, and Tl, preferentially formed HCOO$^-$ species to finally yield methane, ethylene, and alcohols. Depending upon the adsorption of CO_2, CO is produced from stably adsorbed $CO_2^{\bullet-}$ and HCOO$^-$ from free or weakly adsorbed $CO_2^{\bullet-}$. The authors also found that the selectivity for CO was in the following order: Au > Ag > Cu > Zn >> Cd > Sn > In > Pb > Tl > Hg. The electrode potentials of CO_2 reduction are well interconnected with the heat of fusion of metals and the potential of H_2 evolution. The CO selectivity agrees roughly with that of the electrode potential of CO_2 reduction and was explained in terms of stabilization of $CO_2^{\bullet-}$ intermediate species on the electrode surface.

Hori et al.[94] reported the electrochemical reduction of CO_2 in a 0.1 M $KHCO_3$ aqueous solution at constant current density of 5 mA cm^{-2}, and in the presence of various copper single-crystal electrodes. The employed electrodes were Cu(S)-[n(100) × (111)], Cu(S)-[n(100) × (110)], Cu(S)-[n(111) × (100)], Cu(S)-[n(111) × (111)], and Cu(S)-[n(110) × (100)]. The authors found that the electrodes based on (100) terrace surface produce ethylene as the major product, which was further promoted by the introduction of (111) or (110) steps to the basal (100) plane. Methane formation on the n(111) × (111) electrodes decreases when the (111) step atom density augments. The n(111)–(100)

surfaces produce more gaseous products, with methane the favored compound among other C_{2+} species.

Yano et al.[95] reported the electrocatalytic reduction of CO_2 on a copper foil electrode. Catalytic deactivation caused by the deposition of graphitic carbon and other poisons on the electrode was observed. To solve this problem, the Cu-mesh electrode was modified beforehand by copper (I) halides and the CO_2 was then reduced with a constant potential in a potassium halide acidic medium. In addition, the authors developed an electrolysis system in which CO_2 is reduced at the three-phase interface (gas/liquid/solid) on the Cu-mesh electrode. The authors found that the FE for ethylene was significantly increased, whereas that for H_2 was reduced. In the case of CuBr, the conversion of CO_2 to ethylene increased and that for the HER decreased, with an FE of 80% and 9%, respectively. The contribution of CuBr to the CO_2 reduction is related to the reversible combination with CO and C_2H_4. The authors explained that at the three-phase interface, CO_2 is first reduced to CO which is ready to adsorb to copper(I) halide through the π-bond perpendicular to the surface and the CO is then subjected to electron addition from the electrode to be reduced to the methylene radical. This radical is then coupled to form C_2H_4, which is stabilized parallel to the surface through its π-bond.

In another research work, Le et al.[96] reported the electrocatalytic reduction of CO_2 on CuO surfaces. A remarkable yield of CH_3OH of 43 µmol cm^{-2} h and FEs of 38% were measured using an electro-deposited cuprous oxide thin film compared to the results using air-oxidized and anodized Cu electrodes. The authors also observed that CH_3OH yields were dynamic and that CuO is reduced to metallic copper in a simultaneous process.

Kuhl et al.[97] reported an experimental methodology with high sensitivity for the identification and quantification of the CO_2 electroreduction products. The authors studied copper across a range of potentials and detected 16 different C1–C3 products, including ketones, alcohols, and carboxylic acids. A reaction mechanism was proposed.

In another publication, Chen et al.[98] reported the synthesis of Au catalyst and the performance on the electrocatalytic reduction of CO_2. The Au nanoparticles were stable for at least 8 h and exhibited high selectivity toward CO in water using overpotentials of about 140 mV. Other nanostructured Au electrodes rapidly deactivate and require at least an overpotential of 200 mV to achieve comparable CO_2 reduction activities.

Hatsukade et al.[99] studied the electroreduction of CO_2 on metallic silver surfaces. The results indicate that the reaction products and selectivities are potential-dependent. CO, H_2, formate, methane, methanol, and ethanol were identified and quantified. Among them, H_2 and CO were the major products.

Chi et al. also studied the electroreduction of CO_2 using CuO nanoparticles of five different morphologies.[100] The particles were *in situ* reduced and then tested for the CO_2 transformation. The authors reported various alcohols as reaction products and observed high selectivities to ethanol. The nanoparticle morphology plays an important role in the product selectivity and catalytic activity.

Min and Kanan[101] studied the electroreduction of CO_2 using palladium supported on carbon as electrodes. With the metals currently used as cathodic catalysts, the process for the transformation of CO_2 to formate is energetically inefficient because more than 1 V of overpotential for the reduction of CO_2 to formate is required; however, Pd which reduced CO_2 to $HCOO^-$ with no overpotential was neglected as the catalyst due to low reaction rates and an unidentified deactivation pathway. In this work, the authors found that dispersing Pd nanoparticles on a carbon support reaches high mass activities (50–80 mA HCO^{2-} synthesis per one milligram of catalyst) when using less than 200 mV of overpotential in aqueous bicarbonate solutions. The experiments indicate that the rate-determining step is the addition of the generated surface-adsorbed hydrogen to CO_2. The authors also found that deactivation of the catalyst occurred after several hours of reaction, due to the formation of small quantities of CO, however, the catalytic activity was recovered after a brief exposure of the catalyst to air.

Mistry et al.[102] reported oxidized copper catalysts obtained through plasma treatments which displayed lower overpotentials for carbon dioxide electroreduction and measured a selectivity of 60% toward ethylene. Results from characterization indicate that copper oxides are surprisingly resistant to reduction and remain on the surface during reaction. Such species are important for lowering the onset potential and for ethylene selectivity.

In another research work, Cave et al.[103] studied the electrochemical reduction of CO_2 using polycrystalline Au surfaces which show high activity and selectivity to CO. Methanol, CO, H_2, and formate were detected during the electroreduction CO_2 experiments. The authors observed that Au surfaces do not favor the C–O scission and are therefore selective to oxygenates over hydrocarbons.

Tang et al.[104] reported the electrocatalytic reduction of CO_2 on Cu–hydride nanoclusters. It is known that at high overpotentials, copper electrocatalysts can reduce CO_2 to hydrocarbons. By applying density functional theory (DFT), the authors studied this transformation and found that at low overpotentials, copper–hydride nanoclusters with dithiophosphate ligands ($Cu_{32}H_{20}L_{12}$, where L stands for ligands), offer great selectivity toward HCOOH instead of CO. According to the findings, the surface hydrides first transform CO_2 to HCOOH species and then the hydride vacancies are regenerated by the electrochemical proton reduction. At low overpotential, the DFT calculations also predicted that the HER is less competitive than the HCOOH formation. The authors also performed electrocatalytic experiments and found that the studied clusters indeed favored the formation of HCOOH over CO at low overpotentials and that the selectivity for hydrogen is favored at high overpotentials.

In another publication, Peng et al.[105] prepared copper-nanopore catalytic structures, by an alloying–dealloying process, which were then tested in the electrochemical reduction of CO_2. The results indicate that the new nanoporous copper structure reduced the methane formation to less than 1% compared to that observed using a Cu layer with no pores, while keeping the formation of ethylene at a high level of 35%. The experiments were performed in aqueous 0.1 M KHCO$_3$ solution and under −1.3 V (vs. reversible hydrogen electrode, RHE). The authors highlighted the remarkable selectivity of this catalyst toward ethylene, which was ascribed to the exposed specific crystalline orientations.

Zeng et al.[106] prepared and tested Ag$_2$S nanoparticles loaded on N- and S-doped reduced graphene oxide (rGO) as a catalyst in the ECR of CO$_2$. The catalyst showed a remarkable selectivity to CO of 87.4% and good stability over 40 h of continuous operation at ∼70 μA cm^{-2}.

De Luna et al.[107] presented a study related to the electro-redeposition, dissolution, and redeposition of copper from a sol–gel, to influence the copper morphology, oxidation state, and performance of catalysts during the ECR of CO$_2$. The authors utilized in situ soft X-ray absorption spectroscopy to determine the oxidation state of copper under the CO$_2$ reduction conditions with time resolution. The results revealed the beneficial interplay between sharp morphologies and Cu$^+$ oxidation state. The catalyst exhibits a partial ethylene current density of 160 mA cm^{-2} (−1.0 V versus reversible hydrogen electrode) and an ethylene/methane ratio of 200. Figure 7.8 supports the results.

In another research report, Qin et al.[108] prepared and tested a CdS-CNTs electrocatalyst in the ECR of CO$_2$ to CO and noticed an FE greater than 95%. In situ infrared absorption spectroscopy and density functional theory calculations indicate that the increase in the S-vacancies with reaction time changes the electron density of the catalyst surface and decreases the energy barrier for the conversion of COOH* to CO*. In the 10-h test, with the increase of S-vacancy, the yield of CO increased from 199.0 to 243.1 μmol h^{-1} cm^{-2}.

Dinh et al. also studied CO$_2$ electroreduction using copper electrodes.[109] The authors reported that a copper electrocatalyst at an abrupt reaction interface in alkaline electrolyte reduces CO$_2$ to ethylene with 70% FE at a potential of −0.55 V vs. the RHE. It was also commented that the hydroxide ions on or near the copper surface lower the CO$_2$ reaction and the CO–CO coupling activation energy and as a consequence, the onset of ethylene evolution at −0.165 V vs. RHE in 10 M KOH occurs almost simultaneously with the CO production. In addition, the operational stability of the electrode was enhanced by introducing a polymer-based gas-diffusion layer that packed the reaction interface between separate hydrophobic and conductive supports, providing constant ethylene selectivity over 150 h.

Hori[110] and Nitopi[17] indicated that metals such as Cd, Hg, In, Pb, Sn, and TL as electrodes favor the production of formate with FEs above 75% of −5 mA cm^{-2} but require higher overpotentials, compared to other metals as electrodes, whereas metals such as Fe, Ni, and Pt showed FEs above 89% of −5 mA cm^{-2} for H$_2$ requiring low overpotentials than the previous set of metals. The above references also show that Cu as electrode at −1.05 V vs. RHE

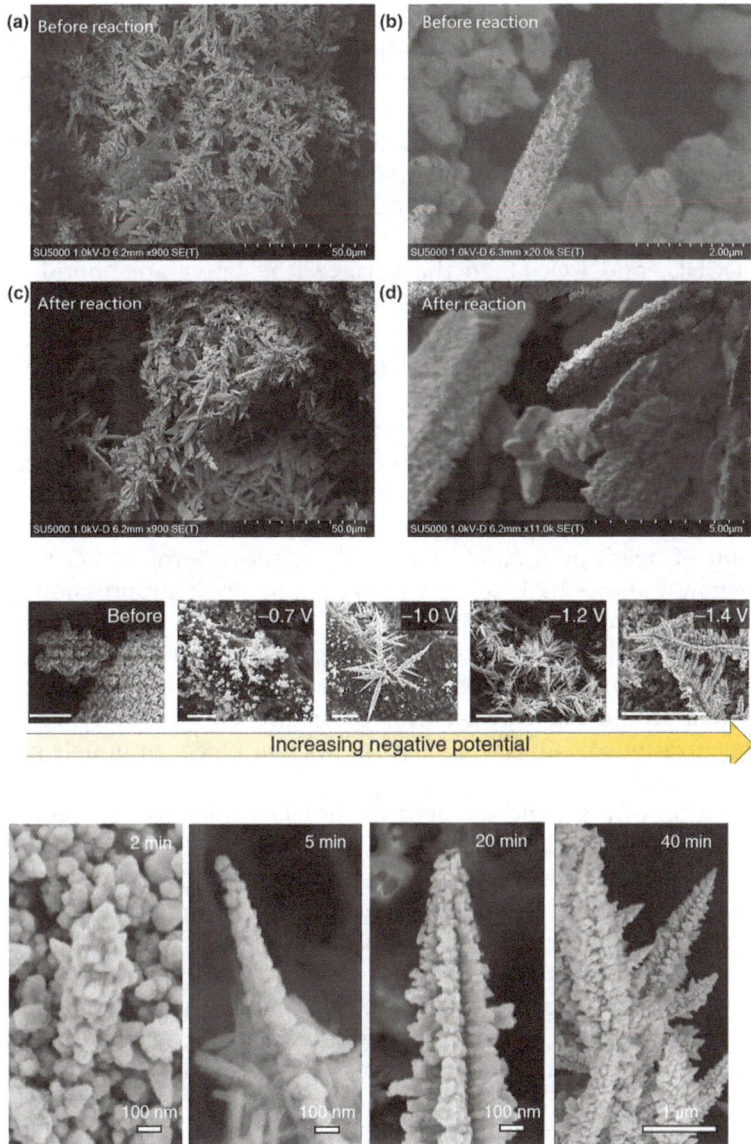

Figure 7.8 Porous copper nanoneedle (NN-Cu) controls before (a) and (b) and after (c) and (d) reaction at −1.2 V *vs.* RHE over at least 1 hour of operation. Below SEM images of the key structure features at their specific applied potentials after at least 1 h of reaction. Evolution of nanoclusters (−0.7 V), nanoneedles (−1.0 V), nanowhiskers (−1.2 V), and dendrites (−1.4 V) at increasing negative potential. The scale bars on these images are 5 μm. At the bottom, SEM images of electroredeposited Cu under −1.0 V *versus* RHE taken at 2, 5, 20, and 40 min from left to right. Reproduced from ref. 107 with permission from Springer Nature, Copyright 2018.

for the reduction of CO_2 is very attractive since it favors the production of hydrocarbons and hydrogen, presenting FEs of 33%, 26%, and 21% of -5 mA cm^{-2} for CH_4, C_2H_4, and H_2, respectively. The authors also emphasized that this information is only a snapshot for the CO_2 reduction since the selectivity for each electrode changes as a function of the experimental conditions and the electrode potential used during the experiment.

Bagger et al.[111] explained the difference in selectivity of some metals considering the binding energies of formate and HER intermediates, such as H, CO, OCHO, and COOH, to the surface (the latter are bound through oxygen and carbon, respectively). The activity of Cu on the electroreduction of CO_2 may be due to the fact that it is the only material that presents negative adsorption energy for CO and positive adsorption energy for H, compared to other metal electrodes.

Zhu et al.[112] commented that due to the impurities generated during the electrochemical CO_2 or CO reduction to C_{2+} products in liquid electrolytes, the practical application is overshadowed. The authors used instead Cu nanocube catalysts and a porous solid electrolyte and observed a continuous production of acetic-acid solutions *via* the electrochemical CO reduction. The Cu catalyst shows high selectivity to acetate, great suppression of other liquid products, an FE of 43%, a partial current of 200 mA cm^{-2}, high relative acetate purity of up to 98 wt%, and great catalyst stability over 150 h. DFT calculations highlighted the role of stepped sites along the nanocube edge of Cu in promoting the formation of acetate. Pure acetic acid solutions, with concentrations of up to 2 wt% (0.33 M), can be continuously produced by using Cu catalysts.

The addition of a second element in the electrodes for the reduction of CO_2 could change the activity and selectivity to H_2, CO, or $COOH^-$. For instance, monometallic transition metal electrodes from the group 8–10 favor the production of H_2. A second element influences the geometric structure and alters the electronic state of the surface and, as a consequence, affects the binding straight of the reaction intermediates on the active sites.[113,114] CO adsorbs on the surface through the carbon atom in a upright position, whereas on bimetallic catalysts it is expected to form carbon and oxygen atomic sites allowing the adsorption of *COOH and *CHO intermediate species on the surface.[115–117] Finally, the catalytic activity and selectivity are expected to change as mentioned above.

7.4.2 Hybrid Catalyst

Zhang et al.[118] presented a study about the electrocatalytic transformation of CO_2 to syngas in the presence of various catalysts. The catalysts are carbon-based with Ni–Fe–N species. It was observed that, depending on the Ni/Fe composition, the catalysts exhibit a wide range of CO/H_2 ratios from 0.14 to 10.86 which is very attractive for various processes. The catalytic samples were evaluated over 8 h with insignificant deactivation. The variation of the FE as a function of the Ni/Fe content was also studied by applying DFT calculations and the determination of the reaction barriers for the HER and CO_2 reduction reactions.

Li et al.[119] synthesized and tested M–N–C catalysts (M = Mn, Fe, Co, Ni, and Cu) in the electro-reduction of CO_2. The difference in activity for CO was attributed to the nature of the metal in the MN_x moieties. A volcano trend between their activity toward CO and the nature of the metal in MN sites was noticed, with Fe and/or Co at the top of the volcano, depending on the electrochemical potential. The FeNC, NiNC, and MnNC showed a FE for CO >80%. By experimental operando X-ray absorption near-edge structure spectroscopy (XANES), it was found that Co and Mn did not change oxidation state with potential, whereas Fe and Ni were partially reduced, and Cu was largely reduced to Cu(0). $Fe^{2+}N_4$–H_2O, $Co^{2+}N_4$–H_2O, and $Ni^{1+}N4$ were identified by XANES as the most active centers in FeNC, CoNC, and NiNC, respectively, at the potentials of −0.5 and −0.6 V vs. RHE. The experimental activity and selectivity could be explained from DFT calculation by considering the difference between the binding energies for CO_2*^- and H* as a descriptor of selectivity toward CO.

In another publication, Pršlja and López[120] computationally analyzed the electroreduction of CO_2 on Ni supported on N-doped carbon catalysts considering both nanoparticles and single atoms. The optimal activity and selectivity are found for the NiN_3 model, in which Ni is in Ni^{1+} oxidation state. NiN_3 exhibits theoretically low onset potential as observed experimentally, and a high reaction rate toward CO. However, saturated N-doped catalysts are more selective toward CO. Theoretical atomic dispersion was also analyzed through computing simulations.

Hossain et al.[121] investigated the effect of temperature on the CO_2 conversion and product selectivity using cobalt(II)-tetraphenyl porphyrin (CoTPP)/multiwalled carbon nanotube (MWCNT) composite as catalyst at reaction temperatures from 20–50 °C. The FE of products changes with temperature and potential. Contrarily to the hydrogen and methane formation, CO and methanol are enhanced at low potential and temperatures. DFT simulations were also performed and the results indicate the differences between the binding energies of CH_2O and CHOH, the binding strength of CO, and the protonation of CHO intermediate.

7.4.3 Metal-free Catalyst

Several research studies concerning carbon materials doped with B,[122,123] N,[124–126] P,[127,128] Si–N,[129] B–N,[130] N–S[131], and N–P[132] catalysts for the electroreduction of CO_2 have been reported.

Ikemiya et al.[122] studied the electrochemical conversion of CO_2 to formic acid using boron-doped diamond electrodes and found that although the FE decreased during the experiment, it can be easily recovered by the electrochemical oxidation of the electrodes using solutions of H_2SO_4, Na_2SO_4, or K_2SO_4. The authors also reported a production rate for formic acid of 328 µmol h^{-1} cm^{-2}, at a current density of −20 mA cm^{-2}, which was the highest production rate ever reported using plate electrode. By successive polarity reversal of plus and minus terminals, this alternative could be attractive for practical application.

Souza et al.[123] reported a review on the electrochemical transformation of CO_2 using boron-doped diamond electrodes. The authors emphasized the enhancement of the catalyst performance by producing surface defects and/or surface functionalization with metal-base materials. The challenges and future research directions for this reaction are also discussed. The authors also mentioned that the main drawbacks of the metal-based electrocatalysts used for the electroreduction of CO_2 are related to the high overpotentials, low stability, and undesirable selectivity.

Chai and Guo[124] published a theoretical study performing DFT calculations and *ab initio* molecular dynamics on the electroreduction of CO_2 using graphene/carbon nanotubes as catalysts and found that the interplay of N-doping and curvature can effectively tune the activity and selectivity of these catalysts. They also reported that the activation barrier for CO_2 is reduced to 0.58 eV in the presence of graphitic-N doped graphene edges compared to 1.3 eV required for the un-doped counterpart. Catalytic curvature also plays an important role in selectivity: 1.5–2.0 V for CO and 1.29–0.49 V for methanol production.

In another research work, Jhong et al.[125] studied the electroreduction of CO_2 using N-doped carbon electrodes. The sample showed high catalytic activity and greater selectivity for CO (98%) than H2 (2%). The authors also reported that the CO partial current density at intermediate cathode potentials ($V = 1.46$ V vs. Ag/AgCl) was up to 3.5 higher than the state-of-the-art Ag nanoparticle-based catalysts, and the maximum current density was 90 mA cm^{-2}. The mass activity and energy efficiency (up to 48%) were also greater than those for the Ag nanoparticle reference.

Kumar et al.[126] reported the electrochemical transformation of CO_2 to CO in the presence of polyacrylonitrile-based heteroatomic carbon nanofibers (CNFs). These catalysts exhibit negligible overpotential of 0.17 V for this reaction and a greater current density compared to silver catalysts under similar experimental conditions. The activity is attributed to the reduced carbons rather than to the electronegative nitrogen atoms. The XPS results, the deconvoluted N1s spectra, the atomic structure, and the proposed reaction mechanism for the ECR of CO_2 in the presence of N-CNF catalysts are shown in Figure 7.9.

Fernandes et al.[133] reported a review on the ECR of CO_2 on metal-free nitrogen-doped carbon electrodes. This kind of material shows some advantages, for instance, low cost, high overpotential for hydrogen production, relatively large catalytic surface area, high availability, and environmentally friendly compared to metal catalysts. In addition, the introduction to heteropolyatoms would create more defects and, as a consequence, increase the catalytic activity. Nitrogen could be located on the carbon surface or inside the carbon structure. More specifically, depending upon the nitrogen source and the methodology used during the synthesis, three nitrogen configurations are reported in the literature: pyrrolic, pyridinic, and graphitic

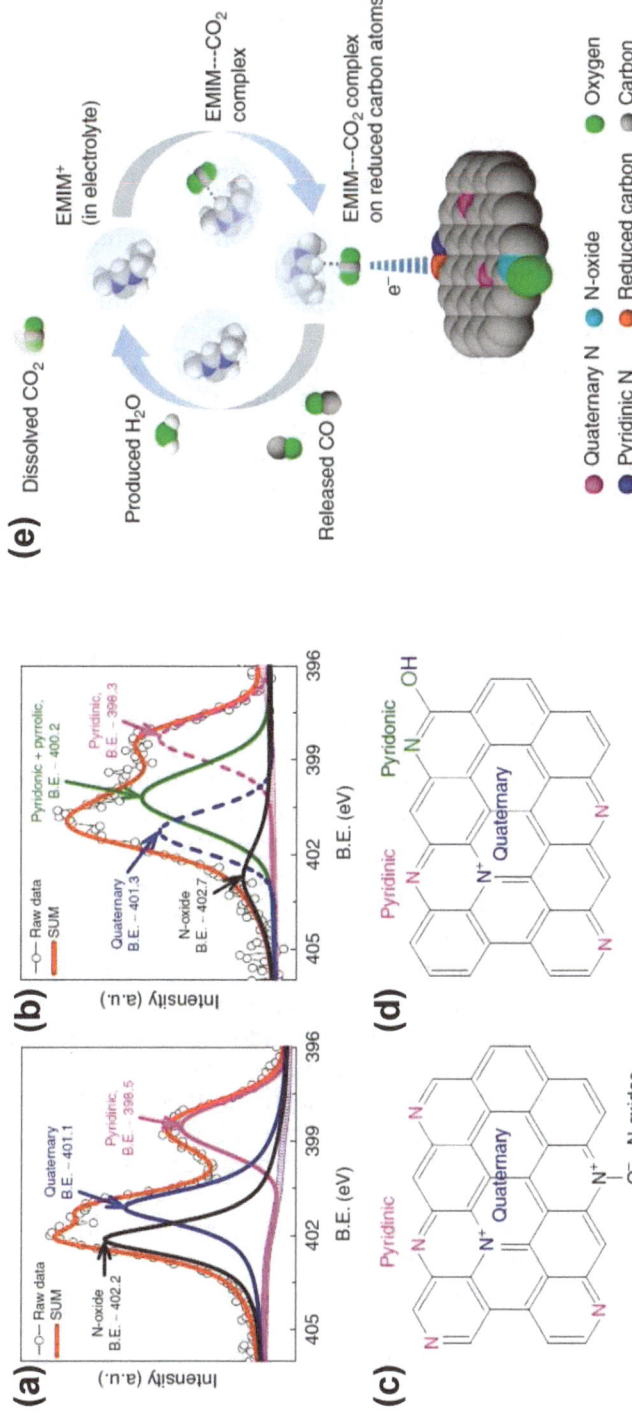

Figure 7.9 XPS results and deconvoluted N1s spectra for N-CNFs before (a) and after the electrochemical experiments (b). The results indicate that the N-oxide signal is drastically reduced during reaction and a new peak (in green) at 400.2 eV (pyridonic N) appears in the spent catalyst. The corresponding atomic structure based on XPS analysis (c) and (d). Proposed mechanism for the ECR of CO_2 (e). Adapted from ref. 126 with permission from Springer Nature, Copyright 2013.

nitrogen (or quaternary nitrogen). The pyridinic and pyrrolic nitrogen are located on defect or edge sites and alter the electronic configuration in the delocalized π-system, whereas nitrogen atoms could also substitute carbon atoms within the graphitic carbon structure keeping the same carbon configuration but adding extra electrons and, as a result, altering the delocalized π-system.[134,135] Due to the excess of electrons resulting from the addition of nitrogen atoms to the carbon structure, the basicity or the electron-donor capacity of the resulting carbon material is improved.[136] In addition, nitrogen presents higher electronegativity than carbon. The consequences in an N-doped carbon material are the reduction in the workfunction, a higher surface energy, an increment in the n-type carrier concentration, and a greater tunable polarization, compared to those properties of the pure graphitic carbon material.[137] The incorporation of nitrogen will finally affect the adsorption–desorption characteristics of the carbon material and promote the catalytic activity in the electroreduction of CO_2. Concerning the mechanism for the transformation of CO_2 using this type of catalyst, it is generally suggested that the rate-limiting step is the transfer of the first electron to the reactant forming CO_2^-. In the case of nitrogen-doped carbon, the catalytic activity would be different depending upon the nitrogen configuration, for instance, pyridinic, pyrrolic, and graphitic nitrogen. Some authors consider the pyridinic nitrogen arrangement as the primary active sites due to the low free energy barrier of these sites when forming key intermediates of COOH,[138,139] which is the rate-limiting step to form CO, whereas some others consider that the pyrrolic nitrogen sites would favor the reduction of CO_2.[140] In this reference based on DFT calculations, it was reported that the conversion of CO_2 to HCOOH would be favored on pyrrolic-nitrogen carbon catalysts and that the hydrogenation of COOH species is the rate-limiting step for the transformation of CO_2 on these materials.

In another research work, Liu et al.[141] studied the electroreduction of CO_2 in the presence of carbon doped with P as catalysts and reported that phosphorus showed a key effect on the catalytic performance. Phosphorus is preferentially present as P–C or P–O species, depending on the synthesis procedure. The electroreduction of CO_2 under a current density of 4.9 mA cm^{-2} and low potential (−0.90 V vs. SHE) on the electrode with abundant P–C bonds showed greater catalytic activity (compared to the P–O sample), an FE of 81%, and an excellent durability (27 h). Density functional theory calculations also demonstrate that the binding energy of the key COOH* intermediates on the P–C sites is greater than that on P–O sites, lowering the energetic barrier of the rate-limiting step.

The research progress, future research, and perspective on the electroreduction of CO_2 using non-metal P-doped catalysts have been reviewed by Zhai et al.[128] In addition, the environment of phosphorus is also addressed.

In another study, Ghausi et al.[142] reported a research work on the electroreduction of CO_2 in the presence of carbon electrodes codoped with silicon and nitrogen (SiNC) as facile pyrolyzation. The catalytic activity of the SiNC sample was two orders of magnitude greater than those observed for single-doped carbon catalysts. The electroreduction of CO_2 on SiNC sites imitates photosynthesis yielded and overall solar-to-chemical efficiency of 12.5% at only 650 mV overpotential. The authors suggested that the elastic-electron structure of –Si(O)C–N– sites is responsible for the high catalytic activity by lowering the adsorption energy of the intermediate species during the electroreduction of CO_2 and the OER.

Xinyue et al.[143] studied the electroreduction of CO_2 in the presence of boron–nitrogen doped carbon catalysts. Using a low overpotential of 423 mV, the authors observed good catalytic performance with a high FE of 95% toward CO during the transformation of CO_2.

In another study, Wang et al.[144] reported the electrochemical reaction of CO_2 using carbon electrodes doped with nitrogen and sulfur. The authors reported the synthesis of carbon nanosheets codoped with nitrogen and sulfur and the catalytic performance of this material in the reduction of CO_2. The catalyst showed good stability, high catalytic activity, and an FE of about 85.4% for the production of CO. The high performance was attributed to the N–S–C sites and the large surface area of the catalyst.

In another research report, Zhou et al.[132] tested carbon electrodes codoped with nitrogen and phosphorus of various contents to explore the relationship between these dopants on the reduction of CO_2. The addition of P to the N-doped system improves the catalytic activity. The optimum sample achieved an FE of 80.8% towards CO under an overpotential of 0.44 V. X-ray photoelectron spectroscopy (XPS) results indicate a suitable pyridinic/graphitic-nitrogen ratio and a P–N content for the enhancement in catalytic activity. In addition, DFT calculations indicate that the graphitic N and pyridinic N sites are important in the formation of *COOH and *CO species.

Wu et al.[145] reported a study related to the electroreduction of CO_2 using carbon nanotubes doped with N to form electron-rich defects which are highly efficient and impart high stability to the catalyst. The authors observed high selectivity to CO of 80% under an overpotential of −0.18 V, which was attributed to high electrical conductivity, preferentially formation of pyridinic N defects as catalytic sites, low free energy for CO_2 activation, and a high barrier for the HER. DFT calculations show a low free energy barrier for the potential-limiting step to form key COOH intermediates and a weak binding energy for the adsorbed CO. The authors reported good electrode stability over 10 h of continuous operation.

In another publication, Sharma et al.[146] studied the active sites on the N-doped carbon nanotubes and the role of the defect density on the selectivity of the electrocatalytic reduction of CO_2. The authors found that the catalytic activity depends on the structural nature of N and the defect density. Compared with pristine carbon nanotubes, the presence of graphitic and pyridinic nitrogen significantly reduces the overpotential, of about −0.18 V, and increases the CO selectivity about 80%. The DFT calculations indicate that pyridinic defects retain a lone pair of electrons capable of binding the CO_2 molecule. However, in the graphitic-like nitrogen, the electrons are located in the π^* antibonding orbital, making them less accessible for CO_2 binding. The theoretical calculations are in agreement with the experimental results.

Wu et al.[147] reported a study related to nitrogen defects on three-dimensional graphene foam incorporated with nitrogen defects for the electroreduction of CO_2. The catalyst required negligible onset overpotential of −0.18 V for CO formation and exhibited superior activity than Ag and Au, achieving similar maximum FE toward CO of about 85% at a lower overpotential of −0.47. The electroreduction of CO_2 was conducted over 5 h with good catalytic stability. The authors also reported that DFT calculations indicate that pyridinic N are the most active sites for the electroreduction of CO_2.

In another study, Lu et al.[148] reported the reduction of CO_2 over graphitic carbon nitride/carbon nanotube composite electrocatalyst. The authors observed that the catalyst was stable during 50 h of operation and presented a very high selectivity toward CO with a maximum FE of 60%, which was attributed to the formation of carbon–nitrogen bonds, high conductivity of the resulting material, and high specific surface area.

A DFT theoretical study performed by Liu et al.[149] for the electroreduction of CO_2 on N-doped graphenes reported that pyrrolic-N graphene sites were the most active, with an overpotential of 0.24 V and HCOOH as the product. The high activity was due to the modification of the graphene electronic properties by the incorporation of N. In addition, the authors commented that doping with nitrogen reduces the free energy barrier to form the COOH intermediate, which is the reaction-limiting step and augments the adsorption energy of the adsorbed COOH with 0.18 eV for undoped graphene and between −1.27 and −3.24 eV for N-doped graphenes.

In another report, Wu et al.[150] tested nanometer-size N-doped graphene quantum dots (NGQD) as a catalyst in the electroreduction of CO_2. The authors highlighted that among the catalysts tested so far (metals, alloys, organometallics, layered materials, and carbon nanostructures) for the electrotransformation of CO_2, only Cu exhibits selectivity toward hydrocarbons and multi-carbon oxygenates, whereas the rest of the catalysts favor CO or formate. The present study shows that the reduction of CO_2 on NGQD exhibited a high total FE of up to 90%, high current densities and low overpotentials, and high selectivity for ethylene and ethanol. The production rates for C2 and C3 are comparable to those obtained using Cu nanoparticle-based electrocatalysts. The authors also concluded from XPS characterization of the catalyst before and after electroreduction of CO_2 that CO_2 adsorbs preferentially onto pyridinic N sites likely located at the edge sites of the NGQD as suggested in another research work.[151]

Jhong et al.[152] studied the electroreduction of CO_2 using a composite formed with pyrolyzed carbon nitride and multiwall carbon nanotube composite as catalyst. The catalyst showed high activity and greater selectivity toward CO compared to H_2, 98% CO and 2% H_2. The partial current density at intermediate cathode potentials ($V = -1.46$ V vs. Ag/AgCl) is up to

3.5 fold higher than that observed for state-of-the-art Ag nanoparticle-based catalysts and the maximum current density was 90 mA cm^{-2}.

Sun et al.^{153} reported the use of N-doped carbon material/carbon paper electrodes on the electroreduction of CO_2 in the presence of an ionic liquid as electrolyte. The authors commented that the FE for CH_4 could be as high as 93.5%, and the current density was about 6-fold higher than that of a Cu electrode under similar experimental conditions. The addition of a trace amount of water in the electrolyte improves the current density effectively, keeping the selectivity for CH_4 constant.

In another study, Li et al.^{154} synthesized S-doped and S,N-doped polymer-derived carbons and tested them as catalysts in the electroreduction of CO_2. The surface basicity favors the reduction of CO_2 compared to the HER. Higher FEs were observed for the dual-doped catalyst than that of the S-doped sample. The S,N-doped catalyst was better at decreasing the overpotential of the reduction process, and showed a maximum FE of 11.3% and 0.18% for CO and CH_4, respectively. The authors reported that the pyridinic nitrogen groups were active in binding the CO_2; however, the quaternary N and thiophenic groups were also involved in the CO_2 reduction. In addition, the formation of CO would be promoted by the positively charged sites on the carbon atom adjacent to pyridinic N, which stabilized the $CO_2^{\cdot-}$ and COOH* intermediates.

Li et al.^{155} also reported a study related to the electroreduction of CO_2 using wood-based activated carbon which was used directly or modified by the introduction of N and/or oxygen. The structures of the materials are similar; however, the surface chemistry differs significantly. At -0.66 V vs. HER, the FEs were 40% and 1.2% for CO and H_2, respectively. The high selectivity for CO was mainly related to the positively charged carbon next to the pyridinic nitrogen which stabilizes the $CO_2^{\cdot-}$ intermediate within the pores. They also commented that the role of quaternary N is less influential, whereas the N oxides (C–N$^+$–O$^-$) are active sites for both the reduction of CO_2 and HER. Their results also indicate that the acidity of the surface increases the overpotential of the maximum FE of either CO or CH_4 formation.

Li et al.^{156} tested an N-doped porous carbon material as catalyst for the electrochemical reduction of CO_2 and found high catalytic activity to produce CO with a maximum FE of 83.7% and a partial current density of CO of 6.6 mA cm^{-2} at an overpotential of 0.71 V in aqueous bicarbonate medium as electrolyte. The mechanistic study indicates that the pyridinic N is responsible for the good catalytic performance. The catalyst was synthesized by a facile one-step pyrolysis approach using wheat flour as carbon source and KOH as starting material. The final catalytic sample shows a large surface area with hierarchical porous structures. The nitrogen content and functional species in the catalytic sample can be controlled to a certain extent by selecting the temperature of pyrolysis. Although most of the papers reported in the literature related to the electroreduction of CO_2 present great selectivities to CO, relatively few favor the production of formate and C2 compounds.

Wang et al.[157] synthesized nitrogen-doped nanoporous carbon (NDNC)/carbon nanotube (CNT) membranes and applied these materials as free-binder electrodes in the catalytic electroreduction of CO_2 in aqueous media. The authors reported that even though the carbon content is similar in both samples their activities significantly differ. The HNCM/CNT presented an FE of 81% toward formate at 0.90 V vs. RHE compared to 40% for the HNCM sample. This was explained by noting that the HNCM/CNT presented a higher quantity of pyridinic-N compared to the NDNC. In addition, both catalytic samples favor the formation of formate and present good stability during reaction.

In another research report, Zhang et al.[158] studied the catalytic electroreduction of CO_2 using nitrogen-doped carbon nanotubes. This metal-free catalyst was robust and selective for the production of formate in aqueous media. The authors also commented that polyethylenimine (PEI) functions as a co-catalyst by significantly reducing the catalytic overpotential and augmenting current density and efficiency. This co-catalysis seems to stabilize $CO_2^{\bullet-}$ and concentrate the CO_2 in the PEI overlayer.

Wang et al.[159] studied the transformation of CO_2 in the presence of N-C electrode and an electrolyte solution of 0.5 M of $KHCO_3$. The catalyst was prepared using melamine as a nitrogen source and graphene oxide as support. The experiment (12 h) showed good formate selectivity (FE = 73%) at −0.84 V vs. RHE. Although higher FE toward formate of 87% was reported,[160] the overpotential was 1.13 V vs. HER compared to 0.74 V observed in the present study.

In another investigation, Song et al.[161] studied the catalytic electrotransformation of CO_2 over electrodes prepared using nitrogen-doped ordered cylindrical mesoporous carbon. The authors reported almost 100% selectivity toward ethanol and a high FE of 77% at −0.56 V vs. RHE. The great catalytic performance was attributed to the synergetic effect of nitrogen and the cylindrical channel configuration of the catalyst as observed also from DFT calculations.

Li et al.[162] synthesized and applied microporous carbons with high nitrogen content located at different positions, pyrazinic-N and pyridinic-N, as catalysts in the electroreduction of CO_2. It was expected that the location of the nitrogen atoms at different positions in the ring unit would provide active sites of different activities. It was indeed found that pyrazinic-N reacted with oxygen and avoids the oxidation of the pyridinic site and as a consequence, the stability of the catalyst was improved. The observed reaction products were CO, CH_4, CH_3OH, formic acid, acetone, and propanol. The authors suggest that the C3 products are formed when more than one nitrogen atom are located close to each other on the surface of the catalyst.

In another study, Yuan et al.[163] reported the synthesis of graphene oxide (GO) catalysts modified with various nitrogen sources such as pyridoxine (vitamin B, VB_6), 4-hydroxypyridine (X-1), 4-aminopyridine (X-2), 8-hydroxyquinoline (X-3), and 5-amino-1,10-phenanthroline (X-4). The samples were then applied as catalysts for the electroreduction of CO_2. The results showed that the sample synthesized with 19.7 mM of VB_6 was the best catalyst, with an overall FE of ≈45% (36.4% ethanol and 8.9% acetone) followed by X-1 (14.2% ethanol and 3.7% acetone). The authors also commented that the electrocatalytic activity of GO-X samples was not only related to the pyridinic-N content, but also the structure of the grafted pyridine derivatives.

Li et al.[164] reported a study related to the electroreduction of CO_2 in the presence of pyridinic-N sites highly exposed on wrinkled porous carbon nanosheets. The catalyst showed high stability and negligible onset overpotential of −0.19 V for CO formation and a maximum FE of 84% at an overpotential of −0.49 V. This was due to the high exposure of the pyridinic-N active sites which favor the CO_2 adsorption and the nanoporous two-dimensional structures which enhanced fast charge transfer. DFT calculations confirm that the pyridinic-N can offer favorable binding sites to COOH* and the subsequent catalytic reduction.

In another study, Chen et al.[165] reported the electroreduction of CO_2 in the presence of nitrogen-doped fullerene derivative catalysts (see Figure 7.10).

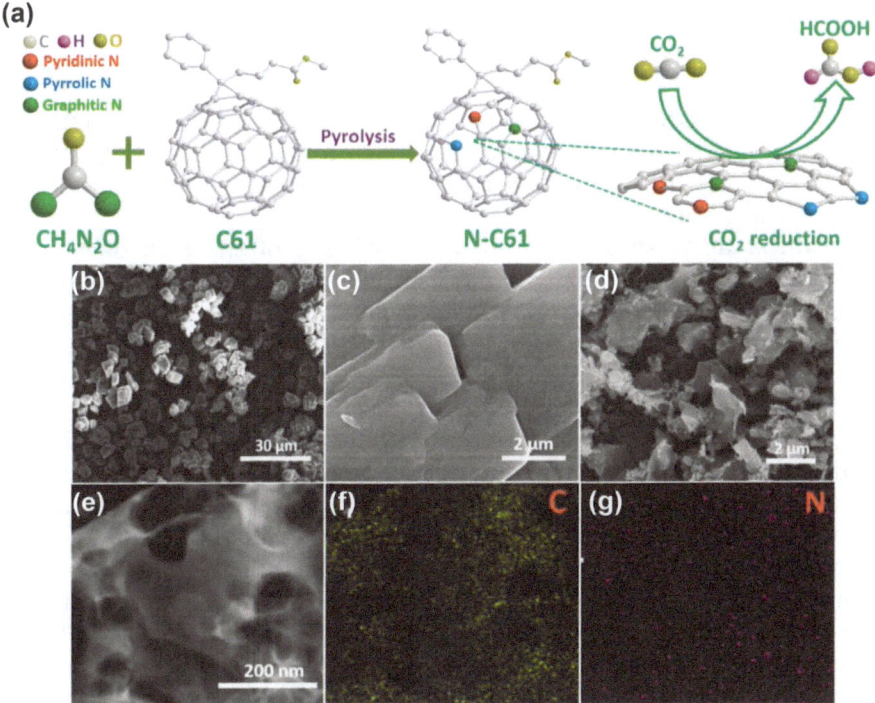

Figure 7.10 Schematic illustration of the fabrication process for the N-C61 electrocatalysts (a). Physical characterization of pristine C61 and N-C61-800. SEM images of pristine C61 (b) and (c). SEM image of N-C61-800 (d). HRTEM images of N-C61-800 (e), and corresponding elemental mappings for carbon (f) and nitrogen (g). Reproduced from ref. 165 with permission from the Royal Society of Chemistry.

The modified fullerene structures are reported for the first time as stable and good catalysts for the transformation of CO_2. The N-doped PC61BM ([6,6]-phenyl C61-butyric acid methyl ester) (N-C61) presents an excellent activity for the CO_2 reduction, reaching a high FE of 91.2% for formate production with good stability at a moderate overpotential of 700 mV. A partial formate current density of 14.7 mA cm^{-2} was observed at -1.1 V vs. RHE, which is greater than most metal-free electrocatalysts. The authors also suggest that graphite-N defects are the most likely active sites for the formation of formate.

Mao and Hatton[166] reviewed non-metal catalysts used in the electroreduction of CO_2 and classified the materials as conducting polymers, pyridinium derivatives, aromatic anion radicals, and heteroatom-doped carbon materials. The required overpotential, the overall catalytic performance, product distribution, catalyst stability, faradaic yield, and reaction mechanism are discussed. In addition, the nature and pH of the electrolyte, the concentration of CO_2, the electrode, and morphology of the catalyst are addressed. The authors also analyzed and identified future research directions.

7.5 Computational Studies

Most of the theoretical studies mentioned in this section are computational and give an idea of the reaction pathway related to the ECR of CO_2 using a specific catalyst. In addition, some other theoretical and experimental studies are also included in this section.

Hussain et al.[167] studied the elementary steps of the ECR of CO_2 on Cu(111) and Pt(111) by density functional theory calculations. By using the nudged elastic band method for applied potentials similar to those used in experimental studies, ranging from -0.7 V to -1.7 V, the minimum sequential protonation by either Tafel or Heyrovsky mechanism was determined. Different detailed mechanisms for the CO_2 reduction on both surfaces are proposed. The theoretical findings are consistent with the experimental observations where copper electrodes lead to a relatively high yield of CH_4, while H_2 is produced almost exclusively at platinum electrodes.

Zou et al.[168] reported a theoretical study of the electroreduction of CO_2 over N-doped graphene quantum dots by performing first-principles simulations. The results demonstrate that the introduction of N atoms into the edges of the graphene quantum dots enhances their bonding with *COOH, which promotes the transformation of CO_2 to CO. The authors also commented that water favors the production of CH_4 instead of CH_3OH, which was due to a much lower kinetic barrier for the transformation of *CH_2OH to *CH_2 via water-mediated proton shuttling. Furthermore, the adsorbed *CH_2 provides sites for the coupling of CO to produce C2 compounds such as C_2H_4 and C_2H_5OH.

In another publication, Prslja and Lopez[169] performed a computational study of the ECR of CO_2 using Ni on N-doped carbon materials, considering nanoparticles and single atoms as catalysts. The authors found that optimal activity and selectivity were noticed for the NiN_3 model in which Ni presents an oxidation state of +1. NiN_3 shows low-onset theoretical potential in agreement with experiments, and a high formation rate for CO. The authors present a mechanism of Ni(211) surface reconstruction that is possible only at high CO coverage of about 8/12 ML by the formation of $Ni(CO)_4$ species which form single active atoms after re-dispersion.

Rodriguez et al.[170] studied the interactions of CO_2 with the (210) surface of brookite TiO_2 by first-principle calculations on cluster and periodic slab systems. The authors found that the charge transfer to CO_2 was negligible, which suggests that this unmodified brookite surface was an unsuitable catalyst to reduce CO_2 as compared to anatase (110). However, when oxygen vacancies were created, the brookite surface was able to activate CO_2. Indeed, CO_2 adsorption experiments on pure brookite surface or with oxygen vacancies, using diffuse reflectance Fourier transform infrared spectroscopy, demonstrate the formation of abundant CO_2^- on the later sample, as found theoretically.

Zhong et al.[171] explored the ECR of CO_2 by computation and machine learning and found that the highest FE to ethylene rate reported so far was over 80% compared to about 66% for pure Cu, which was achieved by using Cu–Al electrocatalysts at a current density of 400 $\mu A\,cm^{-2}$ and an ethylene power conversion efficiency of 55% ± 2 at 150 $\mu A\,cm^{-2}$. The computational results indicate that the Cu–Al alloys provide multiple sites and surface orientation suitable for CO binding for both efficient and selective reduction of CO_2. In addition, theoretical calculations show good agreement with the results from in situ X-ray absorption measurements indicating the enhancement of C–C dimerization.

In another study, Xiao et al.[172] used quantum mechanics to predict the atomistic mechanisms responsible for C1 and C2 products on Cu during the ECR of CO_2. The authors reported the pH-dependent routes to methane and ethylene and identified the key intermediates where branches to methanol, ketene, ethanol, acetylene, and ethane are kinetically blocked. It was found that surface water on Cu plays a key role in the selectivity for hydrocarbon products over the oxygen-containing alcohol products by serving as a strong proton donor.

Hussain et al.[173] studied the ECR of CO_2 in the presence of various close-packed electrode surfaces as catalysts, such as Ag, Au, Cu, Ir, Ni, Rh, and Pt, using a combination of density functional and rate theory. The authors used a realistic model of the electrochemical interface and included the effect of applied potential on the activation energy of the various elemental steps. Cu is the only catalytic metal studied so far in the reduction of CO_2 that produces hydrocarbons and alcohols, in addition to CO. This is due to a crossover in the activation energy at about −0.6 V, where the reduction of CO_2 becomes more facile than HER, until −1.4 V where the activation energy of HER again becomes lower as the mechanism for HER changes from Volmer–Tafel to Volmer–Heyrovsky. Since the on-top sites on the Cu(111) surface cannot easily be occupied by H adatoms, the activation energy for HER remains high for large negative potential. The authors found that changing the applied potential results in shifts in the reaction mechanism. For an applied potential $U > -0.6$ the *CO to *CHO step shows lower activation energy, but when $U < -0.6$ the *CO to *COH step presents lower activation energy. The authors also observed close agreement between the determined current densities and efficiencies, as a function of the applied voltage for various catalysts and products, and the experimental results.

In another study, Cheng et al.[174] applied *ab initio* molecular metadynamics simulations (AIMμD) for the water/Cu(100) system with five layers of the explicit solvent under a potential of -0.59 V at pH 7 and compared this with the experimental results. From these free-energy calculations, the kinetics and pathways for major products (ethylene and methane) and minor products (ethanol, glyoxal, glycolaldehyde, ethylene glycol, acetaldehyde, ethane, and methanol) were determined. For $U > -0.6$ V ethylene is produced *via* the Eley–Rideal (ER) mechanism using $H_2O + e^-$. The rate-determining step is the C–C coupling of two CO, with $\Delta G^{\ddagger} = 0.69$ eV, whereas for $U < -0.6$ eV the rate for ethylene formation diminishes, due mainly to the loss of surface sites which were available for CO adsorption and are now preferentially occupied by H*. The reappearance of C_2H_4 and CH_4 at $U < -0.85$ V results from the *CHO formation produced *via* an ER process of H* with non-adsorbed CO. These findings suggest that for the selective and efficient production of hydrocarbon at a pH of 7 an increase in the CO concentration is required which can be achieved by changing the solvent or alloying the surface.

In another report, Isegawa and Sharma[175] studied the ECR of CO_2 in the presence of Mn modified with $Mg2^+$ which imparts Lewis acidity and improves the reduction of CO_2. DFT calculations demonstrate that the primary role of $Mg(OTf)_2$ (OTf = trifluoromethanesulfonate) is to stabilize a two-electron reduced Mn intermediate through Lewis pair binding, making the reaction thermodynamically and kinetically feasible. In the proposed mechanism, two molecules of CO_2 and Mg(OTf) contribute to the C–O bond cleavage reaction.

Saravanan et al.[176] applied computational quantum chemistry and atomistic thermodynamics to study and identify intermediates formed during the ECR of CO_2 on tin oxide electrodes. The authors found that the hydroxylated and partially reduced SnO species formed during the ECR of CO_2 are the catalytic sites as others have previously concluded. In addition, computational results also indicate that doping this oxide with Nb, Ti, V, or Zn will reduce the overpotential required for the reduction of CO_2.

In another study, Bernstein et al.[177] applied density functional theory to analyze the feasibility of CO_2 electroreduction on a Fe(100) surface. Experimentally, iron as the catalyst for the ECR of CO_2 is nonselective to hydrocarbons; however, the authors analyzed the surface at low coverage and determined the intermediate energies of the paths to produce CH_4 and CH_3OH and found that the Fe(100) surface could be more active than

Cu(111), which is the only catalytically active metal reported so far with high selectivity to hydrocarbons. By considering a series of impediments during the ECR of CO_2 on Fe(100), such as the blockage of the active sites by O*, OH*, the competitive adsorption of H*, CO*, C* and the formation of FeC_x, the results predicted that the Fe(100) surface is covered with C* or CO* species, blocking any C–H bond formation. The authors also found that bulk iron would be unstable, relative to FeC_x at potential relevant to the ECR of CO_2.

In another report, Dean et al.[178] performed a computational study on the ECR of CO_2 in the presence of bimetallic nanoparticles. Cu nanoparticles cannot adsorb and activate CO_2; however, Cu-based nanoparticles can increase the adsorption of CO_2 and the hydrogenation activity. Various heteroatoms were studied by DFT. The authors revealed two descriptors for the CO_2 adsorption on Cu-bimetallic nanoparticles: the local d-band center and the electropositivity of the heteroatom which drive an effective charge transfer from the nanoparticle to the CO_2 molecule. It was found that CuZr nanoparticles can adsorb and activate CO_2 even if Zr is oxidized. The theoretical findings were supported with experiments.

Shen et al.[179] commented that the ECR of CO_2 is an attractive way to store renewable energy in hydrocarbons and other compounds; however, the high overpotential and low efficiency are disadvantages for practical application. A DFT study of the ECR of CO_2 on cobalt porphyrin is presented. The authors found that CO is the main product and the key intermediate formed only when the cobalt center of the complex is in the Co^I oxidation state. Furthermore, formic acid is a minor product generated through a [Co(P)-(OCHO)] intermediate. In addition, CH_4 is also produced in a small quantity and is generated from CO reduction by concerted proton-coupled electron transfers assumed for each electrochemical step. The theoretical findings are consistent with the experimental results previously reported.

In another report, Guo et al.[180] studied the ECR of CO_2 towards C1 products (CO, HCOOH, CH_3OH, and CH_4) on the γ-C_3N_4 surfaces doped with Ni, Co, or Fe by DFT. The structure of the catalysts, the CO_2 adsorption configuration, and the reaction mechanism were systematically studied. The results indicate that CO_2 could be chemically adsorbed on Co–C_3N_4 and Fe–C_3N_4, but physically on Ni–C_3N_4. All catalysts favor the ECR of CO_2 over the HER. It was also found that the ECR of CO_2 proceeds via COOH and OCHO as initial protonation intermediates on Ni–C_3N_4 and Co/Fe–C_3N_4, respectively, resulting in different C1 products. The Co–C_3N_4 catalyst shows great selectivity to methanol and favors the ECR of CO_2, compared to Ni–C_3N_4 and FeC_3N_4.

With the aim to understand the difference in product selectivity observed during the ECR of CO_2 on Cu electrodes, Luo et al.[181] evaluated the thermodynamics and kinetics of this reaction on Cu(100) and Cu(111) surfaces by DFT calculations. The authors suggested that the hydrogenation of CO* to hydroxymethylidyne (COH*) or formyl (CHO*) is a key selective step. The Cu(111) surface favors the formation of COH* species through which methane and ethylene are produced *via* a common CH_2 species when applying a high overpotential (<−0.8 V-RHE), whereas the Cu(100) surface preferentially produces CHO* and the formation of ethylene goes through the coupling of two CHO* species, followed by reduction steps of the C2 intermediates under relatively low overpotential (−0.4 to −0.6 V-RHE). Calculations indicate that the presence of the (111) step sites on the flat (100) terrace can reduce the overpotential for the production of C2, which can be present on Cu(100) due to reconstruction.

Goodpaster *et al.*[182] investigated the pathways for the C–C formation during the ECR of CO_2 on Cu(100) using the periodic Kohn–Sham DFT that incorporates the effects of the electrochemical potential, solvent, and electrolyte. The former was set by relating the applied potential to the Fermi energy and then calculating the number of electrons required by the simulation cell for that specific Fermi energy, whereas the solvent was considered as a continuum dielectric and the electrolyte was described using a linearized Poisson–Boltzmann model. For validating the model, the determined potentials of zero charge for various surfaces were found to agree well with the experimental values presenting a mean average error of 0.09 V. The results indicate that the mechanism for the C–C bond formation occurs at low potential through a CO dimer; however, this pathway is blocked due to a large activation barrier present at high potentials. In the latter case, the C–C bond formation occurs through the reaction between the adsorbed CHO and CO. From the calculations, the determined rate parameters used for the simulation of kinetics related to the formation of ethylene agreed well with previously reported measurements.

In another study, Garza *et al.*[183] proposed a mechanism for the electrochemical transformation of CO_2 to C2 products on Cu electrodes, based on constraints from experimental observations and DFT calculations. The proposed mechanism explains the presence of seven C2 species that have been detected during the CO_2 electrochemical reduction. The results shed light on the difference in activities toward C2 products between Cu(100) and Cu(111) facets. Comparisons with other results in the literature are also addressed.

Todorova *et al.*[184] provided a critical assessment of various proposed scenarios of the initial and post C–C coupling steps for the formation of either ethylene or ethanol. Computational analysis of the parameters controlling product selectivity such as catalyst structure and composition, defects, and the interaction of Cu with a second metal is presented. In addition, reaction conditions, such as pH, applied potential, and electrolyte were also analyzed. A scheme combining the proposed pathways was derived and the issues that are still under debate were highlighted.

In another report, Cheng *et al.*[185] studied the energetics of the electrochemical reduction of CO_2 on metal (M) porphyrin-like motifs incorporated into graphene layers by DFT calculations, with the aim to develop strategies to enhance CO_2 while suppressing the HER. The calculations indicate that the M–H bond is stronger than the M–COOH bond, and consequently, the

HER reaction is favored instead of the reduction of CO_2 to CO. However, when 4f lanthanide or 5f actinide elements are used as the reactive center, the reaction favors the reduction of CO_2 to CO instead of the HER. In addition, there is no scaling relation between the binding energy of H and that of OCHO on the catalyst. The latter species are the key intermediates for the production of formic acid, which is selectively formed whereas the HER is suppressed during the reduction of CO_2. The DFT calculations have also identified several promising electrocatalysts for the reduction of CO_2 to HCOOH with almost zero overpotentials.

Cheng et al.[186] studied the CO_2 reduction reaction in aqueous solution on promising single-atom alloys (SAAs) as electrocatalysts by DFT calculations combined with the Poisson–Boltzmann implicit solvation model. Various SAAs were tested using Au or Ag as a majority element in combination with M (Co, Cu, Ir, Ni, Pd, and Pt). Such SAA species replace surface atoms. The results indicate that CO_2 is reduced to CO on Au or Ag and the CO is further reduced to C1 (alkanes, alkenes, and alcohols) on the second metal. About 14 SAAs, among 28, favor the reduction of CO_2 to C1, whereas the rest of the SAAs prefer the formation of H_2. Rh@Au(100) and Rh@Ag(100) are among the most promising SAAs and were studied in detail. These SAAs reduce CO_2 to methane, through different mechanisms; the minimum applied voltages to drive both Rh@Au(100) and Rh@Ag(100) systems are −1.01 and −1.12 V, respectively; the FEs for the reduction of CO_2 to CO are 60% for gold and 90% for silver SAAs. The results suggest that SAAs can efficiently reduce CO_2 to CH_4 with as low as 40% and 10% losses to HER for Rh@Au(100) and Rh@Ag(100), respectively.

Meng et al.[187] published a review on the theoretical studies related to single- and double-atom catalysts (SAC and DAC) utilized for the electrochemical CO_2 reduction reaction. The attention was focused on the reaction mechanisms resulting from theoretical studies related to SAC and DAC. The current challenges and future development prospects were also summarized.

7.5.1 Final Remarks

Very attractive catalysts for the electrochemical reduction of CO_2 to produce useful hydrocarbons are herein reviewed. The nature of the catalyst, the design of the electrochemical cell, and the experimental conditions, such as temperature, pH, pressure, nature of the electrolyte, and mass transport problems as in the case of using aqueous electrolytes, are important parameters to be considered for the efficient ECR of CO_2. The importance of a solid electrolyte was also addressed. Direct exposition of the gas-vapor reactants to the catalyst can drastically reduce mass transport problems. The catalyst could be pure metals, metal-based, non-metal, or hybrid. Metals such as Au, Ag, Cu, Ga, Sn, In, Zn Tl, and Hg as electrodes preferentially form CO, whereas Cd, Hg, In, Pb, Sn, and Tl favor the formation of $HCOO^-$ species to finally yield methane, ethylene, and alcohols. Other metals, such as Fe, Ni, and Pt, showed high selectivity for H_2 and O_2 produced from the water-splitting reaction. Conversely, pure Cu as electrode favors the production of various hydrocarbons, aldehydes, alcohols, CO, H_2, and O_2, whereas Au surfaces are selective to oxygenates over hydrocarbons. Moreover, the addition of a second element to these metals, the preparation of the electrode, the morphology and crystal orientation of the active species, and the applied potential to the cell, affect the activity and selectivity of the ECR of CO_2.

Compared to planar catalytic surfaces, the nanostructured catalysts lead to improved catalytic activity. Supported active species, such as metal oxides, sulfides, nitrides, and phosphides are also very attractive catalysts for the ECR of CO_2; however, the main drawbacks of the metal-based electrocatalysts are frequently related to the high overpotentials and low stability. Other non-metal catalysts, such as carbon doped with B, N, B–N, S–N, or P are also very promising catalytic materials. Indeed, some reports indicate that non-metal catalysts surpass the catalytic activity showed by metal catalysts under similar experimental conditions. Furthermore, computational studies are also very relevant for searching for highly active catalytic materials suitable for the ECR of CO_2.

Finally, it is important to comment that although the products from the ECR of CO_2 are very attractive and useful, these compounds would eventually end up in water (as in the case of spillover of hydrocarbons), soil (as in the event of irresponsible discarding of stable polymers or compounds into the environment), or as CO_2, polluting the atmosphere again. Natural CO_2 fixation by the vegetal world is insufficient to resolve the contamination problem created by humans, and additional carbon-cycling processes are therefore required. Although the ECR of CO_2 is an interesting alternative for the transformation of CO_2, the optimal procedure needs to be scalable, integrated as one stage of the whole carbon-cycling process and use renewable energy for the electrocatalytic reduction of CO_2. Once the anthropogenic CO_2 emissions are stored, the ECR of CO_2 to produce useful compounds and their utilization can then take place and, finally, the capture of the generated CO_2 is required to repeat the carbon-cycling process.

References

1. Monthly average CO_2 registered at the Global Monitoring Laboratory (GML) of the National Oceanic and Atmospheric Administration (NOAA) for October 2024. Information obtained from https://gml.noaa.gov/ccgg/trends/.
2. Intergovernmental Panel on Climate Change. IPCC Special Report on the Impacts of Global Warming of 1.5 °C-Summary for Policy Makers; Incheon, 2018.
3. A. Majumdar, J. Deutch, R. Bras, S. Benson, E. Carter and Ort, D, Conference of the Parties to the United Nations Framework Convention on Climate Change, Adoption of the Paris Agreement, 2015, pp. 1–32.
4. Q. Zhang, M. Bown, L. Pastor-Pérez, S. Duyar and R. R. Tomas, *Ind. Eng. Chem. Res.*, 2022, **61**(34), 12857.

5. S. Ronsch, J. Schneider, S. Matthischke, M. Schlulter, M. Gotz, J. Lefebvre, P. Prabhakaran and S. Bajohr, *Fuel*, 2016, **166**, 276.
6. S. G. Jadhav, P. D. Vaidya, B. M. Bhanage and J. B. Joshi, *Chem. Eng. Res. Des.*, 2014, **92**, 2557.
7. J. Yang, W. Ma, D. Chen, A. Holmen and B. H. Davis, *Appl. Catal., A*, 2014, **470**, 250.
8. C. Markowitsch and M. Lehner, Impact of the operation conditions on the reverse-water-gas shift reaction, in *Global Challenges for a Sustainable Society. EURECA-PRO 2022. Springer Proceedings in Earth and Environmental Sciences*, ed. J. A. Benítez-Andrades, P. García-Llamas, Á. Taboada, L. Estévez-Mauriz and R. Baelo, Springer, Cham, 2023, DOI: 10.1007/978-3-031-25840-4_10.
9. J. Durst, A. Rudnev, A. Dutta, Y. Fu, J. Herranz, V. Kaliginedi, A. Kuzume, A. A. Permyakova, Y. Paratcha, P. Broekmann and T. J. Schmidt, *Chimia*, 2015, **69**, 769.
10. Q. Fan, M. Zhang, M. Jia, S. Liu, J. Qiu and Z. Sun, *Today Energy*, 2018, **10**, 280.
11. T. Ma, Q. Fan, X. Li, J. S. Qiu, T. B. Wu and Z. Y. Sun, *J. CO2 Util.*, 2019, **30**, 168.
12. Y. Hori, K. Kikuchi, A. Murata and S. Suzuki, *Chem. Lett.*, 1986, **15**, 897.
13. Y. Hori, A. Murata, R. Takahashi and S. Suzuki, *J. Chem. Soc., Chem. Commun.*, 1988, **0**, 17.
14. Y. Hori, A. Murata and R. Takahashi, *J. Chem. Soc., Faraday Trans. 1*, 1989, **85**, 2309.
15. Y. Hori, H. Wakebe, T. Tsukamoto and O. Koga, *Electrochim. Acta*, 1994, **39**, 1833.
16. C. W. Li, J. Ciston and M. W. Kanan, *Nature*, 2014, **508**, 504.
17. S. Nitopi, S. Horch, E. Bertheussen, B. Seger, S. B. Scott, I. E. L. Stephens, X. Liu, A. K. Engstfeld, K. Chan, C. h Hahn, J. K. Nørskov, T. F. Jaramillo and I. Chorkendorff, *Chem. Rev.*, 2019, **119**, 7610.
18. O. S. Bushuyev, P. De Luna, C. T. Dinh, L. Tao, G. Saur, J. van de Lagemaat, S. O. Kelley and E. H. Sargent, *Joule*, 2018, **2**, 825.
19. J. M. Spurgeon and B. A. Kumar, *Energy Environ. Sci.*, 2018, **11**, 15361551.
20. X. W. Mao and T. A. Hatton, *Ind. Eng. Chem. Res.*, 2015, **54**, 4033.
21. G. R. Zhang, S. D. Straub and L. L. Shen, *et al.*, *Angew. Chem., Int. Ed.*, 2020, **59**, 18095.
22. M. Bevilacqua, J. Filippi, H. A. Miller and F. Vizza, *Energy Technol.*, 2015, **3**, 197.
23. B. Endrődi, G. Bencsik, F. Darvas, R. Jones, K. Rajeshwar and C. Janáky, *Prog. Energy Combust. Sci.*, 2017, **62**, 133.
24. L. Fan, C. Xia, F. Yang, J. Wang, H. Wang and Y. Lu, *Sci. Adv.*, 2020, **6**(8), eaay3111.
25. R. Küngas, *J. Electrochem. Soc.*, 2020, **167**, 044508.
26. C. A. R. Pappijn, M. Ruitenbeek, M.-F. Reyniers and K. M. Geem, *Front. Energy Res.*, 2020, **8**, 557466.

27. X. Zhang, S.-X. Guo, K. A. Gandionco, A. M. Bond and J. Zhang, *Mater. Today Adv.*, 2020, **7**, 100074.
28. Y. Quan, J. Zhu and G. Zheng, *Small Sci.*, 2021, **1**, 2100043.
29. I. Brewis, R. -F. Shahzad, R. W. Field, A. Jedidi and S. Rasul, *Discover Chem. Eng.*, 2022, **2**, 2.
30. Z. Kou, X. Li and T. Wang, *et al.*, *Electrochem. Energy Rev.*, 2022, **5**, 82.
31. M. Li, E. Irtem and H. P. I. van Montfort, *et al.*, *Nat. Commun.*, 2022, **13**, 5398.
32. J. Qu, X. Cao and L. Gao, *et al.*, *Nano-Micro Lett.*, 2023, **15**, 178.
33. R. A. Tufa, D. Chanda, M. Ma, D. Aili, T. B. Demissie, J. Vaes, Q. Li, S. h Liu and D. Pant, *Appl. Energy*, 2020, **277**, 115557.
34. D. A. Vermaas and W. A. Smith, *ACS Energy Lett.*, 2016, **1**, 1143.
35. Y. C. Li, D. Zhou, Z. Yan, R. H. Gonçalves, D. A. Salvatore and C. P. Berlinguette, *et al.*, *ACS Energy Lett.*, 2016, **1**, 1149.
36. D. A. Salvatore, D. M. Weekes, J. He, K. E. Dettelbach, Y. C. Li and T. E. Mallouk, *et al.*, *ACS Energy Lett.*, 2018, **3**, 149.
37. M. Ramdin, A. R. T. Morrison, M. de Groen, R. van Haperen, R. de Kler and L. J. P. van den Broeke, *et al.*, *Ind. Eng. Chem. Res.*, 2019, **58**, 1834.
38. D. T. Whipple, E. C. Finke and P. J. A. Kenis, *Solid State Lett.*, 2010, **13**(9), B109–B111.
39. C. Delacourt, P. L. Ridgway, J. B. Kerr and J. Newman, *J. Electrochem. Soc.*, 2008, **155**(1), B42–B49.
40. L. Han, W. Zhou and C. Xiang, *ACS Energy Lett.*, 2018, **3**(4), 855.
41. D. M. Weekes, D. A. Salvatore, A. Reyes, A. Huang and C. P. Berlinguette, *Acc. Chem. Res.*, 2018, **51**(4), 910.
42. D. Higgins, C. Hahn, C. Xiang, T. F. Jaramillo and A. Z. Weber, *ACS Energy Lett.*, 2019, **4**(1), 317.
43. C. M. Gabardo, C. P. O'Brien, J. P. Edwards, C. h McCallum, Y. Xu, C.-T. Dinh, J. Li, E. H. Sargent and D. Sinton, *Joule*, 2019, **3**, 1.
44. Y. Yang and F. Li, *Curr. Opin. Green Sustainable Chem.*, 2021, **27**, 100419.
45. S. Lee, J. D. Ocon, Y. Son and J. Lee, *J. Phys. Chem. C*, 2015, **119**, 4884.
46. J. Shi, F.-X. Shen, F. Shi, N. Song, Y.-J. Jia and Y.-Q. Hu, *et al.*, *Electrochim. Acta*, 2017, **240**, 114.
47. H. Noda, S. Ikeda, A. Yamamoto, H. Einaga and K. Ito, *Bull. Chem. Soc. Jpn.*, 1995, **68**(7), 1889.
48. J. Y. Maeng, S. Y. Hwang, Y. J. Kim, C. K. Rhee and Y. Sohn, *Chem. Eng. J.*, 2023, **470**, 143970.
49. Z. Wei, J. Ding, Z. Wang, A. Wang, L. Zhang, Y. Liu, Y. Guo, X. Yang, Y. Zhai and B. Liu, *Angew. Chem., Int. Ed.*, 2024, **63**(27), e202402070.
50. B. Kumar, M. Llorente, J. Froehlich, T. Dang, A. Sathrum and C. P. Kubiak, Photochemical and photoelectrochemical reduction of CO_2, *Annu. Rev. Phys. Chem.*, 2012, **63**, 541.
51. M. Dunwell, Q. Lu, J. M. Heyes, J. Rosen, J. G. Chen and Y. Yan, *et al.*, *J. Am. Chem. Soc.*, 2017, **139**, 3774.
52. Y. Tomita, S. Teruya, O. Koga and Y. Hori, *J. Electrochem. Soc.*, 2000, **147**, 4164.

53. L. Sun, G. K. Ramesha, P. V. Kamat and J. F. Brennecke, *Langmuir*, 2014, **30**, 6302.
54. Y. Matsubara, D. C. Grills and Y. Kuwahara, *ACS Catal.*, 2015, **5**, 6440.
55. T. C. Berto, L. Zhang, R. J. Hamers and J. F. Berry, *ACS Catal.*, 2015, **5**, 703.
56. Q. Zhu, J. Ma, X. Kang, X. Sun, H. Liu and J. Hu, *et al.*, *Angew. Chem., Int. Ed.*, 2016, **55**, 9012.
57. M. C. Figueiredo, I. Ledezma-Yanez and M. T. M. Koper, *ACS Catal.*, 2016, **6**, 2382.
58. N. E. Mendieta-Reyes, A. K. Díaz-García and R. Gómez, *ACS Catal.*, 2018, **8**, 1903.
59. Y. Tomita, S. Teruya, O. Koga and Y. Hori, *J. Electrochem. Soc.*, 2000, **147**, 4164.
60. Á. Díaz-Duque, A. P. Sandoval-Rojas, A. F. Molina-Osorio, J. M. Feliu and M. F. Suárez-Herrera, *Electrochem. Commun.*, 2015, **61**, 74.
61. A. V. Rudnev, U. E. Zhumaev, A. Kuzume, S. Vesztergom, J. Furrer and P. Broekmann, *et al.*, *Electrochim. Acta*, 2016, **189**, 38.
62. T. Kai, M. Zhou, Z. Duan, G. A. Henkelman and A. J. Bard, *J. Am. Chem. Soc.*, 2017, **139**, 18552.
63. Y. Oh, H. Vrubel, S. Guidoux and X. Hu, *Chem. Commun.*, 2014, **50**, 3878.
64. R. Aydin and F. Köleli, *J. Electroanal. Chem.*, 2002, **535**, 107.
65. S. Kaneco, K. Iiba, M. Yabuuchi, N. Nishio, H. Ohnishi and H. Katsumata, *et al.*, *Ind. Eng. Chem. Res.*, 2002, **41**, 5165.
66. S. Kaneco, H. Katsumata, T. Suzuki and K. Ohta, *Energy Fuels*, 2006, **20**, 409.
67. S. Kaneco, Y. Ueno, H. Katsumata, T. Suzuki and K. Ohta, *Chem. Eng. J.*, 2006, **119**, 107.
68. S. Kaneco, K. Iiba, H. Katsumata, T. Suzuki and K. Ohta, *J. Solid State Electrochem.*, 2007, **11**, 490.
69. M. Murugananthan, M. Kumaravel, H. Katsumata, T. Suzuki and S. Kaneco, *Int. J. Hydrogen Energy*, 2015, **40**, 6740.
70. R. L. Cook, R. C. MacDuff and A. F. Sammell, *J. Electrochem. Sci. Technol.*, 1988, **135**, 1470.
71. D. W. Dewulf and A. J. Bard, *Catal. Lett.*, 1988, **1**, 73.
72. S. Komatsu, M. Tanaka, A. Okumura and A. Kungi, *Electrochim. Acta*, 1995, **40**, 745.
73. Y. Nishimura, D. Yoshida, M. Mizuhata, K. Asaka, K. Oguro and H. Takenaka, *Energy Convers. Manage.*, 1995, **36**, 629.
74. E. J. Dufek, T. E. Lister, S. G. Stone and E. M. McIlwain, *J. Electrochem. Soc.*, 2012, **159**, F514.
75. L. M. Aeshala, R. Uppaluri and A. Verma, *Phys. Chem. Chem. Phys.*, 2014, **16**, 17588.
76. S. D. Ebbesen, S. H. Jensen, A. Hauch and M. B. Mogensen, *Chem. Rev.*, 2014, **114**, 10697.
77. L. M. Aeshala, S. U. Rahman and A. Verma, *Sep. Purif. Technol.*, 2012, **94**, 131.

78. X. Zhang, Y. Song, G. Wang and X. Bao, *J. Energy Chem.*, 2017, **26**, 839.
79. K. Xie, Y. Zhang, G. Meng and J. T. S. Irvine, *Energy Environ. Sci.*, 2011, **4**, 2218.
80. L. Chen, F. Chen and C. Xia, *Energy Environ. Sci.*, 2014, **7**, 4018.
81. Y. Xu, R. K. Miao, J. P. Edwards, S. Liu, C. P. O'Brien, C. M. Gabardo, M. Fan, J. E. Huang, A. Robb, E. H. Sargent and D. Sinton, *Joule*, 2022, **6**, 1333.
82. J. Y. Kim, P. Zhu and F. Y. Chen, *et al.*, *Nat. Catal.*, 2022, **5**, 288.
83. L. Zhang, S. Hu, X. Zhu and W. Yang, *J. Energy Chem.*, 2017, **26**, 593.
84. J. B. Vennekötter, T. Scheuermann, R. Sengpiel and M. Wessling, *J. CO2 Util.*, 2019, **32**, 202213.
85. C. Xia, P. Zhu and Q. Jiang, *et al.*, *Nat. Energy*, 2019, **4**, 776.
86. W. Wu, L. Xu, Q. Lu, J. Sun, Z. Xu, C. Song, J. C. Yu and Y. Wang, *Adv. Mater.*, 2025, **37**, 2312894.
87. B. Wu, B. Wang, B. Cai, C. Wu, W. W. Tjiu, M. Zhang, Z. Aabdin, S. Xi and Y. Lum, *J. Am. Chem. Soc.*, 2024, **146**(43), 29801.
88. N. Chu, Y. Jiang, R. J. Zeng, D. Li and P. Liang, *Environ. Sci. Technol.*, 2024, **58**, 10881.
89. Y. Hori, K. Kikuchi and S. Suzuki, *Catal. Lett.*, 1985, **14**, 1695.
90. R. L. Cook, R. C. MacDuff and A. F. Sammells, *J. Electrochem. Soc.*, 1987, **134**, 2375.
91. R. L. Cook, R. C. MacDuff and A. F. Sammells, *J. Electrochem. Soc.*, 1988, **135**, 1320.
92. Y. Nishimura, D. Yoshida, M. Mizuhata, K. Asaka, K. Oguro and H. Takenaka, *Energy Convers. Manage.*, 1995, **36**(6–9), 620.
93. Y. Hori, H. Wakebe, T. Tsukamoto and O. Koga, *Electrochim. Acta*, 1994, **39**, 1833.
94. Y. Hori, I. Takahashi, O. Koga and N. Hoshi, *J. Mol. Catal. A: Chem.*, 2003, **199**, 39.
95. H. Yano, T. Tanaka, M. Nakayama and K. Ogura, *J. Electroanal. Chem.*, 2004, **565**, 287.
96. M. Le, M. Ren, Z. Zhang, P. T. Sprunger, R. L. Kurtz and J. C. Flake, *J. Electrochem. Soc.*, 2011, **158**, E45.
97. K. P. Kuhl, E. R. Cave, D. N. Abram and T. F. Jaramillo, *Energy Environ. Sci.*, 2012, **5**, 7050.
98. Y. Chen, C. W. Li and M. W. Kanan, *J. Am. Chem. Soc.*, 2012, **134**(49), 19969.
99. T. Hatsukade, K. P. Kuhl, E. R. Cave, D. N. Abram and T. F. Jaramillo, *Phys. Chem. Chem. Phys.*, 2014, **16**, 13814.
100. D. Chi, H. Yang, Y. Du, T. Lv, G. Sui, H. Wang and J. Lu, *RSC Adv.*, 2014, **4**, 37329.
101. X. Min and M. W. Kanan, *J. Am. Chem. Soc.*, 2015, **137**(14), 4701.
102. H. Mistry, A. S. Varela, C. S. Bonifacio, I. Zegkinoglou, I. Sinev, Y.-W. Choi, K. Kisslinger, E. A. Stach, J. C. Yang, P. Strasser and B. R. Cuenya, *Nat. Commun.*, 2016, **7**, 12123.

103. E. R. Cave, T. Hatsukade, J. H. Montoya, C. Shi, K. P. Kuhl, C. Hahn, D. N. Abram, J. K. Nørskov and T. F. Jaramillo, *Phys. Chem. Chem. Phys.*, 2017, **19**, 15856.
104. Q. Tang, Y. Lee, D.-Y. Li, W. Choi, C. W. Liu, C. D. Lee and D. Jiang, *J. Am. Chem. Soc.*, 2017, **139**(28), 9728.
105. Y. Peng, T. Wu, L. Sun, J. M. V. Nsanzimana, A. C. Fisher and X. Wang, *ACS Appl. Mater. Interfaces*, 2017, **9**(38), 32782.
106. L. Zeng, J. Shi, J. Luo and H. Chen, *J. Power Sources*, 2018, **398**, 83.
107. P. De Luna, R. Quintero-Bermudez and C. T. Dinh, *et al.*, *Nat. Catal.*, 2018, **1**, 103.
108. B. Qin, Y. Li, H. Wang, G. Yang, Y. Cao, H. Yu, Q. Zhang, H. Liang and F. Peng, *Nano Energy*, 2019, **60**, 43.
109. C.-T. Ding, *et al.*, *Science*, 2018, **360**, 783.
110. Y. Hori, Electrochemical CO_2 reduction on metal electrodes, in *Modern aspects of Electrochemistry*, ed. C. G. Vayenas; R. E. White and M. E. Gamboa-Aldeco, Springer, New York, NY, 2008; vol. 42, p. 89.
111. A. Bagger, W. Ju, A. S. Varela, P. Strasser and J. Rossmeisl, *Eur. J. Chem. Phys. Phys. Chem.*, 2017, **18**, 3266.
112. P. Zhu, C. Xia, C.-Y. Liu, K. Jiang, G. Gao, X. Zhang, Y. Xia, Y. Lei, H. N. Alshareef, T. P. Senftle and H. Wang, *Proc. Natl. Acad. Sci. U. S. A.*, 2021, **118**(2), e2010868118.
113. J. A. Rodriguez, *Surf. Sci. Rep.*, 1996, **24**, 223.
114. M. Gsell, P. Jakob and D. Menzel, *Science*, 1998, **280**, 717.
115. M. Karamad, V. Tripkovic and J. Rossmeisl, *ACS Catal.*, 2014, **4**, 2268.
116. H. A. Hansen, C. Shi, A. C. Lausche, A. A. Peterson and J. K. Nørskov, *Phys. Chem. Chem. Phys.*, 2016, **18**, 9194.
117. A. A. Peterson and J. K. Nørskov, *J. Phys. Chem. Lett.*, 2012, **3**, 251.
118. M. Zhang, Z. Hu, L. Gu, Q. Zhang, L. Zhang, Q. Song, W. Zhou and S. Hu, *Nano Res.*, 2020, **13**, 3206.
119. J. Li, P. Pršlja, T. Shinagawa, A. J. Martín Fernández, F. Krumeich, K. Artyushkova, P. Atanassov, A. Zitolo, Y. Zhou, R. García Muelas, N. López, J. Perez-Ramírez and F. Jaouen, *ACS Catal.*, 2019, **9**(11), 10426.
120. P. Pršlja and N. López, *ACS Catal.*, 2020, **11**(1), 88–94.
121. M. N. Hossain, P. Pršlja, C. Flox, N. Muthuswamy, H. Jiang, J. Sainio, A. M. Kannan, N. López and T. Kallio, *Appl. Catal., B*, 2022, **304**, 120863.
122. N. Ikemiya, K. Natsui, K. Nakata and Y. Einaga, *ACS Sustainable Chem. Eng.*, 2018, **6**, 8108.
123. L. S. Fernanda, F. L. Osmando, V. S. Elisama and R. Caue, *Curr. Opin. Electrochem.*, 2022, **32**, 100890.
124. G. L. Chai and Z. X. Guo, *Chem. Sci.*, 2016, **7**, 1268.
125. H. R. M. Jhong, C. E. Tornow, B. Smid, A. A. Gewirth, S. M. Lyth and P. J. A. Kenis, *ChemSusChem*, 2017, **10**, 1094.
126. B. Kumar, M. Asadi and D. Pisasale, *et al.*, *Nat. Commun.*, 2013, **4**, 2819.
127. T. F. Liu, S. Ali, Z. Lian, C. W. Si, D. S. Su and B. Li, *J. Mater. Chem. A*, 2018, **6**, 19998.

128. R. Zhai, L. Zhang, M. Gu, X. Zhao, B. Zhang, Y. Cheng and J. Zhang, *Small*, 2023, **19**, 2207840.
129. M. A. Ghausi, J. F. Xie, Q. H. Li, X. Y. Wang, R. Yang, M. X. Wu, Y. B. Wang and L. M. Dai, *Angew. Chem., Int. Ed.*, 2018, **57**, 13135.
130. X. Ma, J. Du, H. Sun, F. Ye, X. Wang, P. Xu, C. Hu, L. Zhang and D. Liu, *Appl. Catal., B*, 2021, **298**, 120543.
131. G. Wang, M. Liu, J. Jia, H. Xu, B. Zhao, K. Lai, C. Tu and Z. Wen, *ChemCatChem*, 2020, **12**(8), 2203.
132. J. Zhou, B. An, Z. Zhu, L. Wang and J. Zhang, *Inorg. Chem.*, 2022, **61**(16), 6073.
133. D. M. Fernandes, A. F. Peixoto and C. Freire, *Dalton Trans.*, 2019, **48**, 13508.
134. D. C. Wei, Y. Q. Liu, Y. Wang, H. L. Zhang, L. P. Huang and G. Yu, *Nano Lett.*, 2009, **9**, 1752.
135. H. B. Wang, T. Maiyalagan and X. Wang, *ACS Catal.*, 2012, **2**, 781.
136. R. Czerw, M. Terrones, J. C. Charlier, X. Blase, B. Foley, R. Kamalakaran, N. Grobert, H. Terrones, D. Tekleab, P. M. Ajayan, W. Blau, M. Ruhle and D. L. Carroll, *Nano Lett.*, 2001, **1**, 457.
137. H. B. Wang, T. Maiyalagan and X. Wang, *ACS Catal.*, 2012, **2**, 781.
138. J. Y. Xu, Y. H. Kan, R. Huang, B. S. Zhang, B. L. Wang, K. H. Wu, Y. M. Lin, X. Y. Sun, Q. F. Li, G. Centi and D. S. Su, *ChemSusChem*, 2016, **9**, 1085.
139. X. L. Zou, M. J. Liu, J. J. Wu, P. M. Ajayan, J. Li, B. L. Liu and B. I. Yakobson, *ACS Catal.*, 2017, **7**, 6245.
140. Y. J. Liu, J. X. Zhao and Q. H. Cai, *Phys. Chem. Chem. Phys.*, 2016, **18**, 5491.
141. T. F. Liu, S. Ali, Z. Lian, C. W. Si, D. S. Su and B. Li, *J. Mater. Chem. A*, 2018, **6**, 19998.
142. M. A. Ghausi, J. F. Xie, Q. H. Li, X. Y. Wang, R. Yang, M. X. Wu, Y. B. Wang and L. M. Dai, *Angew. Chem., Int. Ed.*, 2018, **57**, 13135.
143. X. Ma, J. Du, H. Sun, F. Ye, X. Wang, P. Xu, C. Hu, L. Zhang and D. Liu, *Appl. Catal., B*, 2021, **298**, 120543.
144. G. Wang, M. Liu, J. Jia and Z. Wen, *ChemCatChem*, 2020, **12**(8), 2203.
145. J. J. Wu, R. M. Yadav, M. J. Liu, P. P. Sharma, C. S. Tiwary, L. L. Ma, X. L. Zou, X. D. Zhou, B. I. Yakobson, J. Lou and P. M. Ajayan, *ACS Nano*, 2015, **9**, 5364.
146. P. P. Sharma, J. J. Wu, R. M. Yadav, M. J. Liu, C. J. Wright, C. S. Tiwary, B. I. Yakobson, J. Lou, P. M. Ajayan and X. D. Zhou, *Angew. Chem., Int. Ed.*, 2015, **54**, 13701.
147. J. J. Wu, M. J. Liu, P. P. Sharma, R. M. Yadav, L. L. Ma, Y. C. Yang, X. L. Zou, X. D. Zhou, R. Vajtai, B. I. Yakobson, J. Lou and P. M. Ajayan, *Nano Lett.*, 2016, **16**, 466.
148. X. Y. Lu, T. H. Tan, Y. H. Ng and R. Amal, *Chem. - Eur. J.*, 2016, **22**, 11991.
149. Y. J. Liu, J. X. Zhao and Q. H. Cai, *Phys. Chem. Chem. Phys.*, 2016, **18**, 5491.

150. J. Wu, S. Ma, J. Sun, J. Gold, C. Tiwary, B. Kim, L. Zhu, N. Chopra, I. N. Odeh, R. Vajtai, A. Z. Yu, R. Luo, J. Lou, G. Ding, P. J. Kenis and P. M. Ajayan, *Nat. Commun.*, 2016, **7**, 13869.
151. A. L. Shen, Y. Q. Zou, Q. Wang, R. A. W. Dryfe, X. B. Huang, S. Dou, L. M. Dai and S. Y. Wang, *Angew. Chem., Int. Ed.*, 2014, **53**, 10804.
152. H. R. M. Jhong, C. E. Tornow, B. Smid, A. A. Gewirth, S. M. Lyth and P. J. A. Kenis, *ChemSusChem*, 2017, **10**, 1094.
153. X. F. Sun, X. C. Kang, Q. G. Zhu, J. Ma, G. Y. Yang, Z. M. Liu and B. X. Han, *Chem. Sci.*, 2016, **7**, 2883.
154. W. L. Li, M. Seredych, E. Rodriguez-Castellon and T. J. Bandosz, *ChemSusChem*, 2016, **9**, 606.
155. W. L. Li, B. Herkt, M. Seredych and T. J. Bandosz, *Appl. Catal., B*, 2017, **207**, 195.
156. F. W. Li, M. Q. Xue, G. P. Knowles, L. Chen, D. R. MacFarlane and J. Zhang, *Electrochim. Acta*, 2017, **245**, 561.
157. H. Wang, J. Jia, P. F. Song, Q. Wang, D. B. Li, S. X. Min, C. X. Qian, L. Wang, Y. F. Li, C. Ma, T. Wu, J. Y. Yuan, M. Antonietti and G. A. Ozin, *Angew. Chem., Int. Ed.*, 2017, **56**, 7847.
158. S. Zhang, P. Kang, S. Ubnoske, M. K. Brennaman, N. Song, R. L. House, J. T. Glass and T. J. Meyer, *J. Am. Chem. Soc.*, 2014, **136**, 7845.
159. H. X. Wang, Y. B. Chen, X. L. Hou, C. Y. Ma and T. W. Tan, *Green Chem.*, 2016, **18**, 3250.
160. S. Zhang, P. Kang, S. Ubnoske, M. K. Brennaman, N. Song, R. L. House, J. T. Glass and T. J. Meyer, *J. Am. Chem. Soc.*, 2014, **136**, 7845.
161. Y. F. Song, W. Chen, C. C. Zhao, S. G. Li, W. Wei and Y. H. Sun, *Angew. Chem., Int. Ed.*, 2017, **56**, 10840.
162. W. Li, N. Fechler and T. J. Bandosz, *Appl. Catal., B*, 2018, **234**, 1.
163. J. Yuan, W. Y. Zhi, L. Liu, M. P. Yang, H. Wang and J. X. Lu, *Electrochim. Acta*, 2018, **282**, 694.
164. H. Q. Li, N. Xiao, M. Y. Hao, X. D. Song, Y. W. Wang, Y. Q. Ji, C. Liu, C. Li, Z. Guo, F. Zhang and J. S. Qiu, *Chem. Eng. J.*, 2018, **351**, 613.
165. Z. P. Chen, K. W. Mou, S. Y. Yao and L. C. Liu, *J. Mater. Chem. A*, 2018, **6**, 11236.
166. X. Mao and T. A. Hatton, *Ind. Eng. Chem. Res.*, 2015, **54**(16), 4033.
167. J. Hussain, E. Skúlason and H. Jónsson, *Procedia Comput. Sci.*, 2015, **51**, 1865.
168. X. L. Zou, M. J. Liu, J. J. Wu, P. M. Ajayan, J. Li, B. L. Liu and B. I. Yakobson, *ACS Catal.*, 2017, **7**, 6245.
169. P. Prslja and N. Lopez, *ACS Catal.*, 2021, **11**, 88.
170. M. M. Rodriguez, X. Peng, L. Liu, Y. Li and J. M. Andino, *J. Phys. Chem. C*, 2012, **116**(37), 19755.
171. M. Zhong, K. Tran, Y. Min, C. Wang, Z. Wang and C.-T. Dinh, *et al.*, *Nature*, 2020, **581**, 178.
172. H. Xiao, T. Cheng and W. A. Goddard, *J. Am. Chem. Soc.*, 2017, **139**, 130.
173. J. Hussain, E. Skúlason and H. Jónsson, *Procedia Comput. Sci.*, 2015, **51**, 1865.

174. T. Cheng, H. Xiao and W. A. Goddard, *Proc. Natl. Acad. Sci. U. S. A.*, 2017, **114**, 1795.
175. M. Isegawa and A. K. Sharma, *Sustainable Energy Fuels*, 2019, **3**, 1730.
176. K. Saravanan, Y. Basdogan, J. Dean and J. A. Keith, *J. Mater. Chem. A*, 2017, **5**, 11756.
177. N. J. Bernstein, S. A. M. Akhade and J. Janik, *Phys. Chem. Chem. Phys.*, 2014, **16**, 13708.
178. J. Dean, Y. Yang, N. Austin, G. Veser and G. Mpourmpakis, *ChemSusChem*, 2018, **11**, 1169.
179. J. Shen, M. J. Kolb, A. J. Göttle and M. T. M. Koper, *J. Phys. Chem. C*, 2016, **120**(29), 15714.
180. C. Guo, T. Zhang, X. Deng, X. Liang, W. Guo and X. Lu, *et al.*, *ChemSusChem*, 2019, **12**, 5126.
181. W. Luo, X. Nie, M. J. Janik and A. Asthagiri, *ACS Catal.*, 2016, **6**, 219.
182. J. D. Goodpaster, A. T. Bell and M. Head-Gordon, *J. Phys. Chem. Lett.*, 2016, **7**, 1471.
183. A. J. Garza, A. T. Bell and M. Head-Gordon, *ACS Catal.*, 2018, **8**, 1490.
184. T. K. Todorova, M. W. Schreiber and M. Fontecave, *ACS Catal.*, 2020, **10**, 1754.
185. M.-J. Cheng, Y. Kwon, M. Head-Gordon and A. T. Bell, *J. Phys. Chem. C*, 2015, **119**, 21345.
186. M.-J. Cheng, E. L. Clark, H. H. Pham, A. T. Bell and M. Head-Gordon, *ACS Catal.*, 2016, **6**, 7769.
187. Y. Meng, H. Huang, Y. Zhang, Y. Cao, H. Lu and X. Li, *Front. Chem.*, 2023, **11**, 1172146.

CHAPTER 8

Construction of Active Sites for CO_2 Photoreduction

WA GAO,*[a] YONG ZHUO*[b,c] AND ZHIGANG ZOU[b,c]

[a] School of Physical Science and Technology, Tiangong University, Tianjin, 300387, P. R. China; [b] School of Physics, Jiangsu Key Laboratory of Nanotechnology, Eco-materials and Renewable Energy Research Center (ERERC), National Laboratory of Solid State Microstructures, Collaborative Innovation Center of Advanced Microstructures, Nanjing University, Nanjing, 210093, P. R. China; [c] School of Science and Engineering, The Chinese University of Hongkong (Shenzhen), Shenzhen, Guangdong, 518172, P. R. China
*Emails: gaowa@tiangong.edu.cn; zhouyong1999@nju.edu.cn

8.1 Introduction

The rising concentration of CO_2 in the atmosphere has become a key factor contributing to global climate change, creating an urgent need for innovative technologies to mitigate its impact, achieve carbon neutrality, and drive sustainable development.[1-3] Among the various strategies explored, photocatalytic CO_2 reduction by harnessing solar energy has emerged as a promising approach for converting CO_2 into valuable chemicals and fuels.[4-8] To achieve efficient CO_2 conversion, it is crucial to enhance the light absorption capability of catalysts, promote the separation of photogenerated carriers, and construct effective active sites.[9,10]

Active sites are the specific regions on the surface of photocatalysts where the crucial steps of CO_2 adsorption, activation, and C–C coupling reactions occur.[11-13] Specifically, active sites can enhance the adsorption of CO_2 molecules onto the catalyst surface. A strong and selective interaction between CO_2 and the active site is essential for increasing the local concentration of CO_2, which is a critical first step in the reduction process. In addition, active sites play a vital role in the efficient separation and transfer of photogenerated electron–hole pairs. Well-designed active sites can trap electrons, preventing their recombination with holes, and guiding them toward CO_2 molecules, thus enhancing the reduction reaction. Crucially, the construction of active sites can significantly lower the activation energy required for CO_2 reduction. This is achieved by stabilizing reaction intermediates and transition states, thereby making the reduction process more thermodynamically and kinetically favorable. Additionally, the active sites selectively bind specific intermediates, regulating the reaction pathways to produce target products such as hydrocarbons or alcohols. This selectivity is crucial for achieving high yields of desired products while minimizing undesired by-products.

In recent years, significant research efforts have focused on understanding the nature of active sites and developing strategies to enhance their activity and selectivity.[14] These efforts include the incorporation of various metal and non-metal elements, the creation of surface defects, and the design of heterojunctions that optimize charge separation and transfer.[15] By tuning the atomic and electronic structures of these active sites, more efficient and selective CO_2 capture can be achieved, paving the way for the practical application of photocatalytic CO_2 reduction technology.

This chapter will provide a comprehensive overview of the principles and methods for constructing active sites for photocatalytic CO_2 reduction. It will delve into the impact of active site construction on CO_2 adsorption, activation, and conversion, as well as on C–C coupling, revealing the underlying mechanisms that enhance photocatalytic CO_2 conversion performance. By summarizing relevant research, this chapter aims to elucidate the critical role of constructing active sites with atomic precision in achieving efficient photocatalytic CO_2 conversion and highly selective C_{2+} product formation, while also highlighting the challenges that remain in optimizing performance, providing insights for the future design of efficient photocatalytic energy conversion materials.

8.2 Basic Principles of Photocatalytic CO_2 Reduction

CO_2 is one of the most thermodynamically stable carbon compounds, with a high dissociation energy required to break the C=O bond (750 kJ mol^{-1}), posing a challenge for CO_2 activation and conversion.[16–18] Photocatalytic CO_2 reduction involves three main processes: (1) generation of electron–hole pairs in the catalyst under sunlight, (2) separation and migration of carriers to the surface, and (3) activation and conversion of CO_2 at surface-active sites, which includes C–O bond cleavage, C–H bond formation, and C–C coupling.[19,20] Various factors, such as the electronic band structure, carrier dynamics, and the adsorption and desorption of intermediates, have a significant impact on the reaction pathway.[21,22]

8.2.1 Photogenerated Charge Carrier Separation

The ideal photocatalyst must possess an appropriate band structure for effective light absorption, as well as suitable conduction band (CB) and valence band (VB) positions to drive redox reactions. Specifically, upon exposure to light, the photocatalyst absorbs photons, leading to the excitation of electrons from the VB to the CB, generating electron–hole pairs.[23] These photogenerated carriers migrate to the surface of the catalyst without recombination to participate in the reduction (by electrons) and oxidation (by holes) reactions. Surface redox reactions, which generally occur over a timescale of microseconds, are much slower than rapid charge carrier recombination, which happens within picoseconds to nanoseconds. Consequently, the efficiency of surface catalytic reactions is strongly dependent on the charge dynamics occurring at the surface of the semiconductor. For effective CO_2 reduction, it is crucial that the photogenerated electrons reach the active sites on the catalyst surface before recombination with the holes occurs.[24] Nanostructuring semiconductors imparts unique features, including short charge migration paths, an increased surface-area-to-volume ratio, and tunable electronic properties, which collectively contribute to reducing bulk charge carrier recombination. However, if there are insufficient active sites on the surface to efficiently trap photogenerated carriers, surface recombination becomes significant, resulting in inadequate electron–hole separation. Strategies such as vacancy introduction, doping, and single-atom loading are widely used to reduce surface photogenerated carrier recombination and promote interfacial charge transfer. For example, through a molten salt homogeneous doping strategy, both the bulk and surface structures of the photocatalyst were simultaneously designed. Using two-dimension (2D) BiOCl nanosheets as a prototype model, B_2O_3 served as the molten salt and doping precursor, achieving dual functionality by uniformly doping B from the surface to the bulk.[25] Specifically, bulk B doping significantly reduced the binding energy (E_b) of trapped excitons in 2D BiOCl, thereby accelerating the dissociation of bulk excitons into free carriers. Meanwhile, the surface-doped B was shown to reconstruct the atomic surface of BiOCl and form tightly bound B-oxygen vacancy (B-OV) complexes. These unique B–OV complexes could spontaneously activate CO_2, suppress competitive hydrogen evolution, and facilitate the proton-coupled electron transfer step for selective CO generation by stabilizing *COOH.

8.2.2 CO_2 Adsorption and Activation

The adsorption of CO_2 onto the surface of the photocatalyst is a critical step in the reduction process. This step often involves the formation of intermediates such as $CO_2^{•-}$ (carbon dioxide anion radical), which are essential for further reduction into hydrocarbons or other products. However, the direct reduction of CO_2 through a single-electron transfer to produce the CO_2 anion radical ($CO_2^{•-}$) is highly unfavorable due to the requirement for an extremely negative redox potential, approximately −1.97 V *versus* the standard hydrogen electrode (SHE) in aprotic solvents like N,N'-dimethylformamide and around −1.90 V in water (pH = 7).[26] This significant challenge has driven the development of catalytic pathways that bypass the formation of $CO_2^{\bullet-}$ by utilizing proton-coupled multi-electron transfer processes, thereby enabling CO_2 reduction at more accessible energy levels.

The initial protonation of CO_2 can yield either formate (*OCHO) or carboxyl (*COOH) intermediates, which are subsequently reduced to various C_1-based fuels, with CO and HCOOH being common products of the two-electron reduction process.[27–29] It is important to note that the strength of intermediate adsorption plays a critical role in determining the resulting products. For example, a novel, chemically robust framework, FeTCP-OH-Co, incorporating metalloporphyrin-bridged metal-catechol, was developed to facilitate artificial photosynthesis.[30] Its selectivity for photocatalytic CO_2 reduction to formic acid was 97.8%, and the yield was 4.3 times and 15.7 times higher than that of FeTCP-Co and InTCP-Co, respectively. The primary factors contributing to the excellent performance included the synergistic effects of substituting Fe for In and the attachment of hydroxyl groups at the 2-position of catechol, which enhanced visible-light absorption, promoted charge separation, and led to the uniform distribution of redox-active sites and high CO_2 adsorption affinity. Additionally, the weak interaction between HCOOH and cobalt porphyrin, along with potential hydrogen bonding between HCOOH and hydroxyl groups, aided in the release of HCOOH from the porphyrin ring, thereby preventing its further hydrogenation.

Significant efforts have been dedicated to modifying the surface chemistry of catalysts to facilitate multi-electron reductions that yield products beyond two-electron species.[31–33] To achieve this, it is necessary to design the active site of the catalyst so that the desorption energy of C_1 intermediates such as CO is higher than the free energy required to produce more complex molecules such as CH_3OH and CH_4 by further proton-coupled electron transfer.[34] For instance, single Au atoms anchored on ultrathin $ZnIn_2S_4$ nanosheets ($Au_1/ZnIn_2S_4$) with the Au_1-S_2 structure demonstrated enhanced charge-carrier transfer, contributing to superior catalytic activity.[35] Moreover, these single Au atoms tended to form strong bonds with *CO intermediates, which significantly lowered the energy barrier for *CO protonation and stabilized the *CH_3 intermediate, thereby promoting selective CH_4 generation from CO_2 photoreduction. As a basic industrial raw material, the production of CH_3OH competes with CH_4 during CO_2 photoreduction. Specifically, *CO can be hydrogenated to CH_3OH without undergoing a dehydration step through successive electron and proton additions (*CO + H$^+$ + e$^-$ → *CHO + H$^+$ + e$^-$ → *CH_2O + H$^+$ + e$^-$ → *CH_3O + H$^+$ + e$^-$ → *CH_3OH). Core–shell covalently linked graphitic carbon nitride (C_3N_4)–melamine–resorcinol–formaldehyde microspheres promoted the stepwise hydrogenation of *CO to generate various CH_xO species, ultimately yielding CH_3OH.[36] The further hydrogenation of *CH_2OH to CH_4 and O has a theoretical energy barrier of 2.81 eV (Figure 8.1), which was higher than the

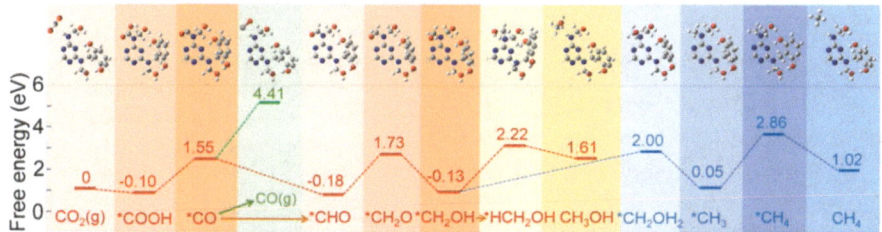

Figure 8.1 DFT calculations of potential intermediates in the reaction. *CO represents surface isocyanate and *COOH carbamic acid. Reproduced from ref. 36 with permission from American Chemical Society, Copyright 2022.

barrier for converting *CH_2OH to CH_3OH (2.35 eV). Therefore, CH_3OH formation was more favorable compared to CH_4 over this structure.

8.2.3 C–C Coupling

Producing C_{2+} fuels with high selectivity is inherently complex, requiring multiple steps, intermediates, and often leading to the formation of by-products.[37,38] Two C_1 species, such as *COOH, *CO, *CH_2, and *CHO radicals, can couple to form crucial *C_2 intermediates like *CO–CO, *CH_2–CH_2, *COOH–COOH, or *CO–CHO, depending on the coupling energy barrier and the accumulation of these intermediates within the catalytic system.[39] The enrichment of key reaction intermediates and the availability of sufficient charge carriers at active sites are essential conditions for successful C–C coupling.[40,41]

To improve the selectivity for the desired C_2 products, it is necessary to control the type and adsorption strength of intermediates, finely tune the catalytic pathways, and regulate the extent of the reduction reaction. For example, the high adsorption capacity and appropriate adsorption strength of CO over $Cu^{\delta+}$/CeO_2–TiO_2 significantly promoted the formation of the *CO–CO intermediate *via* *CO dimerization (as shown in Figure 8.2), thereby enhancing the activity and selectivity for C_2H_4 production.[42] For aldehyde production, the challenges lie in facilitating C–C bond formation while preserving the C=O bond throughout the reaction. For example, modified polymeric carbon nitride (PCN), with its distinctive bubble-like structure comprising alternating amorphous regions and well-crystallized lattice domains, achieved the selective photocatalytic hydrogenation of CO_2 to CH_3CHO.[43] Crucially, on locally crystallized PCN, the energy barrier for converting *CHO to the *OCCHO intermediate was substantially lowered, while the hydrogenation of *CHO to *CH_2O was hindered. This made the formation of the OCCHO intermediate more favorable, leading to the generation of CH_3CHO rather than HCHO through subsequent proton-coupled electron transfer. Additionally, the excessive protonation of CH_3CHO to CH_3CH_2OH was thermodynamically unfavorable on optimized PCN, with an

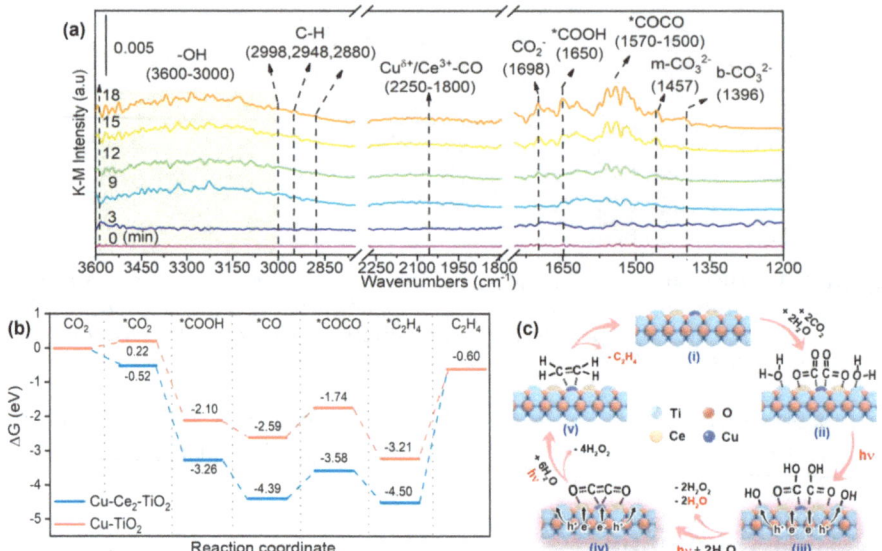

Figure 8.2 (a) *In situ* DRIFTS detections on the $Cu^{\delta+}/CeO_2$-TiO_2 photocatalyst in humid CO_2 atmosphere under irradiation. (b) Calculated free energy diagrams for CO_2 reduction over Cu–TiO_2 and Cu–Ce_2–TiO_2 slabs. (c) Proposed reaction mechanism for photocatalytic CO_2 to C_2H_4 conversion. Reproduced from ref. 42 with permission from American Chemical Society, Copyright 2022.

energy requirement of 1.13 V. The endothermic nature of this step limited further hydrogenation, resulting in highly selective CH_3CHO production.

8.3 Strategies for Creating Active Sites

The creation of active sites is regarded as a key approach commonly used to manipulate the physical and chemical properties of micro- and nanostructures, which in turn affects the efficiency and selectivity of CO_2 photoreduction. This section discusses typical methods for generating active sites, highlighting the effect of active sites on the performance of photocatalytic CO_2 reduction.

8.3.1 Hydro/Solvothermal Route

The hydro/solvothermal technique is a widely used method for synthesizing a variety of materials, particularly in the field of catalysis and nanotechnology. This technique involves conducting chemical reactions in a sealed environment at elevated temperatures and pressures, typically using water (hydrothermal) or organic solvents (solvothermal) as the reaction medium. The unique conditions provided by this method allow for the controlled growth of crystals, the formation of complex structures, and the precise incorporation of various elements into materials. It is considered energy-efficient and versatile, enabling the fabrication of advanced materials with specific properties, such as tailored active sites on catalysts, improved structural stability, and enhanced electronic or optical characteristics. As a typical example, single-atom Ag was doped into $CuInS_2$ using a hydrothermal method, accompanied by the formation of sulfur vacancies (V_s).[44] As a result, dual sites composed of V_s and adjacent In metal atoms were constructed, facilitating the conversion of CO_2 photoreduction products from C_1 compounds to C_2H_4. The silver doping acted as a trap and transfer agent for photogenerated electrons, thereby enhancing the reduction rate. Both experimental results and theoretical calculations confirmed that the key intermediates *CO and *CHO, adsorbed on the V_s and In sites respectively, promote the C–C coupling reaction due to their appropriate distance, asymmetric electron distribution, and strong adsorption capability. As another example, a Cu(I) single-atom modified $W_{18}O_{49}$ nanowire ($Cu_1/W_{18}O_{49}$) photocatalyst was prepared through a one-step hydrothermal synthesis, characterized by asymmetric Cu–W dual active sites.[45] The interaction between W(V) and W(VI) within $W_{18}O_{49}$ helped maintain the stability of Cu(I) during the photocatalytic reaction. The Cu(I) single atom stabilized the *CO intermediate, and the asymmetric Cu–W dual sites effectively reduced the energy barrier for C–C coupling of two neighboring CO intermediates (Figure 8.3), resulting in the highly selective photocatalytic conversion of CO_2 to C_2H_4.

8.3.2 Thermal Annealing

Annealing is a powerful technique for generating vacancies and metal active sites, with the process being heavily influenced by factors such as the gas environment, temperature, and heating rate. These variables play a crucial role in determining the physicochemical properties of the resulting semiconductor photocatalysts. For example, a single-atom catalyst featuring dual-atom sites (DAS) with adjacent Sn(II) and Cu(I) centers embedded in a C_3N_4 framework was synthesized through a one-pot annealing process using urea and metal [Sn(II) and Cu(I)] acetylacetonates.[46] The synergistic effect of the neighboring Sn(II)–Cu(I) diatomic sites stabilized the target intermediate for HCHO formation, resulting in an optimized catalyst that achieved an HCHO yield of 259.1 $\mu mol\,g^{-1}$ with a selectivity of 61% after 24 hours of irradiation. As another example, a dual-metal single-atom photocatalyst with manganese and cobalt active sites (Mn_1Co_1/CN) was synthesized on carbon nitride *via* annealing.[47] Both experimental findings and density functional theory (DFT) calculations suggested that the Mn active site enhanced H_2O oxidation by accumulating photogenerated holes. Concurrently, the Co active site aided in CO_2 activation by elongating the bond length and altering the bond angle of CO_2 molecules. By harnessing the synergistic interaction between these atomic active sites, Mn_1Co_1/CN achieved a CO production rate of 47 $\mu mol\,g^{-1}\,h^{-1}$.

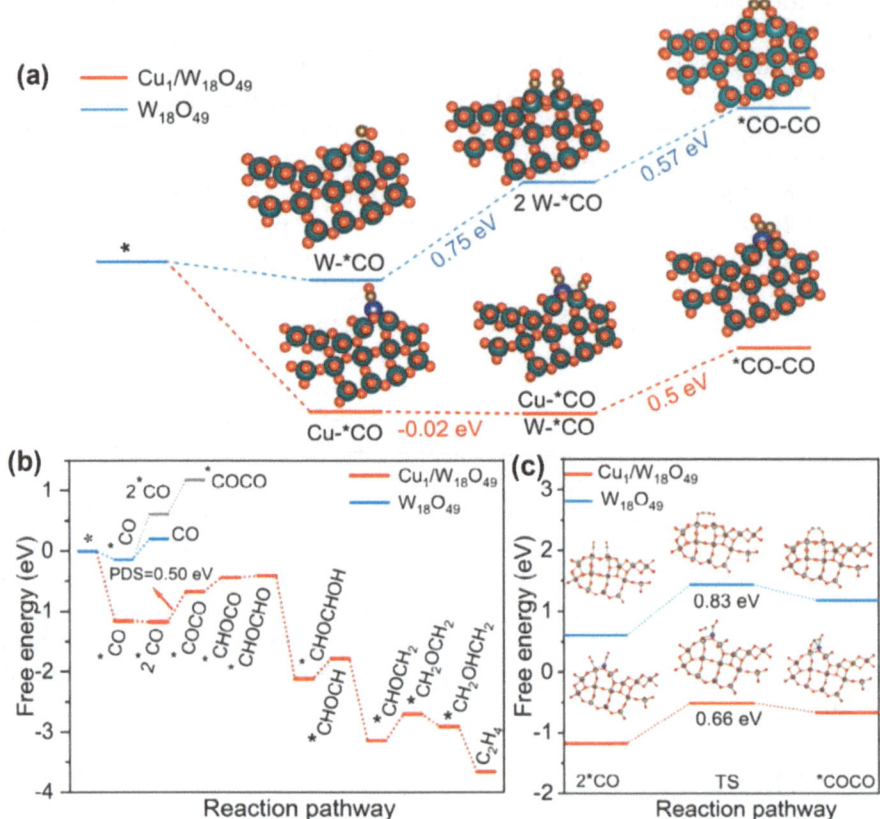

Figure 8.3 (a) Free energy changes of C–C coupling at different active sites. The corresponding structural model consists of Cu (blue), W (dark green), O (red), and C (gray) atoms. (b) The free energy diagram of CO_2 reduction on $Cu_1/W_{18}O_{49}$ and $W_{18}O_{49}$. The blue line shows the more favorable way, while the gray line shows the less favorable way over $W_{18}O_{49}$ catalyst. (c) Transition state energy barriers for C–C coupling on $W_{18}O_{49}$ and $Cu_1/W_{18}O_{49}$. Reproduced from ref. 45, https://doi.org/10.1002/advs.202401933, under the terms of the CC BY 4.0 license, https://creativecommons.org/licenses/by/4.0/.

8.3.3 Etching

Etching serves as an effective method to introduce vacancies into photocatalysts, enhancing photocatalytic CO_2 conversion. A post-etching technique was employed to create an inner-to-outer tandem homojunction in Bi_2WO_6 nanosheets, featuring a tungsten gradient vacancy layer (Vw-BWO).[17] This configuration generated a strong internal electric field, oriented from the exterior toward the interior of the Bi_2WO_6 nanosheets, which drove photoelectrons from the bulk to the surface of the catalyst. Additionally, the introduction of gradient tungsten vacancies modified the coordination environment around O and W atoms, altering the basic sites and shifting the CO_2 adsorption mode on the catalyst surface from weak/strong adsorption at O sites to moderate adsorption at both O and W sites. This adjustment reduced the formation barrier of the crucial *COOH intermediate, thereby improving the thermodynamics of CO_2 conversion. Analogously, an alkali-etching strategy was employed to synthesize a metallic defective $Bi_{19}Br_3S_{27}$ nanowire photocatalyst, V-$Bi_{19}Br_3S_{27}$, featuring abundant Br and S dual vacancies and surface Bi–O bonds.[48] The modified V-$Bi_{19}Br_3S_{27}$ nanowires demonstrated excellent photocatalytic CO_2 reduction performance by converting CO_2 into methanol. The interface reconstruction in V-$Bi_{19}Br_3S_{27}$ nanowires promoted the subsequent protonation of *CO rather than the formation of CO, thereby regulating product selectivity.

8.3.4 Ultrasonication

The ultrasonication technique is widely employed for the creation of active sites. For example, cerium oxide nanostructures with numerous surface defects were synthesized *via* ultrasonic irradiation in an alkaline medium, which enhanced methanol selectivity over CO.[49] Similarly, ultrasonication was used to generate V_s on $ZnIn_2S_4$, which acted as electron traps, thereby improving the separation of photoinduced carriers in the 2D/2D $Ti_3C_2T_x$/$ZnIn_2S_4$ heterostructure (TC/N-ZIS, Figure 8.4).[50] Additionally, low-dimensional materials with a high density of edge sites can also be produced using ultrasonication. For instance, black phosphorus quantum dots (BPQDs) were created through sonication-assisted liquid exfoliation of commercially available black phosphorus crystalline powders, leading to the formation of numerous marginal active sites.[51]

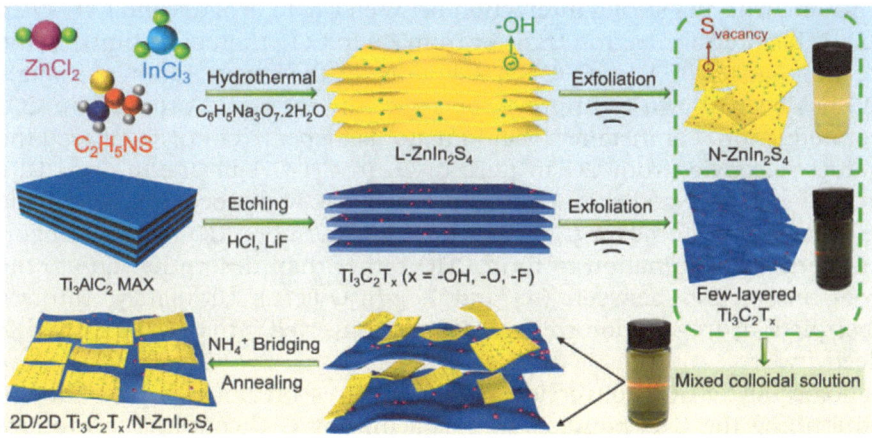

Figure 8.4 Schematic illustration of the synthetic route of layered $ZnIn_2S_4$ (L-$ZnIn_2S_4$), $ZnIn_2S_4$ nanosheet (N-$ZnIn_2S_4$), $Ti_3C_2T_x$, and TC/N-ZIS composite. Reproduced from ref. 50, https://doi.org/10.1002/advs.202103715, under the terms of the CC BY 4.0 license, https://creativecommons.org/licenses/by/4.0/.

8.4 Typical Active Sites for Selective Photocatalytic CO_2 Reduction

The photocatalytic reduction of CO_2 into valuable chemicals and fuels is a crucial technology for addressing global carbon emissions and achieving sustainable energy production. A central challenge in this process is the selective conversion of CO_2 into desired products, such as CO, CH_4, CH_3OH, and C_{2+}, while minimizing the formation of unwanted by-products. The properties and characteristics of active sites constructed on the surface of photocatalysts greatly influence the selectivity of CO_2 reduction products, affecting the formation and adsorption of intermediates, as well as the progression of C–C coupling reactions. This chapter explores the mechanisms by which typical active sites achieve selective control in photocatalytic CO_2 reduction, with a focus on the latest research progress related to the generation of C_{2+} products.

8.4.1 Metal Active Sites (MASs)

MASs are characterized by the exposure of metal atoms on a support material, offering a highly uniform and well-defined environment that is conducive to selective catalysis. Specifically, incorporating individual MAS into catalysts at the atomic scale greatly increases atomic utilization efficiency and enhances the effectiveness of photocatalytic CO_2 conversion. For example, single copper atoms anchored on nitrogen-doped carbon combined with TiO_2 achieved an exceptional 100% CO selectivity in photocatalytic CO_2 reduction.[52] These Cu atoms acted as crucial sites for CO_2 adsorption and activation, with the strong interaction between Cu 3d orbitals and CO_2–O 2p orbitals facilitating electron transfer from Cu to CO_2, thereby optimizing the rate-limiting step ($CO_2^* \rightarrow COOH^*$). Additionally, the incorporation of a single MAS has been shown to improve the selectivity for liquid products in CO_2 photoreduction. For instance, rhenium atoms dispersed in In_2O_3 altered the product selectivity from CO in pure In_2O_3 to CH_3OH in 2Re-In_2O_3.[53] DFT calculations suggested that Re sites promoted H_2O dissociation, supplying the necessary H atoms for CO_2 reduction. Furthermore, CO was more likely to undergo hydrogenation to form CHO rather than desorption due to the strong interaction between CO and Re_1–In_2O_3(111). Ultimately, through subsequent hydrogenation steps, CHO was converted into CH_3OH with high selectivity.

Photocatalytic CO_2 conversion mainly involves breaking the C=O bond and forming the C–H bond, as well as achieving C–C coupling to produce high-value C_{2+} products. MAS plays a crucial role in reducing electrostatic repulsion between C_1 intermediates, thereby facilitating C–C coupling and enabling the highly selective formation of C_{2+} products. An N, S-codoped Fe-MOF MIL-88B, with its well-defined bipyramidal hexagonal prism structure, was engineered to enhance this process.[54] The synergistic effect between Fe–N coordination sites and strategically introduced defects from uncoordinated sulfur increased the electron density disorder around Fe atoms, accelerated photogenerated carrier migration, improved electron storage, and effectively promoted the formation of C–C coupling intermediates, ultimately leading to C_2H_4 production. For liquid products, a Cu single-atom catalyst supported on carbon nitride, featuring a defective low-coordination Cu–N_2 motif (Cu–N_2–V), exhibited superior photocatalytic activity for CO_2 reduction to ethanol compared to Cu–N_3 and Cu–N_4 configurations.[55] In this motif, Cu existed in both Cu^+ and Cu^{2+} oxidation states. Cu^+ sites assisted in CO_2 activation, while the coexistence of Cu^+/Cu^{2+} sites enhanced *CO adsorption and facilitated *CO–*CO dimerization. As a result, ethanol was finally produced from the *CO–*CO dimer through subsequent hydrogenation steps.

Besides a single MAS, precisely designing photocatalysts with dual MASs that can simultaneously enhance both light absorption and catalytic efficiency is a challenging task. Specifically, a critical aspect of hydrocarbon formation is the controlled generation of $Metal_1\cdots C{=}O\cdots Metal_2$ ($M_1\cdots C{=}O\cdots M_2$) intermediates at the photocatalyst interface. This is because the energy required for the concurrent cleavage of $M_1\cdots O$ and $M_2\cdots C$ bonds is significantly greater than that needed to break the C–O bond. For example, $Ag_2Cu_2O_3$ nanowires, which possessed a high density of Cu–Ag Lewis acid–base dual sites on their predominantly exposed (110) surface, served as a model catalyst to achieve 100% selectivity for CH_4 production from CO_2 under light.[56] The $Cu\cdots Ag$ Lewis acid–base dual sites on $Ag_2Cu_2O_3$(110) effectively managed the formation of the $M_1\cdots C{=}O\cdots M_2$ intermediate, converting CO_2 into hydrocarbons. The rate-limiting step, with a Gibbs free energy [ΔG(CHO*)] of 0.75 eV, favored the conversion of CO* to CHO* over desorption from the catalyst surface (1.16 eV). Similarly, V_O-regulated In–Ti dual sites promoted the formation of a stable adsorption configuration of the In–C–O–Ti intermediate, resulting in the highly selective reduction of CO_2 to CH_4.[57]

Catalysts with dual MASs can manipulate electron distribution by forming asymmetric atomic configurations, which profoundly impacts photocatalytic efficiency, especially in multi-electron CO_2 reduction processes. A reverse electron transfer mechanism has been observed in Au and Co bimetallic atom catalysts, where electrons are delocalized from Au and concentrated around Co atoms.[13] This electron-rich environment at the Co sites enhanced their ability to adsorb and activate CO_2 molecules, significantly boosting photocatalytic CO_2 reduction. As a result, Au/Co double single-atom loaded CdS exhibited an increase in CO and CH_4 production by nearly 2800% and 700%, respectively, compared to CdS alone. Moreover, the different electron distributions and valence states of two distinct metal atoms can lead to varied charge distributions in neighboring C_1 intermediates, thereby reducing electrostatic repulsion and facilitating C–C coupling. In Vo-rich Zn_2GeO_4 nanobelts, the asymmetric Zn–O–Ge triatomic sites created unique charge distributions in adjacent C_1 intermediates, promoting C–C coupling and achieving a high conversion rate of CO_2 to CH_3COOH at 29.95%.[58] Similarly, incorporating redox-active Co^{2+}/Ni^{2+} cations (TM) into layered lead iodide hybrids [TJU-39(Pb)] resulted in efficient photocatalytic conversion of CO_2 to C_2H_5OH, with yields ranging from 24.9 to 31.4 μmol g^{-1} h^{-1} and selectivity exceeding 90%.[59] Experimental data indicated that the interlayer TMs were delocalized to the lead iodide layers, forming TM–O–Pd sites with significant asymmetric charge distribution, which lowered the energy barrier for C–C coupling. Specifically, the two-electron reduction of CO_2 to CO* occurred at the charge-rich Pb^{2+} and Ni^{2+} sites, which then underwent C–C coupling to form OC–CO* intermediates. These intermediates, through multiple e^-/H^+ transfer and a dehydration process, ultimately produced C_2H_5OH.

The inherent characteristics associated with the electron states of photocatalysts, such as the spin state of MAS, play a critical role in determining photocatalytic efficiency. For example, by altering the oxidation state of cobalt in the porphyrin center, the spin state of cobalt within covalent organic frameworks (COFs), like COF-367-Co, could be adjusted.[60] The ground states for Co^{II} and Co^{III} in COF-367 were found to be $S = 1/2$ and 0, respectively. As a result, the oxidation state of cobalt had a significant impact on its spin state, which affected the electron distribution in the Co-3d orbital and influenced its interactions with CO_2 and HCOOH. Notably, during CO_2 conversion, the energy barrier for the further conversion of HCOOH was substantially higher for COF-367-Co^{III} compared to COF-367-Co^{II} (1.20 vs. 0.97 eV), thereby preventing further reduction of HCOOH to CO or CH_4, leading to a much higher selectivity for HCOOH in the former. Furthermore, manipulating spin-polarized electrons in $CsPbBr_3$ halide perovskite nanoplates (NPLs) by doping manganese ions (Mn^{2+}) under an external magnetic field improved the photocatalytic CO_2 reduction performance.[61] As a result, Mn-doped $CsPbBr_3$ NPLs showed a 5.7-fold enhancement in CO_2 photoreduction activity under a 300 mT magnetic field from a permanent magnet, compared to undoped $CsPbBr_3$ NPLs. This improvement was attributed to an increase in spin-polarized photoexcited carriers, which extended the carrier lifetime and reduced charge recombination, thereby boosting the photocatalytic performance of Mn-$CsPbBr_3$ NPLs.

8.4.2 Defect Engineering

Defect engineering involves the deliberate introduction of vacancies, dislocations, or other structural defects into the photocatalyst to create active sites that influence selectivity. These engineered defects can alter the electronic band structure, trap photoexcited electrons to prevent recombination with holes, assist in the activation and conversion of CO_2 molecules, and influence the nature of the final products. For example, sulfur-deficient $CuIn_5S_8$ with dual charge-enriched Cu–In sites exhibited high selectivity for the photocatalytic reduction of CO_2 to CH_4.[62] In comparison, pristine $CuIn_5S_8$ single-unit-cell layers produced both CO and CH_4. The low-coordination, charge-enriched Cu and In atoms acted as dual-active sites, providing better stabilization of the rate-limiting COOH* intermediate than the single Cu sites found in pristine $CuIn_5S_8$ layers. This led to a lower activation energy barrier for CO_2 photoreduction in sulfur-deficient $CuIn_5S_8$. Additionally, these low-coordinated Cu and In atoms facilitated the spontaneous and exothermic formation of the CHO* radical (CO* + e^- + H^+ → CHO*), with a Gibbs free energy of CHO* formation less than 0. However, the desorption of CO* from these low-coordinated Cu and In atoms was an endothermic process with a substantial activation energy barrier, favoring further reduction and protonation of *CO to CH_4. As a result, V_S-$CuIn_5S_8$ single-unit-cell layers achieved nearly 100% selectivity for visible-light-driven CO_2 reduction to CH_4 instead of CO. Furthermore, the V_S-$CuIn_5S_8$ layers exhibited minimal decreases in CH_4 production rates even after 10 consecutive cycles, demonstrating their long-term photocatalytic stability.

Vacancies not only improve the generation of C_1 products but also facilitate the formation of C_{2+} products by enriching intermediates, lowering the C–C coupling barrier, and optimizing reaction pathways. In $AgInP_2S_6$, the introduction of V_s led to charge accumulation on the Ag atoms near these vacancies, with the exposed Ag sites efficiently capturing the *CO molecules generated during the reaction.[63] Photogenerated electrons in $AgInP_2S_6$ were drawn to these exposed Ag sites, encouraging the multi-electron conversion of CO_2. The stabilization and accumulation of *CO at the vacancy sites enable further protonation and coupling with other intermediates, rather than the desorption of *CO from the photocatalyst, thereby achieving high selectivity for C_{2+} products (Figure 8.5).

In addition to anion vacancies, cation vacancies are crucial in enhancing CO_2 photoreduction performance. For instance, rose-like BiOCl, rich in bismuth vacancies (V_{Bi}) and consisting of nanosheets with nearly fully exposed active (001) facets, demonstrated highly efficient photocatalytic CO_2 reduction directly from natural air.[64] These rose-like BiOCl structures with V_{Bi} offered numerous adsorption and catalytic sites, improving CO_2 capture and reduction capabilities, and thus accelerating photocatalytic CO_2 reduction with consistent stability over five catalytic cycles. As another example, in the full-spectrum-responsive metallic $ZnIn_2S_4$, which was rich in indium vacancies (V_{In}), the diffusion length of minority carriers was closely related to catalytic efficiency, underscoring the significant role of defects in carrier dynamics.[65] Moreover, the presence of V_{In} reduced the energy barrier for converting CO_2 to CO through the COOH* intermediate, resulting in a high CO production rate in V_{In}-rich $ZnIn_2S_4$ and an almost 28-fold increase compared to $ZnIn_2S_4$ with fewer V_{In}.

The four-electron water oxidation half-reaction can be thermodynamically challenging and may even be the rate-limiting step in CO_2 photoreduction when H_2O is used as the reducing agent. Semiconductors with vacancies demonstrate exceptional ability in modulating water activation and trapping holes, thereby accelerating the H_2O oxidation process and maximizing the use of active hydrogen for CO_2 reduction. Vacancies in SnS_2 atomically thin

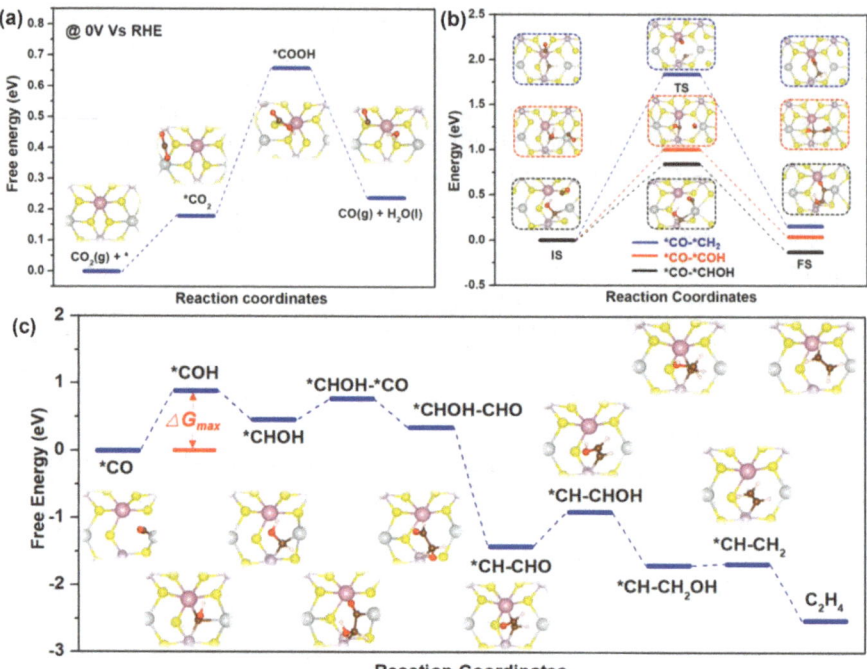

Figure 8.5 (a) Gibbs free energy diagrams for CO_2 reduction to CO over perfect $AgInP_2S_6$. (b) Three kinds of possible C–C coupling pathways over $AgInP_2S_6$ containing V_s. (c) Gibbs free energy diagrams for CO reduction to C_2H_4 over $AgInP_2S_6$ with V_s. The insets show the corresponding optimized geometries for the reaction intermediates during the CO_2 reduction process. Sulfur, phosphorus, indium, silver, carbon, oxygen, and hydrogen atoms are yellow, purple, lilac, gray, black, red, and white, respectively. Reproduced from ref. 63, https://doi.org/10.1038/s41467-021-25068-7, under the terms of the CC BY 4.0 license, https://creativecommons.org/licenses/by/4.0/.

layers (V_s-SnS_2) directly enhanced water oxidation during CO_2 conversion, resulting in improved CO_2 photoreduction efficiency.[66] *In situ* Fourier transform infrared spectroscopy revealed that the peak patterns related to water decomposition shift depending on the concentration of vacancies, highlighting their significant impact on the process. Kinetic studies further confirmed that water decomposition, the rate-limiting step, was indeed accelerated at vacancy sites by lowering the energy barriers for the conversion of $-H_2O$ to $-OH$ and the subsequent release of $-O_2$ from the surface. In another system, Cu single-atom centers and two-coordinated nitrogen vacancies ($N_{2C}V$) were integrated into a carbon nitride matrix ($Cu_1/N_{2C}V$-CN) as dual active sites.[67] The Cu single-atom centers enhanced CO_2 adsorption and activation by accumulating photogenerated electrons, while the $N_{2C}V$ sites promoted H_2O dissociation, facilitating the transformation of COO to COOH.

8.4.3 Edge Configurations

Edge sites, characterized by numerous unsaturated atoms or defects with highly disordered atomic structures, act as active centers for photocatalysis. The dangling bonds at these edges enhance molecular activation and facilitate interfacial reactions. For example, monolayer 2H-WSe$_2$ artificial leaves, featuring reconstructed and imperfect edge atoms, exhibited a strong preference for CO$_2$ adsorption.[68] The internal quantum efficiency of CO$_2$-to-CH$_4$ conversion was found to be inversely proportional to the perimeter of the monolayer 2H-WSe$_2$. Specifically, the edges behaved similarly to metallic regions, which were identified as the primary sites for charge transfer, as confirmed by nanoscale redox mapping of the WSe$_2$–liquid interface. In addition, edge configurations are conducive to the formation of C$_{2+}$ products. The heterojunction of BPQDs and WO$_3$ exhibited efficient solar-driven CO$_2$ conversion to CO, accompanied by a large amount of C$_2$H$_4$ production.[51] On both armchair (AC) and zigzag (ZZ) edges, two adsorbed *CO molecules naturally coupled into a *CO–CO configuration following complete structural optimization, with free energy barriers of -0.20 eV and -0.70 eV, respectively. The formation of *CO–COH species on the AC edge was identified as the potential-limiting step, with an onset potential as low as 0.30 V, which was significantly lower than that of other reported catalysts. These results indicated that the AC and ZZ edges of BP effectively promoted *CO molecular coupling, which then underwent further protonation to produce C$_2$H$_4$. As another example, single-layer CuInP$_2$S$_6$ sheets with a thickness of approximately 0.81 nm were successfully synthesized through simple mechanical exfoliation of the corresponding bulk material for photocatalytic CO$_2$ reduction.[69] The main product generated was C$_2$H$_4$, with a product selectivity of about 56.4% and an electron selectivity as high as 74.6%. The tandem synergistic effect of the charge-enriched Cu–In dual sites at the edges of the CuInP$_2$S$_6$ monolayer (ML) was the primary reason for the efficient conversion and high selectivity of C$_2$H$_4$ production (Figure 8.6). Due to the high barrier for the conversion of CO$_2$ to *COOH at the S atom-exposed surface sites of the ML, it was unable to drive CO$_2$ photoreduction. Under illumination, the edge In sites of the ML first converted CO$_2$ to *CO, which then migrated to the neighboring Cu sites for subsequent C–C coupling reactions, ultimately converting into C$_2$H$_4$, a process that was both thermodynamically and kinetically feasible. Additionally, the ultrathin structure of the ML shortened the distance for photogenerated carriers to transfer from the interior to the surface, suppressing electron–hole recombination, thereby allowing more electrons to survive and accumulate at the exposed active sites, leading to efficient CO$_2$ reduction.

8.4.4 Facet Engineering

The facets that make up the surface of semiconductors are critical factors that must be carefully engineered, as the atomic arrangements on the

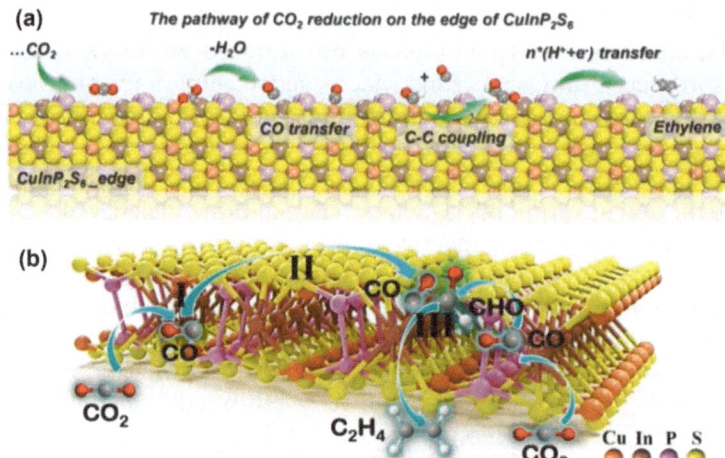
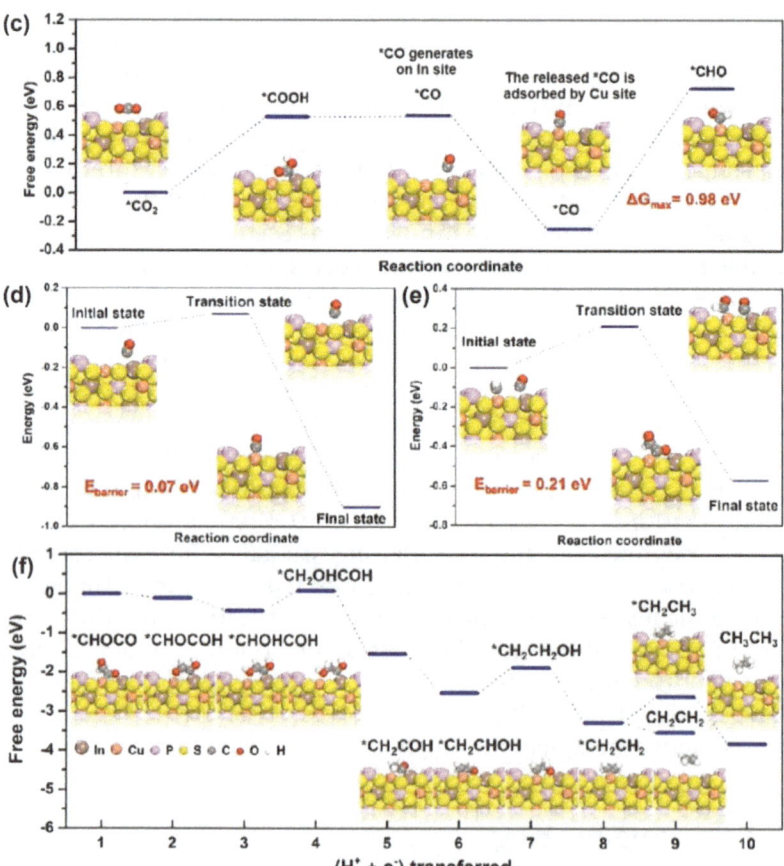

catalyst's surface have a direct impact on the adsorption and activation of reactants. Specifically, surface facets influence the electronic band structures, affecting the efficiency of charge separation and transfer. This may result in photogenerated electrons and holes naturally accumulating on different facets of the semiconductor due to thermodynamically favorable spatial charge separation. Colloidal CdS nanosheets with large basal planes terminated by S^{2-} atomic layers (CdS–S^{2-} NSs) were synthesized using a thermal decomposition approach.[70] These CdS–S^{2-} NSs enabled the migration of photoexcited electrons and holes toward spatially distinct reductive centers (at small edges) and oxidative centers (on the large basal planes), leading to exceptional photocatalytic performance in CO_2-to-CO conversion. This strategic distribution of redox sites not only improved the spatial separation of photoinduced electrons and holes but also enhanced the balanced extraction of these carriers by reducing the hole diffusion distance along the (001) direction of the ultrathin NSs. Additionally, this configuration supported the photostability of CdS–S^{2-}, with minimal changes observed in the absorption spectra. Similarly, polyhedral $BaTiO_3$ nanoparticles featuring (110)/(121) facet junctions exhibited significantly enhanced photocatalytic activity for the production of CO and CH_4 compared to octahedral $BaTiO_3$, which only had (121) facets.[71] DFT and photodeposition studies revealed that the band energy of polyhedral $BaTiO_3$ nanoparticles provided a cascade pathway for more efficient charge flow compared to their octahedral counterparts.

8.5 Conclusions and Perspectives

In the pursuit of sustainable energy solutions, photocatalytic CO_2 reduction offers a promising pathway to convert CO_2 into valuable fuels and chemicals under sunlight. The efficiency and selectivity of this process are profoundly influenced by the design and engineering of active sites on photocatalysts. Over the past decade, significant progress has been made in understanding and developing typical active sites, including MAS, vacancies, edge configuration, and facet exposure. Each of these active sites brings unique advantages in enhancing the selective reduction of CO_2 to desired products,

---

Figure 8.6 Schematic of the pathway for CO_2 reduction on the edge of $CuInP_2S_6$ from (a) top and (b) side view. Photocatalytic CO_2 to C_2H_4 *via* the synergistic effect from the dual sites of In and Cu atoms. The C–C coupling is carried out by *CO and *CHO dimerization. (c) CO_2 is reduced to CO on the edge of $CuInP_2S_6$, and (d) CO subsequently migrates to the Cu site and is reduced to CHO. (e) The kinetic process of CO transfer and C–C coupling. The insets represent reaction intermediates. (f) Gibbs free energy diagrams of the coupled *CO–CHO reduction to C_2H_4. The insets represent reaction intermediates. Reproduced from ref. 69 with permission from John Wiley & Sons, Copyright © 2023 Wiley-VCH GmbH.

such as CO, CH_4, CH_3OH, and C_{2+}. Although substantial progress has been made in CO_2 photoreduction, there remains significant room for growth and a range of challenges ahead.

First, photocatalytic CO_2 reduction yields various carbon-based products. While the generation of C_1 products such as CO, CH_4, and CH_3OH is well-established, fewer successes have been reported in the production of C_2 or even C_3 products. To address this issue, it's necessary to consider how to enrich *C_1 reaction intermediates, reduce C–C coupling barriers, and lessen electrostatic repulsion between *C_1 species. Additionally, producing C_{2+} products involves various competing reactions, multiple steps, and numerous intermediates, making control over the reaction's direction and extent a complex challenge. Therefore, further research is urgently needed on regulating active sites to promote C–C coupling, while preventing over-oxidation or over-reduction, to obtain target C_{2+} products with high selectivity.

Second, the construction of active sites has demonstrated that precise control over the electronic structure, coordination environment, and surface chemistry of the photocatalyst is essential for achieving high selectivity and efficiency in CO_2 reduction. In addition, the development of multifunctional and hybrid photocatalysts that combine the strengths of various active sites could offer new opportunities for enhancing the overall performance of CO_2 reduction systems. The ability to tailor these active sites to stabilize specific intermediates and favor certain reaction pathways has opened up new avenues for improving the performance of photocatalytic systems. However, challenges remain in scaling these technologies for practical applications. Moreover, the stability of active sites under operational conditions, and the optimization of light absorption and charge separation are critical areas that require further research.

Third, numerous *in situ* characterization experiments and theoretical calculations have played a crucial role in identifying active sites and probing the catalytic mechanisms involved. To further advance this field, there is a need to explore more sophisticated atomic-scale characterization techniques capable of monitoring the fundamental steps of catalytic reactions in real time. Additionally, theoretical calculations have mapped out reaction pathways at the active catalytic sites, offering important insights into the photocatalytic process. The reliability of these calculations hinges on accurately simulating real reaction conditions within the theoretical models. Combining theoretical calculations with experimental characterization allows for a comprehensive understanding of the CO_2 conversion mechanisms.

Fourth, machine learning techniques are increasingly being applied to support both experimental and computational studies. By generating predictive models, machine learning can forecast the structure and configuration of active sites for potential photocatalysts. This aids in uncovering the structure–activity relationship, providing guidance for catalyst design, and clarifying the connection between material structure, physicochemical properties, and functional performance. With machine learning, active site identification can be prioritized before active site design, ensuring high precision and accuracy.

In conclusion, the ongoing advancements in the design and understanding of active sites for photocatalytic CO_2 reduction are laying the groundwork for more efficient and selective catalytic systems. Continued research and innovation in this area hold great promise for addressing global carbon challenges and contributing to the development of sustainable energy technologies.

Acknowledgements

The authors wish to acknowledge the support of the NSF of China (22202152 and 224720022), the NSF of Jiangsu Province (No. BK20220006), the Program for Guangdong Introducing Innovative and Entrepreneurial Team (2019ZT08L101), Tianjin Municipal Science and Technology Bureau (No. 24JCQNJC00990), and the University Development Fund (UDF01001159).

References

1. X. D. Li, L. Li, G. B. Chen, X. Y. Chu, X. H. Liu, C. Naisa, D. Pohl, M. Löffler and X. L. Feng, *Nat. Commun.*, 2023, **14**, 4034.
2. J. O. Olowoyo, V. S. Gharahshiran, Y. M. Zeng, Y. Zhao and Y. Zheng, *Chem. Soc. Rev.*, 2024, **53**, d3cs00759f.
3. S. W. Wang, L. G. Wang, D. S. Wang and Y. D. Li, *Energy Environ. Sci.*, 2023, **16**, 2759–2803.
4. Y. Liu, J. H. Sun, H. H. Huang, L. L. Bai, X. M. Zhao, B. H. Qu, L. Q. Xiong, F. Q. Bai, J. W. Tang and L. Q. Jing, *Nat. Commun.*, 2023, **14**, 1457.
5. Y. Wang, Y. Liu, L. Wang, S. Perumal, H. Wang, H. Ko, C. L. Dong, P. P. Zhang, S. J. Wang, T. T. T. Nga, Y. D. Kim, Y. Ji, S. Zhao, J. H. Kim, D. Y. Yee, Y. Hwang, J. Q. Zhang, M. G. Kim and H. Lee, *Nat. Commun.*, 2024, **15**, 6047.
6. J. S. Chen, D. Q. Yuan and Y. B. Wang, *Adv. Funct. Mater.*, 2023, **33**, 202304071.
7. Q. J. Tang, T. H. Li, W. G. Tu, H. Q. Wang, Y. Zhou and Z. G. Zou, *Adv. Funct. Mater.*, 2024, **34**, 202311609.
8. L. Wang, L. Wang, Y. K. Xu, G. X. Sun, W. C. Nie, L. H. Liu, D. B. Kong, Y. Pan, Y. H. Zhang, H. Wang, Y. C. Huang, Z. Liu, H. Ren, T. Wei, Y. Himeda and Z. J. Fan, *Adv. Mater.*, 2024, **36**, 202309376.
9. W. Gao, H. Q. Chi, Y. J. Xiong, J. H. Ye, Z. G. Zou and Y. Zhou, *Adv. Funct. Mater.*, 2024, **34**, 202312056.
10. B. Wang, H. L. Chen, W. Zhang, H. Y. Liu, Z. K. Zheng, F. C. Huang, J. Y. Liu, G. P. Liu, X. W. Yan, Y. X. Weng, H. M. Li, Y. B. She, P. K. Chu and J. X. Xia, *Adv. Mater.*, 2024, **36**, 202312676.
11. Q. Y. Wang, Y. D. Zhang, M. X. Lin, H. W. Wang, Y. Bai, C. Y. Liu, J. L. Lu, Q. Q. Luo, G. M. Wang, H. L. Jiang, T. Yao and X. S. Zheng, *Adv. Energy Mater.*, 2023, **13**, 202302692.

12. H. X. Li, R. J. Li, G. Liu, M. L. Zhai and J. G. Yu, *Adv. Mater.*, 2024, **36**, 202301307.
13. Y. Z. Zhang, B. Johannessen, P. Zhang, J. L. Gong, J. R. Ran and S. Z. Qiao, *Adv. Mater.*, 2023, **35**, 202306923.
14. T. Q. Guo, X. X. Xu, Z. F. Xu, F. F. You, X. Y. Fan, J. Z. Liu and Z. C. Wang, *Adv. Mater.*, 2024, **36**, 202402071.
15. H. M. Liang, C. C. Ye, J. Xiong, G. Z. Hao, J. Lei, W. J. Bai, K. Zhang, W. Jiang and J. Di, *ACS Nano*, 2024, **18**, 21585–21592.
16. X. Y. Deng, J. J. Zhang, K. Z. Qi, G. J. Liang, F. Y. Xu and J. G. Yu, *Nat. Commun.*, 2024, **15**, 4807.
17. Y. H. Wang, J. C. Hu, T. Ge, F. Chen, Y. Lu, R. H. Chen, H. J. Zhang, B. J. Ye, S. Y. Wang, Y. H. Zhang, T. Y. Ma and H. W. Huang, *Adv. Mater.*, 2023, **35**, 202302538.
18. J. Di, C. Chen, Y. Wu, H. Chen, J. Xiong, R. Long, S. Z. Li, L. Song, W. Jiang and Z. Liu, *Adv. Mater.*, 2024, **36**, 202401914.
19. Z. K. Xie, S. J. Xu, L. H. Li, S. H. Gong, X. J. Wu, D. B. Xu, B. D. Mao, T. Zhou, M. Chen, X. Wang, W. D. Shi and S. Y. Song, *Nat. Commun.*, 2024, **15**, 2422.
20. K. Kosugi, C. Akatsuka, H. Iwami, M. Kondo and S. Masaoka, *J. Am. Chem. Soc.*, 2023, **145**, 10451–10457.
21. J. X. Liang, H. Yu, J. J. Shi, B. Li, L. X. Wu and M. Wang, *Adv. Mater.*, 2023, **35**, 202209814.
22. H. W. Huang, J. W. Zhao, H. L. Guo, B. Weng, H. W. Zhang, R. A. Saha, M. L. Zhang, F. L. Lai, Y. F. Zhou, R. Z. Juan, P. C. Chen, S. B. Wang, J. A. Steele, F. L. Zhong, T. X. Liu, J. Hofkens, Y. M. Zheng, J. L. Long and M. B. J. Roeffaers, *Adv. Mater.*, 2024, **36**, 202313209.
23. G. R. Jia, Y. C. Zhang, J. C. Yu and Z. X. Guo, *Adv. Mater.*, 2024, **36**, 202403153.
24. C. H. Chiang, C. C. Lin, Y. C. Lin, C. Y. Huang, C. H. Lin, Y. J. Chen, T. R. Ko, H. L. Wu, W. Y. Tzeng, S. Z. Ho, Y. C. Chen, C. H. Ho, C. J. Yang, Z. W. Cyue, C. L. Dong, C. W. Luo, C. C. Chen and C. W. Chen, *J. Am. Chem. Soc.*, 2024, **146**, 23278–23288.
25. Y. B. Shi, G. M. Zhan, H. Li, X. B. Wang, X. F. Liu, L. J. Shi, K. Wei, C. C. Ling, Z. L. Li, H. Wang, C. L. Mao, X. Liu and L. Z. Zhang, *Adv. Mater.*, 2021, **33**, 202100143.
26. Z. Y. Sun, T. Ma, H. C. Tao, Q. Fan and B. X. Han, *Chem*, 2017, **3**, 560–587.
27. J. M. Zhang, D. X. Shi, J. Y. Yang, L. L. Duan, P. F. Zhang, M. B. Gao, J. L. He, Y. L. Gu, K. Lan, J. W. Zhang, J. Liu, D. Y. Zhao and Y. Z. Ma, *Adv. Mater.*, 2024, **36**, 2409188.
28. S. Hu, P. Z. Qiao, X. L. Yi, Y. M. Lei, H. L. Hu, J. H. Ye and D. F. Wang, *Angew. Chem., Int. Ed.*, 2023, **62**, 202304585.
29. F. Kuttassery, Y. Ohsaki, A. Thomas, R. Kamata, Y. Ebato, H. Kumagai, R. Nakazato, A. Sebastian, S. Mathew, H. Tachibana, O. Ishitani and H. Inoue, *Angew. Chem., Int. Ed.*, 2023, **62**, 202308956.
30. E. X. Chen, M. Qiu, Y. F. Zhang, L. He, Y. Y. Sun, H. L. Zheng, X. Wu, J. Zhang and Q. Lin, *Angew. Chem., Int. Ed.*, 2022, **61**, e202111622.

31. Y. Chai, Y. H. Kong, M. Lin, W. Lin, J. N. Shen, J. L. Long, R. S. Yuan, W. X. Dai, X. X. Wang and Z. Z. Zhang, *Nat. Commun.*, 2023, **14**, 6168.
32. S. H. Yang, W. J. Byun, F. M. Zhao, D. W. Chen, J. W. Mao, W. Zhang, J. Peng, C. Y. Liu, Y. Pan, J. Hu, J. F. Zhu, X. L. Zheng, H. Y. Fu, M. L. Yuan, H. Chen, R. X. Li, M. Zhou, W. Che, J. B. Baek, J. S. Lee and J. Q. Xu, *Adv. Mater.*, 2024, **36**, 202312616.
33. W. Y. Zhang, C. Y. Deng, W. Wang, H. Sheng and J. C. Zhao, *Adv. Mater.*, 2024, **36**, 202405825.
34. J. C. Wu, F. Huang, Q. Y. Hu, D. P. He, W. X. Liu, X. D. Li, W. S. Yan, J. Hu, J. F. Zhu, S. Zhu, Q. X. Chen, X. C. Jiao and Y. Xie, *J. Am. Chem. Soc.*, 2024, **146**, 26478–26484.
35. S. H. Si, H. W. Shou, Y. Y. Mao, X. L. Bao, G. Y. Zhai, K. P. Song, Z. Y. Wang, P. Wang, Y. Y. Liu, Z. K. Zheng, Y. Dai, L. Song, B. B. Huang and H. F. Cheng, *Angew. Chem., Int. Ed.*, 2022, **61**, 202209446.
36. J. Ding, Q. L. Tang, Y. H. Fu, Y. L. Zhang, J. M. Hu, T. Li, Q. Zhong, M. H. Fan and H. H. Kung, *J. Am. Chem. Soc.*, 2022, **144**, 9576–9585.
37. Q. H. Yang, H. Liu, Y. C. Lin, D. S. Su, Y. L. Tang and L. Chen, *Adv. Mater.*, 2024, **36**, 202310912.
38. B. Ni, G. Zhang, H. Wang, Y. Min, K. Jiang and H. Li, *Angew. Chem., Int. Ed.*, 2023, **62**, e202215574.
39. R. Xu, D. H. Si, S. S. Zhao, Q. J. Wu, X. S. Wang, T. F. Liu, H. Zhao, R. Cao and Y. B. Huang, *J. Am. Chem. Soc.*, 2023, **145**, 8261–8270.
40. N. Y. Huang, B. Li, D. Wu, Z. Y. Chen, B. Shao, D. Chen, Y. T. Zheng, W. Wang, C. Yang, M. Gu, L. Li and Q. Xu, *Angew. Chem., Int. Ed.*, 2024, **63**, e202319177.
41. M. R. Zhang, D. Zhang, X. Jing, B. J. Xu and C. Y. Duan, *Angew. Chem., Int. Ed.*, 2024, **63**, 202402755.
42. T. Wang, L. Chen, C. Chen, M. T. Huang, Y. J. Huang, S. J. Liu and B. X. Li, *ACS Nano*, 2022, **16**, 2306–2318.
43. Q. Liu, H. Cheng, T. X. Chen, T. W. B. Lo, Z. M. Xiang and F. X. Wang, *Energy Environ. Sci.*, 2022, **15**, 225–233.
44. Y. Xu, P. Wang, M. Zhang, W. L. Dai, Y. X. Xu, J. P. Zou and X. B. Luo, *Energy Environ. Sci.*, 2024, **17**, 5060–5069.
45. Y. Y. Mao, M. H. Zhang, G. Y. Zhai, S. H. Si, D. Liu, K. P. Song, Y. Y. Liu, Z. Y. Wang, Z. K. Zheng, P. Wang, Y. Dai, H. F. Cheng and B. B. Huang, *Adv. Sci.*, 2024, **11**, 202401933.
46. B. Kim, D. Kwon, J. O. Baeg, P. M. Austeria, G. H. Gu, J. H. Lee, J. Jeong, W. Kim and W. Choi, *Adv. Funct. Mater.*, 2023, **33**, 202212453.
47. H. H. Ou, S. B. Ning, P. Zhu, S. H. Chen, A. Han, Q. Kang, Z. F. Hu, J. H. Ye, D. S. Wang and Y. D. Li, *Angew. Chem., Int. Ed.*, 2022, **61**, 202206579.
48. J. Li, W. F. Pan, Q. Y. Liu, Z. Q. Chen, Z. J. Chen, X. Z. Feng and H. Chen, *J. Am. Chem. Soc.*, 2021, **143**, 6551–6559.
49. G. C. S. Shekar, K. Alkanad, B. Thejaswini, G. Alnaggar, N. Al Zaqri, Q. A. Drmosh, A. Boshaala and N. K. Lokanath, *Surf. Interfaces*, 2022, **34**, 102389.

50. T. M. Su, C. Z. Men, L. Y. Chen, B. X. Chu, X. Luo, H. B. Ji, J. H. Chen and Z. Z. Qin, *Adv. Sci.*, 2022, **9**, 202103715.
51. W. Gao, X. W. Bai, Y. Y. Gao, J. Q. Liu, H. C. He, Y. Yang, Q. T. Han, X. Y. Wang, X. L. Wu, J. L. Wang, F. T. Fan, Y. Zhou, C. Li and Z. G. Zou, *Chem. Commun.*, 2020, **56**, 7777–7780.
52. H. B. Yin, F. Dong, D. S. Wang and J. H. Li, *ACS Catal.*, 2022, **12**, 14096–14105.
53. C. Y. Shen, X. Y. Meng, R. Zou, K. H. Sun, Q. L. Wu, Y. X. Pan and C. J. Liu, *Angew. Chem., Int. Ed.*, 2024, **63**, 202402369.
54. F. Guo, R. X. Li, S. Yang, X. Y. Zhang, H. Yu, J. J. Urban and W. Y. Sun, *Angew. Chem., Int. Ed.*, 2023, **62**, e202216232.
55. H. N. Shi, Y. Liang, J. A. Hou, H. Z. Wang, Z. H. Jia, J. M. Wu, F. Song, H. Yang and X. W. Guo, *Angew. Chem., Int. Ed.*, 2024, **63**, 202404884.
56. S. M. Deng, R. H. Wang, X. Z. Feng, R. J. Zheng, S. K. Gong, X. H. Chen, Y. Z. Shangguan, L. L. Deng, H. Tang, H. Dai, L. L. Duan, C. Y. Liu, Y. Pan and H. Chen, *Angew. Chem., Int. Ed.*, 2023, **62**, 202309625.
57. C. Chen, L. Chen, Y. G. Hu, K. Yan, T. Wang, Y. J. Huang, C. Gao, J. J. Mao, S. J. Liu and B. X. Li, *J. Energy Chem.*, 2023, **86**, 599–608.
58. J. C. Zhu, W. W. Shao, X. D. Li, X. C. Jiao, J. F. Zhu, Y. F. Sun and Y. Xie, *J. Am. Chem. Soc.*, 2021, **143**, 18233–18241.
59. J. L. Yin, D. Y. Li, C. Sun, Y. L. Jiang, Y. K. Li and H. H. Fei, *Adv. Mater.*, 2024, **36**, 202403651.
60. Y. N. Gong, W. H. Zhong, Y. Li, Y. Z. Qiu, L. R. Zheng, J. Jiang and H. L. Jiang, *J. Am. Chem. Soc.*, 2020, **142**, 16723–16731.
61. C. C. Lin, T. R. Liu, S. R. Lin, K. M. Boopathi, C. H. Chiang, W. Y. Tzeng, W. H. C. Chien, H. S. Hsu, C. W. Luo, H. Y. Tsai, H. A. Chen, P. C. Kuo, J. Shiue, J. W. Chiou, W. F. Pong, C. C. Chen and C. W. Chen, *J. Am. Chem. Soc.*, 2022, **144**, 15718–15726.
62. X. D. Li, Y. F. Sun, J. Q. Xu, Y. J. Shao, J. Wu, X. L. Xu, Y. Pan, H. X. Ju, J. F. Zhu and Y. Xie, *Nat. Energy*, 2019, **4**, 690–699.
63. W. Gao, S. Li, H. C. He, X. N. Li, Z. X. Cheng, Y. Yang, J. L. Wang, Q. Shen, X. Y. Wang, Y. J. Xiong, Y. Zhou and Z. G. Zou, *Nat. Commun.*, 2021, **12**, 4747.
64. L. Wang, R. Y. Wang, T. Y. Qiu, L. Q. Yang, Q. T. Han, Q. Shen, X. Zhou, Y. Zhou and Z. G. Zou, *Nano Lett.*, 2021, **21**, 10260–10266.
65. Y. Q. He, C. L. Chen, Y. X. Liu, Y. L. Yang, C. G. Li, Z. Shi, Y. Han and S. H. Feng, *Nano Lett.*, 2022, **22**, 4970–4978.
66. S. K. Yin, X. X. Zhao, E. H. Jiang, Y. Yan, P. Zhou and P. W. Huo, *Energy Environ. Sci.*, 2022, **15**, 1556–1562.
67. Y. Y. Duan, Y. Wang, W. X. Zhang, J. W. Zhang, C. G. Ban, D. M. Yu, K. Zhou, J. J. Tang, X. Zhang, X. D. Han, L. Y. Gan, X. P. Tao and X. Y. Zhou, *Adv. Funct. Mater.*, 2023, **33**, 202301729.
68. M. Qorbani, A. Sabbah, Y. R. Lai, S. Kholimatussadiah, S. Quadir, C. Y. Huang, I. Shown, Y. F. Huang, M. Hayashi, K. H. Chen and L. C. Chen, *Nat. Commun.*, 2022, **13**, 1755.

69. W. Gao, L. Shi, W. T. Hou, C. Ding, Q. Liu, R. Long, H. Q. Chi, Y. C. Zhang, X. Y. Xu, X. Y. Ma, Z. Tang, Y. Yang, X. Y. Wang, Q. Shen, Y. J. Xiong, J. L. Wang, Z. G. Zou and Y. Zhou, *Angew. Chem., Int. Ed.*, 2024, **63**, 202317852.
70. N. Wang, S. Cheong, D. E. Yoon, P. Lu, H. Lee, Y. K. Lee, Y. S. Park and D. C. Lee, *J. Am. Chem. Soc.*, 2022, **144**, 16974–16983.
71. W. H. Cai, Y. B. Wang, L. Zhao, X. Sun, J. Xu, J. Chen, R. C. Shi, P. H. Ma and M. D. Que, *J. Mater. Chem. A*, 2023, **11**, 21746–21753.

CHAPTER 9

Heterogeneous Catalysts in Ammonia Synthesis

MASAAKI KITANO AND HIDEO HOSONO*

MDX Research Center for Element Strategy, Institute of Integrated Research, Institute of Science Tokyo, 4259 Nagatsuta, Midori-ku, Yokohama 226-8503, Japan
*Email: hosono.h.aa@m.titech.ac.jp

9.1 Introduction

Artificial transformation of dinitrogen (N_2) into ammonia (NH_3) is indispensable for life since NH_3 is the key chemical to produce fertilizers and various nitrogen-containing chemicals.[1] The industrial NH_3 synthesis process, the so-called Haber–Bosch (HB) process, was established in 1913. Currently, worldwide NH_3 production reaches approximately 180 million tons per year. In this industrial process, H_2 comes from steam reforming of hydrocarbons such as natural gas, which is conducted at very high temperatures (>800 °C) in a robust and large-scale plant. As a result, HB NH_3 production consumes 1–2% of the global energy demand and emits 1.5 tons of CO_2 per ton of ammonia. As shown in Figure 9.1, low temperature and high pressure are thermodynamically favorable to obtain a high concentration of NH_3, while a high reaction temperature is required because the reaction is kinetically limited by the dissociation of N_2 which possesses a strong triple bond (945 kJ mol^{-1}).[2] As a result, NH_3 synthesis from N_2 and H_2 is typically conducted at high operating temperatures (400–500 °C) and pressures (10–30 MPa) over an Fe-based catalyst.[3,4] However, exothermic NH_3 synthesis combined with endothermic H_2 production makes the HB process energy efficient for mass production of ammonia.

Catalysis Series No. 49
Catalytic Activation of Small Molecules
Edited by Mustafa Yasin Aslan, Angela Daisley, Justin S. J. Hargreaves and José L. Rico
© The Royal Society of Chemistry 2025
Published by the Royal Society of Chemistry, www.rsc.org

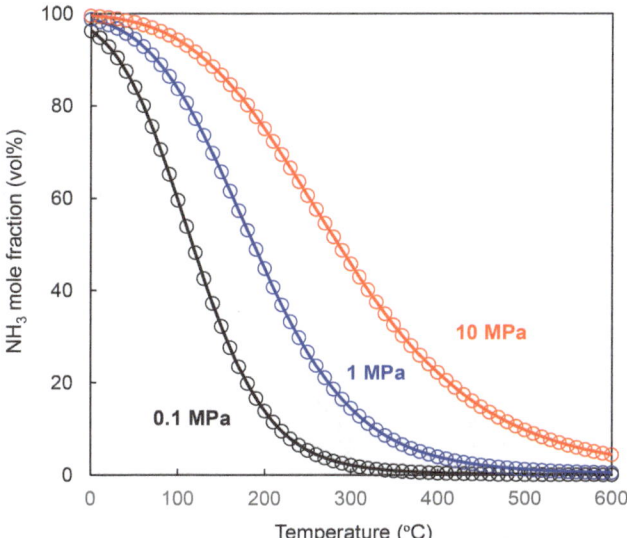

Figure 9.1 Temperature dependence of equilibrium NH_3 production from N_2 and H_2 ($N_2:H_2 = 1:3$).

Recently, NH_3 has attracted significant attention as a promising hydrogen energy carrier because of its high volumetric and gravimetric hydrogen capacity and ease of liquefaction under mild conditions.[5] Accordingly, the development of a green ammonia synthesis process, which utilizes H_2 produced from renewable energies such as wind and solar light, is now becoming a hot topic globally.[6] To realize this new process, the development of new catalysts that can function under lower temperatures and pressures than the Haber–Bosch process is strongly desired.

For typical heterogeneous ammonia synthesis catalysts, a volcano-shaped relationship relating catalytic activity with metal–nitrogen binding energy is a general guide for research since N_2 dissociation is the rate-determining step (RDS).[7–10] According to this relationship, Fe and Ru work as efficient catalysts among the single transition metal (TM) catalysts. NH_3 synthesis over the TM surface takes place *via* the dissociative chemisorption of N_2 and H_2, and subsequent hydrogenation of nitrogen to form NH_x ($x = 1$–3).[11] Thus, the overall reaction is understood in terms of the following elementary steps:

$$H_2(g) \rightarrow H_2(ad), \tag{9.1}$$

$$N_2(g) \rightarrow N_2(ad), \tag{9.2}$$

$$H_2(ad) \rightarrow 2H(ad), \tag{9.3}$$

$$N_2(ad) \rightarrow 2N(ad), \tag{9.4}$$

$$N(ad) + H(ad) \rightarrow NH(ad), \tag{9.5}$$

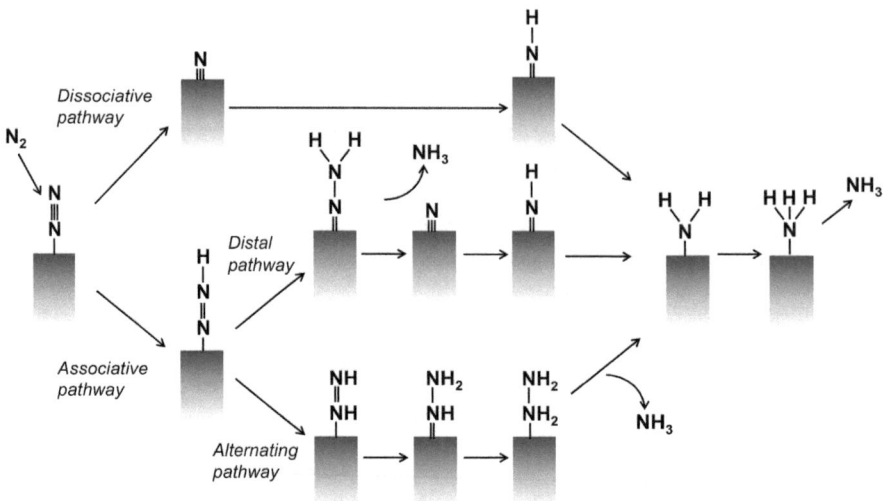

Figure 9.2 Reaction pathway of NH_3 synthesis *via* various N_2 activation processes.

$$NH(ad) + H(ad) \rightarrow NH_2(ad), \tag{9.6}$$

$$NH_2(ad) + H(ad) \rightarrow NH_3(ad), \tag{9.7}$$

$$NH_3(ad) \rightarrow NH_3(g), \tag{9.8}$$

where (g) and (ad) denote gas-phase and adsorbed species, respectively. Various kinetic analysis studies have revealed that step (9.4), N_2 dissociation, controls the overall reaction rate among these steps, and therefore this mechanism is recognized as a "dissociative mechanism" (Figure 9.2). As another reaction mechanism, the "associative mechanism" is proposed, in which the N_2 molecule is not directly dissociated but activated by hydrogenation to form NNH_x intermediates (Figure 9.2). This reaction mechanism is usually observed in organometallic complexes and electrochemical processes, and has been recently reported as an efficient pathway for low-temperature ammonia synthesis on heterogeneous catalysts as well.

To date, various approaches to low-temperature NH_3 synthesis have been conducted using homogeneous and heterogeneous catalysts. In this chapter, we focus on thermal catalytic ammonia synthesis using conventional transition metal (TM)-based catalyst, electride, hydride, and nitride-based materials, and a non-conventional thermal catalytic process. The characteristics of each catalyst in NH_3 synthesis are briefly summarized in Table 9.1.

9.2 Transition Metal-based Catalysts

Transition metal-based catalysts such as Fe and Ru have been developed and practically used in the industrial ammonia synthesis process. In the Fe catalyst, K_2O and Al_2O_3 are included as an electronic promoter and

Table 9.1 Characteristics of various catalysts in NH$_3$ synthesis.[a]

	Conventional TM catalyst	Electride	Hydride	Nitride
Active site	TM surface	TM and electride	TM and hydride	Nitride vacancy
Characteristics	Structure sensitive	Low work function	Active hydrogen (H$^-$)	Highly active without TM particle
RDS	N$_2$ dissociation	N–H formation	N–H formation	N–H formation
Mechanism	L–H mechanism	L–H mechanism	L–H mechanism, MvK mechanism	MvK mechanism, E–R mechanism
Activity	△	○	◎	○
Durability	◎	○	△	△

[a] ◎ excellent, ○ good, △ poor.

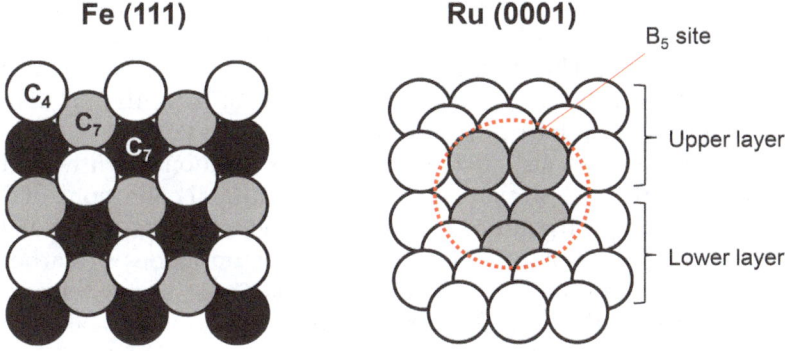

Figure 9.3 Schematic representations of active site on Fe and Ru surfaces.

structural promoter, respectively. In addition, Fe$_3$O$_4$ (magnetite) or Fe$_{1-x}$O (wüstite) is the main component and its content is about 90 wt%.[4] Under ammonia synthesis conditions, these Fe-based oxides are reduced to Fe metal. It is reported that N$_2$ molecules are adsorbed on the Fe surface with side-on orientation (α-state N$_2$), which is regarded as an intermediate for the dissociation of N$_2$.[12] In the side-on N$_2$ geometry, the highest occupied 1π$_u$ orbital of N$_2$ is coupled with the Fe unoccupied d-orbitals, and at the same time, the electron back donation occurs from Fe d-orbitals to the empty 1π$_g$*-antibonding orbital of N$_2$, which in turn facilitates the dissociation of N$_2$ on the Fe surface. The electron back donation is enhanced by electronic promoters such as K$_2$O.[13,14] Detailed surface science experiments and theoretical studies revealed that the 7 coordinate (C7) site which is an unsaturated state of Fe metal (Figure 9.3), and the most highly coordinated surface site on Fe, is the most active site for N$_2$ dissociation on the Fe catalyst surface because the C7 site experiences the largest electronic charge fluctuations within the solid.[15–19] Recently, new strategies have been proposed for Fe-based catalysts by using theoretical simulations. Liu *et al.* predicted through a computation that Fe_3 single-cluster anchored on θ-Al_2O_3(010) works as an efficient ammonia synthesis catalyst.[20] According to this proposal, the N_2 molecule is effectively activated *via* the spin-polarized electron transfer from Fe's 3d orbitals to N_2 π^* orbitals and the first hydrogenation of N_2 to NNH is much faster than the N_2 dissociation on Fe_3/θ-Al_2O_3(010). After the NNH formation, it is more favorable for NNH to dissociate into N and NH rather than the formation of HNNH or NNH_2 species. Such a unique associative mechanism is preferred for Fe_3/θ-Al_2O_3(010) rather than the conventional N_2 dissociation, which is attributed to the large spin polarization, low oxidation state of iron, and multi-step redox capability of Fe_3 cluster. As another example, Si-doping into the industrial Fe catalyst was theoretically predicted to significantly enhance the TOF by 13-fold under typical HB industrial conditions.[21] A hierarchical high-throughput catalyst screening (HHTCS) was used to identify non-transition-metal elements from 18 candidates that can significantly improve the Fe catalyst through surface and subsurface doping. The energy barrier is effectively reduced from 1.68 to 1.38 eV after Si-doping into Fe catalyst, resulting in a 176-fold improvement in the reaction rate. The doped Si significantly decreases the spin polarization difference between 2N and 4N states, stabilizing the 2N state and leading to a decreased barrier in comparison to pure Fe.

Ru catalysts work as an efficient catalyst for ammonia synthesis under lower temperatures and pressures than the operational conditions of the Fe-based catalyst.[22–24] In the Ru-based catalysts, generally less than 10 wt% of Ru is dispersed and supported as nanoparticles on oxides or carbon materials together with an alkali promoter such as Cs, K, and Na (in oxide or hydroxide form). A theoretical study by Nørskov's group revealed that N_2 cleavage on the step surface of Ru preferentially occurs with a lower activation energy than on the terrace.[25] The step site, B_5-type site, consists of five Ru atoms exposing a three-fold hollow site and a bridge site and it is regarded as the active site for the Ru catalyst (Figure 9.3).[26,27] The optimum Ru particle diameter is known to be 1.8–2.5 nm, which maximizes the number of B_5 sites. The N_2 molecule is adsorbed on the Ru surface with end-on orientation (γ-state N_2) through the bonding $3\sigma_g$ and antibonding $1\pi_g^*$ orbitals.[14] The N_2 cleavage is enhanced by the promoter compounds located nearby Ru particles through the charge transfer from the promoter to Ru. Therefore, the ammonia synthesis rate of Ru tends to increase with decreasing electronegativity of the promoter or support materials (*e.g.*, Cs > Rb > K > Na).[28] There have been many reports on Ru catalysts with alkali and/or alkaline earth compounds for ammonia synthesis, among which Cs and Ba-oxides are known as the most efficient electronic promoters.[29–32] Ru supported on high surface area graphite (HSAG) with Cs and Ba promoters was commercialized in the Kellogg Advanced Ammonia Process (KAAP) in 1998.[33] The operation temperature (370–400 °C) and pressure (<9 MPa) are much milder than for the HB process.[32,33] It is well known that the reaction order with respect to ammonia is close to zero for Ru catalyst, which is in contrast to the large negative value for Fe catalyst. Therefore, in the KAAP process, Fe-based catalyst is used in the first catalyst bed, while the following other catalyst beds are filled with the Ru-based catalysts because Ru catalyst works better than Fe catalyst under a high ammonia concentration. To date, various oxide materials such as Al_2O_3, MgO, CaO, BaO, CeO_2, La_2O_3, Sm_2O_3, $BaTiO_3$, $BaCeO_3$, $MgAl_2O_4$, and zeolites have been explored as supports for Ru catalysts. Among oxide-supported Ru catalysts, Ru-loaded Ba-promoted $LaCeO_x$ has been recently reported to exhibit the highest ammonia synthesis rate (52.3 mmol g^{-1} h^{-1}) at 350 °C and 1.0 MPa.[34] A low-crystalline, oxygen-deficient nano-oxide including Ba^{2+}, La^{3+}, and Ce^{3+} accumulated on the Ru surface after very high H_2 reduction treatment (700 °C). The authors of this paper claim the strong electron donation ability of these nano-oxides significantly enhances the N_2 cleavage on Ru surface. In contrast to these studies, Zheng *et al.* reported that Li-promoted Ru catalyst gives the highest ammonia synthesis rate despite its poorest electron-donating ability among alkali promoters.[35] They claim that the Ru step sites are blocked by Li species in this catalyst, and therefore, the adsorbed N_2 molecule is polarized and stabilized by Li^+ on Ru terrace sites. The kinetic energy barrier for the first hydrogenation of N_2 to form NNH on the Li^+-modified Ru terrace site is lower than the classical dissociative pathway on the stepped Ru site.

9.3 Electride-based Catalysts

Inorganic electrides contain electrons at intrinsic lattice sites such as cavities, channels, and interlayers to form individual orbitals and serve as anions individually rather than combining with other atoms in the solid, resulting in very low work functions.[36] Therefore, electride materials can serve as an efficient electronic promoting material in ammonia synthesis. 12CaO·7Al$_2$O$_3$ electride (C12A7:e$^-$) has a cubic structure (space group $I\bar{4}3d$) and its unit cell has a positively charged framework structure composed of 12 subnanometer-sized cages with an inner diameter of 0.4 nm, and its chemical formula is expressed by $[Ca_{24}Al_{28}O_{64}]^{4+}(e^-)_4$.[37] The C12A7:e$^-$ contains a high density of electrons ($N_e = 2.3 \times 10^{21}$ cm^{-3}) in its crystallographic nanocages, resulting in high electrical conductivity (1500 S cm^{-1}) and a very low work function (2.4 eV) comparable to metallic potassium. However, it is thermally stable and resistant to air since the electrons are protected by a stable Ca–O–Al framework, which is in contrast to the metal potassium that is thermally unstable and very sensitive to air.

C12A7:e$^-$ exhibits a strong electron-donating ability toward Ru in ammonia synthesis, leading to an order of magnitude higher turnover frequency (TOF) with almost half the reaction activation energy as compared to conventional alkali-promoted Ru catalysts.[38] This was the first application of an electride material as a catalyst. The catalytic performance of Ru/C12A7:e$^-$ strongly depends on the electron concentration (N_e), *i.e.*, the ammonia synthesis rate is enhanced 10-fold at N_e of 1.0×10^{21} cm^{-3}, and the activation energy (E_a) for ammonia synthesis also reduces to about 50 kJ mol^{-1}

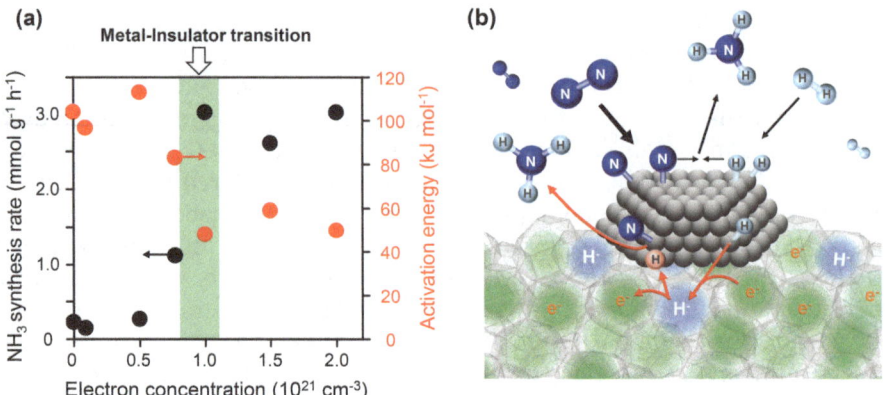

Figure 9.4 (a) Catalytic performance of Ru/C12A7:e$^-$ in NH$_3$ synthesis as a function of electron concentration. Reproduced from ref. 39 with permission from American Chemical Society, Copyright 2015. (b) Possible reaction mechanism of NH$_3$ synthesis on Ru/C12A7:e$^-$. Reproduced from ref. 42, https://doi.org/10.1038/ncomms7731, under the terms of the CC BY 4.0 license, https://creativecommons.org/licenses/by/4.0/.

from about 100 kJ mol^{-1} (Figure 9.4a).[39] As a result, the Ru/C12A7:e$^-$ shows a much higher reaction rate and lower activation energy than other Ru-loaded Ca–Al mixed oxide catalysts as well as Cs–Ru/MgO that is a benchmark catalyst. When the N$_2$ molecule is adsorbed on Ru/C12A7:e$^-$, the N≡N stretching mode of the N$_2$ molecule is observed at 2124–2176 cm^{-1}, which is lower than that of Ru/Al$_2$O$_3$ (2245 cm^{-1}) and Ru/C12A7:O^{2-} (2194 cm^{-1}), suggesting that the N≡N bond of N$_2$ on Ru/C12A7:e$^-$ is weakened by the electron injection from C12A7:e$^-$. Moreover, the C12A7:e$^-$ support can also boost the ammonia synthesis activity of a Co catalyst that has intrinsically low activity for N$_2$ dissociation.[40]

It is worth noting that hydrogen poisoning is one of the most serious issues in low-temperature ammonia synthesis over Ru-based catalysts because Ru metal strongly interacts with hydrogen at low reaction temperatures, which suppresses N$_2$ adsorption on the Ru surface.[41] As a consequence, the ammonia synthesis rate does not increase at elevated pressures. On the other hand, Ru/C12A7:e$^-$ shows a positive reaction order (+0.97) with respect to H$_2$, which is in contrast to the large negative values reported for the conventional Ru catalysts. Ru/C12A7:e$^-$ possesses a unique hydrogen storage capability, *i.e.*, hydrogen atoms can be captured as H$^-$ ions in the cages *via* the reaction with encaged electrons (H^0 + e$^-$ → H$^-$) at the Ru-C12A7:e$^-$ support interface and the reverse reaction (H$^-$ → H^0 + e$^-$) also easily takes place to regenerate electrons in the cages (Figure 9.4b). This reversible hydrogen storage-release reaction in Ru/C12A7:e$^-$ suppresses hydrogen poisoning on the Ru surface. Due to the strong electron-donating ability and unique hydrogen storage properties, ammonia synthesis over Ru-C12A7:e$^-$ is not limited by the N$_2$ dissociation step that is the RDS of conventional catalysts, whereas the N–H bond formation step dominates the overall reaction rate.[42]

The reaction mechanism of ammonia synthesis over Ru/C12A7:e$^-$ catalyst was further investigated by Kammert and co-workers using *in situ* neutron-scattering techniques combined with density functional theory (DFT) calculations and steady-state isotopic transient kinetic analysis (SSITKA).[43] They proposed that the H$^-$ species are unlikely to be reactive under NH$_3$ synthesis conditions because the H$^-$ was entrapped and stabilized in the cage to keep electroneutrality. In addition, the coverage of N-containing intermediates (N$_2$, N, or NH$_x$) on the Ru/C12A7:e$^-$ surface (84%) is much higher than those of Ru/C12A7:O^{2-} (15%) and Ru/MgO (<14%), implying a shift away of the RDS from N$_2$ dissociation to N–H bond formation.

Lu *et al.* reported the first intermetallic 1D-electride, Y$_5$Si$_3$, which showed excellent stability in both air and water. Y$_5$Si$_3$ contains electrons in the small cavity (~4 Å in diameter) surrounded by Y atoms, resulting in the work function of 3.5 eV. While the catalytic activity of Ru/Y$_5$Si$_3$ was not high as compared to Ru/C12A7:e$^-$, Ru/Y$_5$Si$_3$ catalyst exhibited a stable catalytic activity even after exposure to moisture (3% water vapor). The strong orbital hybridization between Y 4d and anionic electrons gives rise to a chemical bond formation that protects the anionic electrons from unintended chemical reactions, leading to high resistance toward water. During ammonia synthesis, Y$_5$Si$_3$ can accommodate hydrogen as H$^-$ to form Y$_5$Si$_3$H, which suppresses the hydrogen poisoning on the Ru surface in a similar manner to Ru/C12A7:e$^-$. Recent theoretical calculations demonstrated that the interstitial electrons on the Y$_5$Si$_3$(0001) surface work as active sites and facilitate N$_2$ cleavage as well as the formation of ammonia in the absence of Ru particles.[44] The N$_2$ dissociation energy barrier on Y$_5$Si$_3$ is much lower than that on the Ru(0001) surface. The eventual ammonia is formed on the top of a surface Y atom and can be desorbed with a low energy barrier of 0.63 eV. Additionally, it was revealed that Y$_5$Ge$_3$ also acts as a catalyst for ammonia synthesis in the same way as Y$_5$Si$_3$.

RTX (R = rare earth, T = transition metal, X = p-block element) ternary intermetallic materials are also reported to have electride-like characteristics and work as good catalyst supports for ammonia synthesis. The RTX intermetallic has sub-nanometer-sized cavities, in which electrons are accommodated, *e.g.*, LaScSi has two types of voids in La$_4$ tetrahedra and La$_2$Sc$_4$ octahedra, both of which are filled with electrons.[45] Ru/LaScSi exhibited high ammonia synthesis activity and chemical stability in air and water. Similarly, other RTX ternary intermetallics such as LaNiSi, LaCoSi, LaFeSi, and LaMnSi promoted the activity of Ru in ammonia synthesis.[46,47] Gong *et al.* investigated a series of ternary intermetallic electrides, LaTMSi (TM = Co, Fe, and Mn) as supports of Ru for ammonia synthesis.[47] All catalysts promoted N$_2$ dissociation with low activation barriers, but the ammonia synthesis rate was not proportional to the number of anionic electrons. Ru/LaMnSi with the highest electron concentration afforded the lowest activity, whereas Ru/LaCoSi showed the highest activity among these catalysts.

DFT calculations revealed that N_2 cleavage is facilitated on the Ru surface by electron donation from the LaTMSi support, but NH_x formations is preferred on La sites. The RDS of ammonia synthesis over Ru/LaTMSi was identified to be NH_2 formation. Croisé et al. reported that the reaction mechanism of ammonia synthesis was switched from a dissociative to an associative pathway on Ru/CeTX (T=Sc, X=Si, Ge) when the reaction temperature decreased from 400 °C to 300 °C.[48] In the TM-supported intermetallic electride catalysts, the intermetallic support injects electrons to the TM site to facilitate N_2 dissociation and the rare-earth metal site works as the second active center to stabilize the intermediate species. Furthermore, the TM site in RTX ternary intermetallics such as LaCoSi and LaRuSi also functions as an active site for ammonia synthesis (Figure 9.5a).[49,50] The Co site in LaCoSi is negatively charged compared with metallic Co. As a result, TOF of LaCoSi is much higher than for the other supported Co catalysts (Figure 9.5b). In addition, a unique "hot atom mechanism" contributes to promote ammonia synthesis over these RTX intermetallic electride catalysts. The strong exothermic adsorption of N_2 on RTX catalyst significantly lowers the energy barrier for N_2 dissociation. Furthermore, H^- ions formed by the reaction of electrons with gas-phase hydrogen provide the activated H to remove the adsorbed N from the metal active site efficiently by forming NH_3. The biggest drawback of these intermetallic catalysts is their very low surface area ($<1\ m^2\ g^{-1}$) and a small number of active sites due to their very high synthesis temperatures. On the other hand, the catalytic performance of LaRuSi can be significantly promoted by selective etching of the surface La and Si atoms using EDTA-2Na solution.[51] Since LaRuSi has a structure in which each element is stacked one atomic layer at a time, there are no B_5 step sites on the exposed Ru surface. Nevertheless, EDTA-treated LaRuSi catalyst exhibited a high ammonia synthesis rate comparable to the Cs-Ru/MgO catalyst with a number of B_5 step sites. This result indicates the active site of Ru is an unsaturated Ru atom including the B5 site on Ru-metals.

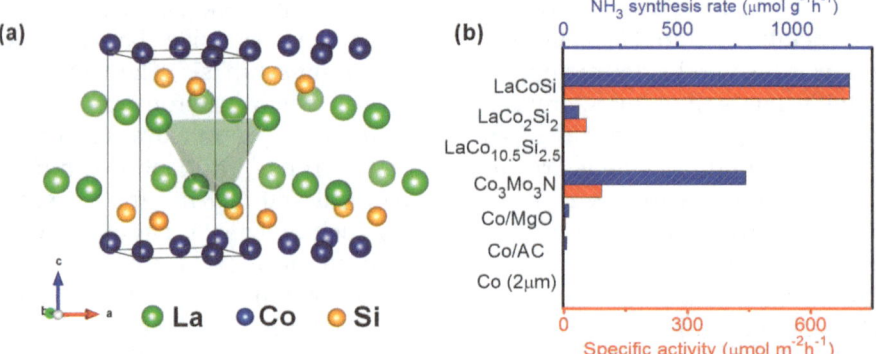

Figure 9.5 (a) Crystal structure of LaCoSi. (b) NH_3 synthesis rate over various Co catalyst at 400 °C and 0.1 MPa. Adapted from ref. 49 with permission from the Authors, Copyright 2019.

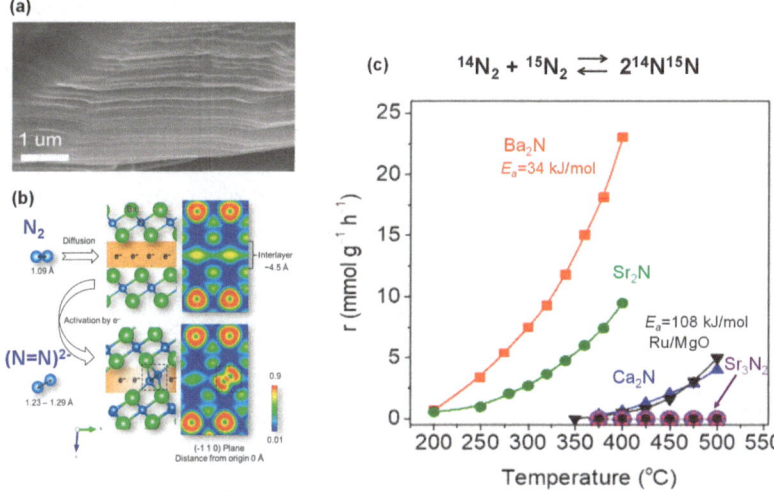

Figure 9.6 (a) SEM image of Ba$_2$N. (b) Schematic illustration of N$_2$ activation by interlayer electron of Ba$_2$N. (c) N$_2$ isotopic exchange reaction over various catalysts. Adapted from ref. 54, https://doi.org/10.1021/jacs.3c09362, under the terms of the CC BY 4.0 license, https://creativecommons.org/licenses/by/4.0/.

AE$_2$N (AE = Ca, Sr, and Ba) with layered structure is known as a 2D electride (Figure 9.6a), in which anionic electrons are confined between the [AE$_2$N]$^+$ layers as counter anions.[52] The work functions (WFs) for in-plane (100) directions of Ca$_2$N, Sr$_2$N, and Ba$_2$N are calculated to be 2.87, 2.61, and 2.38 eV, respectively.[53] Therefore, the interlayer electrons can activate the N$_2$ molecule to form N$_2^{2-}$, diazenide ion, in the interlayer space, which is confirmed by Raman spectroscopy and DFT calculations (Figure 9.6b).[54] Furthermore, Ba$_2$N without any TM sites works as an efficient catalyst for the N$_2$ isotope exchange reaction (N$_2$-IER) and its activity is much higher than that of Ca$_2$N, Sr$_2$N, and conventional Ru catalyst (Figure 9.6c). The activation energy for N$_2$-IER over Ba$_2$N is exceptionally low (35 kJ mol^{-1}), but Ba$_2$N does not work as a catalyst for ammonia synthesis due to facile decomposition into BaH$_2$ in the presence of H$_2$. However, this study clearly demonstrates that electron donation from low WF-Ba$_2$N into N$_2$ is very effective to cleave N$_2$ molecules even in the absence of a TM site.

9.4 Hydride-based Catalysts

Hydride materials have been recently well studied as efficient catalysts for ammonia synthesis because H$^-$ ions serve as both electron and hydrogen donors. The reactivity of H$^-$ in ammonia synthesis is largely influenced by the crystal structure of the solids. For example, most of the electrons in C12A7:e$^-$ can be replaced by H$^-$ ions in the cages to form C12A7:H$^-$. After the full replacement of e$^-$ by H$^-$, the H$^-$ ions are stabilized in the positively charged cage structure and cannot contribute to the catalytic reaction. As a consequence, Ru/C12A7:H$^-$ shows much lower catalytic activity with higher activation energy (154 kJ mol^{-1}) than Ru/C12A7:e$^-$.[42] On the other hand, Ca$_2$NH, which is obtained by the hydrogenation of 2D electride Ca$_2$N, effectively promotes the activity of supported Ru in ammonia synthesis, in which H$^-$ ions act as electron and hydrogen donors.[55] In Ca$_2$NH, H$^-$ ions are located at the interlayer space in place of anionic electrons, and hydrogen can easily go in and out through the interlayer space, allowing a reversible reaction between hydrogen and electrons. This study also revealed the effect of the valence state of hydrogen on the catalytic performance by comparing Ca$_2$NH and CaNH. Both of these materials consist of the same elements, but Ca$_2$NH has H$^-$ ions and CaNH has H$^+$ (Figure 9.7a). Ru/Ca$_2$NH showed above 10 times higher catalytic activity with less than half the activation energy than Ru/CaNH. In addition, Ca$_2$NH suppresses the hydrogen poisoning on the Ru surface *via* the reversible exchange reaction between H$^-$ ions and electrons, whereas the activity of Ru/CaNH decreases with increasing hydrogen partial pressure in a similar way to other common Ru catalysts (Figure 9.7b).

Ca(NH$_2$)$_2$ contains H$^+$ rather than H$^-$, but Ca$_2$NH with H$^-$ is formed at the Ru/support interface when Ru-loaded Ca(NH$_2$)$_2$ is used as a catalyst for ammonia synthesis. Interestingly, Ru flat-shaped nanoparticles with a narrow size distribution (2.1 ± 1.0 nm) are formed on Ca(NH$_2$)$_2$ by the epitaxial growth of Ru on it due to the strong Ru–N interaction and lattice matching between them. In addition, the activity of Ru/Ca(NH$_2$)$_2$ is significantly enhanced by doping a small amount of Ba (3 at%) into the Ca(NH$_2$)$_2$. The

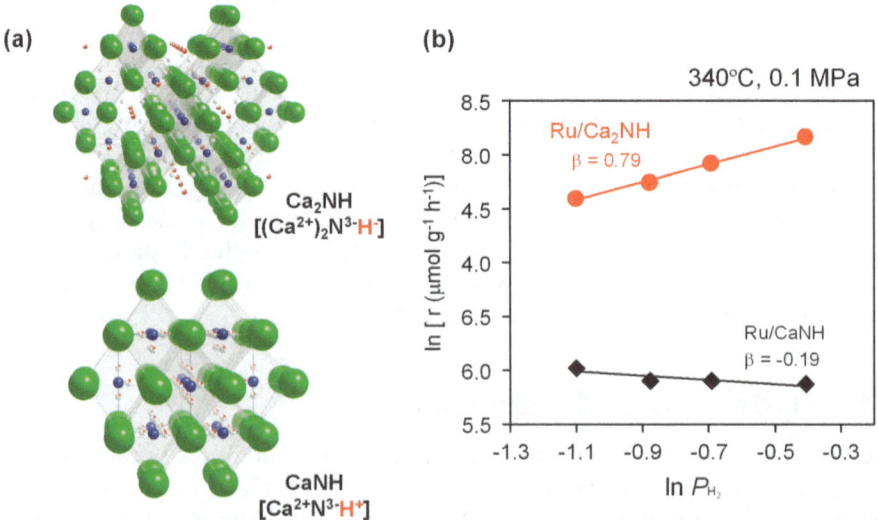

Figure 9.7 (a) Crystal structure of Ca$_2$NH and CaNH. (b) Dependence of NH$_3$ synthesis rate on the partial pressure of H$_2$. Adapted from ref. 55 with permission from the Royal Society of Chemistry.

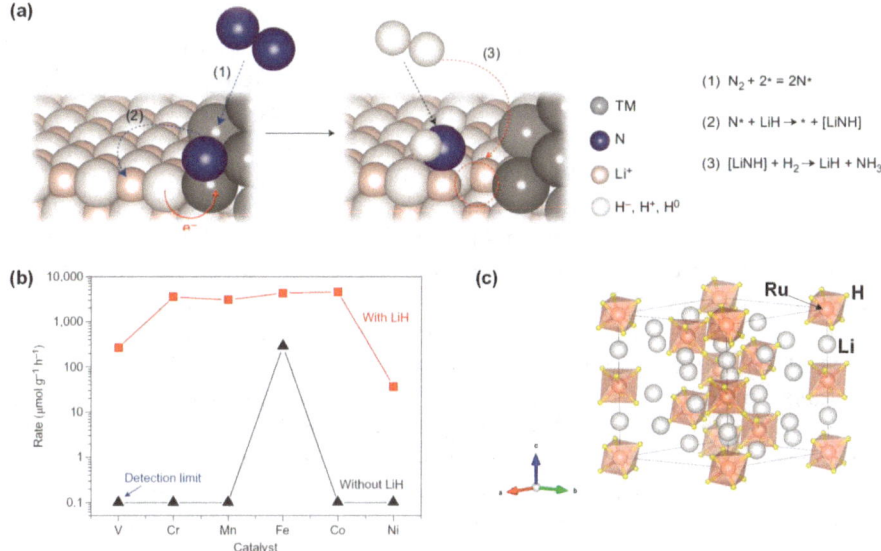

Figure 9.8 (a) Reaction mechanism of NH$_3$ synthesis over TM–LiH composite. (b) NH$_3$ synthesis rate of various TMs with or without LiH at 300 °C. (c) Crystal structure of Li$_4$RuH$_6$. Adapted from ref. 56 with permission from Springer Nature, Copyright 2016.

surface area of Ru/Ba–Ca(NH$_2$)$_2$ increases to 101 m^2 g^{-1} from that of the original Ba–Ca(NH$_2$)$_2$ (17 m^2 g^{-1}). This increase originates from partial decomposition of Ca(NH$_2$)$_2$ by the Ru nanoparticle catalyst to form a mesoporous structure. At the same time, a Ru–Ba core–shell structure is formed on the catalyst surface. The formation of such a unique structure and hydride species is the key factor for the high catalytic performance of the Ru/Ba–Ca(NH$_2$)$_2$ catalyst.

Chen *et al.* have developed composite catalysts of LiH (AH$_x$, A = Li, Na, K, Ba, Ca) with 3d transition metals such as V, Cr, Mn, Fe, Co, and Ni. In this catalyst system, LiH works as a second active site to remove nitrogen from TM to form Li$_2$NH and/or LiNH$_2$ as intermediates, which in turn promotes the overall reaction and results in circumventing the scaling relations (Figure 9.8a).[56] The intermediate Li–N–H compound is hydrogenated to form NH$_3$ and regenerate LiH, and the activation energy of this step is similar to that of the overall ammonia synthesis, which means that the hydrogenation of the Li–N–H intermediate is the RDS. In fact, the ammonia synthesis rates of TM catalysts are enhanced by 3–4 orders of magnitude after combination with LiH (Figure 9.8b). Not only LiH but also other alkali and alkaline earth metal hydrides such as NaH, KH, CaH$_2$, and BaH$_2$ can increase the activity of TMs or their nitrides by several orders of magnitude.[57,58] For instance, the order of the promotion effect on Mn nitride is BaH$_2$ > LiH > KH > CaH$_2$ > NaH, which is different from that of conventional alkali or alkaline earth oxide/hydroxide promoters. In these TM–AH$_x$

composite catalyst systems, the balance between nitridation of AH_x into imide ($A_{2/x}NH$) and hydrogenation of $A_{2/x}NH$ to AH_x is important for the overall catalytic activity. Based on Gibbs free energy change (ΔG) calculation in the temperature range 0–450 °C and 1.0 bar, NaH and KH hardly fix nitrogen to form imides, whereas ΔG for $MgH_2 \rightarrow MgNH$ is strongly negative (<-100 kJ mol^{-1}) and the hydrogenation of MgNH is difficult.[59] On the other hand, Li and Ba have thermodynamically stable hydride and imide forms, which allows both nitridation of AH_x and hydrogenation of $A_{2/x}NH$ to proceed easily. As another possible factor governing catalytic activity, the formation of ternary hydride species at the TM–A_x interface is proposed. The presence of a series of complex hydride clusters such as $[Li_4FeH_6]^-$ and $[Li_5FeH_6]^-$ at the interface of the Fe–LiH composite catalyst is confirmed by gas-phase optical spectroscopy coupled with mass spectrometry and quantum chemical calculations.[60] Although these complex hydrides are typically synthesized under very harsh conditions (900 °C, 6.1 GPa), nitrogen-containing complex intermediates are also identified, which suggests that a small amount of complex hydride species formed at the catalyst surface serves as the active site for ammonia synthesis.

It was demonstrated by the same group that Ru-based ternary hydrides (Li_4RuH_6 and Ba_2RuH_6) can be easily synthesized under mild conditions (400 °C and 1.0 MPa) and function as highly active ammonia synthesis catalysts.[61] When the Ru-based ternary hydrides are supported on MgO, the ammonia synthesis rate is enhanced by ~400 times from the bulk Ru-based ternary hydrides. In particular, MgO-supported Ba_2RuH_6 is one of the most active catalysts under mild conditions reported to date. These catalysts comprise the $[RuH_6]$ anion surrounded by Li or Ba cations as the active site for ammonia synthesis (Figure 9.8c). In these catalysts, the Ru–Ru distance is much longer (>5.0 Å) than that of Ru metal and there is no B_5-type step site. Hence, the dissociative adsorption of N_2 is unlikely to occur. Instead, non-dissociative N_2 reduction is proposed based on theoretical calculations, i.e., N_2 adsorbed on the unsaturated site of RuH_6 is hydrogenated to form N_xH_y ($x=0-2$, $y=0-3$) intermediates that are stabilized by neighboring Li or Ba cations.

Lanthanide hydrides, LnH_{2+x} (Ln = La, Ce, Y), with a fluorite-type structure may accommodate an electron in the central cavity site surrounded by eight H^- ions and exhibit high ammonia synthesis activity at low reaction temperatures (<260 °C) when Ru nanoparticles are supported on their surface.[62] However, LnH_{2+x} is easily nitrided to form LnN_{1-x} during the catalytic reaction, especially above 300 °C. On the other hand, lanthanide oxyhydrides such as $LaH_{3-2x}O_x$ and $CeH_{3-2x}O_x$ have high nitridation resistance and a strong promoting effect on Ru in ammonia synthesis.[63] $LaH_{3-2x}O_x$ is prepared by the partial replacement of H^- site of LaH_3 with O^{2-} and is known as a high H^- ion conductor.[64] The operational temperature of $Ru/LnH_{3-2x}O_x$ is lower by 100 °C than that of corresponding oxide-supported Ru catalysts (Ru/La_2O_3 and Ru/CeO_2). In the lanthanide hydride-based catalysts, the catalytic performance is strongly correlated with the surface H^- ion mobility, i.e., the H^- ion in the support serves as an electron source to promote ammonia synthesis on Ru catalyst. However, high H^- ion mobility also leads to easing of nitridation of the hydride surface, resulting in serious deactivation. Ru/LaH$_3$ is easily deactivated due to the preferred nitridation during ammonia synthesis because the ΔG for LaN formation on LaH$_3$ is much more negative than that on LaH$_{3-2x}$O$_x$. In contrast, LaH$_{3-2x}$O$_x$ has high resistance toward nitridation, which results in high and stable catalytic activity.

Kobayashi et al. demonstrated that Ti-based hydrides, TiH$_2$ and BaTiO$_{2.5}$H$_{0.5}$, worked as efficient catalysts for ammonia synthesis with and without supported TM particles.[65,66] N$_2$ dissociation readily takes place on the Ti metal surface to form TiN because of its high nitrogen-binding energy, but the hydrogenation of TiN is difficult. On the other hand, some of the H^- ions of TiH$_2$ and BaTiO$_{2.5}$H$_{0.5}$ react with nitrogen to form a nitride–hydride surface and continuously produce ammonia under a N$_2$ and H$_2$ gas mixture at 400 °C and 5 MPa. Their catalytic performance is comparable to conventional supported Ru catalysts (Cs–Ru/MgO and Ru/BaTiO$_3$). Furthermore, the catalytic activity of BaTiO$_{2.5}$H$_{0.5}$ was significantly enhanced after Ru, Fe, and Co particles were supported on it. Among them, Ru/BaTiO$_{2.5}$H$_{0.5}$ showed the highest activity. In addition, the catalytic activity is also influenced by the A site cation (Ca, Sr, Ba) in the oxyhydride and the ammonia synthesis rate increases in the order of Ru/CaTiO$_{3-x}$H$_x$ < Ru/SrTiO$_{3-x}$H$_x$ < Ru/BaTiO$_{3-x}$H$_x$. ZrH$_2$ is also reported to work as an efficient promoting support of Ru catalyst in ammonia synthesis. However, the ammonia synthesis rate of ZrH$_2$ itself is about one-tenth or less than that of Ru/ZrH$_2$ at 400 °C and 1 MPa. Thus, transition metal hydrides such as TiH$_2$ and ZrH$_2$ are less active for ammonia synthesis at low temperatures and pressures than alkali, alkaline earth, and rare earth metal hydrides. This result means the metal hydrides which are capable of the formation of stable nitride do not work for efficient NH$_3$ synthesis.

Chang et al. reported that potassium hydride-intercalated graphite (KH$_{0.19}$C$_{24}$) functions as a TM-free catalyst in ammonia synthesis and exhibits similar activity to Ru/MgO catalyst.[67] In typical ammonia synthesis catalysts, the TM site is necessary to activate the N$_2$ molecule because N$_2$ adsorption proceeds via the electron donation from N$_2$ occupied σ orbitals to the TM empty d orbital and the back donation from TM occupied d orbitals to the empty antibonding π^* orbitals of N$_2$. DFT calculations revealed that N$_2$ molecules are adsorbed on the K$_4$H cluster intercalated between the two graphene layers and H$_2$ molecules are activated at the H vacancy sites of the K$_4$H. Due to the large energy barrier for direct N$_2$ dissociation, N$_2$ dissociation is unlikely to occur. Instead, the adsorbed N$_2$ species are hydrogenated to form H$_2$NNH$_2$ intermediates through the alternating associative pathway. This is the first example of a TM-free heterogeneous catalyst for ammonia synthesis.

9.5 Nitride-based Catalysts

Highly active TM catalyst can be predicted by a scaling relationship established by Nørskov and co-workers' theoretical studies.[7] In this relationship, it has been shown that ammonia synthesis activity follows a volcano-shaped plot using nitrogen adsorption energy (E_N) as a descriptor. Co–Mo bimetallic catalyst has an optimum E_N, although the E_N of Co is smaller and that of Mo is larger than the optimum. Cobalt molybdenum nitride, Co_3Mo_3N, was experimentally confirmed to have a higher ammonia synthesis rate than Mo_2N, Fe_3Mo_3N, and Ni_2Mo_3N.[68] In addition, its activity is significantly enhanced by adding Cs promoter and the Cs-Co_3Mo_3N surpasses the activity of commercial Fe-based catalyst (KM-1) at 400 °C and 5 MPa. Hunter et al. revealed the exchangeability of lattice nitrogen species in Co_3Mo_3N by $^{14}N/^{15}N$ isotopic exchange reactions.[69] Co_6Mo_6N is formed after heat treatment of Co_3Mo_3N with H_2 : Ar (3 : 1) at 700 °C. When the Co_6Mo_6N is heated in $^{15}N_2$ atmosphere at 600 °C, Co_3Mo_3N phase is regenerated accompanying $^{14}N/^{15}N$ formation. While this reaction temperature is much higher than that for ammonia synthesis, the presence of H_2 could accentuate the lattice N reactivity. This result implies the possibility of a lattice nitrogen mediated Mars–van Krevelen type mechanism.[70] DFT calculations were conducted to examine whether the reaction mechanism of ammonia synthesis over Co_3Mo_3N was the Langmuir–Hinshelwood (dissociative) or an Eley–Rideal/Mars–van Krevelen (associative) mechanism.[71] In the presence of surface defect sites, the Eley–Rideal mechanism proceeds via the formation of hydrazine and diazene intermediates, where gas-phase hydrogen reacts directly with surface-activated nitrogen in a low barrier process (Figure 9.9). This study clearly suggests that surface defect sites on nitride-based catalysts play a key role in ammonia synthesis under milder conditions.

A novel perovskite oxynitride-hydride, $BaCeO_{3-x}N_yH_z$, was also demonstrated to facilitate N_2 reduction into ammonia via a lattice N^{3-} and H^- mediated Mars-van Krevelen mechanism.[72] The conventional oxide materials such as $BaCeO_3$ do not exhibit any activity for ammonia synthesis, but $BaCeO_{3-x}N_yH_z$ exhibits a stable activity for ammonia synthesis, which suggests that lattice N^{3-} and H^- sites function as active sites. In this catalyst, not only the surface but also bulk N^{3-} and H^- sites contribute to the reaction mechanism as confirmed by ammonia synthesis from isotopically labeled nitrogen ($^{15}N_2$) and hydrogen (D_2). Compared to lattice O^{2-} sites, lattice N^{3-} and H^- sites easily form anion vacancy (V_a) sites with low work function electrons at low reaction temperatures, which in turn facilitate N_2 and H_2 activation into ammonia.

In 2020, a high-performance Ni-based catalyst was achieved by utilizing the nitrogen vacancy sites of LaN as active sites (Figure 9.10a).[73] According to the scaling relation, N_2 cleavage on the Ni surface is too slow because of its very low nitrogen adsorption energy, and therefore Ni is known to be an inactive catalyst for ammonia synthesis. On the other hand, Ni(12.5 wt%)-loaded LaN nanoparticles effectively produce ammonia as high as 5.5 mmol g^{-1} h^{-1}, which is comparable to conventional Ru-based catalysts. Ni supported on C12A7:e$^-$ with a very low work function showed almost no catalytic activity, indicating that the electron donation is not effective for Ni-based catalyst in ammonia synthesis. Kinetic experiments and DFT calculations confirm that the nitrogen vacancy (V_N) sites on LaN function as N_2 activation sites. In addition, the supported Ni particles dissociate H_2 to

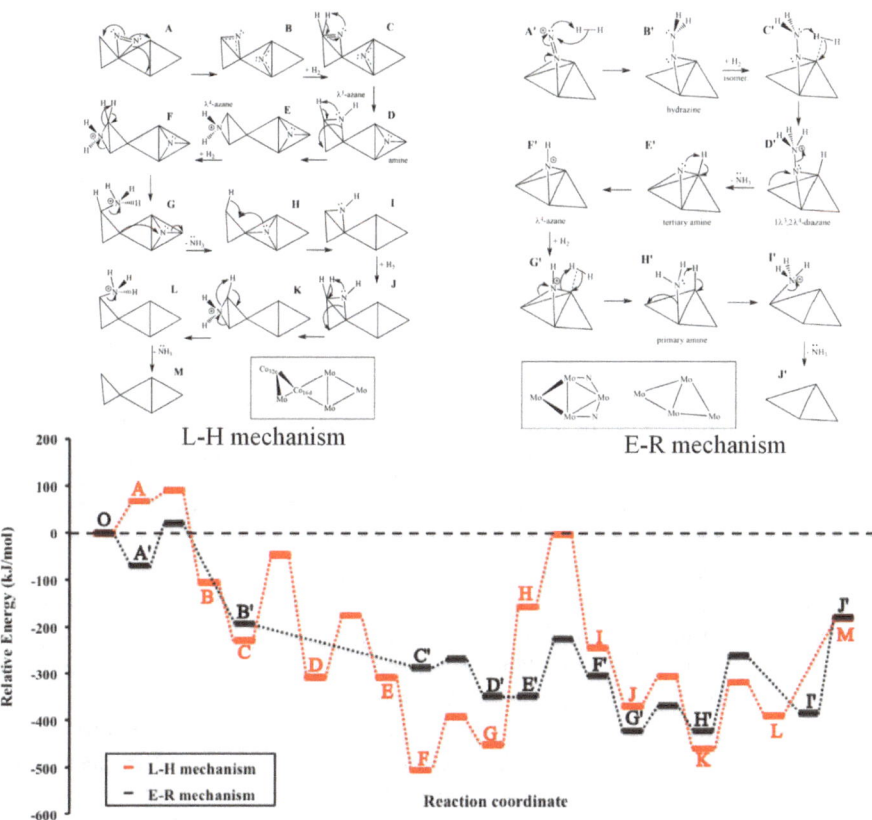

Figure 9.9 Reaction mechanism and energy diagram of Langmuir–Hinshelwood (L–H) and Eley–Rideal/Mars–van Krevelen (E-R/MvK) mechanism for ammonia synthesis on Co_3Mo_3N. Reproduced from ref. 71, https://doi.org/10.1021/acs.jpcc.7b12364, under the terms of the CC BY 4.0 license, https://creativecommons.org/licenses/by/4.0/.

promote the formation of the V_N sites on the LaN surface through the hydrogenation of lattice N sites. The key role of the V_N site on the catalytic performance was elucidated by investigating the relationship between the catalytic activity and V_N formation energy for various binary nitride materials. Although Ni-loaded early-transition metal nitrides such as ZrN, TiN, TaN, and NbN exhibited negligibly low catalytic activity, Ni-loaded rare-earth metal nitrides ReN (Re = Ce, La, and Y) exhibited a high ammonia synthesis rate.[74] The ammonia synthesis rate increased with decreasing the V_N formation energy among the Ni/ReN catalysts, and the order of catalytic performance was Ni/CeN > Ni/LaN > Ni/YN > Ni/ScN (Figure 9.10b). This result strongly supports the proposal that the V_N site on the nitride surface is the main active center for Ni-loaded nitride catalyst systems. DFT calculations revealed that the N_2 molecule is captured at the V_N site and then hydrogenated by spillover hydrogen from Ni particles to form NNH_x species as

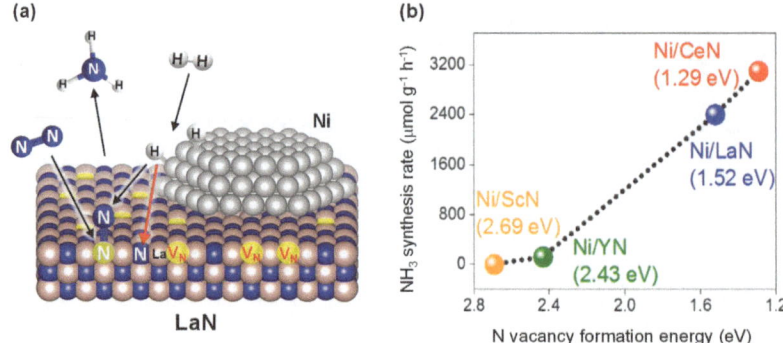

Figure 9.10 (a) Reaction mechanism of NH_3 synthesis over Ni/LaN. Adapted from ref. 73 with permission from the authors, Copyright 2020. (b) Relationship between NH_3 synthesis rate and N vacancy formation energy of various nitride-supported Ni catalysts. Adapted from ref. 74 with permission from American Chemical Society, Copyright 2020.

intermediates. Such an associative reaction mechanism is favored over a direct N_2 dissociation mechanism, which is experimentally supported by the fact that a large energy barrier is required for the N_2 isotope exchange reaction to proceed on Ni/LaN. The V_N site-mediated reaction mechanism is also effective for the supported Co catalysts.

On the other hand, Nørskov and Chorkendorff *et al.* proposed a new mechanism on the Co/LaN catalyst, where the VN site is not an active site. Their surface science techniques combined with theoretical calculations suggested that the spin-mediated promotion effect is significant for La-promoted Co surface.[75,76] They used La atom deposited on Co single crystals as model catalysts and compared the catalytic activity of a stepped Co($10\bar{1}15$) and a terraced Co(0001). From this experiment, both Co surfaces are inactive in the absence of La, whereas the stepped Co($10\bar{1}15$) with La showed ammonia synthesis activity. In this case, La atoms are dispersed on the Co surface and do not form La-nitride or -hydride. Furthermore, He^+ low-energy ion-scattering (LEIS) measurements have also revealed that La is dispersed on the Co surface of Co/LaN catalyst after the ammonia synthesis reaction. DFT calculation revealed that La quenches the spin moment of the nearby Co atoms when the La atom is deposited on the Co metal surface. Also, the adsorption energy of N intermediates and the N–N transition state are more negative on nonmagnetic than on magnetic surfaces. They thus concluded that the local spin-quenching effect of La enhances the N_2 activation on Co surface.

9.6 Non-conventional Ammonia Synthesis Approaches

In addition to the conventional thermal ammonia synthesis process using solid catalysts, several new ammonia synthesis processes have been devised in recent years. Sekine et al. reported that 9.9 wt% Cs/5.0 wt% Ru/SrZrO$_3$ catalyst exhibits a high ammonia synthesis rate (30.1 mmol g^{-1} h^{-1}) at 360 °C and 0.9 MPa with an electric field, which is about six times higher than that without an electric field.[77] In this process, the catalyst bed is directly sandwiched between two stainless steel rod electrodes and an applied external DC electric field (6 mA, 470 V), which contrasts with the conventional electrocatalytic system (Figure 9.11a). In addition, the apparent activation energy decreased from 121 kJ mol^{-1} to 37 kJ mol^{-1} upon applying the electric field. However, N$_2$ dissociation is not promoted by the electric field in the absence of H$_2$. Based on the *in situ* DRIFTS and DFT calculations, they proposed that proton hopping on the catalyst surface is enhanced by the electric field, which in turn facilitates N$_2$ activation into ammonia *via* the associative mechanism.

Zhang et al. reported an efficient NH$_3$ synthesis method using an electromagnetic field.[78] The catalyst employed was bulk iron, which is used in the HB process, and the apparatus consisted of a stainless-steel rod inside a silica glass tube reactor and stainless steel mesh outside the tube. NH$_3$ synthesis was examined under a flow of N$_2$ + H$_2$ applying an RF electromagnetic field (EMF) of 7800 V (root mean square) and 8.6 kHz between these two electrodes. They confirmed no plasma generation under set

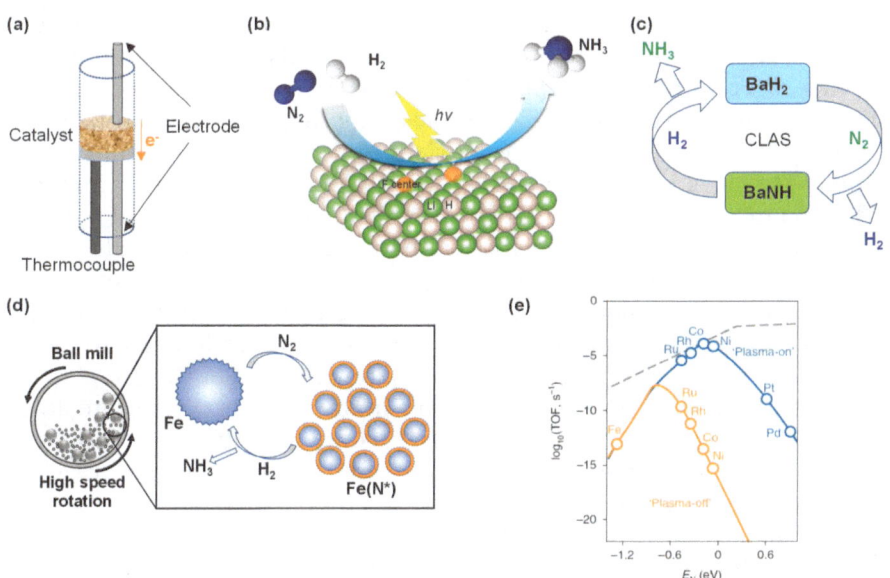

Figure 9.11 Various approaches to ammonia synthesis. (a) Thermal NH$_3$ synthesis with an electric field. (b) Light-driven NH$_3$ synthesis by LiH. (c) Chemical looping NH$_3$ synthesis (CLAS) using BaH$_2$ and BaNH. (d) Mechanochemical NH$_3$ synthesis using Fe powder. (e) Comparison of thermal and plasma-assisted NH$_3$ synthesis. Adapted from ref. 84 with permission from Springer Nature, Copyright 2018.

conditions by monitoring the time course of the voltage, current, and magnetic field (if plasma is generated, a current ripple appears). The results obtained demonstrate excellent performance, such as an NH_3 yield of ~5 times and an energy efficiency of ~2.7 times at 200 °C and 1 MPa compared with the EMF-free condition. They explain this EMF effect in terms of the activation of N_2 molecules adsorbed on metallic iron by a pure electromagnetic field using DFT calculations.

Chen et al. demonstrated light-driven ammonia synthesis by LiH without any external heating (Figure 9.11b).[79] LiH produces H_2 under UV light (<400 nm) irradiation ($2LiH + hv \rightarrow 2Li^+ + 2e^- + H_2(g)$) since it has a wide band gap of 3.68 eV. As a result, photon-generated electrons were formed in the surface H^- vacancies (V_H), which was confirmed by EPR as an F center signal at $g = 2.004$. N_2 molecules were reduced at the V_H sites to form NH and NH_2 species at room temperature as confirmed by FT-IR analysis. When a N_2 and H_2 mixture with a low H_2 content ($N_2 : H_2 = 10 : 1$) gas was introduced at 0.1 MPa, LiH catalyst produced ammonia with a rate of 75 ± 14 µmol g^{-1} h^{-1} only under UV illumination (300–420 nm). In contrast, thermal catalytic ammonia production over LiH proceeds only above 450 °C. It is proposed by DFT calculation that the photocatalytic ammonia synthesis over LiH takes places via the formation of NNH intermediates and subsequent breaking of the N–N bond to produce NH + NH and H_2.

Chemical looping ammonia synthesis (CLAS) is known to have advantages such as operation at atmospheric pressure, circumventing the equilibrium limitation, higher efficiency, and reduced CO_2 emissions.[80] In CLAS, the following two processes are separately conducted: (1) N_2 activation to form nitrogen carriers such as nitrides, imides, or nitrogen-rich intermediates; (2) hydrogenation of the nitrogen carriers to form ammonia. Yang et al. explored Mo_2N as a nitrogen carrier for CLAS.[81] The average ammonia production rate reached ~4.6 mmol g^{-1} h^{-1} by the hydrogenation of Mo_2N at 450 °C, which was three times higher than that via thermo-catalytic ammonia synthesis by Mo_2N. On the other hand, the regeneration of the nitrogen carrier requires 600 °C because of the slow nitrogen fixation kinetics of Mo. As another CLAS approach, an alkali or alkaline earth metal hydrides/imides mediated process was proposed (Figure 9.11c).[82] For this CLAS, N_2 is fixed by hydride to form imide, and the imide is hydrogenated to form ammonia and regenerate hydride. Ni–BaH_2 composite effectively produced ammonia via CLAS below 300 °C and the production rate reached ~6.0 mmol g^{-1} h^{-1} at 300 °C at atmospheric pressure.

A mechanochemical ball-milling method was applied by Han et al. to ammonia synthesis, achieving a high ammonia concentration (82.5 vol%) at 45 °C and 1 atm by using iron powders (Figure 9.11d).[83] This process is divided into two steps: a nitrogen dissociation step and a hydrogenation step. N_2 is mainly adsorbed and dissociated on the defect sites of the Fe(100) surface to form iron nitride (FeN). Then, the FeN is hydrogenated into NH_x ($x = 1$–3), and ammonia is released from the Fe surface with the aid of mechanical energy. The later hydrogenation step is accelerated by increasing the rotation speed and is regarded as the RDS in the mechanochemical ammonia synthesis. Reichle et al. demonstrated one-step continuous ammonia (0.26~0.35 vol%) production at room temperature by a mechanochemical method using Cs-promoted Fe catalyst. The elemental Cs promoter is converted into CsH and $CsNH_2$ during the reaction, which are active Cs promoter species. On the other hand, CsOH formation by the reaction with oxygen passivation layer impurities is the main reason for catalyst deactivation.

Non-thermal plasma-assisted ammonia synthesis has been well studied due to its advantages of mild operation conditions and thermodynamically non-equilibrium process. In this process, the gas temperature is maintained at room temperature, whereas the electron temperature is as high as 1.0×10^4–1.0×10^5 K, which corresponds to 1.0–10 eV in energy. Such high-energy electrons collide with molecular N_2 and H_2, improving NH_3 formation efficiency through the production of excited state atoms and reactive radicals. Mehta et al. predicted that by using a microkinetic model an optimum ammonia synthesis catalyst is shifted from Ru in the thermal catalytic system toward other metals with weaker nitrogen binding energies such as Co and Ni in the plasma-catalytic system (Figure 9.11e).[84] In addition, the optimal reaction rate in the plasma-catalytic system is several orders of magnitude higher than that of thermal catalysis.

9.7 Outlook

Much progress has been made in catalysts for green ammonia synthesis, making it possible to shift the rate-determining step from N_2 dissociation to N–H bond formation. As a result, the next technical challenges to be overcome are as follows:

a. Catalyst for N–H bond formation

   It is now clear that the rate-determining step in NH_3 formation is the creation of N–H bonds, in particular $NH + H \rightarrow NH_2$ for advanced catalysts. The next step is to design and create an appropriate catalyst suitable for acceleration of this process.

b. Ru-free and chemically stable catalyst with high activity

   For industrial application, Ru-free and chemically stable catalysts are required from cost and catalyst-shaping points of view. While rather active catalysts have been reported to date, almost all of them are sensitive to humidity/air. We need to innovate a strategy to break this activity–stability trade-off relationship.

c. Direct synthesis of nitrogen-bearing fine chemicals from N_2

   Apart from NH_3 synthesis, new catalysts discovered to date can greatly accelerate N_2 activation in mild conditions. There are many nitrogen-bearing molecules which are value-added chemicals. If such chemicals could be directly synthesized from N_2 using mild conditions, the benefits should be huge.

Acknowledgements

This study was supported by the FOREST Program (No. JPMJFR203A) from the Japan Science and Technology Agency (JST) and Kakenhi Grants-in-Aid (No. JP22H00272, JPH02204) from the JSPS. Part of this work was supported by the JST-Mirai Program (JPMJMI21E9) from JST.

References

1. J. M. Thomas and W. J. Thomas, *Principles and practice of heterogeneous catalysis*, John Wiley & Sons, 2014.
2. J. A. Pool, E. Lobkovsky and P. J. Chirik, *Nature*, 2004, **427**, 527–530.
3. T. Kandemir, M. E. Schuster, A. Senyshyn, M. Behrens and R. Schlögl, *Angew. Chem., Int. Ed.*, 2013, **52**, 12723–12726.
4. H. Liu, *Ammonia synthesis catalysts: innovation and practice*, World Scientific, 2013.
5. S. Giddey, S. P. S. Badwal, C. Munnings and M. Dolan, *ACS Sustainable Chem. Eng.*, 2017, **5**, 10231–10239.
6. M. Ravi and J. W. Makepeace, *Chem. Sci.*, 2022, **13**, 890–908.
7. C. J. H. Jacobsen, S. Dahl, B. S. Clausen, S. Bahn, A. Logadottir and J. K. Nørskov, *J. Am. Chem. Soc.*, 2001, **123**, 8404–8405.
8. E. Skulason, T. Bligaard, S. Gudmundsdottir, F. Studt, J. Rossmeisl, F. Abild-Pedersen, T. Vegge, H. Jonsson and J. K. Nørskov, *Phys. Chem. Chem. Phys.*, 2012, **14**, 1235–1245.
9. A. J. Medford, A. Vojvodic, J. S. Hummelshoj, J. Voss, F. Abild-Pedersen, F. Studt, T. Bligaard, A. Nilsson and J. K. Nørskov, *J. Catal.*, 2015, **328**, 36–42.
10. J. K. Nørskov, T. Bligaard, J. Rossmeisl and C. H. Christensen, *Nat. Chem.*, 2009, **1**, 37–46.
11. P. Stoltze and J. K. Norskov, *Phys. Rev. Lett.*, 1985, **55**, 2502–2505.
12. M. Grunze, M. Golze, W. Hirschwald, H. J. Freund, H. Pulm, U. Seip, M. C. Tsai, G. Ertl and J. Kuppers, *Phys. Rev. Lett.*, 1984, **53**, 850–853.
13. G. Ertl, M. Weiss and S. B. Lee, *Chem. Phys. Lett.*, 1979, **60**, 391–394.
14. C. N. R. Rao and G. R. Rao, *Surf. Sci. Rep.*, 1991, **13**, 221–263.
15. F. Bozso, G. Ertl, M. Grunze and M. Weiss, *J. Catal.*, 1977, **49**, 18–41.
16. N. D. Spencer, R. C. Schoonmaker and G. A. Somorjai, *J. Catal.*, 1982, **74**, 129–135.
17. D. R. Strongin, J. Carrazza, S. R. Bare and G. A. Somorjai, *J. Catal.*, 1987, **103**, 213–215.
18. G. Ertl, *Angew. Chem., Int. Ed.*, 2008, **47**, 3524–3535.
19. L. M. Falicov and G. A. Somorjai, *Proc. Natl. Acad. Sci. U. S. A.*, 1985, **82**, 2207–2211.
20. J. C. Liu, X. L. Ma, Y. Li, Y. G. Wang, H. Xiao and J. Li, *Nat. Commun.*, 2018, **9**, 1610.
21. Q. An, M. McDonald, A. Fortunelli and W. A. Goddard 3rd, *J. Am. Chem. Soc.*, 2020, **142**, 8223–8232.

22. A. Ozaki, K. Aika and H. Hori, *Bull. Chem. Soc. Jpn.*, 1971, **44**, 3216.
23. K. Aika, A. Ozaki and H. Hori, *J. Catal.*, 1972, **27**, 424–431.
24. K. Aika and A. Nielsen, *Ammonia : catalysis and manufacture*, Springer, 1995.
25. S. Dahl, A. Logadottir, R. C. Egeberg, J. H. Larsen, I. Chorkendorff, E. Tornqvist and J. K. Norskov, *Phys. Rev. Lett.*, 1999, **83**, 1814–1817.
26. T. W. Hansen, P. L. Hansen, S. Dahl and C. J. H. Jacobsen, *Catal. Lett.*, 2002, **84**, 7–12.
27. C. J. H. Jacobsen, S. Dahl, P. L. Hansen, E. Tornqvist, L. Jensen, H. Topsoe, D. V. Prip, P. B. Moenshaug and I. Chorkendorff, *J. Mol. Catal. A: Chem.*, 2000, **163**, 19–26.
28. K. Aika, *Catal. Today*, 2017, **286**, 14–20.
29. O. Hinrichsen, *Catal. Today*, 1999, **53**, 177–188.
30. F. Rosowski, A. Hornung, O. Hinrichsen, D. Herein, M. Muhler and G. Ertl, *Appl. Catal., A*, 1997, **151**, 443–460.
31. H. Bielawa, O. Hinrichsen, A. Birkner and M. Muhler, *Angew. Chem., Int. Ed.*, 2001, **40**, 1061–1063.
32. N. Saadatjou, A. Jafari and S. Sahebdelfar, *Chem. Eng. Commun.*, 2015, **202**, 420–448.
33. G. Prieto and F. Schüth, *Angew. Chem., Int. Ed.*, 2015, **54**, 3222–3239.
34. K. Sato, S.-i Miyahara, Y. Ogura, K. Tsujimaru, Y. Wada, T. Toriyama, T. Yamamoto, S. Matsumura and K. Nagaoka, *ACS Sustainable Chem. Eng.*, 2020, **8**, 2726–2734.
35. J. Zheng, F. Liao, S. Wu, G. Jones, T. Y. Chen, J. Fellowes, T. Sudmeier, I. J. McPherson, I. Wilkinson and S. C. E. Tsang, *Angew. Chem., Int. Ed.*, 2019, **58**, 17335–17341.
36. H. Hosono and M. Kitano, *Chem. Rev.*, 2021, **121**, 3121–3185.
37. S. Matsuishi, Y. Toda, M. Miyakawa, K. Hayashi, T. Kamiya, M. Hirano, I. Tanaka and H. Hosono, *Science*, 2003, **301**, 626–629.
38. M. Kitano, Y. Inoue, Y. Yamazaki, F. Hayashi, S. Kanbara, S. Matsuishi, T. Yokoyama, S. W. Kim, M. Hara and H. Hosono, *Nat. Chem.*, 2012, **4**, 934–940.
39. S. Kanbara, M. Kitano, Y. Inoue, T. Yokoyama, M. Hara and H. Hosono, *J. Am. Chem. Soc.*, 2015, **137**, 14517–14524.
40. Y. Inoue, M. Kitano, M. Tokunari, T. Taniguchi, K. Ooya, H. Abe, Y. Niwa, M. Sasase, M. Hara and H. Hosono, *ACS Catal.*, 2019, **9**, 1670–1679.
41. S. E. Siporin and R. J. Davis, *J. Catal.*, 2004, **225**, 359–368.
42. M. Kitano, S. Kanbara, Y. Inoue, N. Kuganathan, P. V. Sushko, T. Yokoyama, M. Hara and H. Hosono, *Nat. Commun.*, 2015, **6**, 6731.
43. J. Kammert, J. Moon, Y. Cheng, L. Daemen, S. Irle, V. Fung, J. Liu, K. Page, X. Ma, V. Phaneuf, J. Tong, A. J. Ramirez-Cuesta and Z. Wu, *J. Am. Chem. Soc.*, 2020, **142**, 7655–7667.
44. Y. Cao, P. Liu, J. Li, J. Wang, Y. Sun and X.-Q. Chen, *J. Phys. Chem. C*, 2023, **127**, 2953–2962.
45. J. Wu, Y. Gong, T. Inoshita, D. C. Fredrickson, J. Wang, Y. Lu, M. Kitano and H. Hosono, *Adv. Mater.*, 2017, **29**, 1700924.

46. H. Mizoguchi, S. W. Park, K. Kishida, M. Kitano, J. Kim, M. Sasase, T. Honda, K. Ikeda, T. Otomo and H. Hosono, *J. Am. Chem. Soc.*, 2019, **141**, 3376–3379.
47. Y. T. Gong, H. C. Li, J. Z. Wu, X. Y. Song, X. Q. Yang, X. B. Bao, X. Han, M. Kitano, J. J. Wang and H. Hosono, *J. Am. Chem. Soc.*, 2022, **144**, 8683–8692.
48. C. Croisé, K. Alabd, S. Tencé, E. Gaudin, A. Villesuzanne, X. Courtois, N. Bion and F. Can, *ChemCatChem*, 2024, **16**, e202400403.
49. Y. Gong, J. Wu, M. Kitano, J. Wang, T.-N. Ye, J. Li, Y. Kobayashi, K. Kishida, H. Abe, Y. Niwa, H. Yang, T. Tada and H. Hosono, *Nat. Catal.*, 2018, **1**, 178–185.
50. J. Wu, J. Li, Y. Gong, M. Kitano, T. Inoshita and H. Hosono, *Angew. Chem., Int. Ed.*, 2019, **58**, 825–829.
51. J. Li, J. Z. Wu, H. Y. Wang, Y. F. Lu, T. N. Ye, M. Sasase, X. J. Wu, M. Kitano, T. Inoshita and H. Hosono, *Chem. Sci.*, 2019, **10**, 5712–5718.
52. K. Lee, S. W. Kim, Y. Toda, S. Matsuishi and H. Hosono, *Nature*, 2013, **494**, 336–340.
53. S. Liu, W. Li, S. W. Kim and J.-H. Choi, *J. Phys. Chem. C*, 2019, **124**, 1398–1404.
54. Z. Zhang, Y. Jiang, J. Li, M. Miyazaki, M. Kitano and H. Hosono, *J. Am. Chem. Soc.*, 2023, **145**, 24482–24485.
55. M. Kitano, Y. Inoue, H. Ishikawa, K. Yamagata, T. Nakao, T. Tada, S. Matsuishi, T. Yokoyama, M. Hara and H. Hosono, *Chem. Sci.*, 2016, **7**, 4036–4043.
56. P. K. Wang, F. Chang, W. B. Gao, J. P. Guo, G. T. Wu, T. He and P. Chen, *Nat. Chem.*, 2017, **9**, 64–70.
57. W. B. Gao, P. K. Wang, J. P. Guo, F. Chang, T. He, Q. R. Wang, G. T. Wu and P. Chen, *ACS Catal.*, 2017, **7**, 3654–3661.
58. F. Chang, Y. Q. Guan, X. H. Chang, J. P. Guo, P. K. Wang, W. B. Gao, G. T. Wu, J. Zheng, X. G. Li and P. Chen, *J. Am. Chem. Soc.*, 2018, **140**, 14799–14806.
59. W. B. Gaol, J. P. Guo, P. K. Wang, Q. R. Wang, F. Chang, Q. J. Pei, W. J. Zhang, L. Liu and P. Chen, *Nat. Energy*, 2018, **3**, 1067–1075.
60. P. K. Wang, H. Xie, J. P. Guo, Z. Zhao, X. T. Kong, W. B. Gao, F. Chang, T. He, G. T. Wu, M. S. Chen, L. Jiang and P. Chen, *Angew. Chem., Int. Ed.*, 2017, **56**, 8716–8720.
61. Q. R. Wang, J. Pan, J. P. Guo, H. A. Hansen, H. Xie, L. Jiang, L. Hua, H. Y. Li, Y. Q. Guan, P. K. Wang, W. B. Gao, L. Liu, H. J. Cao, Z. T. Xiong, T. Vegge and P. Chen, *Nat. Catal.*, 2021, **4**, 959–967.
62. H. Mizoguchi, M. Okunaka, M. Kitano, S. Matsuishi, T. Yokoyama and H. Hosono, *Inorg. Chem.*, 2016, **55**, 8833–8838.
63. K. Ooya, J. Li, K. Fukui, S. Iimura, T. Nakao, K. Ogasawara, M. Sasase, H. Abe, Y. Niwa, M. Kitano and H. Hosono, *Adv. Energy Mater.*, 2021, **11**, 2003723.
64. K. Fukui, S. Iimura, T. Tada, S. Fujitsu, M. Sasase, H. Tamatsukuri, T. Honda, K. Ikeda, T. Otomo and H. Hosono, *Nat. Commun.*, 2019, **10**, 2578.

65. Y. Kobayashi, Y. Tang, T. Kageyama, H. Yamashita, N. Masuda, S. Hosokawa and H. Kageyama, *J. Am. Chem. Soc.*, 2017, **139**, 18240–18246.
66. Y. Tang, Y. Kobayashi, N. Masuda, Y. Uchida, H. Okamoto, T. Kageyama, S. Hosokawa, F. Loyer, K. Mitsuhara, K. Yamanaka, Y. Tamenori, C. Tassel, T. Yamamoto, T. Tanaka and H. Kageyama, *Adv. Energy Mater.*, 2018, 1801772.
67. F. Chang, I. Tezsevin, J. W. de Rijk, J. D. Meeldijk, J. P. Hofmann, S. Er, P. Ngene and P. E. de Jongh, *Nat. Catal.*, 2022, **5**, 222–230.
68. C. J. H. Jacobsen, *Chem. Commun.*, 2000, 1057–1058.
69. S. M. Hunter, D. H. Gregory, J. S. J. Hargreaves, M. Richard, D. Duprez and N. Bion, *ACS Catal.*, 2013, **3**, 1719–1725.
70. C. D. Zeinalipour-Yazdi, J. S. J. Hargreaves and C. R. A. Catlow, *J. Phys. Chem. C*, 2015, **119**, 28368–28376.
71. C. D. Zeinalipour-Yazdi, J. S. J. Hargreaves and C. R. A. Catlow, *J. Phys. Chem.*, 2018, **122**, 6078–6082.
72. M. Kitano, J. Kujirai, K. Ogasawara, S. Matsuishi, T. Tada, H. Abe, Y. Niwa and H. Hosono, *J. Am. Chem. Soc.*, 2019, **141**, 20344–20353.
73. T. N. Ye, S. W. Park, Y. Lu, J. Li, M. Sasase, M. Kitano, T. Tada and H. Hosono, *Nature*, 2020, **583**, 391–395.
74. T. N. Ye, S. W. Park, Y. Lu, J. Li, M. Sasase, M. Kitano and H. Hosono, *J. Am. Chem. Soc.*, 2020, **142**, 14374–14383.
75. A. Cao, V. J. Bukas, V. Shadravan, Z. Wang, H. Li, J. Kibsgaard, I. Chorkendorff and J. K. Nørskov, *Nat. Commun.*, 2022, **13**, 2382.
76. K. Zhang, A. Cao, L. H. Wandall, J. Vernieres, J. Kibsgaard, J. K. Norskov and I. Chorkendorff, *Science*, 2024, **383**, 1357–1363.
77. R. Manabe, H. Nakatsubo, A. Gondo, K. Murakami, S. Ogo, H. Tsuneki, M. Ikeda, A. Ishikawa, H. Nakai and Y. Sekine, *Chem. Sci.*, 2017, **8**, 5434–5439.
78. B. S. Zhang, Y. Wang, N. Zhang, J. Zhu, W. Ji, F. Chen, X. Chen, Y. Yu and B. Zhang, *Sci. Bull.*, 2023, **68**, 1871–1874.
79. Y. Guan, H. Wen, K. Cui, Q. Wang, W. Gao, Y. Cai, Z. Cheng, Q. Pei, Z. Li, H. Cao, T. He, J. Guo and P. Chen, *Nat. Chem.*, 2024, **16**, 373–379.
80. L. Zhou, X. Li, Q. Li, A. Kalu, C. Liu, X. Liu and W. Li, *ACS Catal.*, 2023, **13**, 15087–15106.
81. S. Yang, T. Zhang, Y. Yang, B. Wang, J. Li, Z. Gong, Z. Yao, W. Du, S. Liu and Z. Yu, *Appl. Catal., B*, 2022, **312**, 121404.
82. W. B. Gao, J. P. Guo, P. K. Wang, Q. R. Wang, F. Chang, Q. J. Pei, W. J. Zhang, L. Liu and P. Chen, *Nat. Energy*, 2018, **3**, 1067–1075.
83. G. F. Han, F. Li, Z. W. Chen, C. Coppex, S. J. Kim, H. J. Noh, Z. Fu, Y. Lu, C. V. Singh, S. Siahrostami, Q. Jiang and J. B. Baek, *Nat. Nanotechnol.*, 2021, **16**, 325–330.
84. P. Mehta, P. Barboun, F. A. Herrera, J. Kim, P. Rumbach, D. B. Go, J. C. Hicks and W. F. Schneider, *Nat. Catal.*, 2018, **1**, 269–275.

CHAPTER 10

Electrocatalytic and Photocatalytic Conversion of Nitrogen

A. DAISLEY

School of Chemistry, Joseph Black Building, University of Glasgow, Glasgow, G128QQ, UK
Email: angela.daisley@glasgow.ac.uk

10.1 Introduction

Ammonia is one of the most important chemicals produced due to its use in making synthetic fertilisers and potential use as a fuel.[1] The industrial route for producing ammonia is *via* the Haber–Bosch process, where N_2 and H_2 are reacted over a doubly promoted iron catalyst at high temperatures and pressures. The hydrogen required for this process is typically generated by the steam reforming of hydrocarbons. Due to these conditions, ammonia production accounts for 1–2% of the world's energy demand and 1.6% of manmade CO_2.[2]

As a result of the negative effects of the Haber–Bosch process, there has been a focus on developing processes that are more sustainable for ammonia production. In order to facilitate localised production, a process would need to be able to operate under more moderate conditions. One possible way to produce ammonia more sustainably is by using electrocatalysis. The electrocatalytic nitrogen reduction reaction (NRR) uses water as the hydrogen source and can be performed at low temperatures and pressures.[3] However, this is a challenge due to the hydrogen evolution reaction (HER) being more favourable and the nitrogen triple bond being

Catalysis Series No. 49
Catalytic Activation of Small Molecules
Edited by Mustafa Yasin Aslan, Angela Daisley, Justin S. J. Hargreaves and José L. Rico
© The Royal Society of Chemistry 2025
Published by the Royal Society of Chemistry, www.rsc.org

difficult to break.[4] As a result, there is generally a low faradaic efficiency for ammonia synthesis at ambient conditions and the small rates achieved limit the commercial potential of this process. Therefore, there has been interest in improving the electrocatalyst, in order to achieve high activity and selectivity as will be discussed in the below sections.

Another possible way to produce ammonia more sustainably is by using photocatalysis. Photocatalytic NRR uses solar energy to perform the reaction under ambient conditions, where a semiconductor uses photo-generated electrons to activate nitrogen.[3] The main challenge for this process is the low efficiency created by the difficulty in nitrogen surface adsorption and activation, hole-pair recombination and appropriate band gap excitation. To solve these issues, defect and band gap engineering and facet and morphology control have been used to design a suitable photocatalyst.

In the below sections, the production of ammonia by electrocatalytic and photocatalytic processes will be discussed in more detail.

10.2 Electrocatalytic Conversion of N_2

In the electrocatalytic reaction, water is oxidised to produce oxygen at the anode and protons and electrons are formed as shown in eqn (10.1):

$$2H_2O \rightarrow O_2 + 4H^+ + 4e^- \tag{10.1}$$

The protons pass through the electrolyte to the cathode, where nitrogen is reduced to form ammonia:[5]

$$N_2 + 6H^+ + 6e^- \rightarrow 2NH_3 \tag{10.2}$$

There have been two main proposed mechanisms for the NRR: a dissociative and associative mechanism.[5] In the dissociative mechanism, the first step involves breaking the nitrogen triple bond, before hydrogenation can occur in the following steps (see Figure 10.1). For the associative mechanism, the nitrogen molecule is adsorbed onto the surface without breaking the nitrogen bond. Then, the reaction can occur *via* two pathways depending on the addition of the hydrogen atoms. In the alternating pathway, the two N atoms are hydrogenated in rotation whereas in the distal pathway, the nitrogen farthest from the surface is hydrogenated first to form ammonia and then, the second N is hydrogenated.

To describe the activity of the catalysts for NRR, the rate and the faradaic efficiency (FE) are reported. The rate is provided as either the catalyst loading mass or the electrode surface area. The FE defines the selectivity of the electrocatalytic process and compares the charge used for the nitrogen reduction to the total charge passed.

In order to accurately determine the ammonia synthesis activity, reliable detection methods and control experiments have to be conducted. Ammonia and nitrogen oxide in air, as well as contamination from the electrolyte and membrane, can result in false findings.[6] Therefore, these sources of

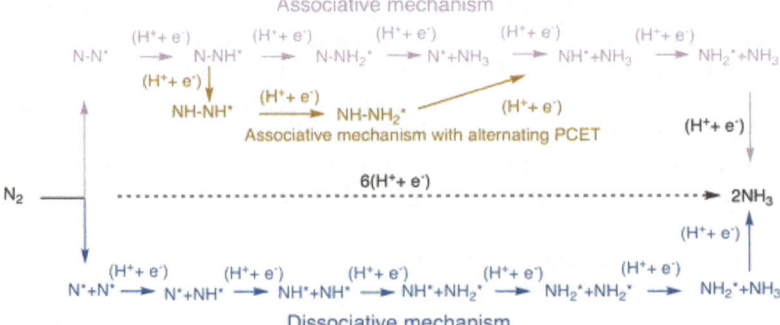

Figure 10.1 Diagram of the possible mechanisms for the electrocatalytic conversion of N_2 to NH_3. The top pathway shows the two associative mechanisms and the bottom pathway shows the dissociative mechanism. Reproduced from ref. 32 with permission from the Royal Society of Chemistry.

ammonia must be removed before accurate results can be recorded. Background control experiments using Ar and $^{15}N_2$ isotopic experiments are necessary to confirm the origin of any detected ammonia. Nitrogen-containing materials such as metal nitrides can also decompose during the reaction and therefore, the nitrogen mass balance must be measured, and ideally longer reactions should be performed to verify the stability of any materials. Furthermore, depending on the pH of the solution, the most reliable ammonia detection methods can vary.[7] Similar concerns have to be taken into consideration during photocatalytic NRR.

10.2.1 Electrocatalytic Conversion of N_2 by a Metal Cathode

Density functional theory (DFT) calculations have been performed on a range of metal surfaces to investigate their potential as electrochemical ammonia synthesis catalysts *via* either the associative or dissociative mechanisms.[4,5] Noble metals such as Ru, Rh and Ir are reported to be potentially the most active metals for NRR, as shown in Figure 10.2. It was also concluded that the flat surfaces of the transition metals Sc, Y, Ti and Zr would be able to reduce N_2 to form ammonia as the nitrogen should bind more strongly to the electrode than hydrogen.[5] However, for other metals, such as Ru or Pt, the HER reaction would be competing with NRR as the protons would cover the metal surfaces, which would prevent nitrogen adsorption and ammonia formation and result in low faradaic efficiencies (FE). Therefore, there is a direct relationship between the adsorption of nitrogen on the catalyst and ammonia production activity.

Two of the most widely studied noble metal catalysts for NRR are gold and ruthenium. Ru/C has been reported to have a low rate of 1.3 µg h^{-1} cm^{-2} at −1.02 V *vs.* Ag/AgC at 1 atm and 90 °C.[8] The anode used was Pt, and Nafion and 2 M KOH were used as the electrolyte. Hydrogen formation was found to

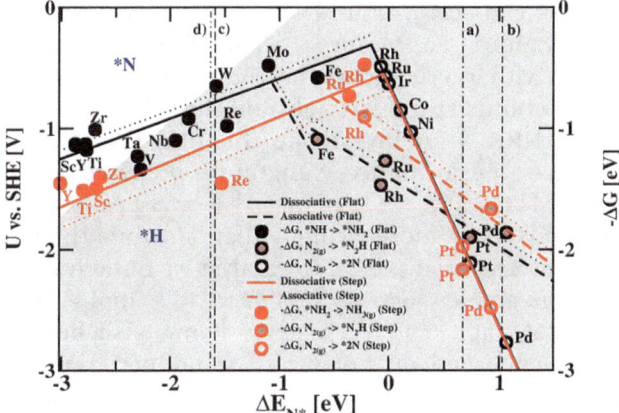

Figure 10.2 Volcano plot of NRR *via* a dissociative mechanism or associative mechanism for metals. Reproduced from ref. 5 with permission from the Royal Society of Chemistry.

be the main limiting factor. It has been stated that particle size can have an effect on the activity. Therefore, Ru nanoparticles were deposited on carbon fibre paper and this material obtained a rate of 5.5 mg h^{-1} m^{-2} at −0.1 V *vs.* reversible hydrogen electrode (RHE) at 20 °C.[9] A faradaic efficiency of 5.4% at 0.01 V *vs.* RHE was observed. The N$_2$ adsorption was suggested to occur rapidly on the edge sites of Ru(001) surfaces through a dissociative mechanism. It has been found that alloying Ru[10] or doping Ru on oxides[11,12] can also improve the activity. For Au, it has been shown that both the morphology and support can have an effect on the performance. Gold with a flower-like structure has a rate of 25.57 μg h^{-1} mg^{-1} at −0.2 V *vs.* RHE and an FE of 6.05% under ambient conditions.[13] This is higher than Au spheres (15.15 μg h^{-1} mg^{-1} [13]) and Au nanorods (6.04 μg h^{-1} mg^{-1} [14]), with the higher activity being attributed to the increase of exposed active sites for NRR. Hollow gold nanocages have also been reported to have better activity than Au nanoparticles of other shapes (rods, spheres and cubes), with a rate of 3.9 μg h^{-1} cm^{-2} at −0.5 V *vs.* RHE and FE of 30.2% at −0.4 V *vs.* RHE[15] due to the increase in surface area and a cage effect that results in more contact of the reactants with the Au. Au nanoparticles with high-index facets reportedly have an FE of 73.32% and a rate of 9.22 μg h^{-1} cm^{-2}.[16] DFT calculations suggest that the high-index surfaces preferentially adsorp *NNH.

To further improve the activity, supporting amorphous Au nanoparticles on cerium oxide and reduced graphite oxide was performed and a rate of 8.3 μg h^{-1} mg^{-1} and FE of 10.10% at −0.2 V *vs.* RHE was reported.[17] It was suggested that the cerium oxide transforms the Au NPs into an amorphous form, resulting in this material having a high concentration of active sites due to the structural distortion. Au has also been supported on TiO$_2$,[18] β-FeOOH nanotubes,[19] Bi nanosheets[20] and C$_3$N$_4$.[21] The support is suggested to provide a synergistic effect that promotes nitrogen adsorption.

Nitrogen fixation can occur naturally by nitrogenase enzymes, which have active centres containing Fe, Mo or V.[22] Therefore, catalysts have been developed for NRR with these metals as the focus. DFT calculations have suggested that Mo binds H too strongly and therefore, HER would be more competitive than NRR.[23] To investigate this and the effect crystal phase orientations have on performance, Yang et al. prepared Mo catalysts with different ratios of (110) and (211) planes.[24] It was proposed that the (110) plane of Mo can adsorb N more strongly than H, while the opposite is true for the (211) plane. As a result, the Mo catalyst with the highest ratio of the (110) plane had the greatest activity of 3.09×10^{-11} mol s^{-1} cm^{-2} and FE of approximately 0.25% at −0.49 V vs. RHE. It has also been reported that single Mo atoms anchored on holey nitrogen-doped graphene (Mo/HNG) and restrained on carbon fibre paper can allegedly obtain a rate of 3.6 μg h^{-1} mg^{-1} and FE of 50.2% at −0.05 V vs. RHE.[25] The catalyst showed good stability and continuously high FE over longer reaction tests of 20 000 s. The active sites for the reaction were suggested to be Mo–N and the vacancies in HNG lower the formation energy of *NNH intermediate. The effect the metal–support interaction has on NRR vs. HER has been investigated for Mo/VO$_2$.[26] This interaction is reported to create electron-deficient sites, which can activate nitrogen and VO$_2$ weakly adsorbs H. The rate of activity for Mo/VO$_2$ is 190.1 μg h^{-1} mg^{-1} and the FE is 32.4% at −0.5 V vs. RHE.

The metal–support interaction of iron has also been investigated for TiO$_2$,[27] carbon nanotubes[28] and Fe$_3$O$_4$.[29] It was found that the form of iron can have an effect on the activity, with Fe/Fe$_3$O$_4$ having an improved rate compared to Fe, Fe$_3$O$_4$ and Fe$_2$O$_3$ nanoparticles. Iron has also been supported on N-doped carbon, which reportedly suppresses HER performance by the structure favourably attracting the access of nitrogen.[30] A rate of 7.48 μg h^{-1} mg^{-1} and an FE of 56.55% were achieved under ambient conditions. The catalyst was shown to be stable during consecutive recycling experiments. Isotopic labelling using ^{15}N$_2$ confirmed that the ammonia produced was from the NRR process. Molybdenum and iron nitrides, carbides and sulphides[31] have also been investigated for NRR and some of these will be discussed in the below sections.

Through DFT calculations, it has been recently suggested that Mn, Ga and In are potential catalysts for NRR as they bind N strongly and therefore, the HER reaction would be limited.[23] As a result, Sinha et al. have investigated the potential of Ga supported on a carbon material as a cathode for the NRR.[32] However, the experimental results showed that the detected levels of ammonia were similar to the concentrations during background ammonia tests. Therefore, it was proposed that Ga is not suitable for NRR as H adsorption was more favourable.

Furthermore, it has been suggested that the scaling relationship between intermediates of N$_2$H* and NH$_2$* results in there being a minimal possible overpotential.[4,33] To overcome this, a catalyst would need to be developed that can strongly bind N$_2$H*, while destabilising NH$_2$*. This could possibly be accomplished by using other catalysts than pure metals, such as metal nitrides.

10.2.2 Electrocatalytic Conversion of N_2 by Metal Nitrides and Carbides

It has been suggested that the metal nitrides can undergo a Mars–van Krevelen mechanism for ammonia synthesis,[34] which is different from the dissociative or associative mechanisms discussed for metals. In this mechanism, the surface lattice nitrogen reacts with hydrogen to form ammonia and this results in a nitrogen vacancy. The vacancy is then replenished by N_2 and the metal nitride is regenerated. However, the nitrogen vacancy could also be filled with oxygen or hydrogen from the electrolyte, which could result in the catalyst becoming inactive.[35] Therefore, the regeneration and stability against poisoning have to be taken into consideration.

DFT calculations on the (100) and (111) facets of the rock salt structure and the (100) and (110) facets of the zinc blende structure have been performed to investigate the stability and activity of transition metal binary nitrides for NRR.[34,36,37] These four facets were chosen as they are the most stable for these nitrides. It was suggested that the (100) facet of the rock salt structure for ZrN, VN, NbN and CrN and the (110) facet of the zinc blende structure for RuN, CrN and WN would be stable and active with low onset potentials as shown in Figure 10.3. However, nitrides will have a multifaceted structure rather than be single-crystalline and hence, all four facets for these nitrides were compared.[38] Although some of the facets of VN, CrN and NbN are highly active, other facets are either inactive or unstable due to ammonia formation leading to decomposition of the nitride. All the facets of

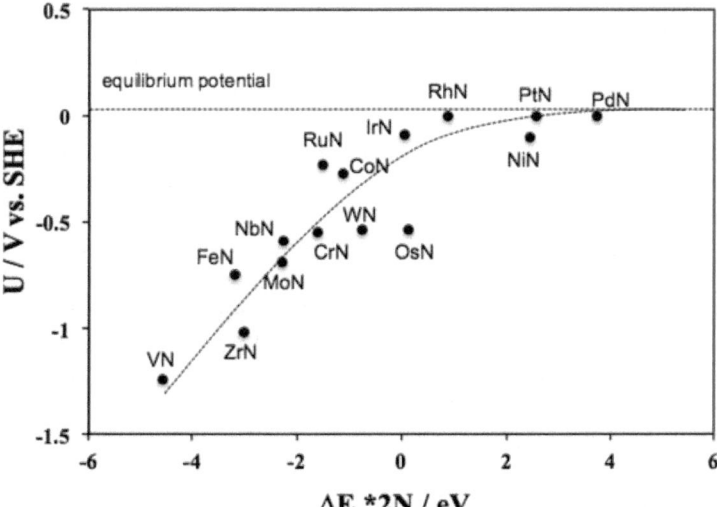

Figure 10.3 Onset potential of ammonia formation *versus* the chemisorption energy of two N-adatoms in the N-vacancy on the (110) facet of zincblende structure of binary nitrides. Reproduced from ref. 34 with permission from American Chemical Society, Copyright 2017.

ZrN are shown to be stable and active for NRR. Another DFT study has also been performed to investigate the structure sensitivity of ZrN through comparison of zirconium oxynitride, zirconium nitride (110) surface and stepped zirconium nitride.[39] The formation of an N vacancy on a step surface was shown to be energetically more unfavourable compared to the pristine surface and O is more likely to bind than N. The (100) facet of the rock salt structure for VN is predicted to be highly active with an FE of 100% at 0.51 V. The (110) facet of RuN in the zinc blende structure is predicted to be active at 0.23 V.

In order to confirm the computational studies, VN nanoparticles, VN nanosheets on Ti mesh (VN/TM) and VN nanowire array on carbon cloth (VN/CC) have been investigated for NRR.[40–42] VN nanoparticles have an initial rate of 3.3×10^{-10} mol s^{-1} cm^{-2} and FE of 6% at -0.1 V vs. RHE, which decreased to 1.1×10^{-10} mol s^{-1} cm^{-2} and 1.6% over 116 h. The active phase for the reaction is suggested to be $VN_{0.7}O_{0.45}$, with surface N sites next to surface O being the active component. The deactivation of the material is due to the conversion of the oxynitride phase to form VN. $^{15}N_2$ was used to confirm the source of nitrogen in the produced ammonia, with both $^{14}NH_3$ and $^{15}NH_3$ being produced, implying that the reaction operates via a Mars–van Krevelen mechanism. Both VN/TM and VN/CC were shown to be selective and stable over 10 consecutive recycling tests. A $^{15}N_2$ isotopic labelling experiment showed that both $^{14}NH_4^+$ and $^{15}NH_4^+$ were also observed over VN/TM. However, some other studies have found that VN and niobium nitride are catalytically inactive for NRR and decompose through the loss of lattice nitride.[43,44]

Chromium nitride has also been investigated experimentally for NRR, with Cr_2N having a rate at 80 °C of 1.4×10^{-11} mol s^{-1} cm^{-2} and FE of 0.58% at -0.2 V.[45] XPS reveals that CrN, oxynitride and oxide phases exist on the surface of Cr_2N. A decrease in activity of 20% was noted after 24 h at -0.2 V. In contrast, the bulk CrN phase was observed to have negligible activity under the same conditions. It is therefore suggested that Cr_2N is the active component and the deactivation observed is due to both leaching of surface lattice nitrogen and conversion of Cr_2N to the CrN phase. However, CrN nanocubes are reported to be active and obtain a rate of 31.11 µg h^{-1} mg^{-1} with an FE of 16.6% at -0.5 V.[46] The CrN nanocubes are also partially oxidised during the reaction and EELS and EDX analyses suggest that an O-rich core and N-rich shell are formed. The material was shown to be stable over five recycling tests. Chromium oxynitride is reported to be more active than CrN and Cr_2O_3 due to the nitrogen on the oxynitride surface being more easily reduced than on CrN due to enhanced charge transfer.[47]

Two-dimensional Ti_2N MXene has been stated to operate via a Mars–van Krevelen mechanism for NRR.[48] However, isotope labelling experiments need to be performed to further support this conclusion. A rate of 11.33 µg h^{-1} cm^{-2} and FE of 19.85% at -0.25 V versus RHE are obtained. $Ti_3C_2T_x$ MXene has also been stated to be active for NRR with a rate of 4.72 µg h^{-1} cm^{-2} and FE of 4.62% at -0.1 V.[49] It was shown that by decreasing the size of the MXene, the FE could be improved and the activity could be enhanced by increasing the number of active edge sites.

As mentioned previously, the NRR activity of molybdenum carbide has been investigated. Mo$_2$N nanorods on a glassy carbon electrode have been shown to be stable and active for NRR, with a rate of 95.1 µg h^{-1} mg^{-1} and FE of 8.13% at −0.3 V.[50] Molybdenum nitride has also been of interest as Mo$_2$N nanorods,[51] MoN nanosheet array on carbon cloth (MoN NA/CC)[52] and MoN nanocrystals (MV-MoN@NC)[53] have been examined. These molybdenum nitrides have been shown to be stable over recycling and long reaction tests. Isotope labelling experiments with ^{15}N$_2$ for MoN NA/CC and MV-MoN@NC suggest that the Mars–van Krevelen mechanism is in operation for these materials. The introduction of Mo vacancies is reported to increase the activity as the rate-limiting step is changed from the reduction of *NH to *NH$_2$, to the reduction of *NH$_2$ to form *NH$_3$, which has a lower reaction barrier.[53] However, there has been a report that Mo$_2$N decomposes into ammonia during the reaction and does not have catalytic activity for NRR.[54] Therefore, it is important that isotope labelling experiments and background checks are performed.

10.2.3 Electrocatalytic Conversion of N$_2$ by Metal Oxides

A range of metal oxides have been investigated for electrocatalytic NRR, including TiO$_2$ and iron oxides. TiO$_2$ nanosheet array supported on Ti plate obtains a rate of 9.16 × 10^{-11} mol s^{-1} cm^{-2} and FE of 2.5% at −0.7 V vs. RHE.[55] A slight decrease in both the rate and FE is observed over 10 cycling experiments, which the authors suggest may be due to the decay of the electrode. Oxygen vacancies formed during the reaction are proposed to increase the adsorption and activation of nitrogen. Ling et al. have looked at regulating the oxygen vacancies by doping TiO$_2$ with Fe, Mn, Co, Ni and Cu metal.[27] Iron on anatase structured TiO$_2$ displays a rate of 30.9 ± 0.4 µg h^{-1} mg^{-1} and an FE of 40.4 ± 1.1% at −0.4 V vs. RHE. XPS and TEM analyses confirm that the number of oxygen vacancies increases after doping with iron. It is suggested that both the oxygen vacancies and iron decrease the barrier for N$_2$ absorption and assist with the formation step for *NNH. It has also been shown that the insertion of Li into TiO$_2$ nanosheets can increase the number of oxygen vacancies, which resulted in higher NRR activity.[56] The presence of Ti^{3+} in TiO$_2$ is another factor that can improve the NRR activity, with Ti^{3+} 3d^1 defect states reportedly increasing the adsorption and activation of nitrogen.[57]

γ-Fe$_2$O$_3$ has been coated on porous carbon paper and used in a membrane electrode assembly (MEA) reactor.[58] The activity was improved compared to when γ-Fe$_2$O$_3$ was used in a half-cell with KOH electrolyte. However, the activity decreased significantly over 25 h, which the authors suggest could be due to degradation of γ-Fe$_2$O$_3$. Conversely, Fe$_2$O$_3$ supported on carbon nanotubes was shown to have a constant rate over 60 h of reaction.[28] The carbon sites at the interface of iron particles and the CNTs are proposed to be the active sites for activating nitrogen.

There has also been a focus on the testing of molybdenum oxides for NRR. MoO_3 nanosheets have a rate of 29.43 $\mu g\,h^{-1}\,mg^{-1}$ and an FE of 0.75% at -0.5 V.[59] DFT calculations show that the active sites are related to the outermost Mo atoms. The importance of oxygen vacancies has also been shown for MoO_3 supported on $Ti_3C_2T_x$ MXene, with increased rates and stabilisation of $*N_2/*N_2H$ intermediates.[60]

Perovskite and phosphate-based materials have also been used for low-temperature and low-pressure NRR. $Ce_{0.8}Sm_{0.2}O_{2-\delta}$ and $SmFe_{0.7}Cu_{0.3-x}Ni_xO_3$ have been tested at temperatures of between 25 to 100 °C and the highest rate was attained for $SmFe_{0.7}Cu_{0.1}Ni_{0.2}O_3$ at 80 °C (1.13×10^{-8} mol s^{-1} cm^{-2}).[61] The effect the morphology has on the activity has been studied for $Ag_2VO_2PO_4$.[62] The material was shown to be active and durable over 24-h reaction, with the rice-grain-like morphology being superior due to the increased surface area.

In Table 10.1, some materials for the electrocatalytic NRR have been given in more detail with the conditions used during the reaction. The recent progress in the development of materials for NRR shows the possibility of electrocatalytic conversion of N_2 to ammonia at ambient conditions. However, there are still some major challenges in this process, including the competition with HER and long-term stability of the cathode. The cathode needs to be highly active, stable and selective to be viable.

10.3 Photocatalytic Conversion of N_2

In the photocatalytic reaction, solar energy is used to convert nitrogen to ammonia over a photocatalyst. The photocatalytic process can be separated into the following two steps:

I The photocatalyst absorbs light causing electrons in the valence band to be excited and jump to the conduction band. This results in the formation of positively charged holes in the valence band and an electron–hole pair is created.

II Either some of the electron–hole pairs will recombine or they will migrate to the surface and participate in the reaction. Water can be oxidised to form oxygen and protons by reaction with the holes in the valence band. Nitrogen can react with the photogenerated electrons and the protons generated from the splitting of water to produce ammonia.

The overall process is provided in Figure 10.4. The first electron transfer to form N^{2-} requires a reduction potential of -4.16 V vs. NHE at pH 0.[63] Therefore, the photocatalyst must have a relatively small band gap and the conduction band should be more negative than the reduction potential of N_2 to achieve efficient ammonia synthesis activity. The expected mechanism of nitrogen activation over photocatalysts is similar to electrocatalysis, with dissociative and associative pathways for metals and the Mars–van Krevelen mechanism in operation for nitrides.[63,64]

Table 10.1 Selected catalysts for electrocatalytic NRR with the stated/used reaction conditions.

Cathode catalyst	Electrolyte	Temperature (°C)	NH$_3$ rate	FE (%)	Potential (V)	Ref.
Ru/C	Nafion and 2 M KOH	90	1.30 μg h^{-1} cm^{-2}	—	−1.02	8
RuPt	1.0 M KOH	RT	5.1 × 10^{-9} g s^{-1} cm^{-2}	13.2	0.123	10
LaFeO-Ru	0.1 M K$_2$SO$_4$	RT	137.5 ± 5.8 μg h^{-1} mg^{-1}	56.9 ± 4.1	−0.7	11
Au flowers	0.1 M HCl	RT	25.57 μg h^{-1} mg^{-1}	6.05	−0.2	13
Au/TiO$_2$	0.1 M HCl	RT	21.4 μg h^{-1} mg^{-1}	8.11	−0.2	18
Mo/VO$_2$	0.05 M H$_2$SO$_4$, pH = 1	RT	190.1 μg mg$_{cat}^{-1}$ h^{-1}	32.4	−0.5	26
Fe/Fe$_3$O$_4$	0.1 M phosphate buffer solution	RT	0.19 μg h^{-1} cm^{-2}	8.29	−0.3	29
VN/CC	0.1 M HCl	RT	2.48 × 10^{-10} mol^{-1} s^{-1} cm^{-2}	3.58	−0.3	42
CrN nanocubes	0.1 M HCl	RT	31.11 μg h^{-1} mg^{-1}	16.64	−0.5	46
MoN NA/CC	0.1 M HCl	RT	3.01 × 10^{-10} mol s^{-1} cm^{-2}	1.15	−0.3	52
Ti$_2$N MXene	0.1 M HCl aqueous solution	RT	11.33 μg h^{-1} cm^{-2}	19.85	−0.25	48
Fe-TiO$_2$	0.1 M Na$_2$SO$_4$	RT	30.9 ± 0.4 μg h^{-1} mg^{-1}	40.4 ± 1.1	−0.4	27
MoO$_{3-x}$/MXene	0.5 M LiClO$_4$	RT	95.8 μg h^{-1} mg^{-1}	20	−0.4	60
Ag$_2$VO$_2$PO$_4$	0.1 M KOH	RT	1.48 mg h^{-1} mg$_{cat}^{-1}$	37.46	−0.2	62

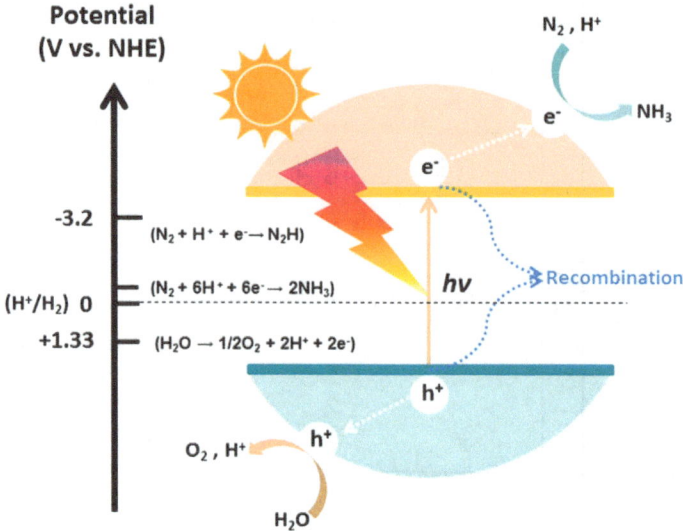

Figure 10.4 Diagram of semiconductor-based photocatalysts used for the reaction of nitrogen to form ammonia. The redox potentials (V vs. NHE at pH = 0) are provided on the left. Reproduced from ref. 63 with permission from the Royal Society of Chemistry.

In order to prevent the recombination of the electron–hole pair, sacrificial donors such as ethanol are used as hole scavengers, which results in an increase in the activity.[65] It has been shown that the type of sacrificial donor can have an effect on the rate and therefore, direct comparisons between materials can be challenging if different scavengers have been used. Furthermore, sacrificial donors can interfere with the detection methods used to determine the amount of ammonia produced.[66] Therefore, control experiments and the type of detection method have to be taken into consideration for accurate ammonia determination.

10.3.1 Photocatalytic Conversion of N_2 by Metal Oxides

Initial studies on soil and sand showed that photocatalytic conversion of nitrogen was possible using nitrogen and water.[67,68] It was suggested that rutile in the samples was responsible for the photocatalytic activity.[69] Therefore, titania materials have been of great interest for photocatalytic nitrogen activation. There is a direct relationship between the rutile content and activity as demonstrated in Figure 10.5. Doping iron, cobalt, molybdenum or nickel on anatase can increase the activity and influences the rutile content of the material.[70] A loading of 0.2% Fe on TiO_2 resulted in a rutile content of 20–30% and the highest activity of 6.0 ± 2.5 µmol. Electron spin resonance spectroscopy and infrared spectroscopy have been used to study the role Fe^{3+} plays in the reaction.[71] Fe^{3+} traps the photogenerated electrons and is reduced to Fe^{2+}, and thus promotes the charge separation of

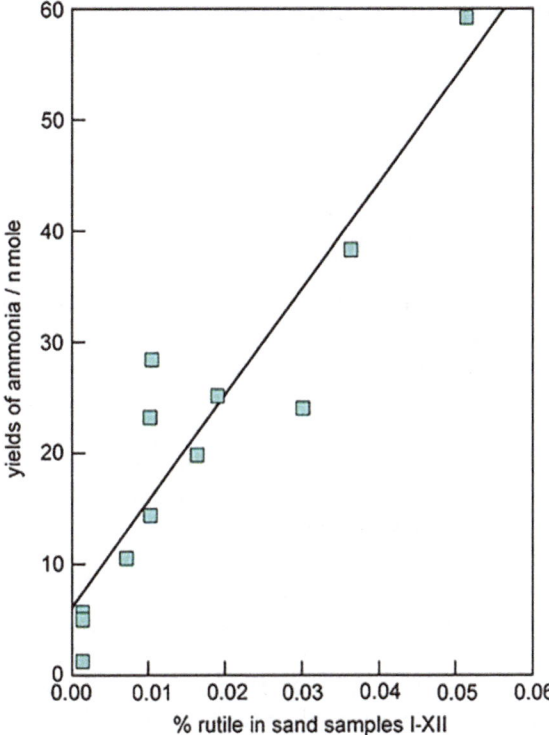

Figure 10.5 Ammonia yield *versus* rutile content for sand samples. Reproduced from ref. 69 with permission from Springer Nature, Copyright 2011.

holes and electrons. Doping TiO_2 with chromium resulted in a lower activity than the iron-doped sample due to the rate of electron–hole recombination being higher.[72] Other metals such as Au, Pt, Pd and Ag were also suggested to lower the photocatalytic activity as the conversion of anatase to the active rutile phase was prevented.[70] However, other research proposes that the loading of Ru, Rh, Pd or Pt on TiO_2 does increase the activity, with 0.24 wt% Ru/TiO_2 having a yield of 1.73 µmol during 1 h of irradiation.[73] It was stated that there was a direct link between the yield and the intermediary M–H bond strength.

It is also noted that a pre-treatment temperature of 1000 °C is required to enhance the activity, with this annealing process expected to produce surface oxygen vacancies.[74] These surface defects are proposed to assist in improving nitrogen adsorption. The active sites are reported to be the Ti^{3+} species in the oxygen vacancies, which can trap the photogenerated electrons and activate nitrogen by electron donation to form Ti^{4+} species.[75] Zhang *et al.* have focused on controlling the number of oxygen vacancies on TiO_2 and found that there is an optimal concentration, with higher vacancy concentrations resulting in increased electron–hole recombination.[76] DFT calculations on rutile TiO_2(110) surface containing oxygen vacancies show that the reaction is driven by water photolysis and photogenerated electrons.[77] Doping of Au on TiO_2 is suggested to increase the number of oxygen vacancies and improve charge separation, with 1 wt% Au/TiO_2 nanotubes having a yield of 5.5 µmol L^{-1} after 5 hours.[78] Single atoms of Ru on TiO_2 nanosheets rich in oxygen vacancies impede HRR and can increase the charge carrier separation.[79] The 1 wt% Ru on TiO_2 is stated to have a rate of 56.3 µmol $h^{-1} g^{-1}$. Meanwhile, single atoms of Ru supported on H_xMoO_{3-y} are suggested to activate H_2 and the Mo^{n+} species trap the localised electrons and activate N_2.[80]

Ternary oxides $SrTiO_3$ and $BaTiO_3$ are able to produce a small amount of ammonia photocatalytically, with yields of 0.41 µmol and 0.87 µmol after 2 h of irradiation, respectively.[81] The activity is shown to increase when these oxides are doped with NiO or RuO_2. However, the activity decreased after 2 h, which the authors suggested could be due to photodecomposition of ammonia. Doping $SrTiO_3$ with iron also increases the activity as the Fe^{3+} sites are stated to activate the N_2.[82] A rate of 30.1 µmol $h^{-1} g^{-1}$ was obtained for $Fe_{0.10}Sr_{0.90}TiO_3$ over 2 hours of reaction.

The photocatalytic performance of iron oxide has also been investigated. Iron oxide has a smaller band gap than TiO_2 (Fe_2O_3: 2.3 eV and TiO_2: 3.2 eV), and thus can be photocatalytically active in the visible light region.[83] A mixture of 90% α-Fe_2O_3 and Fe_3O_4 was initially active, with a yield of approximately 75 µmol after 580 h of irradiation with light of wavelength shorter than 540 nm.[84] However, deactivation was observed after 450 h of illumination. Lashgari et al. have compared the photocatalytic performance of Fe_2O_3 and TiO_2 for NRR.[85] Fe_2O_3 was observed to be more active, which was suggested to be due to iron oxide absorbing more photons in the visible region. As a result, a mixture of Fe_2O_3 and TiO_2 was more active than TiO_2 alone. Hydrazine was observed as a side-product and a decrease in the activity was observed in this study, which the authors suggest is due to the photocatalytic degradation of ammonia. Tennakone et al. have shown that hydrous ferric oxide suspended in aqueous solution can photocatalytically produce ammonia, and the activity can be enhanced by substituting half of the Fe^{3+} with V^{3+}.[86,87] The increase in performance was proposed to be due to the V^{3+} sites capturing the photogenerated holes.

Other metal oxides that are reportedly active for photocatalytic conversion of nitrogen to ammonia include ZnO,[88] tungsten oxide,[89–91] molybdenum oxide[92] and Ga_2O_3.[93] The creation of defects and oxygen vacancies in these oxides is suggested to assist in the activation of nitrogen. ZnO that was synthesised by wet etching is reported to have grain boundaries that resulted in defects.[94] Pt was doped on the ZnO and the material had a yield of 86 µmol after 1 h of irradiation. ZnO has also been doped with MoS_2 and nickel phosphide to investigate the effects these would have on the performance.[95,96] Both the Ni_xP_y co-catalyst and MoS_2 act to increase the photocatalytic performance of ZnO by improving the separation of electron–hole pairs. The Ni_xP_y co-catalyst is also reported to be the active site for nitrogen and proton activation. The oxygen vacancies on MoO_{3-x} nanobelts are reported to be situated on both the (001) and (100) planes and theoretical calculations suggest that the N_2 chemisorbs onto these vacancies.[92] Lu et al. have shown that mixing MoS_{2+x} with MoO_{3-y} results in the material being able to selectively activate N_2 and H_2O and increases the rate to 141 $\mu mol\,h^{-1}g^{-1}$ during 2 h of irradiation.[97] Oxygen vacancies were created in Bi_2MoO_6 by hydrogen treatment and the resulting material was able to convert nitrogen from air.[98] It is clear from these results that oxygen vacancies play an important role in photocatalytic conversion of nitrogen.

10.3.2 Photocatalytic Conversion of N_2 by g-C_3N_4

The role of nitrogen vacancies in the photocatalytic conversion of nitrogen has been investigated for g-C_3N_4.[99] It is suggested that these vacancies can activate N_2 and trap the photogenerated electrons, preventing the recombination of electron–hole pairs. The nitrogen vacancies can be created by either heat treatment with nitrogen or KOH etching.[64] Furthermore, carbon vacancies in C_3N_4 can also activate nitrogen and improve the activity compared to bulk C_3N_4.[100,101] The carbon defect material produces 54 $\mu mol\,L^{-1}$ of ammonia over 100 min.

Doping g-C_3N_4 with Fe_2O_3,[102] boron,[103] Ga_2O_3,[65] metal sulphides[104] and other oxides[105] has been shown to improve the nitrogen photocatalytic activity. The combination of nitrogen vacancies in C_3N_4 and the interface contact between the dopant and C_3N_4 promote photoreduction of nitrogen.[65] It was found that by varying the ratio of Fe^{3+}, a honeycomb structure could be synthesised, resulting in a larger surface area.[106] The Fe^{3+} was also suggested to trap the photogenerated electrons, activate N_2 and promote electron transfer from the catalyst to N_2. The Z-scheme also plays an important role for some of these materials, with Fe_2O_3/g-C_3N_4 having a rate of 47.9 $mg\,h^{-1}L^{-1}$ during 4 h of irradiation.[102,107]

Quaternary sulphides of g-C_3N_4/ZnSnCdS (7.543 $mg\,L^{-1}h^{-1}g^{-1}$) and g-C_3N_4/ZnMoCdS (3.5 $mg\,L^{-1}h^{-1}g^{-1}$) both have a high NH_4^+ rate under visible light and are stable over 20 hours of reaction.[104,108] Coupling between the sulphide and C_3N_4 resulted in separation of electron–hole pairs and improved charge transfer with the distribution of electrons in ZnMoCdS and the holes in C_3N_4. Thus, the sulphur vacancies act as the active sites for nitrogen activation. These results show the importance of doping and defect engineering on improving the performance of the photocatalytic activity.

10.3.3 Photocatalytic Conversion of N_2 by Oxyhalides

Bismuth oxyhalides, with the general formula BiOX, where X can be Cl, Br or I, have been of interest for photocatalytic conversion of N_2. Oxygen vacancies created in bismuth oxybromide, BiOBr, can activate N_2, transfer the electrons between the photocatalyst to the adsorbed nitrogen and act as trapping sites for photogenerated electrons.[109] The average charge carrier lifetime was almost double with oxygen vacancies present. Li et al. also explored the effect the oxygen vacancies on either the (001) or (010) facets have on the mechanism for BiOCl.[110] It is suggested that nitrogen activation occurs by the distal pathway for the oxygen vacancies on the (001) surface and by an alternating pathway on the (010) surface. Therefore, the oxygen vacancies can have a large influence on the N_2 activation pathway. The effects different facets have on the activity have also been explored for Bi_5O_7I nanosheets.[111] It was found that the (001) surface of Bi_5O_7I had a greater activity, apparent quantum efficiency and carrier separation than the (100) surface. The conduction band of the (001) facet was also more negative, which is stated to relate to the improved performance. Both facets were stable over five cycles of 100 minutes. These results suggest that both oxygen vacancies and facet engineering are important for the development of active photocatalysts.

Formation of bismuth vacancies can control the number of oxygen vacancies as there is a reduction in the required formation energy for the oxygen vacancies.[112] As a result, Bi_3O_4Br nanosheets have improved charge separation and photocatalytic activity when the bismuth vacancies are introduced. Unfortunately, oxygen vacancies on the surface of bismuth oxyhalides are susceptible to being oxidised and therefore, the activity decreases. However, the generation of oxygen vacancies during the reaction is stated to improve the stability.[113] Bi_5O_7Br nanotubes can generate oxygen vacancies under visible light and the material had a rate of 1.38 mmol g^{-1} over 1 h of irradiation and quantum efficiency of 2.3% at 420 nm. It was shown that the oxygen vacancy concentration of this material directly relates to the performance.[114] Doping of the oxyhalides with either Fe,[115] Br[116] or MoO_2[117] can also improve the performance. When iron is doped on BiOBr nanosheets, oxygen vacancies are generated, the band gap is reduced, and the valance and conduction bands are negatively shifted. Fe is stated to be the active site for N_2 activation and the material produces 382 µmol g^{-1} of ammonia over 1 hour of irradiation.

10.3.4 Photocatalytic Conversion of N_2 by Metal Sulphides

Sulphur vacancies have also been explored as active sites for adsorbing N_2 and ammonia formation.[118,119] Hu *et al.* have shown that there is a direct relationship between the number of sulphur vacancies and the formation of ammonium, as shown in Figure 10.6. CdS is of interest as it has the potential to be active under visible light due to the appropriate band edge positions.[83] Sulphur vacancies can be introduced into CdS by doping with another metal such as Mo, Ni, Zn and Sn, which causes crystal lattice distortions. $Zn_{0.11}Sn_{0.12}Cd_{0.88}S_{1.12}$ with a sulphur vacancy concentration of approximately 0.09 has an ammonium rate of 4.75 mg L^{-1} h^{-1} g^{-1} and is stable over 4 hours of reaction.[119] A similar promoted activity is observed for $Mo_{0.12}Ni_{0.13}Cd_{0.86}S_{1.1}$ when compared to CdS. This increase in activity is proposed to be due to the sulphur vacancies increasing the N_2 adsorption, trapping the photogenerated electrons and encouraging the charge transfer between the sulphide to the N_2.

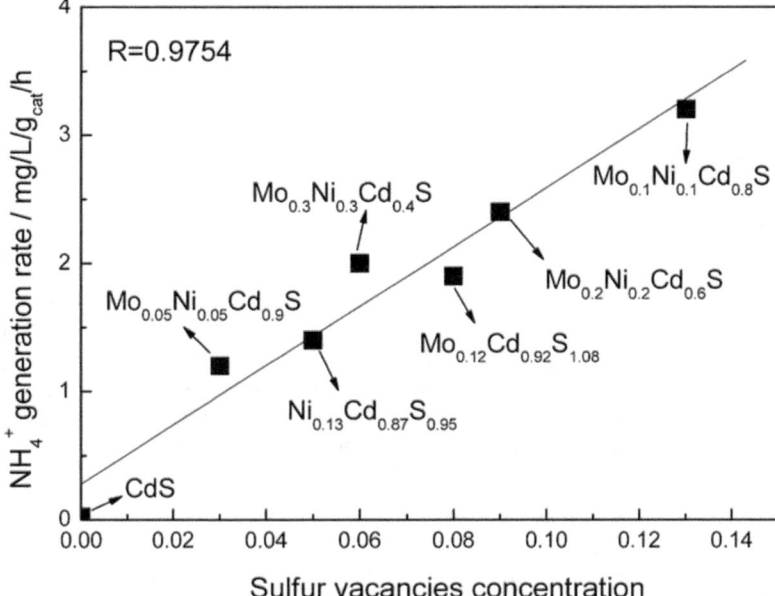

Figure 10.6 The effect of sulphur vacancy concentration on the ammonium formation rate of metal sulphides. Reproduced from ref. 118 with permission from the Royal Society of Chemistry.

There has also been interest in the photocatalytic activity of MoS_2 for NRR. Sun et al. have reported that trions, a high concentration of localised electrons, are the active sites in ultrathin MoS_2.[120] The charged excitons can facilitate the six-electron reduction process to activate N_2 to form NH_3, reducing the production of intermediates. The material has a rate of 325 μmol g^{-1} h^{-1} over 10 hours and was stable over 10 runs. As stated previously, supporting MoS_2 on C-ZnO can result in improvement in the separation of the electron–hole pairs.[95] The authors suggest that the mechanism for NRR changes depending on the wavelength of the light source. Under visible light, the carbon layer transfers electrons to the semiconductor, while under simulated sunlight, it acts as an electron trap. This results in improved performance under simulated sunlight irradiation.

10.3.5 Alternative Materials for the Photocatalytic Conversion of N_2

Layered double hydroxides (LDH) are suitable materials for photocatalytic reactions as they can have narrow band gaps, high surface areas and undercoordinated metal sites.[121] LDH are comprised of positively charged layers with charge balancing anions located in the interlayer space, with the anions being exchangeable. The thickness and cation composition can also be changed, which results in the creation of oxygen vacancies and the tuning

Table 10.2 Selected catalysts for photocatalytic NRR with the stated/used reaction conditions.

Catalyst	Reaction conditions	Light source	NH_3 rate	Scavenger	Ref.
0.24 wt% Ru/TiO$_2$	H$_2$O	150 W Xe arc lamp, UV-vis	1.75 µmol over 1 h	None	73
Reduced TiO$_2$	10:90 mL of methanol and deionised water, 25 °C	300 W Xe lamp, UV-vis	324.9 µmol h^{-1} g^{-1}	Methanol	76
Ru-SA/H$_x$MoO$_{3-y}$	H$_2$, 25 °C	300 W Xenon lamp, λ > 420 nm	4.00 mmol h^{-1}	None	80
BaTiO$_3$	H$_2$O, 40 °C	450 W high-pressure mercury lamp, UV-vis	0.87 µmol over 2 h	—	81
Fe$_2$O$_3$–TiO$_2$	H$_2$O, 25 °C	500 W power-tuneable Xe-lamp	2900 µmol L^{-1} g^{-1} over 2 h	Ethanol	85
Fe$_2$O$_3$	H$_2$O, 25 °C	500 W power-tuneable Xe-lamp	5400 µmol L^{-1} g^{-1} over 2 h	Ethanol	85
1% Pt-ZnO (wet etching)	H$_2$O, RT	450 W Hg lamp, UV	86 µmol/(h/0.1 g)	Na-EDTA solution	94
Mo-doped W$_{18}$O$_{49}$	H$_2$O, RT	300-W Xe lamp, UV-vis	61.9 µmol h^{-1} g^{-1}	None	91
a-MoS$_{2+x}$–MoO$_{3-y}$	H$_2$O, RT, 2.5 mbar	60 W Xe lamp	86.2 µmol h^{-1} g^{-1} over 2 h	—	97
H-Bi$_2$MoO$_6$	H$_2$O, 20 °C, air	300 W Xenon lamp, UV-vis	1.3 mmol h^{-1}	None	98
S-doped g-C$_3$N$_4$ nanosheets	H$_2$O	500-W Xenon lamp	5.99 mM h^{-1} g^{-1} over 4 h	Methanol	100
Fe$_2$O$_3$-g-C$_3$N$_4$	H$_2$O, RT	300 W high pressure Xenon lamp, UV-vis	47.9 mg L^{-1} h^{-1} over 4 h	20 vol% ethanol	102
20% g-C$_3$N$_4$/ZnSnCdS	H$_2$O, 30 °C	250 W high-pressure sodium lamp, λ: 400–800 nm	7.543 mg L^{-1} h^{-1}	0.789 g L^{-1} ethanol	104
BiOBr	H$_2$O, 25 °C	300 W Xenon lamp, λ > 420 nm	104.2 µmol h^{-1} over 1 h	None	109
50 wt% MoO$_2$/BiOCl	H$_2$O	Xenon lamp	35 µmol h^{-1} g^{-1} over 1 h	None	117
Mo$_{0.1}$Ni$_{0.1}$Cd$_{0.8}$S	H$_2$O, 30 °C	250 W high-pressure sodium lamp, λ: 400–800 nm	3.2 mg L^{-1} h^{-1} over 4 h	Ethanol	118
MoS$_2$/C-ZnO	H$_2$O, RT, air	300 W Xe lamp, UV-vis	245 µmol L^{-1} g^{-1} h^{-1}	Ethanol	95
CuNiAl	H$_2$O	100 W LED lamp, visible light	99 µmol h^{-1} g^{-1} over 1 h	—	123
LiH	H$_2$, near ambient conditions	300W xenon lamp, UV λ: 300–420 nm	75 ± 4 µmol h^{-1} g^{-1}	—	125

of the band gap.[122] Zhao et al. studied the photocatalytic NRR activity of ultrathin LDH nanosheets of the type $M^{II}M^{III}$-LDH (M^{II} = Mg, Zn, Ni or Cu and M^{III} = Al or Cr). The oxygen vacancies were proposed to enhance the activation of N_2 and the insertion of Cu^{2+} resulted in further structural distortions and strain, leading to increased ammonia formation. The CuCr-LDH was observed to be the most active of the materials investigated, with a rate of 142.9 µmol L^{-1} over 1 hour of reaction under visible light. More recently, $M^{I}(II)M^{II}(III)$ (M^{I} = Cu or CuNi and M^{II} = Cr or Al) LDHs have been tested for their photocatalytic activity.[123] Partial substitution of Cu with Ni increases the activity as the N_2 adsorption energy is lowered and defects are created. Therefore, by changing the metal cation composition of the LDHs, the oxygen vacancy concentration and photocatalytic performance can be controlled.

Metal–organic frameworks (MOFs) have been of interest for photocatalytic NRR as they are porous and exhibit an adaptable structure, allowing structural engineering. During the synthesis of MOFs, acetic acid can be used to control the growth of the (101) facet, resulting in different ratios of the (100) to (101) facets.[124] DFT calculations suggest that for HMOF (Fe_{III}/Fe_{II}), the (100) facet has more Fe active sites than the (101) facet and that N_2 activation occurs on Fe^{2+} sites. An optimal activity is obtained when the Fe^{3+} to Fe^{2+} ratio is 1 : 1 and the (100) to (101) ratio is 4.84, which is 20 times higher than when the (100) facet is not exposed.

The role of hydrogen vacancies for photocatalytic nitrogen conversion has recently been explored for lithium hydride.[125] When LiH is under UV illumination, the lattice hydrogen reacts to form H_2, leaving behind electron-rich vacancies, called F centres. The activation of nitrogen and N–H bond formation can occur at these F centres under mild conditions and a low H_2 partial pressure. A rate of 75 ± 14 µmol $h^{-1} g^{-1}$ can be obtained over 15 hours of reaction. In contrast, a much higher temperature of 450 °C is required for LiH to be active under thermo-catalytic conditions. These results suggest that hydrides have the potential to be alternative materials for the photocatalytic conversion of N_2.

The results for a selected range of materials for the photocatalytic NRR process are provided in Table 10.2. The examples provided in this section show that this process has promise for converting N_2 under ambient conditions. Nonetheless, the development of materials for nitrogen activation and the accurate detection of ammonia remains a challenge. Furthermore, the use of different scavengers complicates the comparability of different systems, and thus future research needs to focus on a more systematic approach.

10.4 Conclusions

This chapter has detailed the importance of defects, facets, doping, morphology and the use of supports for improving the activity of catalysts for the electrocatalytic and photocatalytic NRR. The incorporation of vacancies into the materials can greatly improve the activity by altering either the rate-limiting step or the mechanism and increasing the N_2 adsorption. Facet control also plays a major role as some facets are more active than others.

In order to improve the activity and selectivity of catalysts for NRR, the material needs to be able to easily activate N_2 and desorb ammonia, while reducing the competing hydrogen production process. The use of non-aqueous electrolytes in electrocatalytic NRR, for example lithium-mediated nitrogen reduction systems,[126] could aid with reducing the access of hydrogen and improve the selectivity.[127] PCET mediators are also an option for suppressing the HER pathway.[128]

Another major challenge for electrocatalytic and photocatalytic NRR is accurately determining the production of ammonia and the reports of false positives in the literature.[129,130] The inclusion of control experiments and isotope labelling studies is a requirement for confirming the origin of the produced NH_3.[131]

Unfortunately, the efficiency and activity of the reported catalysts are too low to be of value.[132] However, the combination of electrocatalytic and photocatalytic processes is a possibility for an alternative and sustainable method of producing ammonia at a higher rate. In the photoelectrochemical cell, the activation of nitrogen takes place at the photocathode and water splitting occurs at the counter electrode. An external bias is applied to aid with the reaction and increases the ammonia formation activity.[3] The development of this process is a possible avenue that could be explored in the future. Further studies are also needed to develop the understanding of the mechanism of N_2 activation and the key intermediate steps of the reaction. Therefore, the combination of theoretical and experimental investigations will allow the development of more stable and active materials for NRR under ambient conditions.

References

1. V. Smil, *Ambio*, 2002, **31**, 126–131126.
2. P. H. Pfromm, *J. Renewable Sustainable Energy*, 2017, **9**, 034702.
3. M. Ismael and M. Wark, *Appl. Mater. Today*, 2024, **39**, 102253.
4. J. H. Montoya, C. Tsai, A. Vojvodic and J. K. Nørskov, *ChemSusChem*, 2015, **8**, 2180–2186.
5. E. Skúlason, T. Bligaard, S. Gudmundsdóttir, F. Studt, J. Rossmeisl, F. Abild-Pedersen, T. Vegge, H. Jónsson and J. K. Nørskov, *Phys. Chem. Chem. Phys.*, 2012, **14**, 1235–1245.
6. H. Liu, N. Guijarro and J. Luo, *J. Energy Chem.*, 2021, **61**, 149–154.
7. Y. Zhao, R. Shi, X. Bian, C. Zhou, Y. Zhao, S. Zhang, F. Wu, G. I. N. Waterhouse, L.-Z. Wu, C.-H. Tung and T. Zhang, *Adv. Sci.*, 2019, **6**, 1802109.
8. V. Kordali, G. Kyriacou and C. Lambrou, *Chem. Commun.*, 2000, 1673–1674.
9. D. Wang, L. M. Azofra, M. Harb, L. Cavallo, X. Zhang, B. H. R. Suryanto and D. R. MacFarlane, *ChemSusChem*, 2018, **11**, 3416–3422.

10. R. Manjunatha and A. Schechter, *Electrochem. Commun.*, 2018, **90**, 96–100.
11. Z. Han, D. Tranca, F. Rodríguez-Hernández, K. Jiang, J. Zhang, M. He, F. Wang, S. Han, P. Wu and X. Zhuang, *Small*, 2023, **19**, 2208102.
12. H. Tao, C. Choi, L.-X. Ding, Z. Jiang, Z. Han, M. Jia, Q. Fan, Y. Gao, H. Wang, A. W. Robertson, S. Hong, Y. Jung, S. Liu and Z. Sun, *Chem*, 2019, **5**, 204–214.
13. Z. Wang, Y. Li, H. Yu, Y. Xu, H. Xue, X. Li, H. Wang and L. Wang, *ChemSusChem*, 2018, **11**, 3480–3485.
14. D. Bao, Q. Zhang, F.-L. Meng, H.-X. Zhong, M.-M. Shi, Y. Zhang, J.-M. Yan, Q. Jiang and X.-B. Zhang, *Adv. Mater.*, 2017, **29**, 1604799.
15. M. Nazemi, S. R. Panikkanvalappil and M. A. El-Sayed, *Nano Energy*, 2018, **49**, 316–323.
16. L. Tan, N. Yang, X. Huang, L. Peng, C. Tong, M. Deng, X. Tang, L. Li, Q. Liao and Z. Wei, *Chem. Commun.*, 2019, **55**, 14482–14485.
17. S.-J. Li, D. Bao, M.-M. Shi, B.-R. Wulan, J.-M. Yan and Q. Jiang, *Adv. Mater.*, 2017, **29**, 1700001.
18. M.-M. Shi, D. Bao, B.-R. Wulan, Y.-H. Li, Y.-F. Zhang, J.-M. Yan and Q. Jiang, *Adv. Mater.*, 2017, **29**, 1606550.
19. H. Sun, H.-Q. Yin, W. Shi, L.-L. Yang, X.-W. Guo, H. Lin, J. Zhang, T.-B. Lu and Z.-M. Zhang, *Nano Res.*, 2022, **15**, 3026–3033.
20. Y. Xu, T. Ren, S. Yu, K. Ren, M. Wang, Z. Wang, X. Li, L. Wang and H. Wang, *Sustainable Energy Fuels*, 2020, **4**, 4516–4521.
21. X. Wang, W. Wang, M. Qiao, G. Wu, W. Chen, T. Yuan, Q. Xu, M. Chen, Y. Zhang, X. Wang, J. Wang, J. Ge, X. Hong, Y. Li, Y. Wu and Y. Li, *Sci. Bull.*, 2018, **63**, 1246–1253.
22. X. Zhang, B. B. Ward and D. M. Sigman, *Chem. Rev.*, 2020, **120**, 5308–5351.
23. E. Draževič and E. Skúlason, *iScience*, 2020, **23**, 101803.
24. D. Yang, T. Chen and Z. Wang, *J. Mater. Chem. A*, 2017, **5**, 18967–18971.
25. C. Zhang, Z. Wang, J. Lei, L. Ma, B. I. Yakobson and J. M. Tour, *Small*, 2022, **18**, 2106327.
26. M. Xie, F. Dai, H. Guo, P. Du, X. Xu, J. Liu, Z. Zhang and X. Lu, *Adv. Energy Mater.*, 2023, **13**, 2203032.
27. Y. Ling, Q. Feng, H. Xie, X. Zheng, X. Chen, Z. Zou, A. Liu, J. Tang, Y. Li and Q. Wang, *ACS Sustainable Chem. Eng.*, 2023, **11**, 12345–12354.
28. S. Chen, S. Perathoner, C. Ampelli, C. Mebrahtu, D. Su and G. Centi, *Angew. Chem., Int. Ed.*, 2017, **56**, 2699–2703.
29. L. Hu, A. Khaniya, J. Wang, G. Chen, W. E. Kaden and X. Feng, *ACS Catal.*, 2018, **8**, 9312–9319.
30. M. Wang, S. Liu, T. Qian, J. Liu, J. Zhou, H. Ji, J. Xiong, J. Zhong and C. Yan, *Nat. Commun.*, 2019, **10**, 341.
31. L. Zhang, X. Ji, X. Ren, Y. Ma, X. Shi, Z. Tian, A. M. Asiri, L. Chen, B. Tang and X. Sun, *Adv. Mater.*, 2018, **30**, 1800191.
32. V. Sinha, F. Rezai, N. E. Sahin, J. Catalano, E. D. Bøjesen, F. Sotoodeh and E. Draževič, *Faraday Discuss.*, 2023, **243**, 307–320.

33. C. J. M. van der Ham, M. T. M. Koper and D. G. H. Hetterscheid, *Chem. Soc. Rev.*, 2014, **43**, 5183–5191.
34. Y. Abghoui and E. Skúlason, *J. Phys. Chem. C*, 2017, **121**, 6141–6151.
35. J. G. Howalt and T. Vegge, *Phys. Chem. Chem. Phys.*, 2013, **15**, 20957–20965.
36. Y. Abghoui and E. Skúlason, *Catal. Today*, 2017, **286**, 69–77.
37. Y. Abghoui and E. Skúlason, *Catal. Today*, 2017, **286**, 78–84.
38. Y. Abghoui, A. L. Garden, J. G. Howalt, T. Vegge and E. Skúlason, *ACS Catal.*, 2016, **6**, 635–646.
39. A. Banerjee, B. M. Ceballos, C. Kreller, R. Mukundan and G. Pilania, *J. Mater. Sci.*, 2022, **57**, 10213–10224.
40. X. Yang, J. Nash, J. Anibal, M. Dunwell, S. Kattel, E. Stavitski, K. Attenkofer, J. G. Chen, Y. Yan and B. Xu, *J. Am. Chem. Soc.*, 2018, **140**, 13387–13391.
41. X. Zhang, R.-M. Kong, H. Du, L. Xia and F. Qu, *Chem. Commun.*, 2018, **54**, 5323–5325.
42. R. Zhang, Y. Zhang, X. Ren, G. Cui, A. M. Asiri, B. Zheng and X. Sun, *ACS Sustainable Chem. Eng.*, 2018, **6**, 9545–9549.
43. R. Manjunatha, A. Karajić, H. Teller, K. Nicoara and A. Schechter, *ChemCatChem*, 2020, **12**, 438–443.
44. H.-L. Du, T. R. Gengenbach, R. Hodgetts, D. R. MacFarlane and A. N. Simonov, *ACS Sustainable Chem. Eng.*, 2019, **7**, 6839–6850.
45. J. Nash, X. Yang, J. Anibal, M. Dunwell, S. Yao, K. Attenkofer, J. G. Chen, Y. Yan and B. Xu, *J. Phys. Chem. C*, 2019, **123**, 23967–23975.
46. Z. Ma, J. Chen, D. Luo, T. Thersleff, R. Dronskowski and A. Slabon, *Nanoscale*, 2020, **12**, 19276–19283.
47. Y. Yao, Q. Feng, S. Zhu, J. Li, Y. Yao, Y. Wang, Q. Wang, M. Gu, H. Wang, H. Li, X.-Z. Yuan and M. Shao, *Small Methods*, 2019, **3**, 1800324.
48. D. Johnson, B. Hunter, J. Christie, C. King, E. Kelley and A. Djire, *Sci. Rep.*, 2022, **12**, 657.
49. Y. Luo, G.-F. Chen, L. Ding, X. Chen, L.-X. Ding and H. Wang, *Joule*, 2019, **3**, 279–289.
50. X. Ren, J. Zhao, Q. Wei, Y. Ma, H. Guo, Q. Liu, Y. Wang, G. Cui, A. M. Asiri, B. Li, B. Tang and X. Sun, *ACS Cent. Sci.*, 2019, **5**, 116–121.
51. X. Ren, G. Cui, L. Chen, F. Xie, Q. Wei, Z. Tian and X. Sun, *Chem. Commun.*, 2018, **54**, 8474–8477.
52. L. Zhang, X. Ji, X. Ren, Y. Luo, X. Shi, A. M. Asiri, B. Zheng and X. Sun, *ACS Sustainable Chem. Eng.*, 2018, **6**, 9550–9554.
53. X. Yang, F. Ling, J. Su, X. Zi, H. Zhang, H. Zhang, J. Li, M. Zhou and Y. Wang, *Appl. Catal., B*, 2020, **264**, 118477.
54. B. Hu, M. Hu, L. Seefeldt and T. L. Liu, *ACS Energy Lett.*, 2019, **4**, 1053–1054.
55. R. Zhang, X. Ren, X. Shi, F. Xie, B. Zheng, X. Guo and X. Sun, *ACS Appl. Mater. Interfaces*, 2018, **10**, 28251–28255.
56. R. Zhao, G. Wang, Y. Mao, X. Bao, Z. Wang, P. Wang, Y. Liu, Z. Zheng, Y. Dai, H. Cheng and B. Huang, *Chem. Eng. J.*, 2022, **430**, 133085.

57. T. Wu, H. Zhao, X. Zhu, Z. Xing, Q. Liu, T. Liu, S. Gao, S. Lu, G. Chen, A. M. Asiri, Y. Zhang and X. Sun, *Adv. Mater.*, 2020, **32**, 2000299.
58. J. Kong, A. Lim, C. Yoon, J. H. Jang, H. C. Ham, J. Han, S. Nam, D. Kim, Y.-E. Sung, J. Choi and H. S. Park, *ACS Sustainable Chem. Eng.*, 2017, **5**, 10986–10995.
59. J. Han, X. Ji, X. Ren, G. Cui, L. Li, F. Xie, H. Wang, B. Li and X. Sun, *J. Mater. Chem. A*, 2018, **6**, 12974–12977.
60. K. Chu, Y. Luo, P. Shen, X. Li, Q. Li and Y. Guo, *Adv. Energy Mater.*, 2022, **12**, 2103022.
61. G. Xu, R. Liu and J. Wang, *Sci. China, Ser. B: Chem.*, 2009, **52**, 1171–1175.
62. D. Gupta, A. Kafle and T. C. Nagaiah, *Faraday Discuss.*, 2023, **243**, 339–353.
63. X. Chen, N. Li, Z. Kong, W.-J. Ong and X. Zhao, *Mater. Horiz.*, 2018, **5**, 9–27.
64. X. Li, X. Sun, L. Zhang, S. Sun and W. Wang, *J. Mater. Chem. A*, 2018, **6**, 3005–3011.
65. V. Devthade, A. Gupta and S. S. Umare, *ACS Appl. Nano Mater.*, 2018, **1**, 5581–5588.
66. X. Gao, Y. Wen, D. Qu, L. An, S. Luan, W. Jiang, X. Zong, X. Liu and Z. Sun, *ACS Sustainable Chem. Eng.*, 2018, **6**, 5342–5348.
67. N. Dhar, E. Seshacharyulu and N. Biswas, *Proc. Natl. Inst. Sci.*, 1941, **7**, 115–131.
68. G. N. Schrauzer, N. Strampach, L. N. Hui, M. R. Palmer and J. Salehi, *Proc. Natl. Inst. Sci.*, 1983, **80**, 3873–3876.
69. G. N. Schrauzer, in *Energy Efficiency and Renewable Energy Through Nanotechnology*, ed. L. Zang, Springer London, London, 2011, pp. 601–623.
70. G. N. Schrauzer and T. D. Guth, *J. Am. Chem. Soc.*, 1977, **99**, 7189–7193.
71. J. Soria, J. C. Conesa, V. Augugliaro, L. Palmisano, M. Schiavello and A. Sclafani, *J. Phys. Chem.*, 1991, **95**, 274–282.
72. L. Palmisano, V. Augugliaro, A. Sclafani and M. Schiavello, *J. Phys. Chem.*, 1988, **92**, 6710–6713.
73. K. T. Ranjit, T. K. Varadarajan and B. Viswanathan, *J. Photochem. Photobiol., A*, 1996, **96**, 181–185.
74. S. Bourgeois, D. Diakite and M. Perdereau, *React. Solids*, 1988, **6**, 95–104.
75. H. Hirakawa, M. Hashimoto, Y. Shiraishi and T. Hirai, *J. Am. Chem. Soc.*, 2017, **139**, 10929–10936.
76. G. Zhang, X. Yang, C. He, P. Zhang and H. Mi, *J. Mater. Chem. A*, 2020, **8**, 334–341.
77. X.-Y. Xie, P. Xiao, W.-H. Fang, G. Cui and W. Thiel, *ACS Catal.*, 2019, **9**, 9178–9187.
78. S. Chang and X. Xu, *Inorg. Chem. Front.*, 2020, **7**, 620–624.
79. S. Liu, Y. Wang, S. Wang, M. You, S. Hong, T.-S. Wu, Y.-L. Soo, Z. Zhao, G. Jiang, Q. Jieshan, B. Wang and Z. Sun, *ACS Sustainable Chem. Eng.*, 2019, **7**, 6813–6820.
80. H. Yin, Z. Chen, Y. Peng, S. Xiong, Y. Li, H. Yamashita and J. Li, *Angew. Chem., Int. Ed.*, 2022, **61**, e202114242.

81. Q.-S. Li, K. Domen, S. Naito, T. Onishi and K. Tamaru, *Chem. Lett.*, 2006, **12**, 321–324.
82. Z. Ying, S. Chen, T. Peng, R. Li and J. Zhang, *Eur. J. Inorg. Chem.*, 2019, **2019**, 2182–2192.
83. A. G. Tamirat, J. Rick, A. A. Dubale, W.-N. Su and B.-J. Hwang, *Nanoscale Horiz.*, 2016, **1**, 243–267.
84. M. M. Khader, N. N. Lichtin, G. H. Vurens, M. Salmeron and G. A. Somorjai, *Langmuir*, 1987, **3**, 303–304.
85. M. Lashgari and P. Zeinalkhani, *Appl. Catal., A*, 2017, **529**, 91–97.
86. K. Tennakone, S. Wickramanayake, C. A. N. Fernando, O. A. Ileperuma and S. Punchihewa, *J. Chem. Soc., Chem. Commun.*, 1987, 1078–1080.
87. K. Tennakone, C. T. K. Thaminimulla and J. M. S. Bandara, *J. Photochem. Photobiol., A*, 1992, **68**, 131–135.
88. H. Miyama, N. Fujii and Y. Nagae, *Chem. Phys. Lett.*, 1980, **74**, 523–524.
89. E. Endoh, J. K. Leland and A. J. Bard, *J. Phys. Chem.*, 1986, **90**, 6223–6226.
90. T. Hou, Y. Xiao, P. Cui, Y. Huang, X. Tan, X. Zheng, Y. Zou, C. Liu, W. Zhu, S. Liang and L. Wang, *Adv. Energy Mater.*, 2019, **9**, 1902319.
91. N. Zhang, A. Jalil, D. Wu, S. Chen, Y. Liu, C. Gao, W. Ye, Z. Qi, H. Ju, C. Wang, X. Wu, L. Song, J. Zhu and Y. Xiong, *J. Am. Chem. Soc.*, 2018, **140**, 9434–9443.
92. Y. Li, X. Chen, M. Zhang, Y. Zhu, W. Ren, Z. Mei, M. Gu and F. Pan, *Catal. Sci. Technol.*, 2019, **9**, 803–810.
93. W. Zhao, H. Xi, M. Zhang, Y. Li, J. Chen, J. Zhang and X. Zhu, *Chem. Commun.*, 2015, **51**, 4785–4788.
94. C. M. Janet, S. Navaladian, B. Viswanathan, T. K. Varadarajan and R. P. Viswanath, *J. Phys. Chem. C*, 2010, **114**, 2622–2632.
95. P. Xing, P. Chen, Z. Chen, X. Hu, H. Lin, Y. Wu, L. Zhao and Y. He, *ACS Sustainable Chem. Eng.*, 2018, **6**, 14866–14879.
96. A. Ray, S. Sultana, S. P. Tripathy and K. Parida, *ACS Sustainable Chem. Eng.*, 2021, **9**, 6305–6317.
97. Z. Lu, S. E. Saji, J. Langley, Y. Lin, Z. Xie, K. Yang, L. Bao, Y. Sun, S. Zhang, Y. H. Ng, L. Song, N. Cox and Z. Yin, *Appl. Catal., B*, 2021, **294**, 120240.
98. Y. Hao, X. Dong, S. Zhai, H. Ma, X. Wang and X. Zhang, *Chem. - Eur. J.*, 2016, **22**, 18722–18728.
99. G. Dong, W. Ho and C. Wang, *J. Mater. Chem. A*, 2015, **3**, 23435–23441.
100. S. Cao, B. Fan, Y. Feng, H. Chen, F. Jiang and X. Wang, *Chem. Eng. J.*, 2018, **353**, 147–156.
101. Y. Zhang, J. Di, P. Ding, J. Zhao, K. Gu, X. Chen, C. Yan, S. Yin, J. Xia and H. Li, *J. Colloid Interface Sci.*, 2019, **553**, 530–539.
102. S. Liu, S. Wang, Y. Jiang, Z. Zhao, G. Jiang and Z. Sun, *Chem. Eng. J.*, 2019, **373**, 572–579.
103. S. Ajmal, A. Rasheed, N. Q. Tran, X. Shao, Y. Hwang, V. Q. Bui, Y. D. Kim, J. Kim and H. Lee, *Appl. Catal., B*, 2023, **321**, 122070.

104. S. Hu, Y. Li, F. Li, Z. Fan, H. Ma, W. Li and X. Kang, *ACS Sustainable Chem. Eng.*, 2016, **4**, 2269–2278.
105. H. Liang, H. Zou and S. Hu, *New J. Chem.*, 2017, **41**, 8920–8926.
106. S. Hu, X. Chen, Q. Li, F. Li, Z. Fan, H. Wang, Y. Wang, B. Zheng and G. Wu, *Appl. Catal., B*, 2017, **201**, 58–69.
107. S. Cao, N. Zhou, F. Gao, H. Chen and F. Jiang, *Appl. Catal., B*, 2017, **218**, 600–610.
108. Q. Zhang, S. Hu, Z. Fan, D. Liu, Y. Zhao, H. Ma and F. Li, *Dalton Trans.*, 2016, **45**, 3497–3505.
109. H. Li, J. Shang, Z. Ai and L. Zhang, *J. Am. Chem. Soc.*, 2015, **137**, 6393–6399.
110. H. Li, J. Shang, J. Shi, K. Zhao and L. Zhang, *Nanoscale*, 2016, **8**, 1986–1993.
111. Y. Bai, L. Ye, T. Chen, L. Wang, X. Shi, X. Zhang and D. Chen, *ACS Appl. Mater. Interfaces*, 2016, **8**, 27661–27668.
112. J. Di, J. Xia, M. F. Chisholm, J. Zhong, C. Chen, X. Cao, F. Dong, Z. Chi, H. Chen, Y.-X. Weng, J. Xiong, S.-Z. Yang, H. Li, Z. Liu and S. Dai, *Adv. Mater.*, 2019, **31**, 1807576.
113. S. Wang, X. Hai, X. Ding, K. Chang, Y. Xiang, X. Meng, Z. Yang, H. Chen and J. Ye, *Adv. Mater.*, 2017, **29**, 1701774.
114. P. Li, Z. Zhou, Q. Wang, M. Guo, S. Chen, J. Low, R. Long, W. Liu, P. Ding, Y. Wu and Y. Xiong, *J. Am. Chem. Soc.*, 2020, **142**, 12430–12439.
115. Y. Liu, Z. Hu and J. C. Yu, *Chem. Mater.*, 2020, **32**, 1488–1494.
116. D. Wu, R. Wang, C. Yang, Y. An, H. Lu, H. Wang, K. Cao, Z. Gao, W. Zhang, F. Xu and K. Jiang, *J. Colloid Interface Sci.*, 2019, **556**, 111–119.
117. C. Xiao, H. Wang, L. Zhang, S. Sun and W. Wang, *ChemCatChem*, 2019, **11**, 6467–6472.
118. Y. Cao, S. Hu, F. Li, Z. Fan, J. Bai, G. Lu and Q. Wang, *RSC Adv.*, 2016, **6**, 49862–49867.
119. S. Hu, X. Chen, Q. Li, Y. Zhao and W. Mao, *Catal. Sci. Technol.*, 2016, **6**, 5884–5890.
120. S. Sun, X. Li, W. Wang, L. Zhang and X. Sun, *Appl. Catal., B*, 2017, **200**, 323–329.
121. Y. Fu, Y. Liao, P. Li, H. Li, S. Jiang, H. Huang, W. Sun, T. Li, H. Yu, K. Li, H. Li, B. Jia and T. Ma, *Coord. Chem. Rev.*, 2022, **460**, 214468.
122. Y. Zhao, Y. Zhao, G. I. N. Waterhouse, L. Zheng, X. Cao, F. Teng, L.-Z. Wu, C.-H. Tung, D. O'Hare and T. Zhang, *Adv. Mater.*, 2017, **29**, 1703828.
123. L. Rizzato, J. Cavazzani, A. Osti and A. Glisenti, *Faraday Discuss.*, 2023, **243**, 388–401.
124. Z. Zhao, H. Ren, Y. Shi, J. Tan, X. Xin, D. Yang and Z. Jiang, *Chem. Eng. J.*, 2022, **443**, 136559.
125. Y. Guan, H. Wen, K. Cui, Q. Wang, W. Gao, Y. Cai, Z. Cheng, Q. Pei, Z. Li, H. Cao, T. He, J. Guo and P. Chen, *Nat. Chem.*, 2024, **16**, 373–379.
126. A. Tsuneto, A. Kudo and T. Sakata, *Chem. Lett.*, 2006, **22**, 851–854.

127. O. Westhead, J. Barrio, A. Bagger, J. W. Murray, J. Rossmeisl, M.-M. Titirici, R. Jervis, A. Fantuzzi, A. Ashley and I. E. L. Stephens, *Nat. Rev. Chem.*, 2023, **7**, 184–201.
128. J. C. Peters, *Faraday Discuss.*, 2023, **243**, 450–472.
129. S. Licht, B. Cui, B. Wang, F.-F. Li, J. Lau and S. Liu, *Science*, 2014, **345**, 637–640.
130. J. Choi, B. H. R. Suryanto, D. Wang, H.-L. Du, R. Y. Hodgetts, F. M. Ferrero Vallana, D. R. MacFarlane and A. N. Simonov, *Nat. Commun.*, 2020, **11**, 5546.
131. *Nat. Commun.*, 2022, **13**, 4642.
132. N. G. Mohan and K. Ramanujam, *Curr. Opin. Electrochem.*, 2024, **45**, 101520.

CHAPTER 11

Catalytic Combustion of Methane in Low-concentration Gas Streams

M. BLIGH, M. DREWERY, L. HARVEY, E. M. KENNEDY AND M. STOCKENHUBER*

Chemical Engineering, University of Newcastle, University Drive, Callaghan, NSW, Australia
*Email: Michael.Stockenhuber@newcastle.edu.au

11.1 Background

11.1.1 Natural Gas Engines

Natural gas power generation has shown prevalence worldwide due to its lower carbon content and lessened impact on climate change. Natural gas consists primarily of methane with small amounts of heavier alkanes, trace CO_2, water, and sulphur-based compounds. Methane, being its primary constituent, makes the hydrogen/carbon (H/C) ratio very high. Converting to a natural gas engine from a diesel engine increases the H/C from 1.8 to approximately 3.9.[1,2] The higher hydrogen and lower carbon equate to less CO_2 produced per mol reacted. So overall, the CO_2 emissions are much lower for natural gas engines. Engines utilising natural gas can be utilised for smoothing peak loads, supporting renewable energy sources, such as solar and wind power. Natural gas engines especially show the most versatility in operation, only taking 100 seconds to start, unlike turbines that take up to 2 hours for the system to operate and produce enough energy for peak loads.[3]

Table 11.1 Transition from solid and liquid fuels to gaseous and renewables. Adapted from ref. 3 with permission from Elsevier, Copyright 2017.

Hydrogen/carbon ratio	0:1	2:1	4:1	C-free
State	Solids	Liquids	Gases	Renewables
Examples	Coal	Diesel gasoline	Natural gas alkanes	Hydrogen

Natural gas engines have been found to have increasing prevalence in transport and power generation, being a more energy-efficient and cleaner alternative to conventional diesel and gasoline-based engines. Natural gas can replace solid and liquid-based fuels in power generation due to its decreased output of greenhouse gases and toxic pollutants, as illustrated in Table 11.1. However, one of the downsides of power generation in gas engines is methane slip, which is described below.

11.1.2 Natural Gas for Power Generation

Coal and liquid-based fossil fuels are the primary sources of power worldwide. The greenhouse gas emissions associated with the use of these fuels have resulted in the need for strict regulations, such as those outlined by the European Commission.[4] The regulation sets three greenhouse gas emissions goals: 40% lower greenhouse emissions than in the 1990s, increased share of renewable energy by at least 27%, and an increase in energy efficiency by 27%. Natural gas shows advantages as an alternative to solid and liquid fossil fuels, such as the carbon production or the overall greenhouse gas emissions being much lower than that of the higher carbon-based fuels.

Utilisation of natural gas as a lower emission energy source currently primarily uses gas turbines for generating electricity. However, using turbines has a significant disadvantage in the dynamic environment required to support renewable sources. An operating turbine can withstand loads associated with peak energy demands, but cannot do so unless operating continuously, as a turbine can take up to 2 hours to achieve steady state associated with full-scale operation. Engines, on the other hand, provide a more readily available energy source, and are able to be switched on and operated within minutes.

Of natural gas and diesel engines, diesel has a faster operational time. A diesel engine can be operational within 60 seconds, while natural gas engines will take up to 100 seconds to operate.[3] Although diesel engines have advantages concerning flexibility, they also have significant disadvantages concerning greenhouse emissions and NO_x production. Due to future NO_x emissions regulations, it has been noted that diesel engine efficiency will hardly improve.[3] Natural gas can run in a lean-burn regime, lowering NO_x production due to its lower temperature, thus addressing future regulations by improving operation efficiency.[3] Due to the lessened NO_x emissions, natural gas engines operating in the lean burn are now the recommended solution able to address operating requirements.[3]

11.1.3 Natural Gas in Transport

Current anthropogenic emissions are an ongoing concern and considered a significant detriment to air quality. This is especially true for internal combustion engines used for transport. Several regulations have been implemented to reduce exhaust emissions of greenhouse gases and toxic pollutants. The Kyoto Protocol, which spans 38 industrialised countries, seeks to regulate greenhouse gas emissions by issuing emissions 5.2% below the levels found in the 1990s.[1] The current regulations in Europe for engine pollutants are the EURO 6 emissions standards implemented in 2015 to reduce toxic pollutants.[5] This reduces nitrogen oxide, carbon monoxide, hydrocarbons, and particulate matter from diesel and gasoline-based engines.[5] The push for greener, lower pollutant engines has led many industrialised countries to use natural gas as their primary fuel for transport.

Burning any fossil fuel will inevitably produce carbon dioxide, but a fuel with a higher hydrogen-to-carbon ratio, such as natural gas, will significantly reduce CO_2 emissions and improve fuel economy. As mentioned, natural gas can operate in a lean burn with a higher air-to-fuel ratio, lowering the burn temperature and increasing overall fuel efficiency due to the reduced harmful side product production.[1]

11.2 Methane Slip

11.2.1 Lean *versus* Stoichiometric

Natural gas engines either operate under lean-burn conditions or stoichiometric conditions.[1] Lean-burn conditions operate with a higher air concentration than methane concentration. In contrast, stoichiometric burn operates with the exact amount of air for methane (17.1 grams of air for 1 gram of methane).[6] When burning under lean conditions, the operating temperature is much lower than that of diesel or gasoline; this inhibits the formation of harmful products such as NO_x, which forms at high temperatures.[1,2,7,8] Under lean conditions, the burning rate of the engine is reduced; this results in a decrease in the overall engine temperature, lowering NO_x production and marginally increasing engine efficiency.[1,9] However, the time required to combust the natural gas is increased, leading to significant heat losses and decreased thermal efficiency, which results in instability and causes misfire cycles and incomplete combustion.[1,2,9] The lower conversion of hydrocarbons means more uncombusted methane exiting in the exhaust. Operating at stochiometric conditions results in a much faster burn, so fast it can self-ignite in the propagating flame front, causing knocking and leading to decreased engine efficiency and even engine destruction.[1,10] The high burn also induces a higher temperature, causing a higher concentration of NO_x^2. The stoichiometric conditions will also increase the likelihood of incomplete combustion, forming soot and CO^2. The effect that lambda (the air-to-fuel ratio, where $\lambda = 1$ is stoichiometric and $\lambda > 1$ is lean)

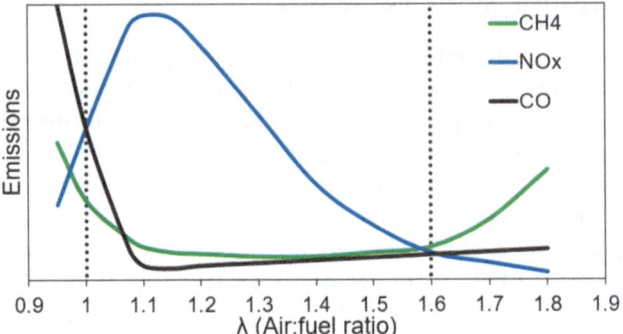

Figure 11.1 The effect of lambda on engine performance, where stoichiometric $\lambda = 1$ and lean burn $\lambda > 1$. Adapted from ref. 1 with permission from Elsevier, Copyright 2007.

has on the engine's performance, the NO_x production, and total hydrocarbon (THC) production is shown in Figure 11.1.

The graph shows that at the more stoichiometric regions $1 < \lambda < 1.2$, the production of NO_x is the highest per horsepower-hour. As the lambda value increases, the NO_x production decreases rapidly, and the engine efficiency rises marginally – consequently, the total hydrocarbon production increases as the air-to-fuel ratio increases. So, to limit the THC and NO_x while increasing the engine efficiency, the best region to be within is $1.6 < \lambda < 1.8$. This leads to a strong preference for engine operation under lean-burn conditions.

11.2.2 Exhaust Methane

Decreasing the overall NO_x output consequently increases the concentration of hydrocarbons in the exhaust, specifically methane, known as slip, as evidenced in Figure 11.1. The amount of methane from the exhaust varies significantly depending on the engine's performance, ranging from 1000 to 7000 ppm.[11] Methane slip is an issue as methane is a potent greenhouse gas, 84–86 times more powerful than carbon dioxide over 20 years and 28–34 times more potent over 100 years.[12,13] Even small concentrations leaving the exhaust can result in considerable environmental concerns, and with the prevalence of natural gas engines increasing, this greenhouse gas effect becomes more significant. In the United States, the EPA has issued a regulation to restrict the amount of methane produced in natural gas vehicles to a limit of 30 mg per mile. This regulation was implemented as a pre-emptive method to reduce future emissions from natural gas engines. In Europe, methane has been directly regulated in heavy-duty gas engines since the Euro 3 standard limits the methane output to 0.5 g kW h^{-1}.[14] The new Euro 7 seeks to restrict methane output to 10 mg km^{-1} for light-duty vehicles, while for stationary engines (such as those used in power generation), the EUROMOT position states a higher limit of 560 mg N^{-1} m^{-1}.[3,15]

The incomplete oxidation of natural gas (methane) in lean-burn gas engines means that post-treatment is necessary to reduce the concentration of methane emitted. Methane has a lower flammability limit between 4–15% at lower temperatures (above 600 °C), so the concentration of the methane slip is too low to flare conventionally.[16,17] If the temperature is increased beyond 1000 °C, thermal oxidation can occur.[18] This will fully combust all the methane without any supplementary pathways assisted by a catalyst. High-temperature operation is not feasible for engine exhausts as heating the exhaust to that temperature will not be an economical solution for engines used for power generation. Additionally, heating the exhaust gas to these temperatures also increases the formation of NO_x. It is therefore proposed that the primary solution for the total oxidation of methane in gas engine exhausts should be through catalytic methods. Compared to other hydrocarbons, methane is very stable, making it much harder to oxidise, especially compared to gasoline and diesel exhaust.[2,6,12,19] Post-treatment of gasoline engine exhaust usually employs a three-way catalyst, which converts toxic pollutants such as NO_x, CO, and hydrocarbons into benign compounds such as CO_2, H_2O, and N_2.[20] Most post-exhaust catalysts use selective catalytic reduction to reduce NO_x to nitrogen gas, however this is not a priority in lean-burn natural gas engines, as the NO_x levels are low, so a specific methane oxidating catalyst is required.

11.2.3 Ventilation Air Methane (VAM)

A similar emitter to natural gas engine exhaust (NGEE) is the exhaust found in coal mine shafts known as ventilation air methane (VAM). VAM has a similar methane concentration range to NGEE,[11] with both having oxygen gas in excess and high water vapour content. VAM research has shown that palladium on alumina (Pd/Al_2O_3) catalysts have high activity but tend to deactivate rapidly under exhaust streams containing water vapour. It was also found that these catalysts required an increased temperature above 500 °C to maintain methane conversion above 90%.[21] Experiments using transition metals exhibited some success in converting VAM streams; cobalt and iron oxides were used, showing high methane conversion activity and cobalt oxide showed little change in conversion in the presence of water compared to dry gas streams.[22] However, adding the noble metal gold to the catalyst did not improve the activity.[22] Even greater stability was found using Pd on zeolite support, which has been shown to have high VAM conversions for long periods.[23,24] Palladium on a TS-1 (Pd/TS-1) catalyst has exhibited long-term stability, with high levels of oxidation activity being observed for 2000 hours, as shown in Figure 11.2.

The main difference between VAM and NGEE is the pre-combustion in NGEE. Thus, the oxygen levels will be reduced from approximately 19% to 8%. Another consequence of combustion is water formation at a concentration greater than 10%, almost five times that present in VAM.[2,6,25] A significant difference is the trace amount of sulphur-containing compounds

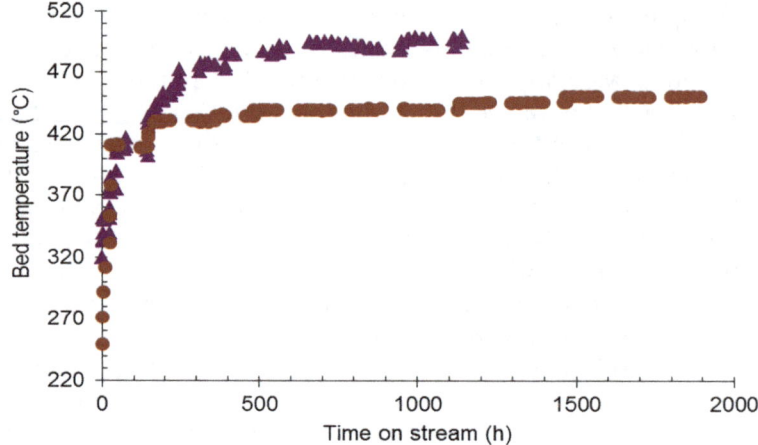

Figure 11.2 Bed temperature as a function of time on stream over (●) Pd/TS-1 and commercial (▲) Pd/Al$_2$O$_3$, maintaining the level of methane conversion above 90%. The feed composition is 7000 ppm CH$_4$, 10 000 ppm CO$_2$, and 30 000–40 000 ppm H$_2$O balanced in air. Reproduced from ref. 23 with permission from the Royal Society of Chemistry.

in natural gas. Most sulphur comes from regulations requiring natural gas to be odorised for safety using methyl mercaptans, with a small percentage of sulphur also found in the gas stream naturally as hydrogen sulphide. Both methyl mercaptans and hydrogen sulphide readily oxidise in engine conditions to form SO$_2$, which can lead to catalyst poisoning.[26] The differences between VAM and NGEE result in the need for tailored technologies to treat methane slip, with catalyst stability a key concern in the viability of catalytic exhaust treatment.

11.3 Active Metal Catalysts

11.3.1 Noble Metals

Concerning methane oxidation, the active metals that have gained the most attention are noble metals, with palladium, platinum, and rhodium catalysts showing high activity and turnover frequency of methane at relatively low temperatures.[27,28]

11.3.1.1 Palladium

Palladium is seen as the most active noble metal for methane combustion at low temperatures.[27] It is well-researched and disputed that palladium's active site is palladium oxide (PdO) at lower temperatures, where thermodynamic stability is found in an oxidative atmosphere.[29] At these conditions, the PdO active site shows the highest activity in methane oxidation compared to other noble metal catalysts.[30] However, it has been suggested that both palladium oxide and palladium metal play a role in the oxidation of methane.[31,32] It was, for example, shown that a Pd/PdOx ratio of 2.5 increases the turnover frequency, for a methane oxidating catalyst.[32] The palladium oxide sites show instability at temperatures beyond 1053 K, while Pd sites are more stable at higher temperatures resulting in a change in activity.[29] These temperatures are not feasible for gas engines, requiring significant additional energy input as the exhaust temperature ranges between 400–550 °C, so operating in the PdO favourability range is the more realistic option.[33]

Under catalytic oxidation conditions, it is suggested that methane interacts with the surface-bound oxygen in PdO rather than gaseous oxygen, with four PdO active sites required to fully oxidise one methane molecule based on stoichiometry.[34] Temperature-programmed reductions of palladium on a zirconia catalyst using methane show a decrease in PdO intensity partnered with methane consumption and carbon dioxide production.[34] Therefore, the reduced active site needs to be re-oxidised, which is usually done through the support.[35] Issues arise as the oxygen concentration within the exhaust is low, between 6% and 10%, and is lower than the water vapour concentration, with the formation of surface hydroxyl species on the support lowering overall oxygen transfer.[35] To withstand the constant reduction cycles, catalyst supports must exhibit a high level of oxygen transfer and hydrophobicity.

A consequence of palladium's high activity is its vulnerability to several deactivating agents, specifically water and sulphur dioxide. Water has been suggested to interact with the PdO active site and form the less active Pd–OH, with further deactivation by the formation of extended palladium hydroxide [Pd(OH)$_2$] also proposed.[25,26,35,36] This theory assumes that the Lewis acid site of Pd^{2+} accepts the Lewis base OH$^-$.[37] While the water-deactivated palladium site activity is substantially reduced compared to PdO, this deactivation has been shown to be reversible, requiring a dry stream to induce water desorption.

Palladium oxide shows a high level of interaction with sulphurous species, exhibiting competition between the methane, deactivating the catalyst as palladium sulphate (PdSO$_4$) forms.[6,38,39] It is suggested that SO$_2$ deactivates palladium through the catalytic oxidation of SO$_2$ to SO$_3$ by palladium, with the SO$_3$ forming palladium sulphate at a high rate.[40] The activity of the sulphated palladium is significantly reduced and shows some reversibility, with palladium sulphate compounds starting to decompose in dry conditions at temperatures greater than 460 °C, releasing SO$_2$.[41] Decomposition stops at approximately 600 °C, where the catalyst will be partially reactivated.[41] In humid conditions, however, the level of deactivation could be far more significant.[41] Depending on whether Pd(OH)$_2$ formation occurs, the water could also compete with methane for the active site and the SO$_2$.[25,35]

11.3.1.2 Platinum

Platinum is another active, noble metal catalyst with high methane oxidation activity. It is used extensively in the oxidation of higher alkanes and alkenes, primarily as a three-way automotive catalyst to oxidise unburnt hydrocarbons in gasoline/diesel engine exhaust.[42] Platinum's activity in methane oxidation is high when the concentration of oxygen and methane is close to stoichiometric or oxygen rich.[42] The oxygen in natural gas exhaust is depleted, but so is the methane, so the exhaust conditions are still lean. In these conditions, palladium shows a higher activity.[42]

It was found that adding platinum to a pallidum catalyst can increase the resistance to inhibition of water and sulphur dioxide. Platinum is a semi-sulphur-resistant catalyst, as it doesn't tend to form an active oxide site.[39] The bimetallic catalyst increases the stability due to the adsorption of SO_2 to the metallic platinum, inhibiting palladium's ability to oxidise the SO_2 to SO_3, preserving the PdO active site.[40,42]

11.3.1.3 Rhodium

Rhodium-based catalysts have gained some increasing interest as automotive exhaust catalysts. Rhodium is an essential component of three-way catalysts, but it has been less researched in specified methane combustion, as palladium and platinum have higher activities.[41] As stated previously, palladium's associated activity makes it prone to the inhibitors of water and SO_2. Palladium has a higher sulphur resistance in dry conditions than rhodium, however, in wet conditions, the level of active palladium sites is depleted faster and shows permanent deactivation.[41] Rhodium's active site is rhodium oxide (Rh_2O_3), and in dry conditions, it readily forms rhodium sulphate [$Rh_2(SO_4)_3$].[41] However, in wet conditions, the hydrophilic sulphate forms a hydrated rhodium sulphate complex ([$Rh(H_2O)_x$]$_2(SO_4)_3 \cdot yH_2O$), which leads to SO_2 release at low temperatures around 300–400 °C.[41] So, a rhodium catalyst shows the potential to be an active catalyst that can withstand the inhibitors prevalent in NGEE.

11.3.2 Non-noble Metal Oxides

Transition metal oxide catalysts are used extensively as they have multiple valence states, allowing redox cycles between different oxidation states. Unlike noble metals, metal oxides are readily available, making them much cheaper.

11.3.2.1 Copper

Research has found that one of the most efficient metal oxide catalysts is copper. It has been reported that the effectiveness of a copper catalyst is strongly dependent upon the support, which is required to have a higher surface area and promote the metal to be in the elemental form, as CuO shows a reduced activity.[43] The commonly used supports for copper catalysts are Al_2O_3, ZrO_2, SiO_2, and $LaMo_xV_{1-x}O_n$.[43–45] The support surface area and the choice of loading are paramount in increasing the concentration of isolated copper species compared to non-isolating, with the isolated copper species being of highest activity owing to the ability of the support to disperse copper on the surface and not form bulk copper species, increasing the number of active sites.[28,43,46] The most active copper catalysts were found to be CuO/ZrO$_2$, which was greater than CuO/Al$_2$O$_3$ and greater than or equal to CuO/SiO$_2$.[28,43] The copper supported by zirconia has the highest activity due to zirconia being able to form stable and dispersed copper species. Meanwhile, silicon-supported copper forms bulk copper oxides with significantly lower activity.

11.3.2.2 Cobalt

Cobalt oxide (Co$_3$O$_4$) catalysts have shown high activity for the oxidation of methane, with combustion activity on the transition metal starting at around 250 °C and reaching 100% conversion at about 450 °C.[22] The activity of cobalt oxide for the total oxidation of methane is not as high as that seen over palladium catalysts, but it has shown a higher resistance to reaction inhibition by water. This is of particular importance when considering NGEE, which contains elevated water concentrations. At 455 °C, when 3% of water is fed into a 0.6% methane in an air feed stream, there was little to no change in cobalt activity, as seen in Figure 11.3. The fact that little change was seen shows that the cobalt oxide does not suffer the same loss in activity as other alternative catalysts and might withstand water conditions greater than 10%.

Certain preparation methods allow a cobalt oxide catalyst to have a high surface area and high comparative activity in methane combustion. Using a cobalt hydroxide precursor and then exposing it to direct thermal decomposition forms cobalt oxide nanoparticles; including nanobelts, nanosheets, and nanocubes with different crystallographic planes.[28,47] The nanosheets with an exposed plane [112] are more active than the others, despite the lower surface area. This is proposed to be a result of the nanosheets' well-defined crystal structure, which was more important than other interspaces within nanocrystals.[28,47]

Cobalt is a suitable catalyst for methane conversion, but unfortunately, it suffers from sintering above 550 °C, significantly reducing catalyst activity. Using a suitable support can, however, remedy this problem. It was found that a magnesium oxide support is capable of forming stable cobalt species, where a CoO–MgO was formed that can disperse Co.[2,28,48,49] The choice of support for a cobalt oxide catalyst significantly affects the performance. It was found that a cobalt catalyst, with a weighting of 10%, performed well with a zirconia and titanium support.[49]

11.3.2.3 Cerium

Oxides of rare earth metals are used extensively to stabilise the support and active phase of automotive catalysts. Cerium oxide has attracted attention in

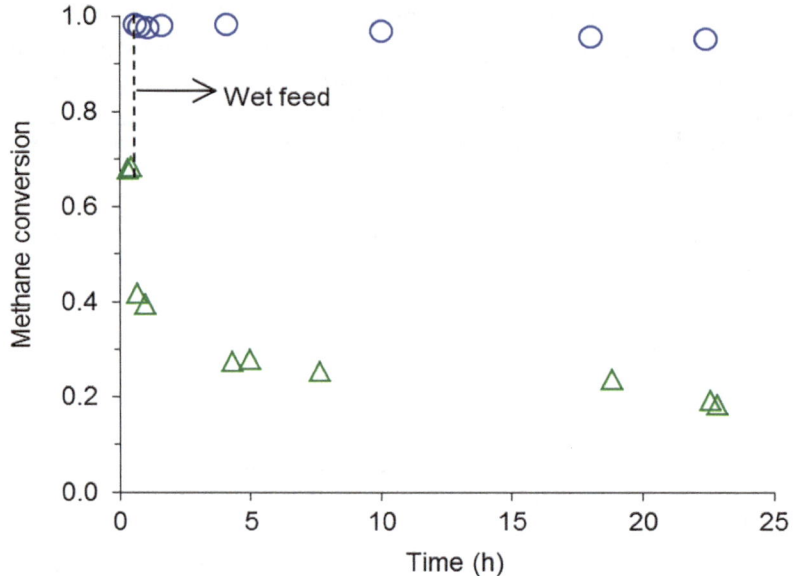

Figure 11.3 Time on stream behaviour under dry and wet feed over Co_3O_4 (○) and Fe_2O_3 (△) catalysts. Feed: 6000 ppm CH_4, 3 vol% H_2O (for wet feed) balance air. GHSV = 100 000 h^{-1}, and bed temperature = 455 °C. Reproduced from ref. 22 with permission from Elsevier, Copyright 2015.

automotive catalysts due to its high oxygen storage capacity.[50,51] Cerium catalysts can be prepared to form nanocrystals with much more activity than ceria microcrystals. Cerium oxide nanocrystals can be formed *via* precipitation using hydrogen peroxide as an oxidiser.[52] This nanocrystalline form of cerium oxide is able to oxidise methane much more efficiently than microcrystalline, reducing the methane oxidation temperature by 100 °C.[52] This is due to the significant increase in surface area, with crystallite sizes of 5 nm, mesopores of 28.5 nm in diameter, and a specific surface area of 158 $m^2 g^{-1}$ after calcination[52]

When CeO_x is combined with PdO in a bimetallic catalyst, the active site reduction is hindered, promoting constant reoxidation of the active metal and increasing the catalyst activity.[50,51] The high amount of oxygen stored in the lattice of the cerium oxide structure readily spills over to the depleted palladium. This increased oxygen mobility level has been useful in the dual support catalyst $Pd/CeO_2/Al_2O_3$,[50,51] which also allows the reoxidation of metallic palladium at higher temperatures. Adding cerium oxide to a zeolite support was found to significantly improve activity; palladium ceria on HZSM-5 lowered the temperature required to reach 100% conversion by approximately 50 °C, resulting in a T_{100} of 375 °C, with the addition of cerium oxide providing higher hydrothermal stability.[53]

Cerium can also be combined with cobalt to increase catalyst activity. A Co_3O_4/CeO_2 catalyst has been found to have increased catalyst activity and thermal stability compared to the cobalt oxide counterpart.[54] A mixed metal ceria and cobalt catalyst with 30% cobalt oxide has been shown to strongly influence the catalyst's redox properties, where the cobalt oxide is dispersed and promotes the efficiency of the Co^{3+}–Co^{2+} redox cycle.[54]

11.3.2.4 Manganese Oxide

Manganese oxide catalysts have shown some promise as methane oxidation catalysts. They are both active metals and additives for mixed metal oxide catalysts. A manganese oxide catalyst paired with alumina showed high catalytic activity, especially when the loading was increased to 15 wt%.[55] These catalysts showed high variability in performance with choice of support, with the alumina support having significantly higher activity than its silicon and titanium oxide counterparts.[55] Manganese oxide has also been used as a promoter in palladium cobalt on alumina catalysts, where altering the ratio between cobalt and manganese demonstrated an improved oxygen transfer ability, therefore increasing the total activity.[56] This high oxygen transfer led to the catalyst having a greater stability under high water concentrations, where it showed high activity at water concentrations of 5 vol%.[56]

Manganese oxide with nickel oxide proved to be an effective catalyst for methane combustion.[57] Utilising composite oxide greatly increases the activity of the nickel catalyst with the appropriate amount of manganese, leading to the formation of small Mn–Ni–O particles and inhibiting the growth of less active NiO crystallite.[57] The atomic ratio for manganese required to achieve this effective state was 0.13.[57] These results indicate that altering the compositions of some mixed metal oxides can greatly affect the overall activity.

11.3.2.5 Chromium Oxide

Another metal oxide highly utilised in mixed metal oxide catalysts is chromium oxide, which has been proven effective in a cobalt–chromium oxide methane oxidising catalyst.[58] Chromium oxides showed high methane oxidation activity in the presence of water vapour and sulphur dioxide. The literature suggests that a cobalt–chromium ratio of 1:2 is the most effective, exhibiting little change in activity under humid conditions.[58] It was proposed that the Cr^{6+} ions cause disorder in the structure of the cobalt chromites, increasing the adsorption of oxygen species and hence increasing the overall catalytic performance.[58]

11.3.2.6 Iron Oxide

Iron oxide has been utilised as an alternative to the more expensive noble metals and has shown relatively high sintering temperature and effective cation diffusion.[59] It has also been shown that iron oxide catalysts are active on silica support in the total combustion of methane to CO_2, with methane conversion shown to have a linear dependence on the iron oxide content in the catalyst.[60,61] Since iron oxide has a low cost, this has been optimised in bulk iron oxide catalysts formed through citrates and precipitation methods with iron contents of 66% and 61%, respectively, showed some activity in methane conversion at a temperature of 600 °C despite the low relative surface area compared to other traditional supported catalysts.[62]

Conversely, nanoparticles from iron oxide were prepared onto an alumina support. These nanocatalysts showed reasonable activity in methane combustion, fully converting 10 000 ppm of methane at 700 °C.[63] The temperature needed to achieve this conversion is relatively high compared to palladium-based catalysts that can accomplish the same activity at a much lower temperature, however it should be highlighted that the iron oxide catalyst could withstand these temperatures and proved stable. This indicates that iron oxide-based catalysts have some potential in these applications because of their abundance and therefore, lower price, but they do not exhibit the same overall activity compared to their noble metal counterparts.

11.3.2.7 Perovskite

Perovskites are crystalline mixed oxides that have a formulae ABX_3, where A and B are cations and X is an anion. The size of the cations can vary significantly, which makes them very flexible in terms of designing compositions. In the following we will focus on O as anion but note that there is a number of other anions possible. The perovskites are versatile and have been applied to oxide catalyst technologies since they have favourable attributes such as flexibility in the choice of cations, very high thermal stability, and resistance to sintering, as well as a unique nature of lattice oxygen that can be applied to specific technologies.[64] Flexibility is a key feature of these catalysts, as around 90% of metallic elements can be partially substituted into the perovskite crystalline structure as the A and B.[65,66] Noble metals and transition metals can be incorporated into the structure, allowing for a catalyst with the active metal within the structure, instead of clusters that will form on support Lewis acid sites in the conventional dry impregnation method. Therefore, perovskites can be utilised as both a support and an active metal.

A lanthanum manganese and zirconia perovskite supported by a palladium-deposited catalyst has been utilised for methane combustion.[67] Even though palladium was deposited *via* standard impregnation methods instead of incorporating it within its structure, the preparation method still proved sufficient, having very high activity in methane combustion.[67] A non-noble metal perovskite of lanthanum and manganese, as well as lanthanum strontium (as the A oxide) and manganese (as the B oxide), was applied to lean methane combustion.[68] The study focused specifically on the effect of the preparation method on the overall activity of the catalyst. It was found that the preparation methods affected the physiochemical properties, including the catalyst activity, with the highest-activity catalyst examined prepared *via* the solvothermal and chemical combustion preparation methods, which was able to achieve an activity comparable to 5 wt% Pt/Al$_2$O$_3$ catalysts.[68]

11.4 Catalyst Supports

A catalyst support should provide an appropriate substrate to allow for the reactant particles to access catalytically active sites. With respect to a combustion catalyst, the support should also have high thermal stability, as well as enough sites to anchor and disperse the active species to ensure that the catalyst does not sinter under the high thermal conditions. Regarding methane oxidation using a palladium catalyst, the other role of the support is the continuous exchange of oxygen to the active metal sites.[35] This oxygen mobility is needed to continue methane oxidation by re-oxidising the active site after each reaction.[35] Water is a known inhibitor of oxygen mobility, absorbing on the support surface, significantly limiting the exchange of oxygen.[35] For NGEE with a water content of greater than 10%, deactivation by water can be significant and hydrophobicity of the support can help alleviate deactivation. Water can act as a Lewis base, so a Lewis acidic support will show hydrophilic properties and thus be significantly affected by the high water content.[37] Regardless of the support associated with hydrophobicity, if palladium is chosen as the active metal, its proposed subtle interaction with water may be a consequence at proposed low operating temperatures.[25] A different active metal must be selected or operated at the high end of the NGEE temperature range. So, another essential property should be a high level of hydrothermal sintering resistance. The temperature for NGEE would not normally exceed 550 °C.[33] Ideally, the active phase of a catalyst should thus have a Tamman temperature above 550 °C.

11.4.1 Alumina

Alumina (Al$_2$O$_3$) with an active palladium metal is the most used support in methane oxidation.[30] Alumina shows good chemical and physical stability, mechanical resistance, a high surface area, and thermal stability.[37] However, alumina shows low oxygen mobility in the presence of water, allowing the active palladium metal to be readily inhibited.[35] The Al$_2$O$_3$ acts as a Lewis acid, allowing hydroxyl groups to form bonds to its surface.[37] Through this acidity, the accumulation of hydroxyls onto the support has been suggested to be the primary way water inhibits an Al$_2$O$_3$-supported catalyst.[35] Water blocks oxygen transfer from the Al$_2$O$_3$ support to the active site.[35] In a palladium catalyst utilised in a natural gas engine exhaust stream, under assumption of a Mars–van Krevelen mechanism the active site will be PdO, which needs oxygen replenishment with every turnover of the catalyst.[35]

Alumina is a sulphating support, as it readily adsorbs SO$_2$ to form aluminium sulphate [Al$_2$(SO$_4$)$_3$].[69] The interaction between sulphur and Al$_2$O$_3$

reduces the concentration of sulphur dioxide and in turn can protect the active metal. Additionally, the oxygen exchange will be limited if the support is sulphated. The presence of water will also deplete Al_2O_3's usefulness as a sulphating support, as the much higher concentrated water will compete for the Al_2O_3 acid sites.[69] As a result, Al_2O_3 can only be considered as a good sulphur-tolerant support when the feed is dry, which is not feasible in a natural gas engine exhaust stream.

11.4.2 Silicas

Silica is a commonly used support for methane combustion, supporting the active metal palladium. It has hydrophobic surface properties and could be a good choice for an NGEE catalyst.[70] However, silica-based catalysts are known to deactivate through the reduction in metal dispersion.[71] The lower binding energy between silica and active metal results in high mobility at temperatures beyond 400 °C. The poor thermal stability and tendency to sinter can cause permanent deactivation.[71]

Silica is also deactivated partially by exposure to carbon dioxide.[71] Carbon dioxide is a weak inhibitor and is not generally considered as affecting the stability of a catalyst. Interestingly, it was reported that over SiO_2-supported catalysts, the catalyst activity was seen to rapidly decrease by 33% with exposure to high concentrations of CO_2.[71]

11.4.3 Zirconia

Zirconia is useful in catalytic processes in both the active phase and as a support due to its good thermal and chemical stability, coupled acid–base properties, and high oxygen storage capacity.[72] The thermal stability of zirconia has been proven to be better than that of alumina by comparison between Pd/Al_2O_3 and Pd/ZrO_2 catalysts.[73] The palladium on the alumina catalyst showed a sudden deactivation at 850 °C, while the zirconia showed little change in activity.[73]

11.4.4 Zeolites

Using zeolites as a support has found recent attention in methane oxidation catalysts. A zeolite is a microporous, crystalline framework that consists of silicon, aluminium, and oxygen. Zeolites as a support offer high thermal stability and high surface area. Over 250 different framework types exist, which can be selected for the best pore dimensions that support metals introduced into extra framework positions.

One of the most important properties of zeolites as supports for metal catalysts for combustion is their ability to bond metal cations and tune their framework charge and in turn acidity. This can be done by replacing Si framework atoms with other elements such as aluminium, titanium, iron, boron, and others. The acidity of the zeolite can not only be altered by changing the type of substitute, but also its concentration.[70]

Low silicon-to-aluminium (Si/Al) ratio zeolites contain higher aluminium content which increases the concentration of binding sites for the metal cations.[70] However, the higher framework charge also introduces more Lewis acidity and consequently low Si/Al zeolite frameworks exhibit low activity in the presence of water due to increased hydrophilicity.[70] Therefore, a higher Si/Al ratio like Pd/ZSM-5 or Pd/Beta, should be chosen for methane oxidation under humid conditions. These catalysts are known to exhibit increased stability and hydrophobicity. However, the reduced number of binding sites consequentially lowers the amount of active metal on the catalyst.[70]

Titanium silicalite-1 (TS-1) is a MFI catalyst with titanium substituted in the framework.[23,24] Titanium substitution leads to a neutral zeolite framework due to Ti being 4-valent. The lack of framework charge reduces the overall affinity of water, increasing the stability in high water conditions.[23,24,74] Temperature-programmed desorption of a TS-1 catalyst showed low retention of polar groups, suggesting a neutral framework. Titanium has also shown some ability to anchor the active metal, increasing the dispersion of the catalyst.[23,24,74] This results in a high stability catalyst for methane combustion with low framework charge but anchoring sites for the active metal.

Similarly, silicalite-1, which has the same zeolite MFI structure but ideally contains no aluminium, lowers Lewis acidity and framework charge. However, without any potential binding sites (like titanium) could restrict the ability of the active metal to anchor onto the support, creating larger active metal clusters and decreasing the overall dispersion. This means specific preparation methods must be employed to achieve adequate formation of active sites. Palladium on silicalite-1 catalysts were prepared *via* a one-pot hydrothermal method, creating palladium nanoparticles along the support structure.[75] This preparation method created a catalyst with 0.6 wt% palladium loading that remained stable for 150 hours under high water concentrations up to 10 vol%.[75] Comparing the same catalyst prepared through the standard incipient wetness method shows a vast difference in activity, with the catalyst prepared *via* incipient wetness deactivating quickly under high water concentrations.[75] Therefore, special considerations must be made when preparing the catalyst for certain supports that do not have the sites that facilitate dry impregnation. Silicalite in different forms has also been utilised in methane oxidation catalysts *via* silicalite-2 (S-2).[76] S-2 zeolite has a MEL-type structure consisting of two identical sets of 10-membered ring channels. An *in situ* synthesis method with a palladium-active metal was used to make a catalyst using this silicalite support.[75] By forming nanocrystals along the support structure, no peak in the water temperature programmed desorption results were observed.[76] This means that the silicalite doesn't absorb any water. However, a Pd/S-2 catalyst prepared *via* dry impregnation was tested under H_2O TPD, and a small peak was shown at a temperature of 250 °C.[76] This suggests palladium clusters can adsorb water. The same analysis was done *via* a palladium alumina catalyst, and the desorption peak dwarfs the other peaks, showing the very high Lewis acidity of the alumina catalyst. As a result of this low water adsorption, the Pd/silicalite-2 catalyst showed high stability at 400 °C under humid conditions.[76]

11.5 Catalyst Preparation

Catalyst preparation strongly influences the properties of the catalysts. These properties can change the structure of the active metal, forming nanostructures, increasing mass transfer, and protecting the active site from certain poisons. Also, active metal support interactions strongly influence reactivity and selectivity and can be tailored by preparation methods.

11.5.1 Impregnation

11.5.1.1 Dry Impregnation

Impregnation is the most simple but economical method of catalyst preparation. This method involves contacting a support with a certain amount of the active metal. The active metal is delivered to the dry support in the form of a salt precursor solution. The precursor is slowly added to the dry support until it reaches a wetted state, called the incipient wetness point, resulting in deeper penetration and a more uniform dispersion.[77] This impregnation method is called incipient wetness or dry impregnation and is the most commonly used supported catalyst preparation method. This is because it is cheap, simple, and very reproducible. Often, several repeated dry impregnations must be run in between drying the slurry in the furnace to achieve this high metal loading. So, incipient wetness is useful for low-metal loading catalysts, specifically a noble metal catalyst.

Noble metals platinum and palladium were prepared successfully *via* an incipient wetness dry impregnation method on alumina support for methane combustion. It has high activity in methane combustion and can convert trace methane of 6000 ppm at very low temperatures.[42]

11.5.1.2 Wet Impregnation

Impregnation can be performed by fully immersing the support with the active metal solution and then drying it. This method, called wet impregnation, relies on the support pores to actively adsorb the active sites. This method is straightforward and reproducible as long as the same support structure is used. However, the particle size distribution can be quite variable due to agglomeration of active phase.

11.5.1.2.1 Ion Exchange. Ion exchange can be used for catalyst that have specific exchange sites (cationic) that can be exchanged with transition metals This method operates by suspending the support in an aqueous solution, where the pH of the solution is crucial in order to prevent precipitation and hydration sphere of the transition metal cations.

Palladium catalysts prepared through ion exchange have shown high activity in methane combustion. Using zeolite supports such as ZSM-5, mordenite, and ferrite exchanged with a palladium precursor improved the activity considerably compared to conventional palladium on alumina catalyst prepared *via* dry impregnation, with the ZSM-5 catalyst showing the highest activity in methane conversion.[78] However, this difference in overall activity could be due to the choice of support over the preparation technique. Another study using supports mordenite, beta, and ZSM-5 compared the preparation methods of ion exchange and impregnation using palladium active metal.[79] It was found that the catalysts did have low activity when prepared using conventional dry impregnation and improved activity was found over ion-exchanged zeolites.[79] Usually, ion exchange results in highly dispersed catalysts creating a more reproducible and active catalyst. However, there is very little control of the loading.

11.5.2 Precipitation

Precipitation utilises a basic solution with a metal support salt and a salt of a compound, usually the precursor. The solution is constantly mixed, and a precipitate of hydroxide is formed, transforming into oxides by heating. The critical choices in designing a precipitation method are salt, alkali, and solvent, mainly water. Consideration must occur to avoid specific ions, such as sulphates and chloride ions, which are often catalyst poisons and are difficult to remove. Usually, nitrate salts or organic compounds are used as they usually fully decompose with temperature, forming the metal oxide required. Utilising co-precipitation, a uniform catalyst can be formed using controlled variables. This includes mixing, the procedure, the order of addition of the precursor and support oxide, and the precipitate's temperature and aging time.

Precipitation methods have been used to make a cheaper alternative to noble metal catalysts in a nickel magnesium oxide catalyst.[80] Several nickel magnesium oxide catalysts were prepared with different ratios *via* co-precipitation.[80] These catalysts had lower overall BET surface area than the standard supports used, like zeolites and alumina, but still had substantial activity in methane combustion.[80] A nickel magnesium catalyst with a molar ratio of 9:1 Ni:Mg showed the highest activity with a T_{50} of approximately 464 °C.[80] The catalyst has a lower activity compared to noble metal catalyst, but is a more cost-effective alternative. A perovskite catalyst with lanthanum and strontium as the "A" oxide and copper as the "B" oxide was prepared *via* co-precipitation for methane oxidation. This catalyst showed some activity at temperatures of 600 °C and above.[81] This is much higher than that of palladium or other noble metal catalysts, but can be useful for applications with higher feed temperatures.

11.5.3 Sol–Gel

The sol–gel method has been utilised in catalyst preparation to achieve nanoparticle catalysts on an industrial scale.[82] This catalyst preparation method is a wet chemical method. The precursor is first dissolved in water or alcohol, heated and stirred.[82] A gel is formed through hydrolysis, which is later dried to form a fine powder.

Spinel magnesium chromate and cobalt chromate catalysts were prepared through a sol–gel method for methane combustion. The magnesium chromate catalysts were active, with a T_{90} of 684 °C. This was attributed to the larger surface area, ease of access to surface oxygen species, and appropriate bulk structure.[83]

11.5.4 Passivation of the Support

Preparation techniques before active metal addition can alter the properties of the catalyst tailored to the feed stream to be catalysed. One property that can be altered is hydrophobicity through the addition of silanol groups to silica supports, which affects the behaviour of the catalyst under wet conditions.[84]

11.5.5 Use of Different Precursors

The choice of precursor has been shown to affect the catalyst activity. Significantly, two manganese oxide catalysts supported on titanium oxide were prepared using two different precursors of manganese nitrate and manganese acetate.[85] The manganese nitrate precursor resulted in mainly MnO_2 species. In contrast, the manganese acetate led to Mn_2O_3, with the manganese acetate-prepared catalyst having a higher low-temperature activity due to the surface Mn_2O_3 species.[85] Zinc oxide catalysts prepared *via* thermal decomposition and precipitation with three different precursors showed similar morphologies indicated by XRD analysis, however the crystal size and particle morphology uncovered by SEM showed variance.[86] Because of this, the catalyst's activity was significantly different, owing to the choice of the precursor as the activity increased.[86] More specifically, several palladium on alumina catalysts were prepared with different precursors for methane oxidating catalysts.[87] The palladium precursors were nitrate, chloride, and acetylacetonate. An XRD analysis highlighted the difference in the crystalline structure, with the chloride precursor catalyst having more crystalline PdO particles.[87] The overall dispersion of the catalysts showed that the acetylacetonate had the highest dispersion, with the next highest being nitrate and then chloride.[87] Because of this, the acetylacetonate catalysts had the highest activity in methane combustion, having a T_{10} of 270 °C, while the nitrate and chloride catalysts had a T_{10} of 334 °C and 378 °C, respectively.[87] Therefore, the precursor had a significant effect on catalysts.

11.6 Deactivation

Catalyst deactivation is a significant problem in any industrial chemical process. Catalysts can be deactivated through several different mechanisms, including sintering, poisoning, and active site inhibition.[30]

11.6.1 Sintering

Sintering of a catalyst involves a gradual coalescence of the active metal or support, typically under high-temperature conditions.[88] The temperature at which metals start to show sintering is known as the Tamman temperature and is approximately half the temperature of a metal's melting point.[88] Both the active metal and support should be kept at a temperature to avoid sintering as sintering of the support can adversely affect the catalyst efficiency as much as the sintering of the active metal. As the support takes up most of the catalyst volume, any sintering will heavily influence the structure, leading in most cases to adverse effects on activity. Active metal sintering usually occurs before any support sintering, mostly due to Ostwald ripening. Sintering of the active metal ends in a gradual, usually irreversible decrease in conversion, which occurs *via* a ripening and coalescence of the metal. When a catalyst is sintered to the point where it's not economically viable to run the process, it must be replaced.

Operating a chemical process at a high temperature can alleviate deactivation from poisoning and inhibition. However, introducing it to a higher thermal load can cause a catalyst to deactivate by sintering. The main catalysts used for methane oxidation are palladium, platinum, and rhodium, which have Tamman temperatures of 641 °C, 741 °C, and 856 °C, respectively. However, Tamman temperatures of the oxides are lower, PdO (293 °C), PtO(139 °C), and RhO (550 °C), which can facilitate sintering under low-temperature operation of methane combustion catalysts from VAM and NGEE conditions.

11.6.2 Sulphur Dioxide Poisoning

Catalyst poisoning involves the catalyst reacting with a contaminant species in the feed stream, producing sub-species with lower or no activity in its previous intended use. The species chemically bonds to the catalyst's active metal, lowering overall conversion. A common poison of catalysts in the process industry is sulphur dioxide (SO_2). Sulphur is found naturally in natural gas as H_2S at very low concentrations. Other natural sources are found in carbonyl sulphide and carbon disulphide. No fundamental regulations are set for these molecules as they are non-existent in the natural gas source or at trace concentrations (parts per billion). Other sulphur-containing compounds are added to the natural gas mixture as an odorant since natural gas is usually odourless. This increases the total concentration of sulphur in commercial natural gas significantly. In Australia, the maximum allowed sulphur concentration is 50 mg m^{-1}.[3,89] In Europe, the concentration varies greatly depending on the country. The highest is Hungary, with a maximum allowable limit of 100 mg m^{-3}; the lowest is Germany, with a maximum permissible limit of 30 mg m^{-3}.[90] Table 11.2 lists the permissible sulphur limits in natural gas for different countries/regions. H_2S and sulphur-containing mercaptans readily oxidise in gas engines to produce SO_2.[89,91]

Even though the concentration of SO_2 will be very low in lean-burn natural gas engines (1–4 ppm), these small amounts of SO_2 can still show high levels of deactivation. Regarding palladium-based catalysts, sulphur dioxide deactivation remains a massive hurdle to overcome. Even one ppm of SO_2 introduced to a methane feed stream can cause rapid deactivation of a palladium catalyst, as shown in Figure 11.4. Sulphur readily attaches to the PdO active site, forming stable palladium sulphate ($PdSO_4$).[26,38]

The less active platinum catalyst doesn't show the same level of deactivation that palladium shows in sulphur conditions. In a concentration of 20 ppm of either H_2S or SO_2, the activity of platinum increases slightly.[92] Regardless, the temperature required to reach an adequate conversion

Table 11.2 Sulphur emission data of various countries/regions (units of ppm).[89,91]

Country/region	Total sulphur	H_2S	Odorants		SO_2 leaving NGEE
			Lower	Upper	
Europe	30–100	5–20	5	15	3
Australia	50	5.7	7	14	<1
USA	17	4	1	8	4

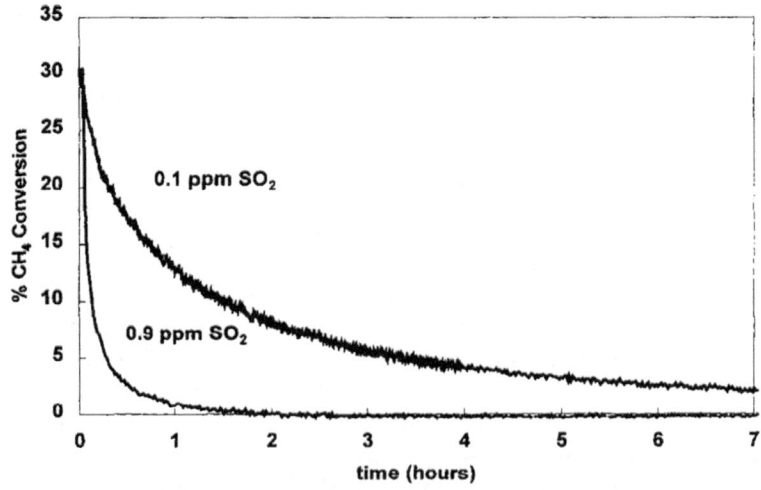

Figure 11.4 Effect of SO_2 on methane activity: 100 g ft^{-3} Pd/ZrO$_2$–SiO$_2$; 320 °C; feed gas: 800 ppm CH_4; 8% O_2, SO_2 as shown, balance N_2; 200 000 h^{-1}. Monolith-supported catalysts, 400 psi. Reproduced from ref. 6 with permission from Elsevier, Copyright 1997.

(>90%) is still too large, 600–650 °C.[92] Rhodium also shows the same relationship as palladium. In dry conditions, the activity decreases significantly.[41,92] But with wet conditions, the formation ($[Rh(H_2O)_x]_2(SO_4)_3 \cdot yH_2O$) was suggested to act as an active site.[41]

The effect sulphur dioxide has on transition metal oxides is similar to that of PdO. Transition metal oxides, cobalt oxide, manganese oxide, copper oxide, and nickel oxide see a significant level of deactivation in methane conversion when in the presence of SO_2. Copper oxide and nickel oxide see little methane oxidation activity even without sulphur dioxide being present.[93] So, with a feed stream containing 40 ppm SO_2, the activity level is so low that conversion is below 10% within 20 hours.[93] The transition metals cobalt oxide and manganese oxide show high activity in the absence of SO_2. But within 20 hours of SO_2 being present, the conversion of a cobalt oxide catalyst was found to be reduced to 65%, and the conversion over the manganese oxide catalyst falls below 20%.[93] Chromium oxide, however, shows formidable activity in methane conversion while showing little effect when SO_2 is present. Temperature-programmed desorption of the chromium catalyst showed zero sulphur dioxide peaks, unlike the cobalt and manganese oxide catalysts, which showed a peak beyond 600 °C.[93] This suggests that sulphates are forming on these catalysts, like palladium.

Using sulphating supports can increase the catalyst's longevity. The SO_2 can spill over onto the support, reducing the concentration of sulphur species on the active metal and restore reactivity.[32] For example, an Al_2O_3 support will react with SO_2 to form aluminium sulphate [$Al_2(SO_4)_3$], which is favoured compared to the formation of $PdSO_4$.[32]

11.6.3 Water Inhibition

Water is a highly effective inhibitor of catalysts. Water inhibition affects noble metal catalysts at low temperatures by forming hydroxyl groups on the active metal and support.[26,94] It is suggested that water converts the PdO active site into $Pd(OH)_2$.[25,26,36,94] This reversible reaction requires either a higher temperature or dry conditions to return the active site.[26] The hydroxides are only thermodynamically stable below 450 °C, showing little inhibition above temperatures of 500 °C in a 1% water stream. Higher temperatures are needed to destroy $Pd(OH)_2$ as the water concentration increases.[25] The support choice can also help reduce the stability of the $Pd(OH)_2$ species. If the support has high oxygen mobility, it can donate some oxygen to the hydrated palladium-producing water, reverting the catalyst to its active metal state.[25] Supports such as Al_2O_3 have low oxygen mobility and show protection from water inhibition.

The hypothesis of stable, hydrated palladium species forming at temperatures around 450 °C is debated.[35] Contrarily, it is suggested that the PdO doesn't interact with the water groups; instead, the support does. Inhibition is then carried out through the lack of oxygen transfer from the support to the active metal sites.[35] Since methane reduces four active palladium sites with every molecule reacted, the level of oxygen transfer between the palladium and support must be high. The formation of hydroxides on the support anchors oxygen species to the support structure, forming mainly metallic inactive palladium.[35]

Both inhibition hypotheses involve the reaction of PdO with water with the support. Thus, water inhibition is not observed with metal catalysts in a zero-valence state such as Pt. This is observed over platinum methane oxidation catalysts.[26] However, the methane oxidation activity of platinum is far less than palladium's.[42]

Synergistic effects can be observed when water and sulphur dioxide are present. Water can attach to the catalyst support, leaving the active metal with the more strongly chemisorbed SO_2.[40,69] Using platinum can help reduce the overall effect of sulphur dioxide poisoning as it does not form very stable sulphates like palladium.[40] Adding NO_x to the feed stream can also increase activation as the NO_x reacts rapidly with the hydroxyl groups that attach to the active metals, producing nitrates.[40]

11.6.4 Coking

The formation of carbonaceous deposits is a significant factor in catalyst deactivation. Carbon coking only shows some prevalence with reactions that involve hydrocarbons. Catalysts that employ carbon chemisorption are especially susceptible to the formation of coke deposits. Coke deposits are unique because any combination can form depending on the reaction conditions, including process conditions, active metal catalyst, and support.[30] Coke on a catalyst blocks the active sites, inhibiting their interaction with the process gas. Further, coking can induce pore blockage and even reactor blockage.

Palladium was found to incorporate carbon and has shown increased activity in transforming amorphous carbon into graphitic carbon.[30] Coking is usually caused by palladium's interaction with a highly concentrated hydrocarbon or several hydrocarbons, so hydrocarbon species comprise the bulk of the process gas. Most well-known instances of palladium coking are found through hydrogenation or reforming.[95] It was found that the carbonaceous deposits formed between the support and the active metal catalyst are different.[30]

Despite the high oxygen concentration in lean methane gas streams, carbonaceous deposits were observed in lean methane oxidation conditions.[96] Most of the species are oxygenated like carbonyls and carboxylic acid groups attached to the Pd. The presence of carbonaceous deposits can occur at low temperatures of around 180 °C and at higher temperatures.

11.7 Regeneration

11.7.1 Temperature-based Regeneration

For water deactivation, an increase in temperature higher than 450 °C will destroy the Pd hydroxide species depending on the water concentration.[25]

The palladium sulphates formed in sulphur dioxide deactivation find instability at temperatures beyond 750 °C.[40] Removing the SO_2 will cause some regeneration but can result in sintering.

11.7.2 Nitrogen Oxide Regeneration

The presence of NO has been found to regenerate catalysts under certain conditions.[40] NO competes for the active sites with methane in dry conditions, just like sulphur dioxide and water.[40] This slowly deactivates the catalyst but to a much lesser extent than the other previously mentioned inhibitors. In wet conditions, water-inhibited NO can reactivate the active site. By adding NO to the feed stream, the NO reacts with the hydroxyl groups to form HNO_2.[40] This species is readily oxidative and can combust methane rapidly. This removes a hydroxyl group inhibiting the catalyst and returns the active site to its metallic form. This interaction has only been recorded to work in bimetallic palladium and platinum.[40]

11.8 Stability

11.8.1 Water

11.8.1.1 Temperature

The formation of hydroxyls on the active metal or the support is the primary deactivation pathway for palladium-based catalyst. The palladium on the Al_2O_3 catalyst shows little inhibition above 500 °C in a 1% water stream.[25] At higher water concentrations, increasing the temperature up to 500 °C can provide a level of stability, as seen in Figure 11.5.

Even though the catalyst performed for 1000+ hours, it still showed a deactivation level at 500 °C, mainly due to the formation of carbonaceous deposits.

11.8.1.2 Catalyst

Water readily affects the activity of palladium on Al_2O_3, but, for example, reduced deactivation is observed using TS-1 as a support (Figure 11.2). It was found that hydroxyl spillover from Al_2O_3 support onto ruthenium in partial methane oxidation is kinetically significant.[97] Hydroxyls forming on Al_2O_3 can quickly transfer hydroxy groups to the active metal promoting deactivation. On hydrophobic supports the hydroxy concentrations are low, resulting in decreased spillover and in turn increased stability.[25,26]

11.8.2 Sulphur Dioxide

11.8.2.1 Consumption

One way to reduce catalyst deactivation by sulphur is to use a sacrificial adsorption catalyst. This method of removing sulphur dioxide is feasible as

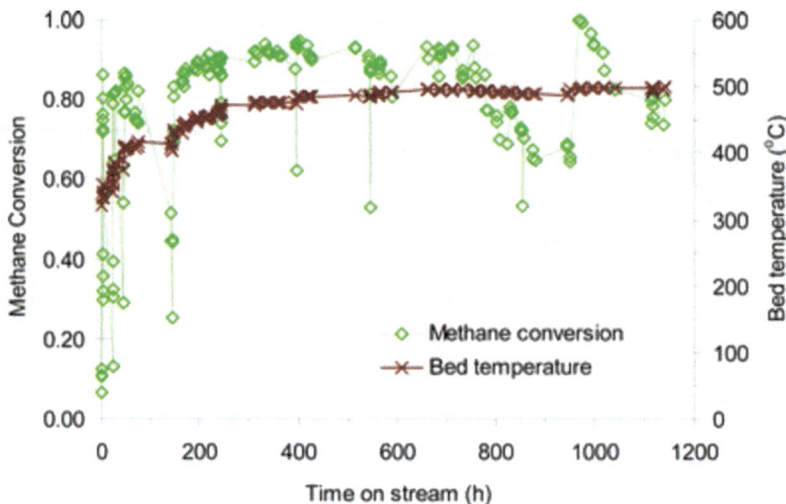

Figure 11.5 Time-on-stream fractional methane conversion and corresponding reactor bed temperature over 1.0 wt% Pd/Al$_2$O$_3$ catalyst. Feed = 7000 ppm CH$_4$, 10 000 ppm CO$_2$, 30 000–40 000 ppm H$_2$O and balance air. GHSV = 75 000–110 000 h^{-1}. ◊ = methane conversion, X = bed temperature. Reproduced from ref. 21 with permission from the Royal Society of Chemistry.

the concentration of sulphur dioxide is very low in the exhaust gas (around 1–2 ppm), so only a relatively low amount of catalyst will be affected.

Many metal oxides can adsorb sulphur dioxide, forming a stable sulphate. The standard metal oxides used for adsorption or reaction with sulphur dioxide are cobalt, chromium, copper, iron, manganese, nickel, zinc, and vanadium.[98,99] Most of these metal oxides form sulphates, but some adsorb SO$_2$ onto their surface. Adsorption of SO$_2$ onto the metal surface is a reversible process at low temperature and thus requires careful control of reaction conditions.

For a metal oxide to be efficient at SO$_2$ removal in the feed stream, it needs a high level of selectivity, a high level of conversion, and a high capacity for sulphate formation.

Capacity of the sulphate compound on the metal oxide is another critical factor in the lifetime of the metal oxide and in turn for the lifetime of the primary catalyst. Metal oxides such as iron oxide have a low storage capacity of 0.03 grams of sulphur dioxide consumed per gram of adsorbent. Other metals such as copper, cobalt, manganese, and nickel have higher capacities of around 0.5 grams of SO$_2$ consumed per gram of metal oxide.[99]

11.9 Reaction Kinetics

11.9.1 Mechanisms

The total oxidation of methane is shown in eqn (11.1).

$$CH_{4(g)} + 2O_{2(g)} \rightarrow CO_{2(g)} + 2H_2O_{(g)} \quad (11.1)$$

This reaction is relatively simple, but once a catalyst is used to promote oxidation under non-flammable conditions, the level of complexity increases. The three main mechanisms proposed for the catalytic oxidation of methane are Langmuir–Hinshelwood, Eley–Rideal, and Mars–van Krevelen. All three of these mechanisms have been suggested to be followed by a palladium-based catalyst.[32,100–102]

11.9.1.1 Langmuir–Hinshelwood

The Langmuir–Hinshelwood mechanism requires the two reactants to adsorb on the catalyst surface before they react to form the products. So, oxygen and methane must both adsorb to the catalyst for methane's catalytic oxidation. The process can be described *via* reaction eqn (11.2) and (11.13).[101]

$$CH_4 + * \rightarrow CH_4^* \tag{11.2}$$

$$O_2 + 2* \rightarrow 2O^* \tag{11.3}$$

$$CH_4^* + O^* \rightarrow CH_3^* + OH^* \tag{11.4}$$

$$CH_3^* + O^* \rightarrow HCHO^* + H^* \tag{11.5}$$

$$H^* + O^* \rightarrow HO^* + * \tag{11.6}$$

$$HCHO^* + O^* \rightarrow CHO^* + HO^* \tag{11.7}$$

$$CHO^* + * \rightarrow C^* + OH^* \tag{11.8}$$

$$C^* + O^* \rightarrow CO^* + * \tag{11.9}$$

$$CO^* + O^* \rightarrow CO_2^* + * \tag{11.10}$$

$$CO_2^* \rightarrow CO_2 + * \tag{11.11}$$

$$OH^* + OH^* \rightarrow H_2O^* + O^* \tag{11.12}$$

$$H_2O^* \rightarrow H_2O + * \tag{11.13}$$

where the "*" represents a surface species, the methane adsorbs to the surface sites in this mechanism, and the hydrogen bonds are broken in two ways. The methane is dehydrated by surface oxygen to form a hydroxyl. Or hydrogen is removed, and oxygen is added to create a C–O bond, leaving a hydrogen surface species that needs to be oxidised *via* another oxygen surface species to form a hydroxyl. The step taken is dependent upon the temperature and conditions. Either step is repeated four times to produce four OH* groups and one C* group. Two oxygen surface species then oxidise the carbon species to form carbon dioxide, which leaves the catalyst surface, leaving a surface vacant site. The four hydroxide species combine to form two water species, two oxygen-adsorbed species, and two vacant surface sites. The overall mechanism is shown in eqn (11.14).

$$CH_4^* + 6O^* + * \rightarrow 2H_2O + CO_2 + 6* + 2O^* \qquad (11.14)$$

So, six oxygen surface species and a vacant surface site must completely oxidise surface methane to form two water and carbon dioxides. Leaving six vacant surface species and two surface oxygens. So, four surface sites must be oxidised for a complete turnover. Depending on the concentration of water in the gaseous stream and the temperature, the pathway of the CHO* species in step 8 can change, as shown below in eqn (11.15)–(11.17).

$$CHO^* \rightarrow CO^* + H^* \qquad (11.15)$$

$$CO^* + OH^* \rightarrow COOH^* + * \qquad (11.16)$$

$$COOH^* \rightarrow CO_2^* + H^* \qquad (11.17)$$

Instead of a hydroxyl leaving the carbon species, hydrogen breaks off, leaving a CO* species. This species interacts with an OH* group to form COOH*, which forms CO_2 and leaves a hydrogen surface species. This mechanism is promoted more in wet conditions as there would be more hydroxyls forming due to the reverse of steps 12 and 13. This mechanism has been accounted for in platinum clusters containing catalysts.[101]

11.9.1.2 Eley–Rideal

The Eley–Rideal mechanism involves a gas species that adsorbs to the catalyst, reacting with another gaseous species. Concerning the total oxidation of methane, oxygen is first adsorbed onto the catalyst and then reacts with gaseous methane. The rest of the mechanism is shown *via* eqn (11.18)–(11.26).[101]

$$O_2 + 2* \rightarrow 2O^* \qquad (11.18)$$

$$2O^* + CH_4 \rightarrow H_2CHO^* + OH^* \qquad (11.19)$$

$$H_2CHO^* + O^* \rightarrow HCHO^* + OH^* \qquad (11.20)$$

$$HCHO^* + O^* \rightarrow CHO^* + OH^* \qquad (11.21)$$

$$CHO^* + O^* \rightarrow CO^* + OH^* \qquad (11.22)$$

$$CO^* + O^* \rightarrow CO_2^* + * \qquad (11.23)$$

$$CO_2^* \rightarrow CO_2 + * \qquad (11.24)$$

$$OH^* + OH^* \rightarrow H_2O^* + * \qquad (11.25)$$

$$H_2O^* \rightarrow H_2O + * \qquad (11.26)$$

Catalytic Combustion of Methane in Low-concentration Gas Streams 369

So, this mechanism is very similar to the Langmuir–Hinshelwood, but methane doesn't need to compete with oxygen for surface sites. This means that the reaction order concerning methane should be 1. The reaction order for oxygen is dependent upon the concentration of oxygen. If oxygen is in far excess, it will be zero, but if it's close to stoichiometric, oxygen could be 1.

11.9.1.3 Mars–van Krevelen

In the Mars–van Krevelen mechanism, methane adsorbs on the catalyst and reacts with lattice oxygen. The lattice oxygen is then replenished *via* gaseous oxygen; the entire mechanism is shown in Figure 11.6. This mechanism is proposed for a palladium oxide-based catalyst.

So, overall, the first two hydrogen bond breakages occur when they interact with lattice oxygen. The formed lattice hydroxides move to a vacant lattice palladium where the last two hydrogen bond breakages arise with the interaction with PdOH species – the movement of lattice hydroxides causes vacant lattice sites. The vacant lattice sites require gaseous oxygen and a high oxygen mobility among palladium for a PdO group to donate their oxygen to the second vacant site. After entirely dehydrating methane, a PdCO group interacts with an oxygen lattice to form *OCO. Both PdH$_2$O and *OCO species desorb to form water and carbon dioxide, respectively. Carbon dioxide desorption will leave a vacant site that will need a second gaseous oxygen to adsorb to complete the mechanism.

So, from this mechanism, lattice oxygen, palladium oxide, and lattice palladium are needed to oxidise the methane.

In the Mars–van Krevelen mechanism, carbon dioxide will interact with vacant lattice sites. This can inhibit the catalyst at a rate dependent upon the overall oxygen mobility and the concentration of gaseous oxygen compared to gaseous CO_2. Under lean conditions oxygen concentration by far exceeds CO_2 so this inhibition will be less relevant under lean than rich conditions.

Based on the mechanism, water could interact with lattice palladium sites to form Pd(H$_2$O) and further interact with palladium oxide to form two palladium hydroxide surface species. This interaction will remove two sites available for methane oxidation. Even though PdOH is necessary for the breakage of the last two C–H bonds, without a PdCH$_2$O species, the rate to remove PdOH will reduce.

11.9.2 Effect of Water

Water inhibits the combustion of methane, which is observed as an increase of the activation energy when water is added to the reactants.[103] Activation energy of combustion changes when additional water is added to the feed stream of lean methane mixtures and the activation energies of Pd on Al$_2$O$_3$ and Pd/TS-1 change from 128 kJ mol^{-1} to 369 kJ mol^{-1} on used catalysts, respectively. The change in activation energy is suggested to result from

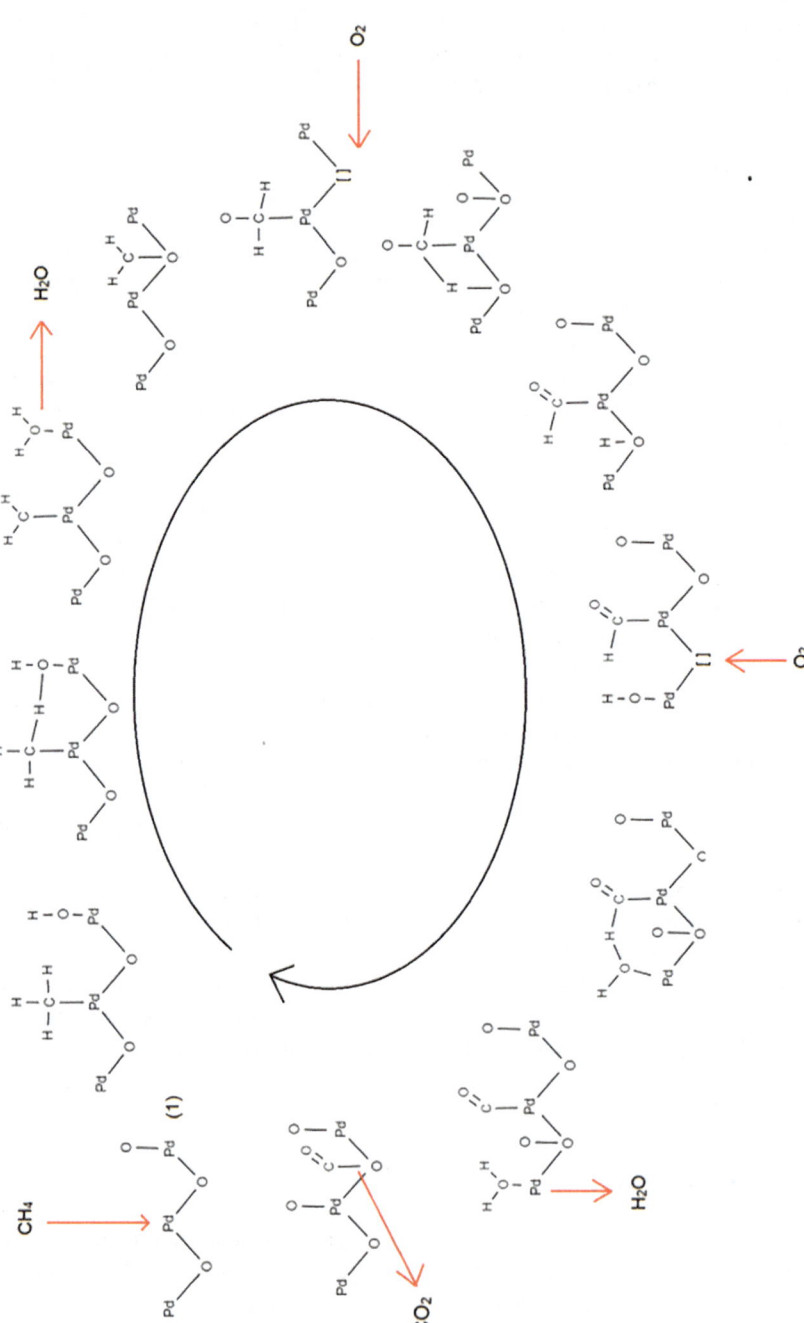

Figure 11.6 Proposed Mars–van Krevelen mechanism for the total oxidation of methane. Adapted from ref. 101 with permission from American Chemical Society, Copyright 2022.

different heat of adsorption depending on the water coverage, which in turn results in a different apparent activation energy.

It has been proposed that the effect of wet conditions on the activation energy changes *via* eqn (11.27).[100]

$$\left[(E_A)_{app}\right]_{Dry} = \frac{1}{1-\gamma}\left[(E_A)_{app}\right]_{Wet} \tag{11.27}$$

where γ represents the reaction order of water concerning the oxidation of methane.

11.9.3 Effect of Support

The support's effect on the catalyst or the active metal is complex. It allows the active metal to be dispersed amongst the support framework and allows particle size, dispersion, and phase variances. It has been suggested that the methane oxidation rate depends on the palladium oxide particle size.[28]

11.9.4 Reaction Order

The equation of methane oxidation is shown in eqn (11.1); from this, the rate equation is shown in eqn (11.28).

$$r = k[CH_4]^\alpha [O_2]^\beta [CO_2]^\gamma [H_2O]^\delta \tag{11.28}$$

where α, β, γ, and δ are the reaction orders for methane, oxygen, carbon dioxide, and water, respectively. In most cases, both products will be present in the feed and will contribute to the overall kinetics.

11.9.4.1 Effect of Water

Water can act as a Lewis base and interact with the palladium oxide Lewis acid sites to form low-activity hydroxides. Water has been suggested to interact only with the catalyst's active metal in oxide form and doesn't interact with the metal form.[100] Therefore, the reaction order with respect to water has been unanimously determined to be negative.[104,105] With a varied water concentration of 0.6% to 3.1%, the reaction order was found to be -0.8 ± 0.2 on a palladium alumina catalyst with a 7.3 wt% loading.[104] This effect may be attributed to the reverse of the relationship shown in eqn (11.13) and (11.26) and the desorption of water shown previously in Figure 11.6. These mechanisms show a step that requires the water to leave the catalyst and complete a full turnover. Water present in the feed could lower the rate at which this desorption occurs by adsorbing to the catalyst, reversing the suggested interaction.

The effect of water has been found to affect catalyst performance at minimal concentrations of 10^{-5} mol.[106] This implies that the water generated *via* the oxidation of methane would affect the catalyst. For VAM conditions, the water concentration is between 1–3%. So, under these conditions, the reaction order of water should be -0.8 ± 0.2, assuming the support and amount of active sites on the catalyst does not play a role in the reaction order. However, for NGEE, the water concentration of approximately 10% could decrease the reaction order further. Regardless, since the concentration is approximately five times that of VAM, the negative reaction order will have a much more significant effect on overall activity.

The effect of temperature has been noted for water's inhibitory properties. As the temperature increases, the palladium hydroxide sites are more prone to decompose to water and palladium metal. It was found that for a water stream of 2.7%, the inhibitory effect of water is null beyond a temperature of 450 °C.[71,103] However, at concentrations past 2.7%, the temperature required to achieve this effect may be increased.

11.9.4.2 Effect of Carbon Dioxide

The effect of carbon dioxide on the reaction rate of the total oxidation of methane has been suggested to be approximately zero under most conditions, with palladium on alumina catalysts.[104] Due to the effect of carbon dioxide on a methane oxidation catalyst, it has been briefly discussed. For the Mars–van Krevelen mechanism, CO_2 must be desorbed removing a lattice of oxygen, leaving a vacant lattice site. This is unlike water, as the water interacts with palladium species. Regardless, this effect can affect catalyst activity, and this has been found under high concentrations of CO_2 (greater than 0.5%) where the reaction order shifts from zero to −2, with a palladium silicon–alumina catalyst with 7.7% palladium loading.[105] This negative effect is pretty much void once the temperature reaches 400 °C,[71] where the proposed PdO-CO_2 sites have a minimal residence time.[71] It should be noted that this negative effect was found on a catalyst with a Si support but there was no effect with an alumina support.[105]

11.9.4.3 Effect of Gaseous Oxygen

The effect of oxygen gas on the reaction rate of methane combustion under similar conditions of VAM and NGEE has been noted to be very close to zero.[104,107] This is likely because methane is the limiting reagent, not oxygen gas. For both VAM and NGEE, the oxygen gas concentration to methane concentration is approximately 27 : 1 and 20 : 1, respectively. Even though the stoichiometric factor of oxygen in the oxidation of methane is 2, at 100% methane conversion, the change in oxygen concentration will be minimal. Thus, the overall effect of oxygen gas concentration between the concentrations of VAM and NGEE should be small. However, at high water levels the Pd hydroxide formation could result in a lowering of the oxygen concentration available on the surface.[104,107] This would be enhanced for Langmuir–Hinshelwood reaction mechanisms.

If the mechanism follows that of a Mars–van Krevelen mechanism, the level of incorporation of oxygen within the lattice support may be a rate-determining step, and oxygen concentration could have an effect on the rates. This is highly dependent on the choice of support. Some transition metal oxide supports have a high oxygen capacity and incorporate oxygen concentration into the rate.[108]

Acknowledgements

The authors acknowledge the support of Australian Coal Association Research Projects (ACARP) and the Australian Government Department of Industry, Science and Resources.

References

1. H. M. Cho and B.-Q. He, *Energy Convers. Manage.*, 2007, **48**, 608–618.
2. P. Lott and O. Deutschmann, *Emiss. Control Sci. Technol.*, 2021, **7**, 1–6.
3. G. Pirker and A. Wimmer, *Energy Convers. Manage.*, 2017, **149**, 1048–1065.
4. D.-G. f. E. European Commission, European Commission, Brussels, 2020.
5. P. Bielaczyc, J. Woodburn and A. Szczotka, *Appl. Energy*, 2014, **117**, 134–141.
6. J. K. Lampert, M. S. Kazi and R. J. Farrauto, *Appl. Catal., B*, 1997, **14**, 211–223.
7. K. Varde, N. Patro and K. Drouillard, Lean burn natural gas fueled SI engine and exhaust emissions 0148-7191, SAE Technical Paper, 1995.
8. A. Ayala, N. Y. Kado, R. A. Okamoto, B. A. Holmén, P. A. Kuzmicky, R. Kobayashi and K. E. Stiglitz, SAE International, 2002.
9. S. K. Chen and N. J. Beck, SAE International, 2001.
10. T. Kato, K. Saeki, H. Nishide and T. Yamada, *JSAE Rev.*, 2001, **22**, 365–368.
11. P. G. Kristensen, B. Karll, A. B. Bendtsen, P. Glarborg and K. I. M. Dam-Johansen, *Combust. Sci. Technol.*, 2000, **157**, 262–292.
12. P. Da, L. Tao, K. Sun, L. M. Golston, D. J. Miller, T. Zhu, Y. Qin, Y. Zhang, D. L. Mauzerall and M. A. Zondlo, *Nat. Commun.*, 2020, **11**, 4588.
13. G. Myhre, D. Shindell, F.-M. Bréon, W. Collins, J. Fuglestvedt, J. Huang, D. Koch, J.-F. Lamarque, D. Lee and B. Mendoza, *Climate Change 2013- The Physical Science Basis*, 2014, pp. 659–740.
14. S. Lee, U. H. Yi, H. Jang, C. Park and C. Kim, *Energy*, 2021, **220**, 119766.
15. EUROMOT, ed. The Europen association of the Internal Combusion Engine and Alternative Powertrain Manufacturers, 2015.
16. B. Vanderstraeten, D. Tuerlinckx, J. Berghmans, S. Vliegen, E. Van't Oost and B. Smit, *J. Hazard. Mater.*, 1997, **56**, 237–246.
17. C. Robinson and D. Smith, *J. Hazard. Mater.*, 1984, **8**, 199–203.
18. R. Liu, Y. Liu and Z. Gao, 2008 2nd International Conference on Bioinformatics and Biomedical Engineering, 2008.

19. H. Yamamoto and H. Uchida, *Catal. Today*, 1998, **45**, 147–151.
20. U. G. Alkemade and B. Schumann, *Solid State Ionics*, 2006, **177**, 2291–2296.
21. J. F. Adi Setiawan, E. M. Kennedy, B. Z. Dlugogorski and M. Stockenhuber, *Catal. Sci. Technol.*, 2014, **4**, 1793–1802.
22. A. Setiawan, E. M. Kennedy, B. Z. Dlugogorski, A. A. Adesina and M. Stockenhuber, *Catal. Today*, 2015, **258**, 276–283.
23. H. Hosseiniamoli, A. Setiawan, A. A. Adesina, E. M. Kennedy and M. Stockenhuber, *Catal. Sci. Technol.*, 2020, **10**, 1193–1204.
24. H. Hosseiniamoli, G. Bryant, E. M. Kennedy, K. Mathisen, D. Nicholson, G. Sankar, A. Setiawan and M. Stockenhuber, *ACS Catal.*, 2018, **8**, 5852–5863.
25. R. Gholami, M. Alyani and K. J. Smith, *Catalysts*, 2015, **5**, 561–594.
26. P. Gélin, L. Urfels, M. Primet and E. Tena, *Catal. Today*, 2003, **83**, 45–57.
27. S. H. Oh, P. J. Mitchell and R. M. Siewert, in *Catalytic Control of Air Pollution*, American Chemical Society, 1992, vol. 495, pp. 12–25.
28. J. Chen, H. Arandiyan, X. Gao and J. Li, *Catal. Surv. Asia*, 2015, **19**, 140–171.
29. Y. Ozawa, Y. Tochihara, M. Nagai and S. Omi, *Chem. Eng. Sci.*, 2003, **58**, 671–677.
30. P. Albers, J. Pietsch and S. F. Parker, *J. Mol. Catal. A: Chem.*, 2001, **173**, 275–286.
31. H. Stotz, L. Maier, A. Boubnov, A. T. Gremminger, J. D. Grunwaldt and O. Deutschmann, *J. Catal.*, 2019, **370**, 152–175.
32. P. Auvinen, N. M. Kinnunen, J. T. Hirvi, T. Maunula, K. Kallinen, M. Keenan, R. Baert, E. van den Tillaart and M. Suvanto, *Appl. Catal., B*, 2019, **258**, 117976.
33. H. Ohtsuka, *Emiss. Control Sci. Technol.*, 2015, **1**, 108–116.
34. J.-H. Park, J. H. Cho, Y. J. Kim, E. S. Kim, H. S. Han and C.-H. Shin, *Appl. Catal., B*, 2014, **160-161**, 135–143.
35. W. R. Schwartz, D. Ciuparu and L. D. Pfefferle, *J. Phys. Chem. C*, 2012, **116**, 8587–8593.
36. C. Cullis, T. Nevell and D. Trimm, *J. Chem. Soc., Faraday Trans. 1*, 1972, **68**, 1406–1412.
37. Q. Dai, Q. Zhu, Y. Lou and X. Wang, *J. Catal.*, 2018, **357**, 29–40.
38. D. L. Mowery, M. S. Graboski, T. R. Ohno and R. L. McCormick, *Appl. Catal., B*, 1999, **21**, 157–169.
39. P. Lott, M. Eck, D. E. Doronkin, A. Zimina, S. Tischer, R. Popescu, S. Belin, V. Briois, M. Casapu, J.-D. Grunwaldt and O. Deutschmann, *Appl. Catal., B*, 2020, **278**, 119244.
40. N. Sadokhina, G. Smedler, U. Nylén, M. Olofsson and L. Olsson, *Appl. Catal., B*, 2018, **236**, 384–395.
41. Y. Zhang, P. Glarborg, K. Johansen, M. P. Andersson, T. K. Torp, A. D. Jensen and J. M. Christensen, *ACS Catal.*, 2020, **10**, 1821–1827.
42. R. Burch and P. K. Loader, *Appl. Catal., B*, 1994, **5**, 149–164.

43. G. Águila, F. Gracia, J. Cortés and P. Araya, *Appl. Catal., B*, 2008, **77**, 325–338.
44. A. Khalesi, H. R. Arandiyan and M. Parvari, *Chin. J. Catal.*, 2008, **29**, 960–968.
45. H. R. Arandiyan and M. Parvari, *J. Nat. Gas Chem.*, 2008, **17**, 213–224.
46. P. W. Park and J. S. Ledford, *Appl. Catal., B*, 1998, **15**, 221–231.
47. L. Hu, Q. Peng and Y. Li, *J. Am. Chem. Soc.*, 2008, **130**, 16136–16137.
48. M. A. Ulla, R. Spretz, E. Lombardo, W. Daniell and H. Knözinger, *Appl. Catal., B*, 2001, **29**, 217–229.
49. T.-C. Xiao, S.-F. Ji, H.-T. Wang, K. S. Coleman and M. L. H. Green, *J. Mol. Catal. A: Chem.*, 2001, **175**, 111–123.
50. L. M. T. Simplício, S. T. Brandão, D. Domingos, F. Bozon-Verduraz and E. A. Sales, *Appl. Catal., A*, 2009, **360**, 2–7.
51. P. O. Thevenin, A. Alcalde, L. J. Pettersson, S. G. Järås and J. L. G. Fierro, *J. Catal.*, 2003, **215**, 78–86.
52. H.-J. Choi, J. Moon, H.-B. Shim, K.-S. Han, E.-G. Lee and K.-D. Jung, *J. Am. Ceram. Soc.*, 2006, **89**, 343–345.
53. C. Shi, L. Yang and J. Cai, *Fuel*, 2007, **86**, 106–112.
54. L. F. Liotta, G. Di Carlo, G. Pantaleo, A. M. Venezia and G. Deganello, *Appl. Catal., B*, 2006, **66**, 217–227.
55. J. Hu, W. Chu and L. Shi, *J. Nat. Gas Chem.*, 2008, **17**, 159–164.
56. Y. Xu, X. Chen, Z. Wang, S. Fan, W. Zhang, H. Liu and Y. Zheng, *Int. J. Hydrogen Energy*, 2021, **46**, 15526–15538.
57. Y. Zhang, Z. Qin, G. Wang, H. Zhu, M. Dong, S. Li, Z. Wu, Z. Li, Z. Wu, J. Zhang, T. Hu, W. Fan and J. Wang, *Appl. Catal., B*, 2013, **129**, 172–181.
58. J. Chen, X. Zhang, H. Arandiyan, Y. Peng, H. Chang and J. Li, *Catal. Today*, 2013, **201**, 12–18.
59. J. G. McCarty, M. Gusman, D. M. Lowe, D. L. Hildenbrand and K. N. Lau, *Catal. Today*, 1999, **47**, 5–17.
60. T. Kobayashi, N. Guilhaume, J. Miki, N. Kitamura and M. Haruta, *Catal. Today*, 1996, **32**, 171–175.
61. V. A. Sazonov, Z. R. Ismagilov and N. A. Prokudina, *Catal. Today*, 1999, **47**, 149–153.
62. A. L. Barbosa, J. Herguido and J. Santamaria, *Catal. Today*, 2001, **64**, 43–50.
63. S. C. Kwon, M. Fan, T. D. Wheelock and B. Saha, *Sep. Purif. Technol.*, 2007, **58**, 40–48.
64. Y.-C. Lin and K. L. Hohn, *Catalysts*, 2014, **4**, 305–306.
65. M. A. Peña and J. L. G. Fierro, *Chem. Rev.*, 2001, **101**, 1981–2018.
66. Goldschmidt V. M. and Holmsen D., in *Geochemische verteilungsgesetze der elemente*, Kommission bei J. Dybwad, 1927.
67. A. Civera, G. Negro, S. Specchia, G. Saracco and V. Specchia, *Catal. Today*, 2005, **100**, 275–281.
68. N. Miniajluk, J. Trawczyński, M. Zawadzki and W. Tylus, *Adv. Mater. Phys. Chem.*, 2018, **8**, 193–215.
69. D. L. Mowery and R. L. McCormick, *Appl. Catal., B*, 2001, **34**, 287–297.

70. A. W. Petrov, D. Ferri, F. Krumeich, M. Nachtegaal, J. A. van Bokhoven and O. Kröcher, *Nat. Commun.*, 2018, **9**, 2545.
71. R. Burch, F. J. Urbano and P. K. Loader, *Appl. Catal., A*, 1995, **123**, 173–184.
72. S. Cimino, R. Pirone and L. Lisi, *Appl. Catal., B*, 2002, **35**, 243–254.
73. K. Narui, K. Furuta, H. Yata, A. Nishida, Y. Kohtoku and T. Matsuzaki, *Catal. Today*, 1998, **45**, 173–178.
74. D. P. Serrano, G. Calleja, J. A. Botas and F. J. Gutierrez, *Sep. Purif. Technol.*, 2007, **54**, 1–9.
75. W. Wang, W. Zhou, W. Li, X. Xiong, Y. Wang, K. Cheng, J. Kang, Q. Zhang and Y. Wang, *Appl. Catal., B*, 2020, **276**, 119142.
76. Y. Sun, G. Xu, Y. Wang, W. Shi, Y. Yu and H. He, *Environ. Sci. Technol.*, 2023, **57**, 20370–20379.
77. F. Pinna, *Catal. Today*, 1998, **41**, 129–137.
78. Y. Li and J. N. Armor, *Appl. Catal., B*, 1994, **3**, 275–282.
79. H. Maeda, Y. Kinoshita, K. R. Reddy, K. Muto, S. Komai, N. Katada and M. Niwa, *Appl. Catal., A*, 1997, **163**, 59–69.
80. G. Caravaggio, L. Nossova and M. J. Turnbull, *Chem. Eng. J.*, 2021, **405**, 126862.
81. L. Zhang, Y. Zhang, H. Dai, J. Deng, L. Wei and H. He, *Catal. Today*, 2010, **153**, 143–149.
82. B. D. Abduladheem Turki Jalil, S. Chupradit, W. Suksatan, M. Javed Ansari, I. H. Shewael, G. H. Valiev and E. Kianfar, *Adv. Mater. Sci. Eng.*, 2021, 5102014.
83. J. Hu, W. Zhao, R. Hu, G. Chang, C. Li and L. Wang, *Mater. Res. Bull.*, 2014, **57**, 268–273.
84. J. Gao, X. Zhang, Y. Yang, J. Ke, X. Li, Y. Zhang, F. Tan, J. Chen and X. Quan, *Chem. – Asian J.*, 2013, **8**, 934–938.
85. J. Li, J. Chen, R. Ke, C. Luo and J. Hao, *Catal. Commun.*, 2007, **8**, 1896–1900.
86. O. W. Perez-Lopez, A. C. Farias, N. R. Marcilio and J. M. C. Bueno, *Mater. Res. Bull.*, 2005, **40**, 2089–2099.
87. L. M. T. Simplício, S. T. Brandão, E. A. Sales, L. Lietti and F. Bozon-Verduraz, *Appl. Catal., B*, 2006, **63**, 9–14.
88. R. J. Liu, P. A. Crozier, C. M. Smith, D. A. Hucul, J. Blackson and G. Salaita, *Appl. Catal., A*, 2005, **282**, 111–121.
89. A.-R. T. Operations, 2017.
90. REDUBAR, *A register of all gas regulations and norms concerning the necessary gas quality for allowing the transport in the natural gas grid*, 2008.
91. EUROMOT, 2017.
92. V. Meeyoo, D. L. Trimm and N. W. Cant, *Appl. Catal., B*, 1998, **16**, L101–L104.
93. S. Ordóñez, J. R. Paredes and F. V. Díez, *Appl. Catal., A*, 2008, **341**, 174–180.
94. K. Persson, L. D. Pfefferle, W. Schwartz, A. Ersson and S. G. Järås, *Appl. Catal., B*, 2007, **74**, 242–250.
95. P. Chen, H.-B. Zhang, G.-D. Lin and K.-R. Tsai, *Appl. Catal., A*, 1998, **166**, 343–350.

96. A. Setiawan, E. M. Kennedy, B. Z. Dlugogorski, A. A. Adesina, O. Tkachenko and M. Stockenhuber, *Energy Technol.*, 2014, **2**, 243–249.
97. D. Wang, Z. Li, C. Luo, W. Weng and H. Wan, *Chem. Eng. Sci.*, 2003, **58**, 887–893.
98. H.-H. Tseng, M.-Y. Wey, Y.-S. Liang and K.-H. Chen, *Carbon*, 2003, **41**, 1079–1085.
99. M. P. D. Thomas E. Koball and, *AIChE Symposium Series* 1977, 73.
100. D. Ciuparu and L. Pfefferle, *Appl. Catal., A*, 2001, **209**, 415–428.
101. Z. Tang, T. Zhang, D. Luo, Y. Wang, Z. Hu and R. T. Yang, *ACS Catal.*, 2022, **12**, 13457–13474.
102. M. Rotko, A. Machocki and B. Stasinska, *Appl. Surf. Sci.*, 2010, **256**, 5585–5589.
103. A. Setiawan, J. Friggieri, H. Hosseiniamoli, E. M. Kennedy, B. Z. Dlugogorski, A. A. Adesina and M. Stockenhuber, *Phys. Chem. Chem. Phys.*, 2016, **18**, 10528–10537.
104. J. C. van Giezen, F. R. van den Berg, J. L. Kleinen, A. J. van Dillen and J. W. Geus, *Catal. Today*, 1999, **47**, 287–293.
105. F. H. Ribeiro, M. Chow and R. A. Dallabetta, *J. Catal.*, 1994, **146**, 537–544.
106. C. F. Cullis and B. M. Willatt, *J. Catal.*, 1984, **86**, 187–200.
107. R. Rudham and M. K. Sanders, *J. Catal.*, 1972, **27**, 287–292.
108. M. Haneda, T. Mizushima and N. Kakuta, *J. Phys. Chem. B*, 1998, **102**, 6579–6587.

CHAPTER 12
Conclusion

MUSTAFA YASIN ASLAN,[*a] **ANGELA DAISLEY,**[*b]
JUSTIN S. J. HARGREAVES[*b] **AND JOSÉ L. RICO**[*c]

[a] Department of Chemical Engineering, Faculty of Engineering and Natural Sciences, Usak University, 64200, Usak, Turkey; [b] School of Chemistry, Joseph Black Building, University of Glasgow, Glasgow, G12 8QQ, UK; [c] Laboratorio de Catálisis, Facultad de Ingeniería Química, Universidad Michoacana de San Nicolás de Hidalgo, Edificio V1, C.U. Morelia Mich., C.P. 58060, México
*Emails: mustafa.aslan@usak.edu.tr; Angela.Daisley@glasgow.ac.uk; Justin.Hargreaves@glasgow.ac.uk; jose.rico@umich.mx

In principle, there are a number of definitions which could be applied for "small molecules". In determining the contents of this book, we have restricted our consideration to methane, nitrogen, carbon dioxide and carbon monoxide. Most of these molecules pose very significant challenges in terms of their activation, particularly in terms of industrial application. However, their successful transformation *via* various pathways is well established, and industrially practiced, in some cases, whereas in others there are routes of significant potential interest which have not progressed to application and are still at the research stage. Some of the general considerations related to the activation of the various small molecules considered within this book are outlined below.

Taking the dinitrogen molecule as an example, the challenge in activation of its very strong triple bond (946 kJ mol^{-1}) is well known. It does, however, react with lithium metal very slowly.[1] The biological activation of N_2 by nitrogenase, which co-produces hydrogen, occurs under ambient conditions[2] – however the timescale on which it ensues whilst being suitable in a biological context would be totally impractical for industrial application. In this context, the development of the Haber–Bosch process was a truly landmark achievement of the 20th century which provided a convenient route to synthetic fertiliser and thereby sustenance of the ever-expanding global population. The process itself is operated at very high pressure and moderate reaction temperature, with such operational parameters being dictated by the combination of thermodynamic limitations and the requirement for an acceptable reaction rate. Recent areas of interest related to the development of new catalysts which would be suitable for the application of small-scale ammonia production, and thus appropriate for the production of fertiliser close to its point of use and employing hydrogen generated from sustainable means (such as by electrolysis of water using electricity derived from solar and/or wind power) are very attractive. Chemical looping approaches for ammonia production involving coupled reaction processes have been the subject of a number of recent reviews.[3,4] One additional approach would be to directly incorporate nitrogen into an organic target product. This has been accomplished to an extent in some systems, for example,[5] and chemical looping routes employing the hydrogenation of metal nitride materials for the production of aniline have been identified as a very desirable, albeit highly challenging, target.[6]

Methane is also a molecule which is non-polar and possesses strong bonds (C–H average bond strength $ca.$ 416 kJ mol^{-1}). Steam reforming is a well-established technology which is practiced widely on an industrial scale. Much of the current interest in dry reforming is related to the possibility of reacting two potent greenhouse gases (CH_4 and CO_2) to yield syngas ($CO + H_2$), thereby opening up pathways to a range of different hydrocarbons and oxygenates such as alcohols by application of the Fischer–Tropsch process. Oxidative coupling has been discussed as a potential route to higher, more valuable, hydrocarbons, with ethylene being a more desirable target than ethane due to the added functionality of the double bond. With this in mind, it is interesting to highlight production routes from methane to acetylene using an electric arc (Huels process) and pyrolytic-based routes which have been practiced historically.[7] Direct cracking of methane to generate CO_x-free H_2 is topical, as is methane dehydroaromatisation – arguably these routes are closely related and the fact that hydrogen is the major product in molar terms for the latter route is not often as fully appreciated as it should be even if it is somewhat neglected. As has been the subject of a number of reviews,[8,9] the direct partial oxidation of methane to produce methanol or formaldehyde attracts much interest as it potentially bypasses the multistage processes of steam reforming and methanol synthesis (and additionally methanol oxidation). Partial oxidation to yield syngas is also a desirable pathway which has been widely investigated.[10]

Whilst carbon monoxide is isoelectronic with dinitrogen, with an even higher bond energy (1072 kJ mol^{-1}), it is a strongly polar molecule which facilitates reactivity. As a component of syngas, it is a very important intermediate for the production of hydrocarbons and alcohols via Fischer–Tropsch-based routes.

Another important application of CO is for carbonylation, such as the large-scale production of acetic acid and methyl acetate by the Monsanto and Cativa-based processes[11] as well as the industrial-scale production of methyl methacrylate *via* the Alpha process.[12] The conversion of carbon dioxide, a non-polar molecule, presents particular challenges related to its high thermodynamic stability. Recently, the conversion of waste CO_2 to methanol has proved to be an area of particular research attention.[13] It is interesting in this context to highlight the role of the water–gas shift reaction in the traditional $Cu/ZnO/Al_2O_3$ catalyst where it is, in fact, CO_2 which is hydrogenated.[14] Other conversion pathways for CO_2 relate to its reaction with epoxides to generate cyclic carbonates, which is a 100% atom efficient alternative to routes involving phosgene and diols.[15] Dry reforming is a particularly desirable reaction. However, an alternative challenging reaction which has been reported from time to time involves the production of acetic acid ($CH_4 + CO_2 \rightarrow CH_3COOH$).[16]

In the above very brief summary, we have mentioned some of the pathways of interest for conversion of the small molecules of interest. A number have not been directly addressed by the contents of the book and accordingly appropriate references to the various different areas have been given. An aspect which can be seen in terms of the future perspective is the increased attention to sustainability and this has spurred significant interest in electrocatalytic pathways which can be driven from sustainably derived electricity and photocatalytic approaches which can be driven from sunlight. Some coverage of these areas is presented within this book. These are very exciting and highly challenging areas which are developing rapidly. Related to this, we emphasise the requirements for rigorous product identification and also the exclusion of potential sources of error which may lead to erroneous conclusions.[17]

References

1. O. V. Ignatenko, V. A. Komar, S. V. Leonchik, N. A. Shempel, A. Ene, A. Cantaragiu, M. V. Frontasyeva and V. N. Shvetsov, *J. Alloys Compd.*, 2013, **581**, 23.
2. T. Chen, P. A. Ash, L. C. Seefeldt and K. A. Vincent, *Faraday Discuss.*, 2023, **243**, 270.
3. E. K. Fu, F. Gong, S. J. Wang and R. Xiao, *Small*, 2024, **20**, 2305095.
4. R. J. L. Pereira, I. S. Metcalfe and W. T. Hu, *Appl. Energy Combust.*, 2023, **16**, 100226.
5. S. Kim, F. Loose and P. J. Chirik, *Chem. Rev.*, 2020, **120**, 5637.
6. A. Daisley and J. S. J. Hargreaves, *Catal. Today*, 2023, **423**, 113874.
7. J. R. Fincke, R. P. Anderson, T. Hyde, B. A. Detering, R. Wright, R. L. Bewley, D. C. Haggard and W. D. Swank, *Plasma Chem. Plasma Process.*, 2002, **22**, 105.
8. N. Agarwal, S. H. Taylor and G. J. Hutchings, *Frontiers of Green Catalytic Selective Oxidations*, ed. K. P. Bryliakov, Springer, Singapore, 2019, p. 37.
9. T. J. Hall, J. S. J. Hargreaves, G. J. Hutchings, R. W. Joyner and S. H. Taylor, *Fuel Process. Technol.*, 1995, **42**, 151.

10. L. Li, N. H. M. D. Dostagir, A. Shrotri, A. Fukuoka and H. Kobayashi, *ACS Catal.*, 2021, **11**, 3782.
11. G. J. Sunley and D. J. Watson, *Catal. Today*, 2000, **58**, 293.
12. S. G. Khokarale and J.-P. Mikkola, *Green Chem.*, 2019, **21**, 2138.
13. N. Lawes, K. J. Aggett, L. R. Smith, T. J. A. Slater, M. Dearg, D. J. Morgan, N. F. Dummer, S. H. Taylor, G. J. Hutchings and M. Bowker, *Catal. Lett.*, 2024, **154**, 1603.
14. M. Bowker, *ChemCatChem*, 2019, **11**, 4238.
15. M. Alves, B. Grignard, R. Mereau, C. Jerome, T. Tassaing and C. Detrembleur, *Catal. Sci. Technol.*, 2017, **7**, 2651.
16. E. M. Wilcox, G. W. Roberts and J. J. Spivey, *Catal. Today*, 2003, **88**, 83.
17. S. Z. Andersen, V. Čolić, S. Yang, J. A. Schwalbe, A. C. Nielander, J. M. McEnaney, K. Enemark-Rasmussen, J. G. Baker, A. R. Singh, B. A. Rohr, M. J. Statt, S. J. Blair, S. t Mezzavilla, J. Kibsgaaard, P. C. K. Vesborg, M. Cargnello, S. F. Bent, T. F. Jaramillo, I. E. L. Stephens, J. K. Nørskov and I. Chorkendorff, *Nature*, 2019, **570**, 504.

Subject Index

^{1}H NMR spectroscopy, 177
2D electride, 303
4-aminopyridine (X-2), 254
4-hydroxypyridine (X-1), 254
5-amino-1,10-phenanthroline (X-4), 254
8-hydroxyquinoline (X-3), 254

α-cristobalite, 111

ab initio molecular dynamics, 248
ab initio molecular metadynamics simulations (AIMμD), 258
activated carbon (AC), 45, 176
active metal catalysts, 348
    noble metals, 348
        palladium, 348–349
        platinum, 349–350
        rhodium, 350
    non-noble metal oxides, 350
        cerium, 351–353
        chromium oxide, 353
        cobalt, 351
        copper, 350–351
        iron oxide, 353–354
        manganese oxide, 353
        perovskite, 354–355
active sites, 271
AECs. *See* alkaline electrolysis cells (AECs)
alcohols, 160
aldehydes, 193
alkali additives, 178
alkali metals, 177, 184
alkaline earth metal oxides, 103–106, 115
alkaline electrolysis cells (AECs), 236
alkaline metal oxides, 130
alkanes, 126
alkoxy mechanism, 191, 192
alloying–dealloying process, 243
alpha process, 380
alumina (Al_2O_3), 355–356
aluminium sulphate [$Al_2(SO_4)_3$], 363
ammonia (NH_3) synthesis, 294, 318
    electride-based catalysts, 299–303
    hydride-based catalysts, 303–307
    nitride-based catalysts, 307–310
    non-conventional ammonia synthesis approaches, 310–313
    transition metal-based catalysts, 296–299
ammonium molybdate, 148
amorphous carbon, 135
anatase, 176
Anderson–Schulz–Flory (ASF) polymerization model, 186
    product distribution, 167
aqueous-based reactors, 232
armchair (AC), 285
Aspen Plus®, 62, 65
associative mechanism, 296
autothermal reforming of methane (ATR), 4–5

Subject Index

balance equations, 63–66
bifunctional catalysts, 211
bimetallic catalysts, 18, 204
binding energy, 273
bismuth oxyhalides, 331
Bjerrum plot, 233
black phosphorus quantum dots (BPQDs), 279, 285
boron nitride (BN), 18
Boudouard reaction, 153
B-oxygen vacancy (B-OV) complexes, 273
BPQDs. See black phosphorus quantum dots (BPQDs)
Brunauer–Emmett–Teller (BET) surface area, 106

calcium-looping technology (CaL), 133
carbide mechanism, 186–188
carbon black (CB), 40, 45, 176
carbon black seeding, 55
carbon capture and storage (CCS), 224
carbon–carbon bond formation, 229
carbon catalyst, 45–48
carbon coking, 364
carbon dioxide (CO_2), 40, 86, 206, 223
    adsorption, 207
    catalysts, 241
        hybrid catalyst, 246–247
        metal-based catalysts, 241–246
        metal-free catalyst, 247–256
    computational studies, 256–262
    electrocatalytic reduction of, 225
        electrocatalytic cells, 229–232
        electrolytes, 232–240
    electronic configuration of, 206–207
    theoretical aspects of, 224–225

carbon dioxide hydrogenation
    catalysts, 202
        active metals, 203–204
        promoters, 205–206
        supports, 204–205
    kinetics of, 211
    mechanism of, 207
        conversion of CO_2 to methanol, 208–209
        Fischer–Tropsch synthesis mechanism, 210
        hydrogenation of CO_2 to methane, 209–210
carbon monoxide (CO), 40, 62, 86, 185, 200
carbon monoxide hydrogenation
    catalysts, 167
        active metals, 169–175
        promoters, 177–179
        supports, 175–177
    kinetics of, 196–199
    mechanism of, 186
        alkoxy mechanism, 191
        carbide mechanism, 186–188
        CO insertion mechanism, 190–191
        enolic mechanism, 188–190
        other alternative mechanisms, 191–196
carbon nanofibers (CNFs), 176, 248
carbon nanotubes (CNTs), 176, 254
catalyst–electrolyte interface, 232
catalysts, 7, 41, 209, 241
    carbon catalyst, 45–47
    catalyst regeneration, 50–51
    co-feeding, 49–50
    deactivation, 152, 361
        coking, 364
        sintering, 361
        sulphur dioxide poisoning, 361–363
        water inhibition, 363–364

catalysts (*continued*)
- dry reforming of methane, 13
  - modifications of Ni/Al$_2$O$_3$ catalyst, 13
  - noble and non-noble metals as promoters, 13–15
  - supported noble metal catalysts, 19–20
  - supported non-noble metal catalysts, 20–21
  - supports, 15–19
- hybrid catalyst, 246–247
- metal-based catalysts, 241–246
- metal catalyst, 42–45
- metal-free catalyst, 247–256
- for N–H bond formation, 313
- preparation, 358
  - impregnation, 358–359
  - passivation of support, 360
  - precipitation, 359
  - sol–gel method, 360
  - use of different precursors, 360
- reaction kinetics, 47–49
- regeneration, 50–51
- steam reforming of methane, 8
  - promoters, 8–9
  - supported noble metal catalysts, 11–12
  - supported non-noble metal catalysts, 12–13
  - supports, 9–10
- supports, 355
  - alumina, 355–356
  - silica, 356
  - zeolites, 356–358
  - zirconia, 356

catalytic dehydroaromatisation
- molybdenum-containing catalysts, 148–154
- other catalytic systems, 154–155

CB. *See* carbon black (CB)

C–C coupling reactions, 271, 275–276

CCS. *See* carbon capture and storage (CCS)

cell reactions, 63

cerium, 351–353

CFD. *See* computational fluid dynamics (CFD)

chemical looping, 27–28, 122
- for CH$_4$ conversion–perspectives, 141–142
- combustion of CH$_4$, 126–129
- concept, history, and fundamentals, 122–125
- dry reforming of methane with chemical looping, 131–135
- with methane, 125–126
- methane cracking with chemical looping, 135–137
- methane to other products *via* chemical looping routes, 138
  - benzene synthesis, 140–141
  - methanol synthesis, 139–140
- for oxidative coupling of methane, 137–138
- partial oxidation of methane with oxygen carriers, 129–131

chemical looping ammonia synthesis (CLAS), 312

chemical looping benzene synthesis, 140–141

chemical looping combustion (CLC), 127

chemical looping–cracking of methane (CL–MC), 127

chemical looping–dehydroaromatisation to benzene (CL–DHA), 127

chemical looping for oxidative coupling of methane (CL–OCM), 137–138

chemical looping–methanol synthesis (CL–MeOH), 127, 139–140

Subject Index

chemical looping–super dry reforming of methane (CL-SDRM), 127, 133
chromium nitride, 324
chromium oxide, 353, 363
chromium oxynitride, 324
CLAS. *See* chemical looping ammonia synthesis (CLAS)
CLC. *See* chemical looping combustion (CLC)
CL-DHA. *See* chemical looping–dehydroaromatisation to benzene (CL-DHA)
CL-MC. *See* chemical looping–cracking of methane (CL-MC); methane cracking with chemical looping (CL-MC)
CL-MeOH. *See* chemical looping–methanol synthesis (CL-MeOH)
CL-OCM. *See* chemical looping for oxidative coupling of methane (CL-OCM)
CLR. *See* coal liquefaction residue (CLR)
CL-SDRM. *See* chemical looping–super dry reforming of methane (CL-SDRM)
CNFs. *See* carbon nanofibers (CNFs)
CNTs. *See* carbon nanotubes (CNTs)
CO. *See* carbon monoxide (CO)
CO_2 hydrogenation (CO_2-H), 61
CO_2 methanation, 202
CO_2 photoreduction
  basic principles of photocatalytic CO_2 reduction, 272
    C–C coupling, 275–276
    CO_2 adsorption and activation, 273–275
    photogenerated charge carrier separation, 273
  strategies for creating active sites, 276
    etching, 278–279
    hydro/solvothermal route, 276–277
    thermal annealing, 277–278
    ultrasonication, 279
  typical active sites for selective photocatalytic CO_2 reduction, 280
    defect engineering, 282–284
    edge configurations, 285
    facet engineering, 285–287
    metal active sites, 280–282
coal liquefaction residue (CLR), 45
cobalt (Co), 42, 351
cobalt-based catalysts, 183, 197
cobalt carbide, 169
cobalt catalysts, 168, 183, 204
cobalt oxide (CoO), 204
cobalt(II)-tetraphenyl porphyrin (CoTPP), 247
co-electrolysis, 61–62
co-feeding, 49–50
COFs. *See* covalent organic frameworks (COFs)
coking, 364
computational fluid dynamics (CFD), 52, 53, 56
conduction band (CB), 273
copper (Cu), 200, 350–351
copper-based catalysts, 203
copper oxide, 363
covalent organic frameworks (COFs), 282
Cu–Al electrocatalysts, 257
Cu-based catalysts, 200, 205
Cu-based nanoparticles, 259
Cu–ZnO, 208
Cu–ZnO–ZrO_2 (CZZ), 208
Cu–ZrO_2, 208

DCFC. *See* direct carbon fuel cell (DCFC)
defect engineering, 282–284
dehydroaromatisation (DHA), 140, 148

dehydrogenative coupling, 89
density functional theory (DFT), 14, 101, 159, 196, 207, 208, 243, 256, 275, 277, 301, 320, 322, 326
DHA. See dehydroaromatisation (DHA)
diatomic species, 97
diffuse reflectance infrared Fourier-transform spectroscopy (DRIFTS), 101
dimethyl ether (DME), 58, 76
direct carbon fuel cell (DCFC), 59, 60, 76
dissociative mechanism, 296
DRM. See dry reforming of methane (DRM)
dry impregnation, 358
dry reforming of methane (DRM), 3–4, 131
    modifications of Ni/Al_2O_3 catalyst, 13
    noble and non-noble metals as promoters, 13–15
    supported noble metal catalysts, 19–20
    supported non-noble metal catalysts, 20–21
    supports, 15–19
dry reforming of methane with chemical looping (CL–DRM), 131–135
dual-atom sites (DAS), 277

ecMR. See electrochemical membrane reactor (ecMR)
edge configurations, 285
electride-based catalysts, 299–303
electrocatalysts, 229
electrocatalytic C–C formation, 227
electrocatalytic cells, 229
    H-cells, 229–230
        membrane-based flow reactors, 230
        microfluidic reactor, 230–232
electrocatalytic reaction, 319
electrochemical cell, 224

electrochemical membrane reactor (ecMR), 239
electrolytes, 232
    aqueous, 233–234
    non-aqueous, 234
    pure water, 232–233
    solid, 234–240
electromagnetic field (EMF), 311
electron acceptors, 185
electron paramagnetic resonance (EPR), 101
electron spin resonance spectroscopy, 328
Eley–Rideal (ER) mechanism, 92, 258, 308
Ellingham diagrams, 124
EMF. See electromagnetic field (EMF)
endothermic reactions, 163
energy efficiency, 66
enhance oil recovery (EOR), 224
enolic mechanism, 188–190
EPR. See electron paramagnetic resonance (EPR)
equilibrium constant, 233
etching, 278–279
ethanol, 50
ethylene, 90, 229
exergy destruction rates, 67, 68, 70
exergy efficiency, 66, 68
exhaust methane, 346–347
exothermic reactions, 163

facet engineering, 285–287
faradaic efficiency (FE), 224, 319, 320
Faraday constant, 227
FE. See faradaic efficiency (FE)
Fe-based catalysts, 43, 203, 294, 298
Fermi energy, 260
Fischer–Tropsch synthesis (FTS), 84, 88, 159, 164, 208, 211, 224
    carbon dioxide adsorption, 207
    carbon dioxide hydrogenation catalysts, 202–206
    carbon monoxide hydrogenation catalysts, 167–179

Subject Index

CO and H_2 adsorption on Fischer–Tropsch catalysts, 179
  carbon monoxide adsorption, 180–183
  co-adsorption and interaction of, 184–185
  electronic configuration of, 179–180
  hydrogen adsorption, 183–184
electronic configuration of CO_2, 206–207
kinetics of carbon monoxide hydrogenation, 196–199
mechanism of carbon dioxide hydrogenation, 207–210
mechanism of carbon monoxide hydrogenation, 186
  alkoxy mechanism, 191
  carbide mechanism, 186–188
  CO insertion mechanism, 190–191
  enolic mechanism, 188–190
  other alternative mechanisms, 191–196
product distribution, 165–167, 202
production of hydrocarbons from carbon monoxide hydrogenation, 160–163
products of carbon dioxide hydrogenation, 200–201
thermodynamics of carbon dioxide hydrogenation reactions, 201–202
thermodynamics of CO hydrogenation reaction, 163
  free energy of reaction, 164–165
  heats of reaction, 163–164
through carbon dioxide hydrogenation, 199–200
flame spray pyrolysis, 12
fluidized-bed reactors, 135, 201
formaldehyde (HCHO), 141
formic acid (HCOOH), 141, 259
Fourier transform infrared spectroscopy, 257
frustrated Lewis pairs (FLPs), 207
FTS. *See* Fischer–Tropsch synthesis (FTS)

gas-diffusion electrodes (GDEs), 230, 231, 239
gaseous fuels, 127
gas hourly space velocity (GHSV), 104, 105
GDEs. *See* gas-diffusion electrodes (GDEs)
GHSV. *See* gas hourly space velocity (GHSV)
Gibbs free energy, 85, 91, 97, 306
glycine nitrate combustion (GNC), 108
graphene, 176
graphene oxide (GO), 254
graphite, 176

Haber–Bosch (HB) process, 88, 294, 295, 318, 379
Hägg carbide, 161
HB process. *See* Haber–Bosch (HB) process
H-cells, 229–230
HER. *See* hydrogen evolution reaction (HER)
Heyrovsky mechanism, 256
hierarchical high-throughput catalyst screening (HHTCS), 298
highest occupied molecular orbital (HOMO), 87
high surface area graphite (HSAG), 298
hot atom mechanism, 302
hybrid catalyst, 246–247
hydride-based catalysts, 303–307
hydride materials, 303
hydrogen (H_2), 1, 147, 200
  adsorption, 183–184

hydrogen evolution reaction (HER), 225, 318, 322
hydrogen fuel cell, 60, 62
hydro/solvothermal technique, 276–277
hydrothermal method, 277
hydroxyapatite (HAp), 12

impregnation, 358
    dry impregnation, 358
    wet impregnation, 358–359
infrared spectroscopy, 328
in situ Fourier transform infrared spectroscopy, 284
in situ infrared absorption spectroscopy, 244
in situ soft X-ray absorption spectroscopy, 244
ion exchange, 358–359
ionic liquids (IL), 234
Ir-based catalysts, 12
iron (Fe), 42
iron-based catalysts, 161, 196
iron catalysts, 168, 171
iron oxide, 353–354

Kellogg Advanced Ammonia Process (KAAP), 298

Langmuir adsorption isotherm, 184
Langmuir–Hinshelwood–Hougen–Watson (LHHW) approach, 196
Langmuir–Hinshelwood (LH) mechanism, 211, 308
lanthanide hydrides, 306
lanthanum oxide (La_2O_3), 106
layered double hydroxides (LDH), 333
Le Chatelier principle, 202
LHHW approach. *See* Langmuir–Hinshelwood–Hougen–Watson (LHHW) approach
linearized Poisson–Boltzmann model, 260
liquefied natural gas (LNG), 40

low-energy ion-scattering (LEIS) measurements, 310
lowest unoccupied molecular orbital (LUMO), 87

machine learning techniques, 288
magnesium oxide (MgO), 206
manganese (Mn), 206
manganese oxide, 353
Mars–van Krevelen mechanism, 128, 129, 139, 141, 308, 323
MEA. *See* membrane electrode assembly (MEA)
mechanochemical ball-milling method, 312
membrane-based flow reactors (MFR), 230
membrane electrode assembly (MEA), 232, 325
mesoporous carbons, 176
metal active sites (MASs), 280–282
metal-based catalysts, 241–246
metal-based electrocatalysts, 248, 262
metal catalyst, 42–45
metal-free catalyst, 247–256
metal nitride, 323
metal–organic frameworks (MOFs), 172, 203, 204, 335
metal promoters, 13–14
methanation, 200
methane (CH_4), 84, 125, 200, 202, 379
    catalysts, 7–21
    chemical looping combustion, 126–129
    conversion routes, 88
    dry reforming of methane, 3–4
    importance of H_2 production and H_2 production technologies, 1–2
    (micro)kinetics and DFT studies, 21–25
    new generation reactors and process intensification, 25

chemical looping reforming of methane, 27–28
other new-generation methane-reforming routes, 28
oxidative (steam or dry) reforming of methane, 27
sorption-enhanced reforming of methane, 25–27
partial oxidation and autothermal reforming of, 4–5
steam reforming of methane, 2–3
thermodynamics of reactions, 5–7
methane cracking, 40
case studies, 58
methodology, 62–67
results and discussion, 67–76
system description, 58–62
catalysts, 41
carbon catalyst, 45–47
catalyst regeneration, 50–51
co-feeding, 49–50
metal catalyst, 42–45
reaction kinetics, 47–49
methodology, 62–67
solar reactors, 51
catalyzed solar reactors, 56–57
directly irradiated solar reactors, 52–53
indirectly irradiated solar reactors, 53–56
molten media reactors, 57–58
solar energy collection, 58
methane cracking with chemical looping (CL–MC), 135–137
methane slip, 345
exhaust methane, 346–347
lean *versus* stoichiometric, 345–346
ventilation air methane, 347–348
methane temperature-programmed surface reaction (CH_4-TPSR), 101
methanol (CH_3OH), 50, 58, 141, 193, 200, 202, 208
production, 61
methanol-to-aromatics (MTA) reactions, 200
methanol-to-gasoline (MTG) process, 199
methanol-to-olefins (MTO) reactions, 200, 211
MFR. See membrane-based flow reactors (MFR)
microfluidic reactors (MR), 230–232
mixed ionic and electronic conductors (MIECs), 239
Mn–Na–W–SiO_2 catalyst, 111–114
Mo-based catalyst, 169
modified Fischer–Tropsch synthesis, 199
MOFs. See metal-organic frameworks (MOFs)
molten media reactors, 57–58
molybdenum-containing catalysts, 148–154
molybdenum nitride, 325
monoatomic species, 97
monomers, 186
multicarbonyls, 183
multiwalled carbon nanotube (MWCNT), 247
MXenes, 325

natural gas
engines, 343–344
for power generation, 344
in transport, 345
natural gas engine exhaust (NGEE), 347
NDNC. See nitrogen-doped nanoporous carbon (NDNC)

N-doped graphene quantum dots (NGQD), 252
NGQD. See N-doped graphene quantum dots (NGQD)
Ni-based catalysts, 9, 24, 43, 44, 308
nickel (Ni), 42
nickel oxide, 363
Ni metal nanoparticles, 14
nitride-based catalysts, 307–310
nitrogen (N_2), 248
    electrocatalytic conversion of, 319
        by metal cathode, 320–322
        by metal nitrides and carbides, 323–325
        by metal oxides, 325–326
    fixation, 322
    oxide regeneration, 365
    photocatalytic conversion of, 326
        alternative materials for, 333–335
        by g-C_3N_4, 331
        by metal oxides, 328–331
        by metal sulphides, 332–333
        by oxyhalides, 331–332
nitrogen-doped nanoporous carbon (NDNC), 254
nitrogen reduction reaction (NRR), 318, 322
noble metals, 348
    palladium, 348–349
    platinum, 349–350
    rhodium, 350
non-aqueous electrolytes, 233
non-conventional ammonia synthesis approaches, 310–313
non-noble metal oxides, 350
    cerium, 351–353
    chromium oxide, 353
    cobalt, 351
    copper, 350–351
    iron oxide, 353–354
    manganese oxide, 353
    perovskite, 354–355
non-thermal plasma (NTP), 202, 208, 209
non-thermal plasma-assisted ammonia synthesis, 313
NRR. See nitrogen reduction reaction (NRR)
NTRL-RK property method, 63

OCM. See oxidative coupling of methane (OCM)
OER. See oxygen evolution reaction (OER)
olefins, 160, 171
one-dimensional model, 57
one-step hydrothermal synthesis, 277
organic solvents, 276
Ostwald ripening, 361
oxidative coupling, 379
oxidative coupling of methane (OCM), 84, 114, 137
    activation of C–H bond and possibilities for methane conversion, 87–89
    active sites for, 100–103
    alternative oxidants for, 95–99
    characteristics and environmental impact, 85–86
    development of catalysts for, 103
        alkaline earth metal oxides, 103–106
        Mn–Na–W–SiO_2 catalyst, 111–114
        perovskites, 109–111
        rare earth oxides, 106–108
    impact of reaction conditions in, 94–95
    reaction mechanism and kinetic aspects, 91–94
    thermodynamic aspects, 90–91
oxidative processes, 137
oxide mechanism, 191
oxygen carrier, 27, 122
oxygen evolution reaction (OER), 225
oxygen vacancies, 100

palladium, 348–349
palladium catalysts, 359
palladium sulphate (PdSO$_4$), 362
paraffins, 160
partial oxidation of methane (POM), 2, 4–5, 130
partial oxidation of methane with oxygen carriers (CL-POM), 129–131
PCN. *See* polymeric carbon nitride (PCN)
PEI. *See* polyethylenimine (PEI)
PEM. *See* proton exchange membranes (PEM)
PEMFC. *See* proton-exchange membrane fuel cell (PEMFC)
Peng–Robinson equation, 62
perovskites, 18, 109–111, 115, 326, 354–355
phase diagrams, 124
phosphate-based materials, 326
photocatalysis, 200, 209, 319
photocatalyst, 273, 288
photogenerated charge carrier separation, 273
photothermal catalysis, 200, 209
photovoltaic (PV) system, 73
plasma-catalytic system, 313
platinum, 349–350
Poisson–Boltzmann implicit solvation model, 261
polyethylenimine (PEI), 236, 254
polymeric carbon nitride (PCN), 275
polyvenyl alcohol (PVA), 236
POM. *See* partial oxidation of methane (POM)
post-etching technique, 278
potassium hydroxide (KOH), 236
power-to-gas technology, 209
precipitation methods, 359
pre-treatment methods, 13
proton-exchange membrane fuel cell (PEMFC), 59, 60, 69
proton exchange membranes (PEM), 230
PVA. *See* polyvenyl alcohol (PVA)

pyrazinic-N, 254
pyridinic-N, 254
pyridoxine, 254

quantum mechanical calculations, 115

radical reactions, 93
Raman spectroscopy, 101, 303
Rankine cycle, 58
rare earth metal oxides, 115
rare earth oxides, 106–108
reaction kinetics, 47–49, 366
    effect of water, 369–371
    Eley–Rideal mechanism, 368–369
    Langmuir–Hinshelwood mechanism, 367–368
    Mars–van Krevelen mechanism, 369
    reaction order, 371–373
    support's effect, 371
reaction order, 371
    effect of carbon dioxide, 372
    effect of gaseous oxygen, 372–373
    effect of water, 371–372
reduced graphene oxide (rGO), 244
reference temperature, 75
regeneration, 364
    nitrogen oxide regeneration, 365
    temperature-based regeneration, 364–365
renewable energy source, 41
reverse Boudouard reaction, 132
reverse water gas shift reaction (rWGSR), 7, 62, 131, 159, 199, 200, 204, 205, 207, 208, 210, 224
reversible hydrogen electrode (RHE), 244, 321
rhodium, 350
rhodium catalysts, 198
Ru–Ba core–shell structure, 305
Ru-based catalysts, 298, 300
Ru nanoparticles, 321

ruthenium, 159
rWGSR. *See* reverse water gas shift reaction (rWGSR)

SAAs. *See* single-atom alloys (SAAs)
Sabatier reaction, 209, 210
SACs. *See* single-atom catalysts (SACs)
Schulz–Flory distribution function, 165
semiconductors, 283
SHE. *See* standard hydrogen electrode (SHE)
silica, 356
single-atom alloys (SAAs), 261
single-atom catalysts (SACs), 199, 203, 207
sintering, 361
slip, 346
small molecules, 378
sodium, 112
SOECs. *See* solid oxide electrolysis cells (SOECs)
solar energy, 41, 326
   collection, 58
solar reactors, 51
   catalyzed solar reactors, 56–57
   directly irradiated solar reactors, 52–53
   indirectly irradiated solar reactors, 53–56
   molten media reactors, 57–58
   solar energy collection, 58
sol–gel method, 10, 104, 360
solid electrolyte, 261
solid oxide electrolysis cells (SOECs), 236
solid polymer electrolyte (SPE), 235
solid proton conducting electrolysis cells (SPCECs), 236
SPE. *See* solid polymer electrolyte (SPE)
SRM. *See* steam reforming of methane (SRM)
stability, 365
   sulphur dioxide, 365–366
   water, 365

standard hydrogen electrode (SHE), 274
statistical optimization tools, 95
steady-state isotopic transient kinetic analysis (SSITKA), 301
steam methane reforming (SMR), 41
steam reforming of methane (SRM), 2–3, 8, 21–23, 129
   promoters, 8–9
   supported noble metal catalysts, 11–12
   supported non-noble metal catalysts, 12–13
   supports, 9–10
structure–activity relationships, 288
subsystems, 59
   CO_2 hydrogenation, 61
   co-electrolysis, 61–62
   DCFC, 60
   MEOH, 61
   PEMFC, 60
sulphur dioxide (SO_2), 361
   consumption, 365–366
sulphur vacancies, 332
support promoters, 14–15
surface basicity, 97
surface redox reactions, 273
syngas, 63, 129
synthetic fertilisers, 318, 379

Tafel mechanism, 256
Tammann temperature, 4, 361
Temkin's model, 21
temperature-based regeneration, 364–365
temperature-programmed desorption of oxygen (TPD-O_2), 101
temperature-programmed reduction with hydrogen (TPR-H_2), 101
thermal annealing, 277–278
thermal catalysis, 209
thermal hydrogenation, 200
thermodynamics of reactions, 5
   dry reforming of methane, 6–7
   steam reforming of methane, 5–6

transition metal-based catalysts, 296–299
transition metal (TM) catalysts, 14, 295
transition metal oxides, 363
turnover frequency (TOF), 168, 299

ultrasonication technique, 279

valence band (VB), 273
ventilation air methane (VAM), 347–348

water (H_2O), 200
    catalyst, 365
    temperature, 365
water–gas shift (WGS) activity, 203
water–gas shift (WGS) reaction, 21, 41, 71, 94, 161, 197, 380
water inhibition, 363–364
wet impregnation, 358
    ion exchange, 358–359
work functions (WFs), 303

X-ray absorption near-edge structure spectroscopy (XANES), 247
X-ray photoelectron spectroscopy (XPS), 101, 251

zeolites, 139, 177, 210, 299, 356–358
zeolitic imidazolate frameworks (ZIFs), 203
zero CO_2 emission, 224
ZIFs. *See* zeolitic imidazolate frameworks (ZIFs)
zigzag (ZZ) edges, 285
zinc, 170, 206
zinc oxide (ZnO), 206
zirconia (ZrO_2), 356
zirconium nitride, 324
zirconium oxynitride, 324